Eutrophication in Planktonic Ecosystems: Food Web Dynamics and Elemental Cycling

Developments in Hydrobiology 127

Series editor
H.J. Dumont

Eutrophication in Planktonic Ecosystems: Food Web Dynamics and Elemental Cycling

Proceedings of the Fourth International PELAG Symposium, held in Helsinki, Finland, 26–30 August 1996

Edited by

T. Tamminen & H. Kuosa

Reprinted from Hydrobiologia, volume 363 (1997/8)

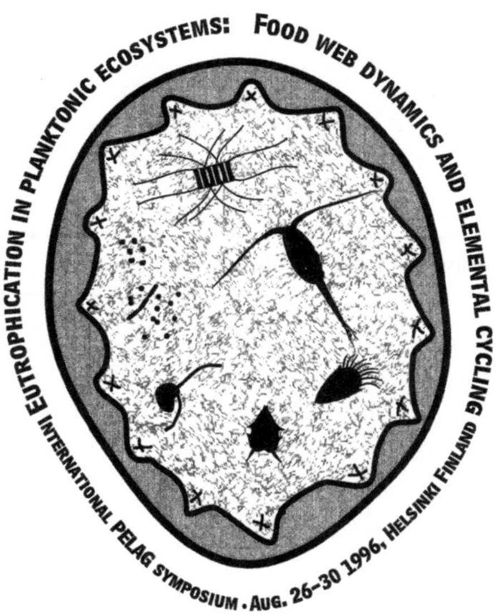

Kluwer Academic Publishers

Dordrecht / Boston / London

Library of Congress Cataloging-in-Publication Data

A C.I.P. Catalogue record for this book is available from the Library of Congress.

ISBN 978-90-481-5041-0

Published by Kluwer Academic Publishers,
P.O. Box 17, 3300 AA Dordrecht, The Netherlands.

Sold and distributed in the North, Central and South America
by Kluwer Academic Publishers,
101 Philip Drive, Norwell, MA 02061, U.S.A.

In all other countries, sold and distributed
by Kluwer Academic Publishers,
P.O. Box 322, 3300 AH Dordrecht, The Netherlands.

Title page illustration based on the symposium logo created by Jan Ekebom

Printed on acid-free paper

Printed in the Netherlands.

Contents

Hydrobiologia 363: vii–viii, 1998.
T. Tamminen & H. Kuosa (eds), Eutrophication in Planktonic Ecosystems: Food Web Dynamics and Elemental Cycling.

Foreword

(so what on earth is a PELAG symposium??)

Soon after the Ancient Romans, in 1982 I think, a major Finnish research-funding foundation received an application, where a bunch of nobodies described experimental teamwork on planktonic processes and nutrient limitation of the Gulf of Finland, the Baltic Sea. No funding was received. After the decision, the research director of the foundation gave encouraging comments on the application as such, recommended resubmitting after a year – and suggested also that the responsible leader of the team might be considered better qualified by the steering committee if he could present himself as something other than 'student' (after 9 years at the university. . .).

As it were, after a year the team had managed to run one experiment – quite literally on a beg-borrow-and-steal basis – to demonstrate their approach in a renewed application, travel around Europe on a 4-congress ramble to display their great results and to spy what was going on, the responsible leader could present himself as 'M.Sc.', and lo and behold! – funding was received for the team both from the foundation and the Academy of Finland. Meanwhile, another private foundation had decided to support one of the participants of the pilot experiment, plus to give grants for certain students. After these sudden twists of fate, no time was wasted in re-interpreting recent history. Much to their surprise, the innocent funding parties learned that instead of funding their relatively-young and relatively-promising picks of choice, they were all now jointly funding research project PELAG (1984–86, with 7 full-time participants).

Whatever they originally might have thought of this interpretation, no hard feelings were nurtured when this period was over, as the same funding parties supported more than twice the original crew in a project described – this time from the start – as PELAG II (1987–89, *'Ecological Plankton Research of the Baltic Sea'*). Instead of the big bang that any competent observer was expecting after this phase, PELAG III started in 1991, and this time for 5 years (1991–95, *'Nitrogen Discharge, Pelagic Nutrient Cycles, and Eutrophication of the Northern Baltic Coastal Environment'*) with ca. 10 scientists.

Since then, we have become tired of counting. Research Team PELAG, with a homebase since early 90's in the Finnish Environment Institute, and an essential outpost at the Tvärminne Zoological Station, University of Helsinki, is currently heavily involved in EU-projects, as the trend in Finland is the same as in most European countries. The heydays of big national-only projects are more or less history, and networking within EU and other international bodies is a necessity to ensure the resources that are needed for effective scientific work.

This is especially true when it comes to the overall topic of the symposium – processes taking place in the planktonic ecosystem, more specifically from the eutrophication point of view. Understanding the short-term and long-term responses of aquatic ecosystems to changes in nutrient inputs requires integrated studies of the planktonic food web, which by definition calls for concurrent focusing on numerous components and their interactions. This in turn translates into relatively resource-hungry operations, and into a need for widening the basis of isolated projects. As the knowledge on the functioning of planktonic ecosystems accumulates, there is a potential threat that the gap between 'basic' aquatic ecology and 'applied', management-oriented research expands.

This would be most unfortunate in several respects. The latter must necessarily be kept simple but definitely not stupid. We believe that 'basic' and 'applied' research cannot be differentiated by, for example, methodologies applied, or by scientific criteria set, but by the way the scientific questions are posed. Relevant 'basic' questions can and should be asked within a plenitude of 'applied' topics in order to get really applicable answers. The symposium was organized according to this view into 8 sessions, each opened by extensive keynote presentations on the field:

- nutrient limitation of phytoplankton growth (keynotes by B. Riemann and U. Larsson)
- nutrient cycles within the planktonic food web (keynote by P. Glibert)
- DOM – sources, composition and uptake (keynote by F. Azam)
- resource limitation vs. shaping of the food web by grazing. I. Pico-microplankton (keynote by P. K. Bjørnsen)

- resource limitation vs shaping of the food web by grazing. II. Micro-mesoplankton (keynote by T. Kiørboe)
- temporal and spatial variability – coupling of physical and biological planktonic processes (keynote by D. Aksnes)
- retention vs export food chains: processes controlling sinking loss from the pelagic system (keynote by P. Wassmann)
- planktonic food web modelling (keynote by T. F. Thingstad)

Our original ambition was that the meeting would add to the difficult art of combining state-of-the-art aquatic ecology and environmental concerns, most of all eutrophication problems. This is a task that easily surpasses mere opportunistic fund-raising schemes (so familiar to all researchers) – it serves well both as a scientific challenge and a source of motivation. Doing decent research is, after all, quite some work. It would be a drag if it were of no use.

The 140 participants showing up from 21 countries of Europe, Asia and northern America, with 48 oral presentations (including the 9 keynote lectures) and 45 posters, took care of our goal even beyond our expectations. A fair share of these insights are now included in these Proceedings. I trust that researchers with widely varying specific fields of interest within the overall topic will find this collection stimulating.

Once again I want to thank the invited lecturers, the sponsors of this symposium (Finnish Environment Institute, Finnish Ministry of the Environment, Academy of Finland, Maj and Tor Nessling Foundation, Nordisk Forskerutdanningsakademi) and the organizing committee for making the meeting possible, and the secretariat (Pirjo Kuuppo, Riitta Autio) for making it happen. And of course all participants who showed up. At least we enjoyed it ourselves!

Helsinki, 12. 3. 1998
on behalf of the Research Team PELAG,

TIMO TAMMINEN

Hydrobiologia **363:** 1–12, 1998.
T. Tamminen & H. Kuosa (eds), Eutrophication in Planktonic Ecosystems: Food Web Dynamics and Elemental Cycling.

Interactions of top-down and bottom-up control in planktonic nitrogen cycling

Patricia M. Glibert
Horn Point Environmental Laboratory, Center for Environmental and Estuarine Studies, University of Maryland, P.O. Box 775 Cambridge, Maryland 21613, USA

Key words: top-down control, bottom-up control, NH_4^+ regeneration, nutrient limitation, trophodynamics

Abstract

Although our understanding of the complexity of the plankton and microbial food webs has increased substantially over the past decade or two, there has been little appreciation to date of the interactions between top-down (grazing) control and bottom-up (nutrient supply) control on the structure and nutrient cycling processes within these webs. The quality of nutrient supply, both in terms of the relative proportion of inorganic: organic nitrogen, as well as the relative proportion of inorganic nitrogen substrates has a direct impact on rates of nitrogen uptake, and ultimately on the relative composition of phytoplankton and bacteria. At the same time, grazing by microzooplankton and macrozooplankton also influences both the composition of the food web and the rate of supply of nitrogen. The impact of macrozooplankton on rates of nitrogen cycling in a microbial community is complex: macrozooplankton release NH_4^+, urea, and amino acids by direct excretion and by 'sloppy feeding', but they also control both the rates of nitrogen regeneration and uptake within the community by grazing the microzooplankton, the primary regenerators of NH_4^+, and the phytoplankton, the primary consumers of nitrogen. Thus, grazing and nitrogen recycling are intricately connected: the presence of large zoooplankton simultaneously provides top-down control of biomass and bottom-up nutrient supply. These relationships vary depending on the scale of interest, and have important consequences for how we measure and model total nitrogen cycling in a natural food web.

Introduction

One need not be a scholar of the history of biological oceanography to appreciate the fact that phytoplankton production in marine systems has traditionally been thought to be limited by *either* nutrients *or* grazing, but rarely both. Most students can recite the important contributions of both Liebig (1855), in terms of the Law of the Minimum, and Lindeman (1942), in terms of Trophic Dynamics, but do not often consider both simultaneously. Nor it is often realized that these contributions were controversial at the time they were introduced. The evolution of thought, and the controversies surrounding the applications of the principles of the Law of the Minimum, have been eloquently reviewed by DeBaar (1994). He describes the contributions of Nathansohn (1908), Gebbing (1910), and others, who challenged the the hypothesis of limitation by a single nutrient, and suggested that multiple limitation must occur. Indeed, Nathansohn (1908; quoted in DeBaar, 1994) viewed the plankton as a system in *dynamic balance* between production and loss – a concept that was decades ahead of its time:

> *'When we now turn to the study of external, especially chemical, conditions for the plankton and its quantitative distribution, we will clearly recognize the great variety of problems that ensue once we consider the continual growing and dying of the plankton, and once we keep in mind that its biomass results from the balance of these two processes.'*

The literature on limitation of primary productivity in marine systems is further confounded by a confusion, or more correctly, imprecision, regarding what is being limited, growth rate of the organisms, or biomass accumulation in the system (Glibert, 1988; Malone et al., 1996). The concept of nitrogen limitation in the sea was crystallized by the work of Dugdale & Goering (1967), in which the distinction between 'new' and 'regenerated' nitrogen was first drawn. In systems for which the rate of supply of 'new' nitrogen is relatively high, the total production of the system is generally higher than in systems for which this supply is small: total production being the product of growth rate and biomass (Eppley & Peterson, 1979; Glibert, 1988; Malone et al., 1996). However, the regeneration of nitrogenous nutrients can support high uptake and growth rates of phytoplankton, even when concentrations are vanishingly low (McCarthy & Goldman, 1979). Interestingly, the distinction between limitation of growth and biomass was noted by Nathansohn (1908; DeBaar, 1994), but not well appreciated until decades later:

> *It makes a great difference whether we study a nutrient solution in respect to its maximum potential of production...or in respect to the rate with which the production in it proceeds, as we will have to do when we want to clarify the conditions of plankton development. The first variable is, in accordance with the repeatedly mentioned Liebig's Law, always dependent on the one nutrient which happens to be in the minimum, while the latter can be affected by all possible components.'*

Whereas there is a physiological limit to how fast a rate can occur in a biological system (i.e. uptake, assimilation or growth), the limitation as to how much biomass can occur is determined by external factors (total nutrient load, grazing). Thus, responses to nutrient perturbations can be very rapid, while responses of a system to total nutrient supply are much slower; physiology responses are faster than those of total system biomass (Malone et al., 1996).

While there are, and will continue to be debates regarding the degree and manner in which nitrogen may be limiting phytoplankton growth and productivity in oceanic waters, my purpose here is to examine some of the more subtle aspects regarding the interactions between grazing control and nitrogen supply. Thus, the theme of this paper is not whether nutrients or grazing regulates the response of the system, but how, and to what degree, the responses to bottom up (i.e. nutrient supply) and top down (i.e. grazing) factors are coupled. There has been little discussion of the scale-dependence of this coupling and its impact on trophodynamics (Legendre & Rassoulzadegan, 1995). This discussion will focus on the impacts of macrozooplankton, although similar arguments could be made for other components of the planktonic food web. Indeed, Wikner & Hagström (1988) and others, have suggested that there are multiple trophic links even in communities of organisms <5 μm in size.

A dynamic balance

As Harvey (1945) noted:

> *'The standing crop of phytoplankton, or breeding stock, at any time is merely a 'momentary balance' between the processes of production and of consumption by animals and bacteria. The concentration of nutrient salts in the water mass at any time is likewise a momentary balance between the regeneration by animals and bacteria and their consumption by the phytoplankton.'*

The notion, therefore, of a dynamic balance between production and loss processes maintaining both biomass and dissolved substrates is clearly not a new one; however, our understanding of the organisms mediating the regeneration of nitrogen has changed tremendously in the past two decades. There is now considerable evidence, from size fractionation studies to controlled laboratory experiments, that microzooplankton play a primary role in the regeneration of nitrogen in the sea (Johannes, 1965, 1968; Fenchel & Harrison, 1976; Caron & Goldman, 1990). The degree to which bacteria serve as regenerators or consumers of nutrients depends in large part on their growth rate and physiological state (Goldman et al., 1987; Goldman & Dennett, 1991; Jørgensen et al., 1993).

The contribution macrozooplankton and larger microzooplankton make to total nutrient regeneration is complex; they can both enhance and reduce the flow of regenerated nitrogen. First, they release nitrogen directly, but the form of nitrogen released (for example, NH_4^+, urea or amino acids) may depend on what they ate and how long ago they ate it (Bidigare, 1983; Miller, 1992). Typically macrozooplankton excretion rates do not provide a significant proportion of the total nitrogen regenerated in a system (Bidigare, 1983; Dam et al., 1996), but one study in Long Island Sound showed that as much as 50% of the total nitrogen required by phytoplankton was by zooplankton excretion alone (Harris, 1959). Second, they graze on both the phytoplankton, the primary consumers of nitrogen, and the protozoa, the primary regenerators of nitrogen (Caron & Goldman, 1990). Thus, whether macrozooplankton will have a net positive impact on the regeneration of nitrogen via their direct release, or will have a negative impact by reducing other regenerators will depend on the trophic interactions within the microbial loop and between the microbial loop and the macrozooplankton. These interactions have been shown experimentally by Glibert et al. (1992), in which increasing numbers of copepods added to natural coastal plankton assemblages resulted in an increase in the rate of NH_4^+ regeneration up to about 20 copepods l^{-1}, but when additional copepods were added, the rate of NH_4^+ regeneration actually decreased. The decrease at higher macrozooplankton densities was hypothesized to be due to trophic interactions, resulting in a decrease in the regeneration by the smaller regenerators. Miller et al. (1995), in summer estuarine experiments, also demonstrated that NH_4^+ regeneration rates were highest at densities of 10–20 copepods l^{-1}, with lower rates observed at copepod densities both above and below this concentration.

Macrozooplankton, and larger microzooplanton, also tend to be 'sloppy' grazers (Dagg, 1974; Lampert, 1978; Eppley et al., 1981). This can result in a release of both inorganic (NH_4^+) and organic (amino acids and other low molecular weight compounds, as well as high molecular weight compounds) nitrogen. The latter may, in turn, stimulate the growth of, and/or nitrogen regeneration by, bacteria, thereby reducing their demand for inorganic nitrogen substrates (Reimann et al., 1986; Roman et al., 1988). In mesocosm studies, Carlsson et al. (1995) observed increases in both tintinnids and ciliates in the presence of both humic substances and copepods, suggesting that bacteria, the microzooplankton food, increased when sufficient organic substrate was available. Thus, through differential grazing and direct and indirect release of nitrogen, macrozooplankton can alter the composition and the rates of uptake and regeneration of nitrogen by the entire community (Roman et al., 1988; Miller et al., 1995; Christaki & Van Wambeki, 1995; Carlsson et al., 1995).

The interactions between the impact of new nitrogen input and macrozooplankton grazing can be formalized in a schematic flow diagram (Figure 1). These interactions are depicted as a series of interconnected ratios, representing parameters that are, for the most part, both easily or regularly measured, and which have known effects on the growth of one or more components in the food web. This approach is similar to that proposed by Legendre & Rassoulzadegan (1995) where links between availability and uptake of nitrogen and carbon and plankton size fractions were shown. They used these relationships to predict a continuum of trophic pathways from those dominated by the herbivorous food web to the microbial loop.

In this scheme, 'bottom up' control is depicted as the input of new nitrogen, typically in the form of NO_3^- (Figure 1). New inorganic nitrogen entering a system changes the ratio of dissolved inorganic to organic nitrogen (DIN/DON). Altering this ratio in turn affects the relative availability of inorganic nitrogen sources (NO_3^-/NH_4^+) or the ratio of dissolved organic carbon to nitrogen (DOC/DON). It is these ratios that have a direct effect on the kinetics of uptake of phytoplankton and/or bacteria, as the rate of uptake is a function of availability of substrate. Furthermore, the kinetics of uptake by phytoplankton of the reduced nitrogen forms are controlled by different physiological and environmental parameters than those of the oxidized forms (McCarthy, 1982; Goldman & Glibert, 1983). For example, the uptake systems of NH_4^+ and urea are usually thought to be less light dependent than that of NO_3^- (MacIsaac & Dugdale, 1972; Fisher et al., 1982), and temperature dependence of NH_4^+ uptake may also be different from that of NO_3^- (Lomas & Glibert, submitted).

The biomass of phytoplankton or bacteria may or may not respond directly to an increased rate of uptake of the nutrient, depending on the degree to which nutrient uptake, or biomass, is limited by that nutrient (Figure 1). There is, nevertheless, growing evidence that the relative availability of different forms of nitrogen can play an important role in shaping the quality of the phytoplankton or bacterial communities. For example, while many factors come into play in determining species composition, in general, systems dominated by oxidized forms of nitrogen often appear to have proportionately greater abundance of large phytoplankton, while systems dominated by regenerated nitrogen appear to have a greater fraction of small phytoplankton and/or bacteria. Episodic increases in phytoplankton in response to increases in proportional availability of NO_3^- are often (but not always) the result of increases in diatom biomass (Goldman, 1993). The relationship between new nitrogen and utilization by phytoplankton of larger size classes (primarily diatoms) has also been demonstrated in nitrogen uptake studies. In coastal and upwelling environments, Glib-

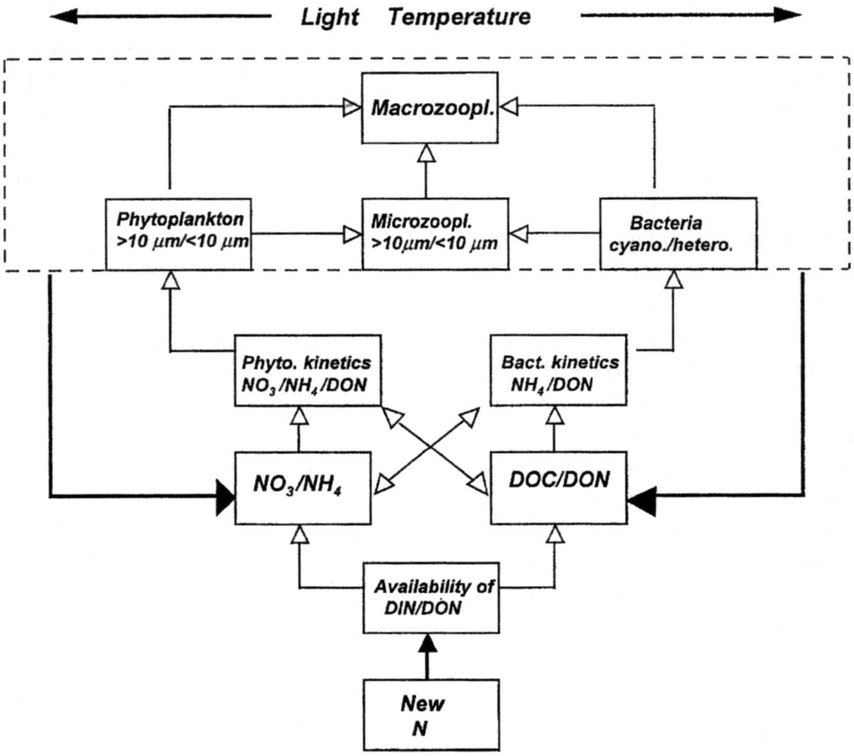

Figure 1. Flow diagram depicting relationships between new nitrogen input ('bottom-up' control), the ratio of ambient nitrogen and carbon availability, phytoplankton and bacterial uptake, plankton size fractions, macrozooplankton grazing ('top-down' control) and the release of nitrogen during grazing. The size fractions of the phytoplankton and microzooplankton boxes are arbitrarily shown here as >10 μm/<10 μm to be consistent with the examples provided in text. See text for further description.

ert et al. (1982), Probyn (1985), and Kokkinakis & Wheeler (1988) have shown that diatom-dominated assemblages use proportionately more NO_3^- than do nano- and picoplankton, but there are some other examples where this relationship is not as apparent (Chisholm, 1992). However, modeling efforts also predict that the input of new nitrogen is related to the biomass production and export, and inversely related to foodweb complexity and nitrogen regeneration (e.g. Vezina & Platt, 1987; Ducklow et al., 1989). Less clear is the regulation of the development of cyanobacteria versus heterotrophic bacteria.

Viewing these relationships from the perspective of regulation at the top, it is readily apparent that grazing can have a direct effect on the biomass of other trophic levels (Figure 1). Macrozooplankton will shape the species composition of the phytoplankton and microzooplankton through selective grazing, and while bacteria and cyanobacteria are not typically grazed directly by macrozooplankton (although there are exceptions), they can be consumed directly when attached to other particles. Selective grazing by the smaller microzooplankton has a similar impact. All of these grazing interactions, in turn, produce a flux of regenerated nitrogen, and the balance of nutrient ratios is again altered. Preferential grazing on different prey will also have differential effects on the flow of regenerated nitrogen. Simplistically, we might predict that grazing on phytoplankton will result in greater production of NH_4^+, as regeneration occurs and as phytoplankton, the primary consumers of NH_4^+ are removed. On the other hand, preferential grazing on microzooplankton will result in a decrease in the production of NH_4^+, as the primary regenerators of NH_4^+ are consumed. The system is not simple, however, as copepods further stimulate the regeneration of NH_4^+ by bacteria through the release of DON during feeding and metabolism (Roman et al., 1988), and by preying on larger microzooplankton, also relieve smaller microzooplankton from predation, in turn resulting in higher NH_4^+ regeneration (Glibert et al., 1992; Miller et al., 1995). Thus, the balance is a delicate one, with numerous feedbacks and impacts on both long and short term scales. Examples of these impacts across various scales are presented here.

Methods

This paper draws on data from several studies, spanning a range of scales. The seasonal study was conducted in the plume of the Chesapeake Bay, USA, during 1985–1986. The estuarine plume is characterized by a horizontal scale of 10–100 km, a vertical scale of 5–20 m, and a time scale of several days (Boicourt et al., 1987). Sampling was conducted while following surface drogues, with measurements of nutrients and biomass, rates of inorganic nitrogen and amino acid uptake and NH_4^+ regeneration, and rates of weight-specific grazing by copepods every few hours. Methods for all of the sampling and analytical techniques are documented in Malone and Ducklow (1990), Fuhrman (1987, 1990), Roman (1990) and Glibert et al. (1991).

The mesocosm study was conducted during summer 1995. The mesocosms consisted of tanks with 2 different sizes, but with a constant surface area: volume ratio. The volumes of the tanks were 0.1 m^3 (B tank) and 10 m^3 (D tank), with diameter of the tanks being 0.5 m and 2.4 m, respectively. At the start of the experiment, the mesocosms were filled with water from the Choptank River, a tributary of the Chesapeake Bay (<2 μm filter). Subsequently, they were diluted with 10% of the tank volume per day with the same water source. The experiments were run under a 12:12 L:D cycle and illuminated with cool white fluorescent light. Both mesocosms were well mixed with paddle stirrers throughout the experiment.

The tanks were run for a period of $\sim$10 days with no supplemental nutrient addition. Subsequently, 25 μmol NH_4^+ were added to each tank. The NH_4^+ was enriched with ^{15}N to a final enrichment of 10 atom %. The ^{15}N enrichment of the particulate nitrogen was monitored in the <10 and <202 μm fractions, as well as in the copepods over a several day period. In addition, biomass and nutrient concentrations were determined along the same time scale. Particulate nitrogen was analyzed using a Control Equipment CHN Analyzer, nutrients were analyzed with an autoanalyzer, and ^{15}N enrichment was determined by mass spectrometry as described in Glibert et al. (1991) and Glibert & Capone (1993).

The data from the smallest scale experiment were collected during a 48 hr carboy experiment conducted in late summer 1993. A complete description of this experiment is provided in Miller et al. (1997). Water from the Choptank River was dispensed into several 20 l polyethylene carboys, and the carboys were subsequently maintained in a constant temperature room under 13.5: 10.5 L:D. One set of carboys received an addition of 10 copepods l^{-1} (collected from the Choptank River). The initial inorganic nitrogen concentrations were <1.5 μg atom N l^{-1}. One control and one copepod-enriched carboy also received 20 μg atom N l^{-1} as arginine (to yield a DOC/DON $\leq$4). Carboys were sampled numerous times over 48 h for the determination of changes in ambient biomass and nutrients. Analytical techniques used are fully described in Miller et al. (1997).

Results and discussion

Example 1: Interactions on a seasonal scale

The plume of the Chesapeake Bay is an interesting site for examining the interactions between top-down and bottom-up control. The plume is a hydrographically distinct parcel of water, which is characterized by a distinct salinity structure and high biological productivity relative to ambient shelf water (Boicourt et al., 1987). There is generally little export of inorganic nutrients from the Bay, thus ambient inorganic nutrient concentrations are generally <5 μg atom N l^{-1}in the plume. The plankton community within the plume tends to shift from one that is dominated by autotrophic processes near the outflow source to one that is dominated by heterotrophic processes away from the source. This shift has been characterized in terms of decreasing chlorophyll concentrations, increasing bacterial and zooplankton abundance, and increasing rates of NH_4^+ regeneration (Boicourt et al., 1987; Malone & Ducklow, 1990; Glibert, et al. 1991).

During our study, a seasonal shift in the relative impact of bottom-up versus top-down effects on the distribution of size classes of chlorophyll was observed (Figure 2). In spring, NO_3^- dominated the nitrogen pool, and there was a positive relationship between the ratio of ambient NO_3^-/NH_4^+ and the relative proportion of chlorophyll in the >10 μm fraction/<10 μm fraction (Figure 2A), suggesting that the high input of NO_3^- appeared to promote growth by the larger phytoplankton. A doubling in the ratio of

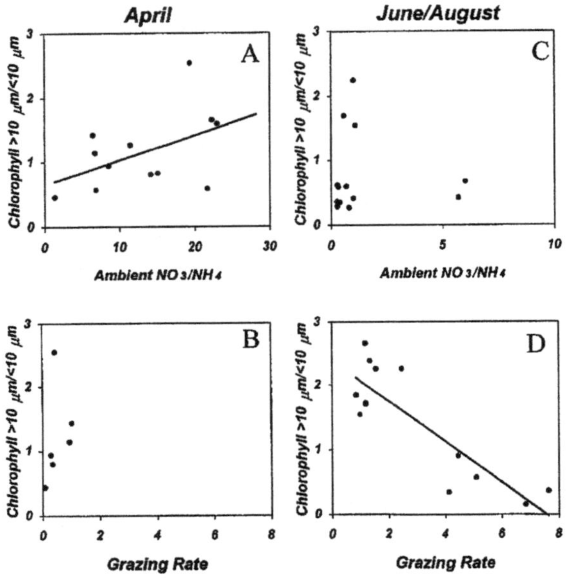

Figure 2. (A) Relationship between the ratio of ambient NO_3^-/NH_4^+ and the size distribution of chlorophyll, as the ratio of $>10\ \mu m/<10\ \mu m$ fractions, for Chesapeake Bay plume, April 1986. (B) Relationship between the weight-specific grazing rate (l (mg C)$^{-1}$ h^{-1}) of copepods on autotrophs (^{14}C -labeled particles) and the ratio of $>10\ \mu m/<10\ \mu m$ chlorophyll fractions, for Chesapeake Bay plume, April 1986. (C) As for (A), except for the months of June and August, 1985. (D) As for (B), except for the months of June and August, 1985. Lines shown are linear regressions.

NO_3^-/NH_4^+ correlated with nearly a 50% increase in the $> 10\ \mu m/<10\ \mu m$ chlorophyll. During this time of year, measurements by M. Roman revealed that the weight-specific grazing rate (l (mg C)$^{-1}$ h^{-1}) on autotrophs (^{14}C-labeled particles) by copepods was low, and there appeared to be little or no effect of grazing on chlorophyll size distribution (Figure 2B). In contrast, during the summer months, the reduced forms of nitrogen represented most of the nitrogen that was available, and no correlation emerged between the ratio of ambient NO_3^-/NH_4^+ and phytoplankton size fractions (Figure 2C). During summer, however, rates of weight-specific grazing (l (mg C)$^{-1}$ h^{-1}) on autotrophs by copepods were considerably higher, and a significant inverse relationship with chlorophyll size fractions was apparent suggesting that copepods were removing the larger phytoplankton ($r^2=0.73$, $P<0.005$; Figure 2D).

The impact of grazer control can also be seen in the direct relationships between NH_4^+ regeneration rates and the weight-specific grazing rates (Glibert et al., 1991). Across all seasons, a direct relationship between grazing and NH_4^+ regeneration was found (Figure 3A); highest grazing rates and highest NH_4^+ production rates were noted in the summer months. Consequently, temperature alone was a good predictor of the mean rate of NH_4^+ regeneration ($r^2=0.93$; Figure 4). Higher grazing rates also result in the production of organic compounds through sloppy feeding, with the rate of regeneration of dissolved free amino acids (DFAA; measured by J. Fuhrman) being roughly 10–15% that of NH_4^+ (Figure 3B). Direct correlations between NH_4^+ uptake and the uptake of dissolved free amino acids have previously been presented for the summer data of this study (Glibert et al., 1991), and Malone & Ducklow (1990) have shown that rates of carbon release in the summer are the direct consequence of higher copepod biomass and grazing rates. Ultimately, in summer, the biomass of the bacteria exceeded that of the phytoplankton (Malone & Ducklow, 1990), largely supported by products of sloppy feeding and grazer-mediated trophic changes.

Example 2: Interactions on a population scale

Mesocosms "*are simplified models of nature and agreements and disagreements with nature continue to be defined in every new experiment*" (Oviatt, 1994). Mesocosms are not true models of nature in a number of ways, such as the artificial light regime, and potential artifacts of surface wall area to volume. Nevertheless, they provide dramatic examples of the interplay between the effects of nutrient uptake leading to increased biomass production and grazing effects leading to decreased biomass.

In the mesocosms studied here, which differed significantly in size, the same nutrient pulse added at the same time resulted in dramatically different rates of incorporation by the respective fractions of the biomass (Figure 5). In Tank B, the ^{15}N atom % of both the $<202\ \mu m$ biomass, as well as the $<10\ \mu m$ biomass increased rapidly within the first few hours, but the ^{15}N atom % of the larger biomass fraction was consistently higher than that of the smaller biomass fraction. No further incorporation of $^{15}NH_4^+$ label, that is, increase in ^{15}N atom %, was observed after the first half day, corresponding with the time at which NH_4^+ concentrations were reduced to $<5\ \mu g$ atom N l^{-1}. On the other hand, in Tank D, the rate of incorporation of ^{15}N label proceeded more slowly than in Tank B, and by 24 h, the ^{15}N atom % enrichment of the $<10\ \mu m$ fraction exceeded that in the $<202\ \mu m$ fraction. Furthermore, differences in the rate of increase in the ^{15}N atom % enrichment in the copepods in the two tanks can also

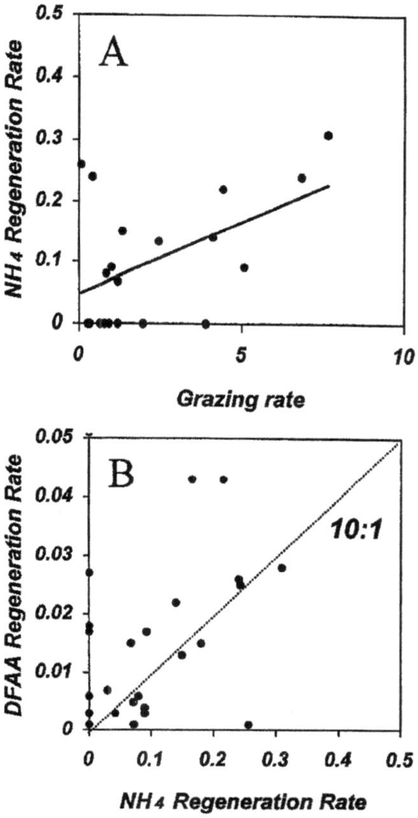

Figure 3. (A) Relationship between the weight-specific grazing rate (l (mg C)$^{-1}$ h^{-1}) of copepods on autotrophs (^{14}C-labeled particles) and the regeneration rate of NH_4^+ (μg atom N l^{-1} h^{-1}) for samples collected in the plume of the Chesapeake Bay across all seasons. (B) Relationship between the regeneration rate of NH_4^+ (μg atom N l^{-1} h^{-1}) and the regeneration rate of dissolved free amino acids (DFAA; μg atom N l^{-1} h^{-1}). Line shown is fit by linear regression.

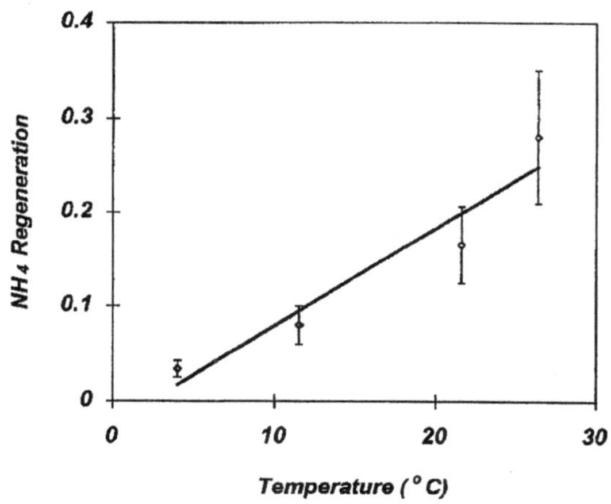

Figure 4. Relationship between mean water temperature and mean NH_4^+ regeneration rate (μg atom N l^{-1} h^{-1}) for four seasons studied in the Chesapeake Bay outflow plume. Line shown is fit by linear regression.

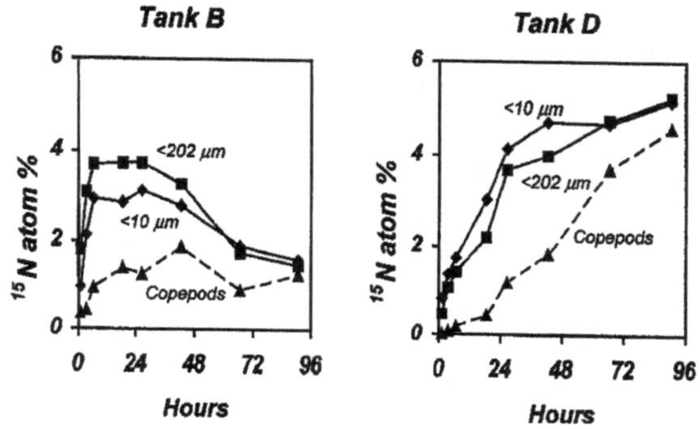

Figure 5. The ^{15}N atom percent enrichment of two size classes of plankton (<202 μm and <10 μm) and copepods with time after a single pulsed addition of $^{15}NH_4^+$ to the mesocosm tanks indicated.

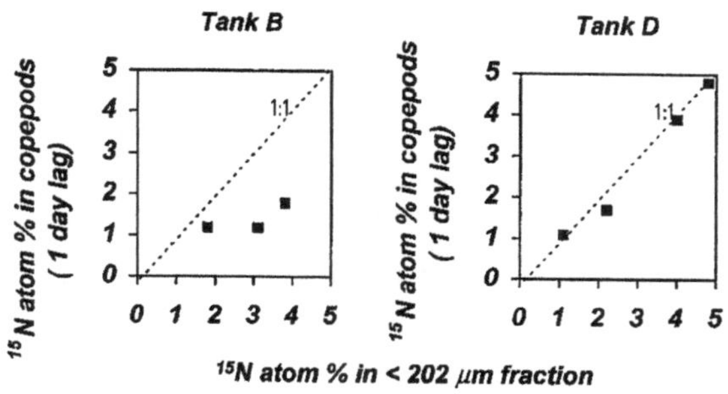

Figure 6. Relationship between the ^{15}N atom percent enrichment in the < 202 μm plankton size class and the ^{15}N atom percent enrichment in the copepods, lagged by one day, in the mesocosm tanks indicated.

be seen. Whereas the rate of increase in ^{15}N atom % enrichment in the copepods in Tank B increased for about 48 hrs, then stopped, in Tank D, the ^{15}N atom % of the copepods increased throughout the 4 days of measurement. By comparing the increase in ^{15}N atom % in the copepod fraction with the increase in atom% of the phytoplankton biomass, with a 1 day lag period, it can be seen that the ^{15}N atom % of the copepods in Tank D mimicked that of the <202 μm phytoplankton the day before, but in Tank B, the enrichment in the copepods after day 2 fell below that of the phytoplankton (Figure 6). This strongly suggests that the copepods in Tank D were grazing the phytoplankton in Tank D throughout the period of this study, but in Tank B, either the copepods stopped feeding after day 2, or chose an alternate (unlabeled) food. The alternate food sources may have been unlabeled microzooplankton in the water column or nonphytoplankton material accumulated at the wall surface or in the sediment. The smaller size of Tank B increased the likelihood that copepods were feeding at the walls or in the sediment rather than on the plankton.

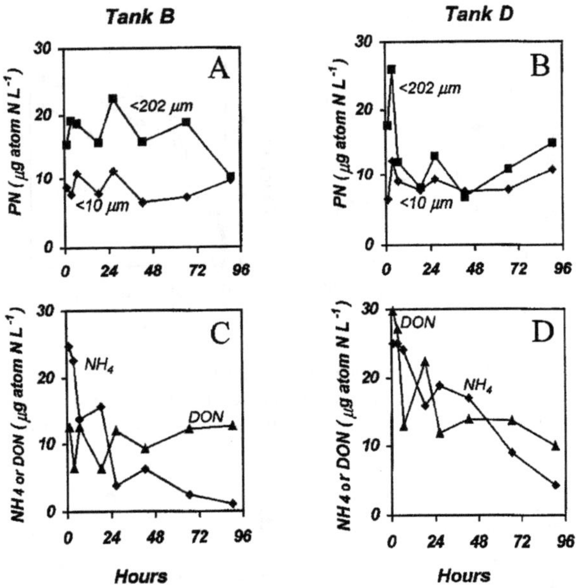

Figure 7. The change in particulate nitrogen (PN) concentration (μg atom N l^{-1} h^{-1}) in two size classes of plankton material (<202 μm and <10 μm) with time for tank B (panel A) and tank D (panel B), and the change in NH_4^+ (♦) and DON (▲) concentration (μg atom N l^{-1} h^{-1}) with time for tank B (panel C) and tank D (panel D).

The effects of the different grazing patterns can be seen in the distribution of biomass and nutrients in these tanks over the several day time course (Figure 7A, B). In Tank D, where the copepods accumulated more ^{15}N label from the phytoplankton, a sharp decrease in the availability of particulate nitrogen <202 μm can be seen, whereas such a sharp decrease was not observed in the other tank. Also, the higher biomass that remained in Tank B depleted the available NH_4^+ pulse considerably sooner (Figure 7C), eventually leading to nitrogen limitation. Finally, the mean concentrations of DON during the first 2 days of sampling in Tank D were about twice those in Tank B (Figure 7C, D), likely due at least in part to zooplankton-mediated processes. In Tank D the high concentrations of NH_4^+ and DON, combined with a high removal rate of the phytoplankton led to a doubling of bacterial abundance by day 5 (data not shown). These results again point to the strong interplay between zooplankton grazing, and the release of substrates utilizable by bacteria as well as by phytoplankton.

Example 3: Interactions on a metabolic scale

It is well recognized that bacteria can act as either net consumers or net producers of NH_4^+, and this is determined largely by the quality of the organic material available to the bacteria (Goldman et al., 1987; Goldman & Dennett, 1991; Jørgensen et al., 1993). One measure of substrate quality is the DOC/DON ratio. Other factors potentially determining whether bacteria will be consumers or regenerators of NH_4^+ include their nutritional state and their growth rate (Goldman & Dennett, 1991). Tupas & Koike (1990) have demonstrated that in natural populations of bacteria not only can assimilation and regeneration of NH_4^+ occur simultaneously (presumably by different members of the population), but also, the nutritional demands of the bacteria are met through the simultaneous uptake of NH_4^+ with amino acids. Hagström et al. (1988) showed that bacterivores and cyanobacteria contribute a significant fraction of organic material from cell excretion or lysis when primary production is high and this material is rapidly incorporated by the heterotrophic bacteria. Thus, bacteria can be a significant source and sink for both inorganic and organic nitrogen.

In our 48-hour carboy experiment, the DOC/DON ratio was manipulated along with the abundance of copepods. In the control carboy, and in the carboy in which the copepod abundance only was increased, there was little change in ambient concentration of NH_4^+ over the time period of study (Figure 8A, B). Processes of uptake and regeneration were in approximate balance under these conditions. In sharp contrast, in the carboy with added arginine, there was a rapid release of NH_4^+ between hours 17 and 23 of incubation (Figure 8C). Furthermore, in the carboy in which both arginine and copepod abundance were manipulated, the net accumulation of NH_4^+ was even larger, exceeding that which was added in organic form (Figure 8D). Approximately 30% more NH_4^+ was produced in this carboy than in the carboy receiving arginine alone. The greater net increase in NH_4^+ concentrations may have resulted from the direct excretion of NH_4^+ by the copepods, but the evidence from the control + copepod treatment suggests that this rate was not large. More likely, due to grazing, there was a production of additional low DOC/DON substrates, thereby providing more substrate for bacterial regeneration (Miller et al., 1997). Also, the copepods, by preying on microzooplankton probably relieved smaller microzooplankton from predation, resulting in higher NH_4^+ regeneration

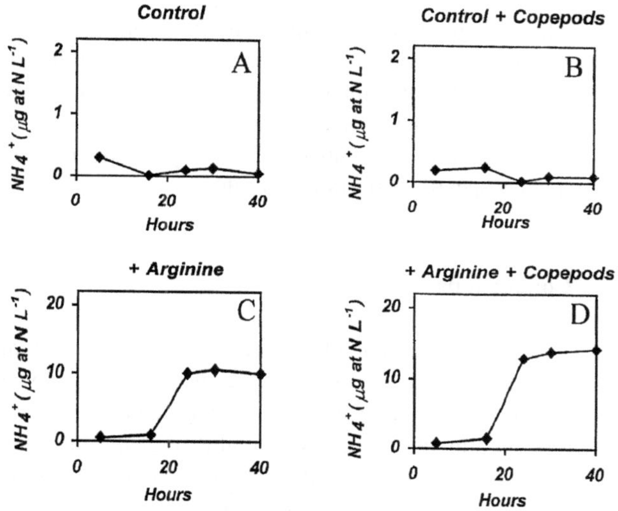

Figure 8. The change in NH_4^+ concentration (μg atom N l^{-1}) with time for four carboys of estuarine water which received no supplemental additions (control; panel A), 10 copepods l^{-1} (control + copepods; panel B), 20 μg atom N l^{-1} as arginine (+ arginine; panel C), or 20 μg atom N l^{-1} as arginine plus 10 copepods l^{-1} (+ arginine + copepods; panel D).

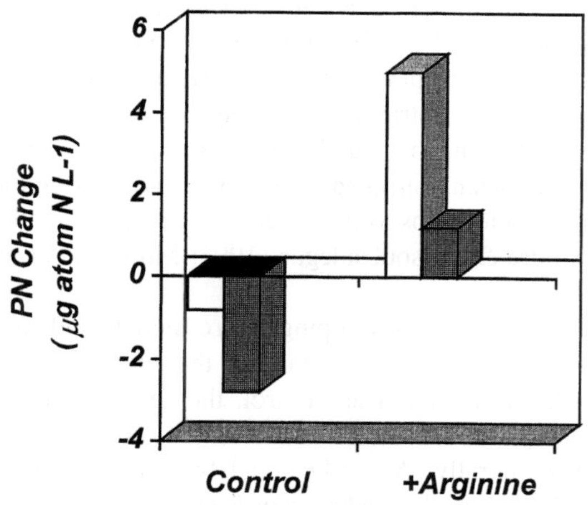

Figure 9. The net change in particulate nitrogen (PN) after 48 hrs of incubation for the carboys receiving the same supplemental additions described in Figure 8. The light bars represent the treatments with no additional copepods, and the dark bars are the + copepod treatments.

by the nanoflagellates (Roman et al., 1988; Miller et al., 1995). The change in particulate nitrogen in these treatments lends support to the notion that there was greater removal of both phytoplankton and microzooplankton in the presence of copepods, and that the added arginine provided was incorporated into biomass (Figure 9).

This result is consistent with other mesocosms experiments (e.g. Roman et al., 1988; Vaqué et al., 1989; Carlsson et al., 1995) in which phytoplankton biomass was found to be lower, but overall productivity rates higher in mesocosms containing grazers and suggests that the presence of copepods results in increased nutrient regeneration by both microzooplankton and bacteria. Roman et al. (1988) also observed a higher ratio of bacteria to phytoplankton in mesocosms with copepods than those without, further underscoring that in controlling microzooplankton and phytoplankton abundance through grazing, inorganic and organic substrates are not only released, but serve to stimulate production by the bacteria. There is no doubt that population dynamics must also influence our measurements of nitrogen regeneration in typical experiments where one group of organisms is artificially separated from another by size fractionation.

Other interactions

These examples only touch on some of the possible interactions between grazing and nutrient cycling. Other interactions between top-down and bottom-up control include viral interactions, heterotrophic nutrition by phytoplankton, mixotrophy, and release of organic nitrogen by phytoplankton.

Viral infection of bacteria and phytoplankton in many ways has similar effects to grazing: the reduction in biomass and the regeneration of nutrients. Viral infection results in the host cell rupturing, releasing all cellular components to the media (Proctor & Fuhrman, 1991; Suttle et al., 1991; Fuhrman, 1992). Proctor & Fuhrman (1991) reported that up to 60% of bacterial mortality and 30% of cyanobacterial mortality in coastal and oceanic systems may be due to viral infection. Viruses can therefore play an important role in both organic and inorganic nitrogen regeneration.

While it has long been recognized that bacteria are the primary sinks for DON, it is becoming increasingly recognized that some phytoplankton can utilize, and indeed compete effectively with bacteria, for organic nitrogen (Antia et al., 1991; Glibert, 1993). Organic nitrogen has been implicated in the outbreak of some nuisance blooms, for example (Paerl, 1988), particularly where the DIN/DON ratio has declined. Some phytoplankton have also been recently recognized to have an extracellular enzyme for scavenging nitrogen from organic substrates, oxidizing the DON to NH_4^+, H_2O_2, and an α-keto-acid. The released NH_4^+ is then available to be transported across the cell membrane (Palenik & Morel, 1990a, b; Pantoja & Lee, 1994; Mulholland et al., in press). This extracellular

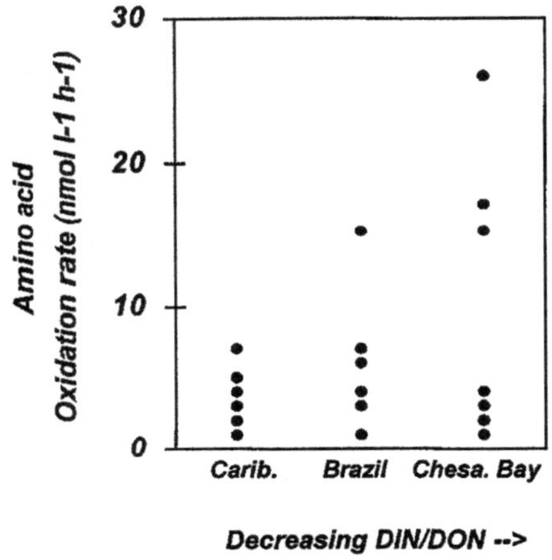

Figure 10. The rate of cell surface oxidation of the amino acid lysine for samples collected covering a span of DIN/DON ratios. All rates were measured when ambient concentrations of DIN were <1 μg atom N l^{-1}. Concentrations of DON in the Caribbean and coastal Brazil waters were <10 μg atom N l^{-1}, while concentrations of DON in summer mesocosm experiments with Chesapeake Bay water were ~ 30 μg atom N l^{-1}. Data are described more fully in Mulholland et al. (In press).

enzyme system appears to be present in certain species of phytoplankton, including cyanobacteria, prymnesiophytes and dinoflagellates. Cell surface oxidation of amino acids has been documented in a wide range of natural waters, from coastal and estuarine to oligotrophic, and relationships are beginning to emerge suggesting that the rate and contribution of this nitrogen source to the phytoplankton is in part a function of the DIN/DON availability (Figure 10; Pantoja & Lee, 1994; Mulholland et al., in press). While considerable variability exists in the rates of cell surface oxidation that have been measured in natural waters, highest rates to date have been documented from summer conditions in coastal and estuarine environments that are low in inorganic nitrogen, but which have relatively high concentrations of organic nitrogen available (Figure 10). Further studies are needed to address how these enzymes are related to the heterogeneous pool of DON in natural systems and how important this process is in overall nitrogen cycling.

Some flagellates also have the ability to combine photosynthesis and nutrient uptake with phagotrophy. Mixotrophy appears to be particularly common among the dinoflagellates, and they can consume prey close to their own size (Schepf & Elbrächter, 1992; Hansen et al., 1994). However, there has little work on the nutrient budget of such algae, and therefore how much of the consumed nitrogen is retained by the cells or released is not well known. Phytoplankton can and do release organic nitrogen under a wide range of nutritional conditions (Williams, 1990; Bronk & Glibert, 1991, 1994), but much controversy and uncertainty remains as to the physiological mechanisms involved, the environmental controls over the process, and the species specific differences in rates.

Conclusions

The examples provided here have served to show that in most systems, across a range of scales, top-down and bottom-up control are in a dynamic balance. Factors such as the relatively availability of the nitrogen substrates, the nutritional status of the component organisms, the number of trophic interactions, along with environmental parameters such as ambient light and temperature, all determine the extent of this balance. The stability of the balance will depend on the composition of the food web or the microbial web. Regeneration of nitrogen promotes stability. Thus, the presence of grazers can enhance the regeneration of NH_4^+ and DON in a system, but the magnitude and time scale of the effect will depend on the interactions of the trophic pathways. Factors driving the input of new nutrients, including physics, and factors controlling the abundance and grazing rates of the macrozooplankton, need to be considered in extending these relationships to food web models (Banse, 1994; Legendre & Rassoulzadegan, 1995; Mousseau et al., 1996).

Therefore, in developing future models and sampling strategies for understanding the relative roles of top-down or bottom-up control, the key is in understanding that these factors are linked, both spatially and temporally. As aptly noted by Hunter & Price (1992), '*In trophic webs, as in Escher paintings, flow can be upward or downward. From every intermediate level in a trophic web there are 'ladders' going up and 'chutes' going down, and the major players in the game are not restricted to the top or the bottom of the web.*'

Acknowledgments

I thank the organizers of the International PELAG Symposium for the opportunity to deliver this keynote address and for organizing an excellent conference. I thank my colleagues in the Chesapeake Bay plume study, particularly M. Roman, T. Malone, J. Fuhrman, and C. Garside for various data used in this analysis. I also thank C. Miller, J. Ludlam, M. Berg, M. Mulholland, and M. Lomas for lively discussions and insights, as well as assistance with analytical analysis and data interpretation associated with the mesocosm, the carboy, and the cell surface oxidase studies. This work was partially funded by the EPA Multiscale Experimental Ecosystem Research Center Project #R819640, and is contribution number 2813 from the Center for Environmental and Estuarine Studies.

References

Antia, N. J., P. J. Harrison & L. Oliveira, 1991. The role of dissolved organic nitrogen in phytoplankton nutrition, cell biology and ecology. Phycologia 30: 1–89.

Azam, F. T., T. Fenchel, J. G. Field, L. A. Meyer-Reil & F. Thingstad, 1983. The ecological role of water-column microbes in the sea. Mar. Ecol. Prog. Ser. 10: 257–263.

Banse, K., 1994. Grazing and zooplankton production as key controls of phytoplankton production in the open ocean. Oceanography 7: 13–20.

Bidigare, R. R., 1983. Nitrogen excretion by marine zooplankton. In Carpenter, E. J. & D. G. Capone (eds), Nitrogen in the Marine Environment. Academic Press: 385–409.

Boicourt, W. C., S.-Y. Chao, H. W. Ducklow, P. M. Glibert, T. C. Malone, M. R. Roman, L. P. Sanford, J. A. Fuhrman, C. Garside & R. W. Garvine, 1987. Physics and microbial ecology of a buoyant estuarine plume on the continental shelf. EOS 68: 666–668.

Bronk, D. A. & P. M. Glibert, 1991. A ^{15}N method for the measurement of dissolved organic nitrogen release by phytoplankton. Mar. Ecol. Prog. Ser. 77: 171–182.

Bronk, D. A. & P. M. Glibert, 1994. The fate of the missing ^{15}N differs among marine systems. Limnol. Oceanogr. 39: 189–195.

Carlsson, P., E. Granéli, P. Tester & L. Boni, 1995. Influences of riverine humic substances on bacteria, protozoa, phytoplankton, and copepods in a coastal plankton community. Mar. Ecol. Prog. Ser. 127: 213–221.

Caron, D. A. & J. C. Goldman, 1990. Protozoan nutrient regeneration. In Capriulo, G. M. (ed.), Ecology of Marine Protozoa. Oxford, 283–306.

Chisholm, S. W., 1992. Phytoplankton size. In Falkowski, P. G. & A. D. Woodhead (eds), Primary Productivity and Biogeochemical Cycles in the Sea. Plenum, 213–238.

Christaki, U. & F. Van Wambeke, 1995. Simulated phytoplankton bloom input in top–down manipulated mesocosms: comparative effect of zooflagellates, ciliates and copepods. Aquat. Microb. Ecol. 9: 137–147.

Dagg, M. J., 1974. Loss of prey contents during feeding by an aquatic predator. Ecology 55: 903–906.

Dam, H. G., X. Zhang, M. Butler & M. R. Roman, 1995. Mesozooplankton grazing and metabolism at the equator in the central Pacific: Implications for carbon and nitrogen fluxes. Deep-Sea Res. 42: 735–756.

DeBaar, H. J. W., 1994. VonLiebig's law of the minimum and plankton ecology (1899–1991). Prog. Oceanogr. 33: 347–386.

Ducklow, H. W., M. J. R. Fasham & A. F. Vezina, 1989. Derivation and analysis of flow networks for oceanic plankton systems. In F. Wulff, J. G. Field & K. H. Mann (eds), Network Analysis in Marine Ecology. Springer 159–205.

Dugdale, R. C. & J. J. Goering, 1967. Uptake of new and regenerated forms of nitrogen in primary productivity. Limnol. Oceanogr. 12: 196–206.

Eppley, R. W., S. G. Horrigan, J. A. Fuhrman, E. R. Brooks, C. C. Price & K. Sellner, 1981. Origins of dissolved organic matter in Southern California coastal water: experiments on the role of zooplankton. Mar. Ecol. Rog. Ser. 6: 149–159.

Eppley, R. W. & B. J. Peterson, 1979. Particulate organic matter flux and planktonic new production in the deep ocean. Nature 282: 677–680.

Fenchel, T. & P. Harrison, 1976. The significance of bacterial grazing and mineral cycling for the decomposition of particulate detritus. In Anderson, J. M. & A. MacFayden (eds), The Role of Terrestrial and Aquatic Organisms in Decomposition Processes. Blackwell, Oxford, 285–299.

Fisher, T. R., P. R. Carlson & R. T. Barber, 1982. Carbon and nitrogen primary productivity in three North Carolina estuaries. Estuar. coast. Shelf Sci. 15: 621–644.

Fuhrman, J. A., 1987. Close coupling between release and uptake of dissolved free amino acids in seawater studies by an isotope dilution approach. Mar. Ecol. Prog. Ser. 37: 45–52.

Fuhrman, J. A., 1990. Dissolved free amino acid cycling in an estuarine outflow plume. Mar. Ecol. Prog. Ser. 66: 197–203.

Fuhrman, J. A., 1992. Bacterioplankton roles in cycling of organic matter: the microbial loop. In Falkowski, P. G. & A. D. Woodhead (eds), Primary Productivity and Biogeochemical Cycles in the Sea. Plenum, 361–383.

Gebbing, J., 1910. Über den Gehalt des Meeres an Stickstoffnährsalzen. Untersuchungergebnisse der von der Deutschen Südpolar-Expedition (1901–1903) gesammelten Meerwasserproben. Internationale Revue der gestamten Hydrobiologie 3: 50–66.

Glibert, P. M., 1988. Primary productivity and pelagic nitrogen cycling. In T. H. Blackburn & J. Sørensen (eds), Nitrogen Cycling in Coastal Marine Environments. SCOPE 33, J. Wiley & Sons 3–31.

Glibert, P. M., 1993. The interdependence of uptake and release of NH_4^+ and organic nitrogen. Mar. Microb. Food Webs 7: 53–67.

Glibert, P. M. & D. G. Capone, 1993. Mineralization and assimilation in aquatic, sediment, and wetland systems. In Knowles, R. & T. H. Blackburn (eds), Nitrogen Isotope Techniques, 243–272.

Glibert, P. M., C. Garside, J. A. Fuhrman & M. R. Roman, 1991. Time-dependent coupling of inorganic and organic nitrogen uptake and regeneration in the plume of the Chesapeake Bay estuary and its regulation by large heterotrophs. Limnol. Oceanogr. 36: 895–909.

Glibert, P. M., J. C. Goldman & E. J. Carpenter, 1982. Seasonal variations in the utilization of ammonium and nitrate by phytoplankton in Vineyard Sound, Massachusetts, USA. Mar. Biol. 70: 237–249.

Glibert, P. M., C. A. Miller, C. Garside, M. R. Roman & G. B. McManus, 1992. NH_4^+ regeneration and grazing: interdependent processes in size-fractionated $^{15}NH_4^+$ experiments. Mar. Ecol. Prog. Ser. 82: 65–74.

Goldman, J. C., 1993. Potential role of large oceanic diatoms in new primary production. Deep Sea Res. 40: 159–168.

Goldman, J. C., D. A. Caron & M. R. Dennett, 1987. Regulation of gross growth efficiency and ammonium regeneration in bacteria by substrate C:N ratio. Limnol. Oceanogr. 32: 1239–1252.

Goldman, J. C. & M. R. Dennett, 1991. Ammonium regeneration and carbon utilization by marine bacteria grown on mixed substrates. Mar. Biol. 109: 369–378.

Goldman, J. C. & P. M. Glibert, 1983. Kinetics of inorganic nitrogen uptake by phytoplankton. In Carpenter, E. J. & D. G. Capone (eds), Nitrogen in the Marine Environment. Academic: 233–274.

Hagström, Å., F. Azam, A. Andersson, J. Wikner, & F. Rassoulzadegan, 1988. Microbial loop in an oligotrophic pelagic ecosystem: Possible roles of cyanobacteria and nanoflagellates in the organic fluxes. Mar. Ecol. Prog. Ser. 49: 171–178.

Hansen, B., P. K. Bjørnsen & P. J. Hansen, 1994. The size ratio between planktonic predators and their prey. Limnol. Oceanogr. 39: 395–403.

Harris, E., 1959. The nitrogen cycle of Long Island Sound. Bull. Bingham, oceanogr. Coll. 17: 31–64.

Harvey, H. W., 1945. Recent Advances in the Chemistry and Biology of Seawater. Cambridge University Press.

Hunter, M. D. & P. W. Price, 1992. Playing chutes and ladders: heterogeneity and the relative roles of bottom-up and top-down forces in natural communities. Ecology 73: 724–732.

Johannes, R. E., 1965. Influence of marine protozoa on nutrient regeneration. Limnol. Oceanogr. 10: 433–442.

Johannes, R. W., 1968. Nutrient regeneration in lakes and oceans. In Droop, M. R. & E. J. F. Wood (eds), Advances in Microbiology of the Sea. Academic Press 203–213.

Jørgensen, N. O. G., N. Kroer & R. B. Coffin, 1993. Dissolved free amino acids, combined amino acids, and DNA as sources of carbon and nitrogen to marine bacteria. Mar. Ecol. Prog. Ser. 98: 135–148.

Kokkinakis, S. A. & P. A. Wheeler, 1988. Uptake of ammonium and urea in the northeast Pacific: comparison between netplankton and nanoplankton. Mar. Ecol. Prog. Ser. 43: 113–124.

Lampert, W., 1978. Release of dissolved organic carbon by grazing zooplankton. Limnol. Oceanogr. 23: 831–835.

Legendre, L. & F. Rassoulzadegan, 1995. Plankton and nutrient dynamics in marine waters. Ophelia 41: 153–172.

Liebig, J. Von (1855) Principles of agricultural chemistry with special reference to the late researches made in England, 17–34. Reprinted in: Cycles of Essential Elements (Benchmark papers in Ecology, Vol. I, L. R. Pomeroy, 1974, Dowden, Hutchinson & Ross, Inc., Straussburg, Pennsylvania, 11–28.

Lindeman, R. L., 1942. The trophic-dynamic aspect of ecology. Ecology 23: 399–418.

MacIsaac, J. J. & R. C. Dugdale, 1972. Interactions of light and inorganic nitrogen in controlling nitrogen uptake in the sea. Deep-Sea Res. 19: 209–232.

Malone, T. C., D. J. Conley, T. R. Fisher, P. M. Glibert, L.W. Harding & K.G. Sellner, 1996. Scales of nutrient-limited phytoplankton productivity in Chesapeake Bay. Estuaries 19: 371–385.

Malone, T. C. & H. W. Ducklow, 1990. Microbial biomass in the coastal plume of Chesapeake Bay: Phytoplankton-bacterioplankton relationships. Limnol. Oceanogr. 35: 296–312.

McCarthy, J. J., 1982. The kinetics of nutrient utilization. In Platt, T. (ed.), Physiological Bases of Phytoplankton Ecology. Can. J. Fish. aquat. Sci 210: 211–233.

McCarthy, J. J. & J. C. Goldman, 1979. Nitrogenous nutrition of marine phytoplankton in nutrient depleted waters. Science 203: 670–672.

Miller, C. A., 1992. Effects of food quality and quantity on nitrogen excretion by the copepod, *Acartia tonsa*, PhD dissertation, University of Maryland, College Park.

Miller, C. A., D. L. Penry & P. M. Glibert, 1995. The impact of trophic interactions on rates of nitrogen regeneration and grazing in Chesapeake Bay. Limnol. Oceanogr. 40: 1005–1011.

Miller, C. A., P. M. Glibert, G. M. Berg & M. R. Mulholland, 1997. The effects of grazer and substrate amendments on nutrient and plankton dynamics in estuarine enclosures. Aquat. Microb. Ecol., 12: 251–261.

Mousseau, L., L. Legendre & L. Fortier, 1996. Dynamics of size-fractionated phytoplankton and trophic pathways on the Scotian Shelf and at the shelf break, Northwest Atlantic. Aquat. Microb. Ecol. 10: 149–163.

Mulholland, M. R., P. M. Glibert, G. M. Berg, L. Van Heukelem, S. Pantoja & C. Lee, in press. Extracellular amino acid oxidation by microplankton: A cross-ecosystem comparison. Aquat. microb. Ecol.

Nathansohn, A., 1908. Über die allgemeinen Produktionsbedingungen im Meere, Beiträge zur Biologie des Planktons, von H. H. Gran und Nathansohn. Internationale Revue der gestamten Hydrobiologie 1: 38–72.

Oviatt, C. A., 1994. Biological considerations in marine enclosure experiments: Challenges and revelations. Oceanography 7: 45–51.

Palenik, B. & F. M. M. Morel, 1990a. Amino acid utilization by marine phytoplankton: a novel mechanism. Limnol. Oceanogr. 35: 260–269.

Palenik, B. & F. M. M. Morel, 1990b. Comparison of cell-surface L-amino acid oxidases from several marine phytoplankton. Mar. Ecol. Prog. Ser. 59: 195–201.

Pantoja, S. & C. Lee, 1994. Cell-surface oxidase of amino acids in sea water. Limnol. Oceanogr. 39: 1718–1725.

Paerl, H. W., 1988. Nuisance phytoplankton blooms in coastal, estuarine, and inland waters. Limnol. Oceanogr. 33: 823–847.

Probyn, T. A., 1985. Nitrogen uptake by size-fractionated phytoplankton population in the southern Benguela upwelling system. Mar. Ecol. Prog. Ser. 22: 249–258.

Proctor, L. M. & J. A. Fuhrman, 1991. Roles of viral infection in organic particle flux. Mar. Ecol. Prog. Ser. 69: 133–142.

Riemann, B., N. O. G. Jørgensen, W. Lampert & J. A. Fuhrman, 1986. Zooplankton induced changes in dissolved free amino acids and in production rates of freshwater bacteria. Microb. Ecol. 12: 247–258.

Roman, M. R., H. W. Ducklow, J. A. Fuhrman, C. Garside, P. M. Glibert, T. C. Malone & G. B. McManus, 1988. Production, consumption, and nutrient cycling in a laboratory mesocosm. Mar. Ecol. Prog. Ser. 42: 39–52.

Roman, M. R., M. J. Furnas & M. M. Mullin, 1990. Zooplankton abundance and grazing at Davies Reef, Great Barrier Reef, Australia. Mar. Biol. 105: 73–82.

Schnepf, E. M. & M. Elbrächter, 1992. Nutritional strategies in dinoflagellates. Eur. J. Protistol. 28: 3–24.

Suttle, C. A., A. M. Chan & M. T. Cottrell, 1991. Use of ultrafiltration to isolate viruses from seawater which are pathogens of marine phytoplankton. Appl. envir. Microbiol. 57: 721–726.

Tupas, L. & I. Koike, 1990. Amino acid and ammonium utilization by heterotophic marine bacteria grown in enriched seawater. Limnol. Oceanogr. 35: 1145–1155.

Vaqué, D., C. Marrasé, V. Iñiguez & M. Alcarez, 1989. Zooplankton influence on phytoplankton-bacterioplankton coupling. J. Plankton Res. 11: 625–632.

Vezina, A. F.& T. Platt, 1987. Small-scale variability of new production and particulate fluxes in the ocean. Can. J. Fish. aquat. Sci. 44: 198–205.

Wikner, J. & Å. Hagström, 1988. Evidence for a tightly coupled nanoplanktonic predator-prey link regulating the bacteriovores in the marine environment. Mar. Ecol. Prog. Ser. 50: 137–145.

Williams, P. J. LeB., 1990. The importance of losses during microbial growth: Commentary on the physiology, measurement and ecology of the release of dissolved organic material. Mar. Microb. Food Webs 4: 175–193.

Hydrobiologia **363:** 13–27, 1998.
T. Tamminen & H. Kuosa (eds), *Eutrophication in Planktonic Ecosystems: Food Web Dynamics and Elemental Cycling.*

Population regulation and role of mesozooplankton in shaping marine pelagic food webs

Thomas Kiørboe
Danish Institute for Fisheries Research, Charlottenlund Castle, DK-2920 Charlottenlund, Denmark

Key words: Copepods, growth, mortality, grazing, vertical flux

Abstract

Copepods constitute the majority of the mesozooplankton in the oceans. By eating and being eaten copepods have implications for the flow of matter and energy in the pelagic environment. I first consider population regulation mechanisms in copepods by briefly reviewing estimates of growth and mortality rates and evidence of predation and resource limitation. The effects of variations in fecundity and mortality rates for the demography of copepod populations are then examined by a simple model, which demonstrates that population growth rates are much more sensitive to variations in mortality than to variations in fecundity. This is consistent with the observed tremendous variation in copepod fecundity rates, relatively low and constant mortality rates and with morphological and behavioral characteristics of pelagic copepods (e.g., predator perception and escape capability, vertical migration), which can all be considered adaptations to predator avoidance. The prey populations of copepods, mainly protozoa (ciliates) and phytoplankton, may be influenced by copepod predation to varying degrees. The highly variable morphology and the population dynamics (e.g., bloom formation) of the most important phytoplankton prey populations (diatoms, dinoflagellates) suggest that predation plays a secondary role in controlling their dynamics; availability of light and nutrients as well as coagulation and sedimentation appear generally to be more important. The limited morphological variation of planktonic ciliates, the well developed predator perception and escape capability of some species, and the often resource-unlimited *in situ* growth rates of ciliates, on the other hand, suggest that copepod predation is important for the dynamics of their populations. I finally examine the implications of mesozooplankton activity for plankton food webs, particularly their role in retarding vertical fluxes and, thus, the loss of material from the euphotic zone.

Introduction

This paper examines the role of mesozooplankton in structuring pelagic food webs. I partly do this in light of the classical question of top-down versus bottom-up control (Hairston et al., 1960; Verity & Smetacek, 1996). Obviously, both predation mortality and resource limitation determine population processes, but traditionally, most emphasis has been put on resource limitation in marine research. There are good reasons for that: the gross distribution of phytoplankton in vast areas of the ocean follows the availability of inorganic nutrients, the seasonality of the phytoplankton often depends on the availability of light, and the abundance of mesozooplankton is highest in seasons and regions with the highest phytoplankton production, where also fisheries yield typically peaks. The significance of top-down control has been repeatedly demonstrated in freshwater systems but rarely in marine environments even though fisheries yield per unit of phytoplankton production is 5–10 times higher in marine than in fresh water systems (Rudstam et al., 1994). Thus, one would expect predation processes and top-down control to be particularly important for structuring marine pelagic food webs.

Copepods constitute the majority of the mesozooplankton biomass in the ocean (ca. 80%; Verity & Smetacek, 1996) and may, thus, be considered a particularly successful group in the pelagic environment. In examining the role of mesozooplankton I shall, therefore, primarily be focusing on the planktonic copepods. The specific role of the planktonic copepods in structuring food webs cannot be evaluated solely from the rate at which they process matter. An assessment of the behaviour, morphology, life history traits, growth rates, and population dynamics of the copepods and of their predator and prey populations may provide cues to the understanding of the evolutionary success of copepods and to their dominance of todays oceans' mesozooplankton. And, hence, to their role in structuring pelagic food webs.

The approach taken in this study is first to examine to what extent (if at all) copepods and their prey populations (phytoplankton and protozoa) have individual growth rates *in situ* that are limited by availability of resources, or whether they grow mainly at maximal, temperature-dependent rates. I do this by compiling all available direct determinations of *in situ* growth rates of copepods, ciliates and phytoplankton. Variations in the sizes of populations whose individuals grow at near maximal rates are governed mainly by predation. The converse is not the case, because populations with individual growth rates that are severely food limited will both vary in accordance with variations in the availability of resources, and be influenced by variation in predation; i.e., a more complex pattern. To further explore the role of resource and predator limitation I next consider behavioral, morphological, and demographic characteristics of copepods and their prey populations. Finally, I examine the implications of mesozooplankton activity for the pelagic food web, particularly their role in modifying vertical fluxes and, thus, the loss of material from the euphotic zone.

In a recent review of the role of copepods in controlling ocean production Banse (1995) was considering those regions of the ocean where the concentration of phytoplankton does not vary seasonally. In this article I will mainly be considering seasonal seas; i.e., most neritic waters and some open oceans (e.g. the North Atlantic) of temperate and polar regions. This includes the Baltic.

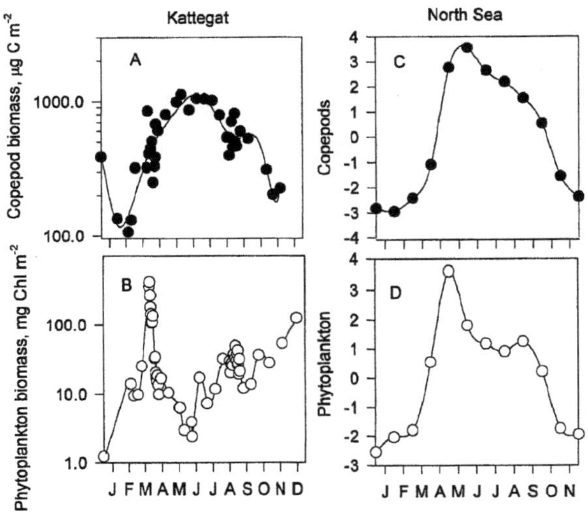

Figure 1. Seasonal variation in the concentration of planktonic copepods and phytoplankton in the Kattegat (Kiørboe & Nielsen 1994) and the North Sea (Colebrook 1979). The units for the North Sea data are standard deviations from the annual mean.

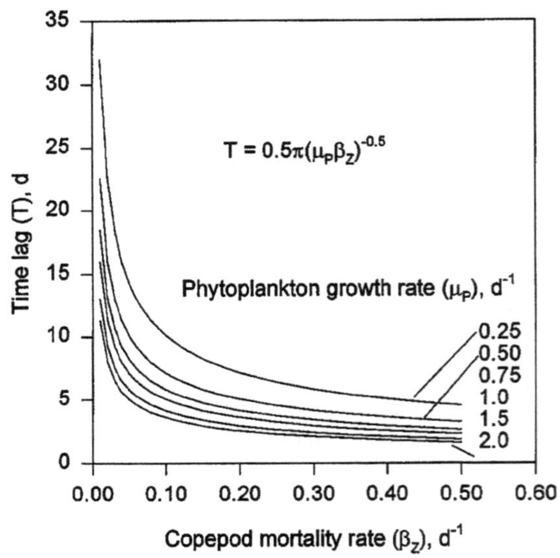

Figure 2. Phase shift between predator (copepods) and prey (phytoplankton) population oscillations estimated from the Lotka-Volterra model as a function of phytoplankton growth rate and copepod mortality rate.

What regulates copepod population sizes?

The seasonal variation of phyto- and mesozooplankton in temperate waters often shows what resembles coupled oscillations (Figure 1) which might suggest a classical Lotka-Volterra predator–prey relationship; i.e., that the prey (phytoplankton) is limited by predation, and the predator (zooplankton) is resource limited. The Lotka-Volterra predator-prey model implies that the two populations oscillate one quarter of a cycle out of phase and with a period $T = 2\pi(\mu_P\beta_Z)^{-0.5}$, where μ_P is the phytoplankton growth rate in the absence of predation and β_Z is the zooplankton (copepod) mortality rate (Pielou, 1969). The time lag between the seasonal (spring) peak of the phytoplankton and that of the zooplankton in temperate waters is typically on the order of 1 to several months (Figure 1). *In situ* growth rates of bulk phytoplankton, μ_P, are in excess of 0.25 d^{-1} in cold (5 °C) water, and much higher (> 1–2 d^{-1}) in warmer water (Banse, 1995 and Figure 6). Reported *in situ* copepod mortality rates, β_Z, range between 0.02–0.3 d^{-1} (Table 2). Realistic values of μ_P and β_z, thus, predict time lags of less than 1 month (Figure 2). Only for values of μ_P and β_Z that are in the extreme lowest range of those reported may the predicted time lag be consistent with that observed in, for example, the North Sea (where the ascent of a large overwintering population may violate the assumptions of the model). However, for any realistic values, the prediction is inconsistent with observations in more shallow seas such as the Kattegat (Figure 1). Traditionally, most emphasis has been put on resource limitation of copepod growth and population dynamics and although the gross seasonal pattern does suggest that the copepods are ultimately resource limited, this simple analysis suggests that mortality (predation) may be equally important in determining the population dynamics of copepods in the oceans. Thus, the question of top-down vs. bottom-up control appears to depend on the scale at which the problem is considered.

Evidence of resource limitation

Copepods and other mesozooplankton feed in a nutritionally dilute environment. However, they have developed techniques to extract sufficient food from this environment to maintain populations. Over the last several decades much research has been devoted to elucidating the food gathering mechanisms in copepods, and to estimate feeding and growth rates in natural environments. This emphasis on feeding rates and gathering of difficult accessible (dilute) food has focused research on food limitation as a critical factor for the maintenance of copepod populations. There is, in fact, sufficient evidence from the field, that copepods are frequently food limited:

Juvenile growth: There are not many direct estimates of juvenile growth rates *in situ*. Huntley & Lopez (1992) compiled field data on cohort development rates and indirectly inferred *in situ* juvenile growth rates from those observations. Their data suggested that juvenile growth was mainly limited by temperature and, thus, not much limited by resources. However, their approach to estimate growth rate can be criticized because it is biased towards the conclusion they arrived at (see e.g. Kiørboe & Nielsen, 1994). What their data *do* suggest, however, is that development rate – which is different from growth rate – is often close to maximum *in situ*. In Figure 3 I have compiled all available direct estimates of *in situ* growth of juvenile copepods. Growth rates have in all cases been estimated more directly by incubating juveniles in ambient water by some variant of the technique suggested by Tranter (1976) and subsequently developed by Kimmerer & McKinnon (1987). All estimates were converted to a temperature of 15 °C by assuming a Q_{10} of 3.0 (Sabatini & Kiørboe, 1994). Laboratory studies have shown that maximum growth in juvenile copepods is independent of body size, and the average of the maximum growth rates estimated in the laboratory have been indicated for nauplii and copepodids (from Kiørboe & Sabatini, 1995). On average, nauplii appear

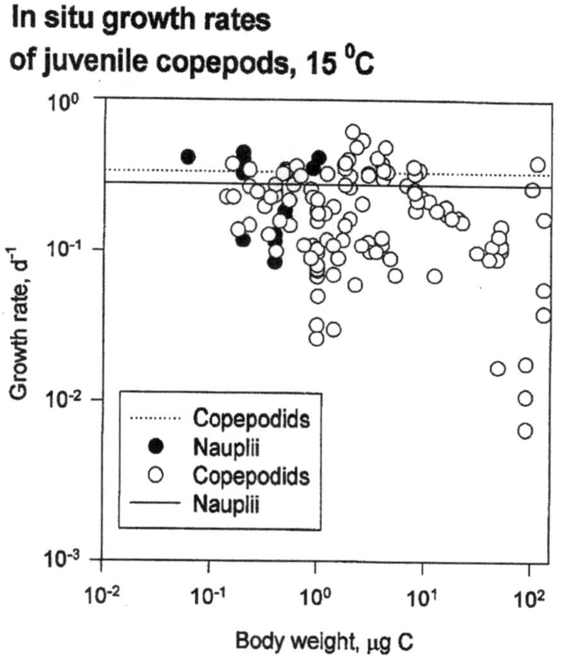

Figure 3. Direct estimates of copepod *in situ* specific growth rates in nauplii (closed symbols)and copepodids(open symbols) as determined by variants of the incubation technique of Tranter (1978). All measurements have been converted to 15 °C by assuming a Q_{10} of 3.0. The field observations have been compared with average maximum growth rates measured in the laboratory for nauplii (full line) and copepodids (dotted line), as derived from the review of Kiørboe & Sabatini (1995). Field data are from Burkill & Kendall (1982), Chrisholm & Roff (1990a,b), Diel & Klein Breteler (1986), Fransz & Diel (1985), Haslund & Fryd (1990), Hutchings et al. (1995), Kimmerer (1983), Kimmerer & McKinnon (1987), Kivi et al. (1993), Miller & Nielsen (1988), Peterson et al. (1991) and Walker & Peterson (1991).

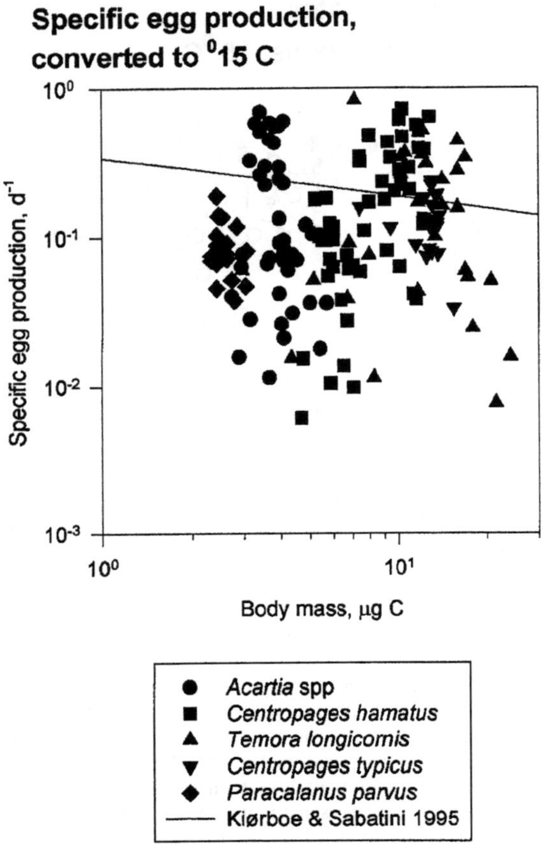

Figure 4. Direct estimates of *in situ* weight-specific egg production rates (incubation method) for 5 species of neritic copepods as recorded during one annual cycle in the Southern Kattegat (Data from Kiørboe & Nielsen, 1994). All data have been converted to 15 °C by assuming a Q_{10} of 3.0. Field data have been compared with average maximum specific egg production in broadcast spawning calanoid copepods measured in laboratory and derived form the review of Kiørboe & Sabatini (1995).

to grow at rates identical to those maximally predicted from laboratory studies (0.28 ± 0.13 and 0.28 ± 0.08 d^{-1} in field and lab.), while copepodids on average grow somewhat slower (0.20 ± 0.13 and 0.34 ± 0.08 d^{-1}), suggesting that the latter are somewhat food limited on average. This lends only partial support for the conclusion of Huntley & Lopez (1992). Further evidence of food limitation stem from simultaneous observations of female fecundity and juvenile growth which suggest a correlation between the two (Peterson et al., 1991; Kiørboe & Nielsen, 1994); this, in turn, suggests that food availability is limiting both. Also, Kimmerer & McKinnon (1987) in a seasonal study showed that *in situ* growth rates of juvenile *Acartia tranteri* were always less than maximal, and could be

Table 1. Comparison of average *in situ* growth rates with those maximally estimated in the laboratory and coefficient of variation of *in situ* growth rates of copepod nauplii, copepodids, females (egg production), pelagic ciliates and bulk phytoplankton. Data are those reported in Figures 3, 4, 6 and 8

	In situ growth/max growth	CV% of growth *in situ*
Copepods		
nauplii	1.00	47%
copepodids	0.59	62%
Females	0.91	116%
Ciliates	0.52	60%
Phytoplankton (0.5 × Eppley)	0.65	49%

enhanced by the addition of phytoplankton to ambient water. Finally, vertical migration may reduce *in situ* growth relative to that measured by incubation techniques, and this may in particular be relevant to larger copepods (cf. Aksnes, 1996). However, overall juvenile growth appears not to be strongly food limited.

Female fecundity: There is abundant evidence that female fecundity *in situ* is very often much less than maximum and frequently correlated to measures of ambient food availability (Dagg, 1978; Checkley, 1980; Durbin et al., 1983; Kiørboe & Nielsen, 1994). This suggests that fecundity is frequently resource limited. Data in Figure 4, from just one sampling site, illustrate the variability in egg production both between and within species. A substantial portion of the observations lie above the line of average maximum fecundity in laboratory experiments (review by Kiørboe & Sabatini, 1995) which may suggest that laboratory measurements have underestimated maximum egg production. In considering the degree of food limitation I will, therefore, rather emphasise the *variability* in fecundity; the more egg production or growth depend on ambient food level, the more variable will it be. If all data are converted to 15 °C and averaged, the coefficient of variation is 116%. This is more than for nauplii (47%) and copepodid (62%) growth rates (Table 1), even though the inherent variability in the measuring techniques is less for egg production than for juvenile growth (cf. Miller et al., 1984), and despite the fact that the fecundity data were collected from one site while the growth data were collected from a variety of sites.

I conclude from the above analysis that resource limitation of copepod growth is unequally distributed among developmental stages. On average, female fecundity *in situ* is more limited by food availability than juvenile growth, and nauplii are less limited (if at all) than copepodids. Food limitation, thus, appears to become more severe with development. I shall later examine the demographic implications of these observations.

Evidence of predation limitation

There are relatively few field estimates of mortality rates (Tables 2 and 3) in copepods and even fewer studies that have elucidated the role of predation-mortality in shaping the seasonal and spatial occurrence of copepod populations (Ianora & Buttino, 1990; Kiørboe & Nielsen, 1994; Mullin, 1991). Post hatching mortality rates in small neritic copepods vary of course between species and sites but are on average modest (0.11 d^{-1} for both nauplii and copepodids; Table 2). In contrast mortality rates of free eggs are high, averaging 3.0 d^{-1} for the species and sites studied (Table 3). However, the significance of mortality rates in limiting population sizes cannot be deduced alone from estimates of their magnitude. There are a number of morphological and behavioral features that suggest that mortality – and predator avoidance – has been a very significant factor in copepod evolution. These include (based partly on Verity & Smetacek, 1996):

Shape, predator perception and escape capability: Most planktonic copepods have very similar morphologies; i.e., a hydrodynamically shaped elongated body with well developed musculature. This allows a rapid escape response, as has been documented in many copepods (Ohman, 1988). Another common feature is the presence of extending antennae with mechanosensory hairs by which approaching predators can be perceived (Yen et al., 1992). The similarity in body shape among copepods, and the fact that this shape cannot be considered an adaptation to food acquisition, suggest predation to be a significant factor in its evolution (Verity & Smetacek, 1996).

Distribution patterns: Enclosed shallow bays and fjords provide 'predation environments' that differ from more open oceans: visual predation and predation due to benthic filter feeders are both more intense. The characteristic zonation of copepod species from the bottom of such bays seawards is thought to be a consequence of different species' adaptation to the different predation environments (Kimmerer, 1991).

Diel vertical migration: Many copepods undertake diurnal vertical migration, and this is often considered an adaption to reduce the risk of predation (Frost, 1988). This is further emphasized by the fact that the vertical migration behaviour can be modified by the presence of predators in a way that reduces predation risk (Bollens & Frost, 1991; Bollens et al., 1994). Also, ovigerous females of several sac-spawning copepods, which pay a particular high demographic price of being eaten by a predator (they loose not only their own life, but also that of their offspring), show vertical distributions and migration patterns that differ from those of co-occurring non-ovigerous females and which further reduces predation risk (Vuorinen, 1987; Bollens & Frost, 1991). These adaptations may have a price in terms of lowered feeding and fecundity (Kiørboe & Sabatini, 1994; Ohman, 1990).

Spawning strategy: Most calanoid copepods, that dominate the marine environment, are broadcast spawners; i.e. they release their eggs in the water column. Available estimates of the mortality rate of such suspended eggs suggest that they suffer very high mortalities (see Table 3), presumably because they are eaten by planktivorous organisms, including the copepods themselves. Brood protection has evolved in some marine copepods (most cyclopoids carry their eggs), but is far from being as common as in freshwater; egg-carrying copepods and cladocerans, which carry their embryos, dominate the mesozooplankton there, and egg-carrying behaviour is even common among microzooplanktonic rotifera. This may suggest that egg mortality is more crucial in freshwater environments, which may be related to the generally much higher density of planktivorous organism here than in ocean waters and, thus, a much higher predation risk of suspended eggs. A possible adaptation to the much higher mortality rates of free as compared to carried eggs is that the former have egg development times that are only 1/3 of the latter, and thus allows predator escape reactions of the offspring earlier (Kiørboe & Sabatini, 1994). Copepod nauplii, like later stages, can perceive approaching predators and feeding currents and have developed relatively efficient escape responses (Williamson & Vanderploeg, 1988; Tiselius & Jonsson, 1990).

Table 2. Reported mortality rates of copepod nauplii and copepodids

Species	Nauplii Mortality rate, d^{-1}	Copepodids Mortality rate, d^{-1}	Reference
Acartia clausii		Avg. 0.16	Landry (1978)
			Myers & Runge (1983)
Acartia tranteri	Avg. 0.16	Avg. 0.02	Kimmerer & McKinnon (1987)
Acartia tonsa		0.10–0.41 (median: 0.16)	Andersson (1996)
Acartia hudsonica	0.05	0.05	Durbin & Durbin (1981)
Paracalanus parvus		0.3	Aksnes & Magnesen (1987)
Centropages hamatus		0.07	Same
Pseudocalanus elongatus		0.11	Same
Temora longicornis		0.15	Same
Temora longicornis	0.14 (0.05–0.22)	0.10 (0.05–0.19)	Bakker & Rijswiljk (1987)
Pseudocalanus sp.		0.04 (females)	Ohman (1986)
Pseudocalanus newmani	Avg. 0.10	0.04–0.16	Ohman (1995)
			Ohman & Woods (1996)
Small calanoids		0.03–0.15 (Avg. 0.08)	Kiørboe & Nielsen (1994)

Table 3. Reported mortality rates of free eggs in neritic copepods

Region	Period	Species	Egg mortality, d^{-1}	Reference
Long Island Sound	All year	*Temora longicornis*	0.5–21 (Avg. 4.6)	Peterson & Kimmerer (1994)
Kattegat		Small calanoids	0–2.3 (Avg. 0.8)	Kiørboe & Nielsen (1994)
Swedish west coast		*Acartia tonsa* *Paracalanus parvus*	3.5–9.6	Kiørboe et al. (1988)
Long Island Sound		*Acartia tonsa*	up to 4.7	Beckmann & Peterson (1986)
Roskilde Fjord		*Acartia tonsa*	up to 5.2 (Median: 2.2)	Andersson (1996)
Various		*Acartia clausii*	0.8–1.3 (Including NI)	Landry (1978) Uye (1982)

Demographic implications of resource limitation and predation mortality

I have examined the demographic implications of the above documented variation in fecundity and mortality rates by means of a simple model that expresses population growth rate as a function of fecundity and mortality. The model is based on the assumption of a population of broadcast spawning copepods with egg hatching time τ, development time δ (time from hatching to adulthood), fecundity m, egg mortality β_e, and posthatching mortality β. For such a population the net reproductive rate, R_0, and the population growth rates, r, are given by (Kiørboe & Sabatini, 1994):

$$R_0 = (m/\beta)e^{-(\beta_e-\beta)\tau-\beta\delta}$$

$$r = \ln R_0/(\delta + \beta^{-1}).$$

Combining the two equations yields

$$r = (\ln(m/\beta) - (\beta_e - \beta)\tau - \beta\delta)/(\delta + \beta^{-1}).$$

In the following simulations I have assumed that both egg hatching time and development time depend only on temperature. In fact, development time may vary somewhat with the availability of food. However, it is much less variable than the growth rate (Berggreen et al., 1988), and the above data suggest limited variability of growth – and hence δ – with environmental factors other than temperature in field populations. Putting development time $\delta = 25$ d and hatching time $\tau = 1.5$ d, which are the averages for free spawning copepods at 15 °C (Kiørboe & Sabatini, 1995), leaves three parameters to determine variation in population growth rate, namely the mortality rates of eggs and post hatching individuals, and the female fecundity. In Figure 5 I have varied one at a time and assumed constant values for the two others. The values assumed are typical values; the conclusions below are robust to realistic variation in the magnitude of these assumed values.

As expected, population growth rate increases with fecundity but, interestingly, it is positive even at very low fecundities ($\sim$1 egg female^{-1} d^{-1}). It then increases very steeply with fecundity and becomes almost independent of fecundity when this exceeds 5–10 eggs female^{-1} d^{-1} (Figure 5a). An increase in fecundity from, for example, 10 to 100% of the maximum possible does not increase the population growth rate much. This would suggest that selection pressure for high fecundity is not very strong: there is not much to achieve.

The initial increase in population growth rate with increasing post hatching mortality is due to the unrealistic implicit assumption in the model that, as β goes towards 0, the copepods get eternal life (Figure 5b). I have not attempted to correct for this. The point I want to illustrate here is that, at a relatively low mortality rate, population growth becomes negative. The threshold value calculated here, 0.15 d^{-1}, is close to typical values found for copepod populations (Table 2). At higher mortality rates, population growth declines rapidly. This would suggest a strong selection pressure to minimize post-hatching mortality. Finally the exercise illustrates that population growth can be maintained at even incredibly high egg mortality rates (Figure 5c).

This model exercise provides a framework for interpreting the above observations of fecundity and mortality *in situ*. It demonstrates that copepod populations can be maintained with very low fecundities and with extremely high egg mortality rates. There would, thus, not be much selection pressure to maximize, respectively, minimize these rates. The variable fecundity and high egg mortality observed in field populations is consistent with this. On the other hand, post hatching mortality rates exceeding a relatively low value are critical for population maintenance and there would, thus,

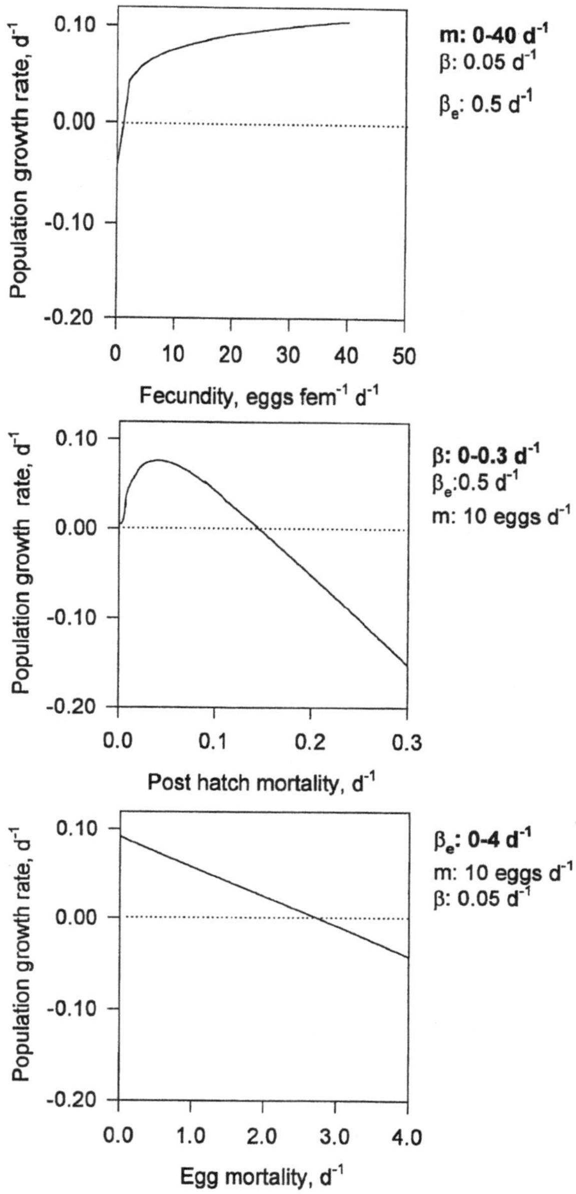

Figure 5. Simulation of population growth rate in a broadcast spawning copepod as function of egg production rate (m), post hatching mortality rate (β), and egg mortality rate (β_e). In each simulation one of the three parameters are allowed to vary while the two others are kept constant. The constant values are $m = 10$ eggs d^{-1}, $\beta = 0.05$ d^{-1}, $\beta_e = 0.5$ d^{-1}.

be strong selection for predator avoidance in copepods. Again, the evidence reviewed above seems to be consistent with this. Thus, population dynamics in planktonic copepods appear to be strongly controlled by top-down processes.

The effect of copepod grazing on phyto- and protozooplankton

The main food of planktonic copepods is phytoplankton and microzooplankton. Most copepods are omnivorous to some degree, i.e., they feed on both types of foods. They may use different hunting strategies for phytoplankton and microzooplankton, and the efficiency by which they capture different types of food depend on both the hunting strategy, the size and escape capability of the prey (Berggreen et al., 1988; Jonsson & Tiselius, 1990) as well as on the ambient levels of turbulence (Saiz & Kiørboe, 1995). Clearing rates on motile microzooplankters are typically higher than clearing rates on phytoplankton (Stoecker & Egloff, 1987; Wiadnyana & Rassoulzadegan, 1989). Below I discuss the implications of copepod grazing to populations of ciliates and phytoplankters.

Phytoplankton

The oceans' phytoplankton production is overall limited mainly by the availability of inorganic nutrients in the euphotic zone. This is why coastal areas and particularly upwelling regions are the most productive. Primary production is the product of biomass and growth rate and these two components may not be regulated the same way.

Growth: Phytoplankton growth is potentially limited both by availability of building blocks and by light. Banse (1995) considered the regulation of phytoplankton growth in environments where grazing and remineralization were balancing (or balanced by) nutrient uptake and phytoplankton growth. He argued that zooplankton grazing (and remineralization) was driving the rate of phytoplankton production, rather than vice versa. As noted by Banse (1995), however, it is difficult to distinguish cause and effect in a truly steady-state situation, and there are equally good arguments that the control is rather opposite, i.e. that phytoplankton growth is mainly limited by light (Goldman, 1987). Let us briefly review the evidence. Banse (1995) compiled all available direct determinations of *in situ* bulk phytoplankton growth rates (Figure 6a). This plot very much resembles that obtained for maximum, nutrient replete growth rates of unicellular algae at continuous, saturating light in the laboratory by Eppley (1972) (Figure 6b). To facilitate comparison, the enveloping curve of Eppley's data as well as 50% of that rate (to approximate the natural light cycle) have also been shown on the graph with the field data. Two points suggest that phytoplankton growth *in situ* is mainly independent of the concentration of nutrients: (i) the scatter below the curves in Figure 6a for *in situ* growth is very similar to that obtained by Eppley for nutrient replete cells, and (ii) there is no difference in growth rates of phy-

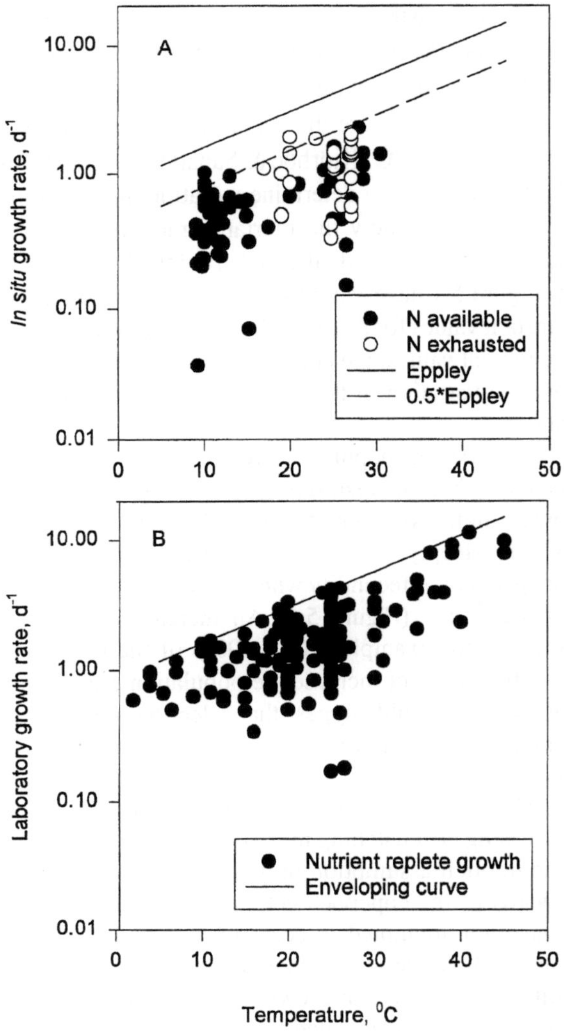

Figure 6. A. Direct determinations of *in situ* bulk phytoplankton growth rates compiled from a variety of environments (replotted from Banse, 1995). Open symbols are environments in which nitrogen has been exhausted and closed symbols are environments in which nitrogen is available. The full line is the curve enveloping nutrient replete growth rates measured in the laboratory for a variety of phytoplankton species in continuous light (from Eppley, 1972), and the dotted line is half that rate (to simulate diurnal variation in light). B. Nutrient replete growth rates of a variety of phytoplankton species in laboratory experiments (replotted from Eppley, 1972; observations of 0 growth have been omitted).

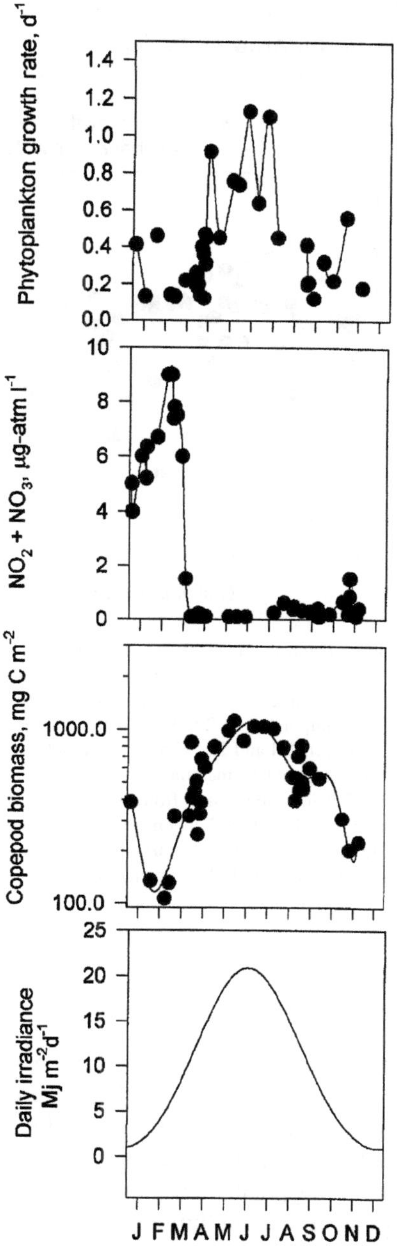

Figure 7. The seasonal variation in phytoplankton growth rate, concentration of NO_2 and NO_3, irradiance, and biomass of planktonic copepods as determined at a permanent station in the Southern Kattegat, Denmark. Modified from Kiørboe (1996).

toplankton between oligotrophic environments (where inorganic nutrients may be below detection level) and growth in eutrophic environments. Furthermore, the coefficient of variation for all measured growth rates (corrected for variation in temperature) is relatively low (49%) and similar to that observed for nauplii above (Table 1). These observations corrobate the suggestion by Goldman (1987) that phytoplankton in the ocean is typically growing at maximum rates, only limited by light. Figure 7, with seasonal data from the Kattegat, further supports this contention. Goldman's (1987) hypothesis of light limitation also implies that changes in light intensity leads to corresponding changes in growth rate. This is confirmed by the common experience that field populations incubated in the laboratory show a positive relation between growth and light – as routinely demonstrated by the P-I-curves obtained in connection with primary production measurements. Evidence to the contrary stems from incubation experiments, in which addition of nutrients may stimulate the growth rate (see e.g. Banse, 1995 for references).

If zooplankton grazing and remineralization rather than light is governing the rate of primary production in steady state systems, as maintained by Banse (1995), then the phytoplankton growth rate should increase with increasing zooplankton grazing rate. The similar patterns of copepod biomass and phytoplankton growth rate in a system which is not in steady state (Figure 7) might be taken as supporting evidence for Banse's thesis. However, one could equally well argue that light was determining the phytoplankton growth rate in this seasonal system, because daily irradiance shows a seasonal pattern similar to that of the phytoplankton growth rate and because light is evidently limiting phytoplankton growth in periods of high ambient nutrient concentrations (spring) (Figure 7). Also, copepod grazing is only a relatively small fraction of phytoplankton production; in this particular environment only about 12% on an annual basis (Kiørboe & Nielsen, 1994).

Biomass and size distribution: Obviously, phytoplankton biomass is limited by the availability of building blocks, i.e. inorganic carbon and nutrient salts, being macronutrients (such as nitrogen) or micronutrients (such as iron). However, it may be modified by grazing and sedimentation, particularly when the ambient resources of nutrients have been exhausted. As noted, grazing due to mesozooplankton typically constitutes only a relatively small fraction of the phytoplankton production in seasonal environments, and blooms, in particular, rather sediment out of the euphotic zone due to aggregation of the cells (e.g. Kiørboe et al., 1994). Mesozooplankton may, however, significantly modify this flux (see below).

This does not imply that zooplankton grazing has no effects on the phytoplankton. First, pico- and nanophytoplankton growth is typically balanced by (micro)zooplankton grazing, both in the oligotrophic, steady-state environments discussed by Banse (1995), but also in periods of a plentiful nutrient supply; i.e., when nutrients are or temporarily become available. This explains why the concentration of small phytoplankton cells in the ocean is relatively constant (not bloom-forming), and is caused by the similar population growth rates of the pico- and nanoalgae and their microzooplankton predators, which enables the microzooplankton to control their phytoplankton prey populations (e.g. Kiørboe, 1993). In contrast, copepods and other mesozooplankton appear unable to check their net-phytoplankton prey populations, because they have very dissimilar realized population growth rates, which implies a significant time-lag in the numerical response of the copepods to variations in net-phytoplankton abundance. Predator-limitation of the copepod's population growth rate reinforces this effect. Thus, when nutrients and light become available, the net-phytoplankton grows rapidly and escapes predator control; it blooms! Because of the lack of predator control, such blooms continue until all inorganic nutrients have become exhausted, and the cells sediment to the seafloor. Paradoxically, a requirement for nutrient limitation of the phytoplankton thus appears to be a preceding nutrient enrichment of the photic zone. Examples include spring blooms in many temperate and arctic seas, that typically consist of relatively large-sized diatoms. Big phytoplankton cells are at a competitive disadvantage relative to their smaller sisters in terms of light harvesting, nutrient uptake kinetics and growth rates; it is therefore the demographic characteristics of their predators that allows for the mere existence of netphytoplankton in the ocean (Munk & Riley, 1952; Kiørboe, 1993). In summary, then, the effects of zooplankton grazing on the phytoplankton are (i) to modify the size distribution by 'allowing' big cells to bloom in periodic environments (mesozooplankton) and (ii) to regenerate nutrients for continued phytoplankton growth in oligotrophic environments (mainly microzooplankton). The close coupling between grazing, remineralization and phytoplankton growth in steady state environments makes it difficult to decide whether the turnover rate here is governed mainly by light intensity or by grazing and remineralization.

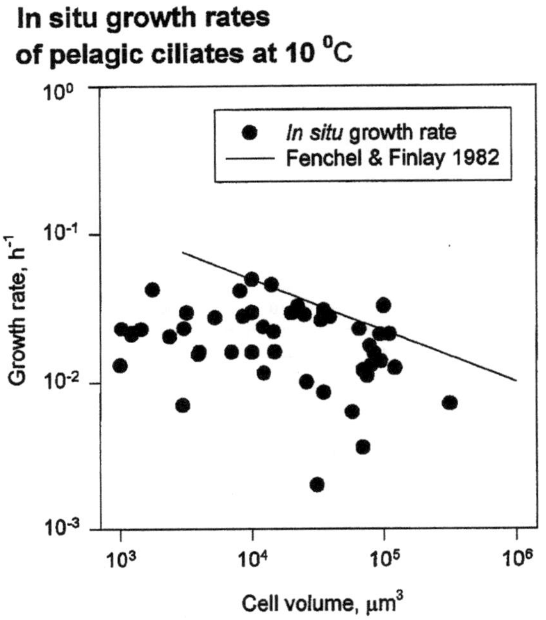

Figure 8. Direct determinations of *in situ* growth rates in pelagic ciliates. All measurements have been made by incubating size-fractionated water samples. All measurements have been converted to 10 °C by assuming a Q_{10} of 2.6. Only data obtained at ambient temperatures between 1.4 and 20 °C have been included. The field data have been compared to maximum growth rates measured in the laboratory (full line; regression from Fenchel & Finlay 1982). Field data are from Carrick & Fahnenstiel (1992), Dolan (1991), Hansen & Christoffersen (1995), Leakey et al. (1994), Levinsen (1995), Nielsen & Kiørboe (1994), Smetacek (1984), Tumantseva & Kopylov (1985).

Ciliates

In a seasonal study in the Kattegat, Nielsen & Kiørboe (1994) found that the oligotrich ciliates were growing at maximum, temperature-dependent individual rates throughout the year, and that variation in copepod grazing was governing population abundances. Thus, there were no signs of nutrient limitation. The capability of copepod populations to check populations of ciliates has been demonstrated several times (e.g. Stoecker & Egloff, 1987; Kivi et al., 1993; Lonsdale et al., 1996). Other studies, on the other hand, have demonstrated that ciliate growth rates *in situ* depend on food availability (e.g. Verity, 1986), thus, suggesting food limitation. In Figure 8, I have compiled all available direct determinations of ciliate *in situ* growth rates, converted to 10 °C by assuming a Q_{10} of 2.6 (Nielsen & Kiørboe, 1994). All measurements have been made by incubating size-fractionated samples at *in situ* conditions. With this approach the growth of the ciliate population is monitored in the (assumed) absence of predation. I have also shown Fenchel & Finlay's (1983) regression between body volume and maximum growth rates for ciliates as determined in laboratory experiments. Evidently, in many cases *in situ* growth rates are identical to those maximally measured in the laboratory, and these populations must, thus, be checked by predation. However, in several environments reported growth is less than maximum, suggesting food limitation in these instances. On average, measured *in situ* growth is 52% of maximum (Table 1). The approach to directly determine growth has some inherited biases, however, that will tend to underestimate growth: size fractionation may not exclude all predators, and food may become limiting during incubation due to reduced remineralization rates. Thus, resource limitation may be less than suggested by Figure 8, which is further supported by a modest coefficient of variation (60.4%, after correction for variation in cell volume; Table 1).

I conclude that pelagic ciliates in seasonal environments often are growing at maximum rates, but at times show resource-limitation. In environments in which the ciliates do grow at maximum unlimited rates and the resident populations of copepods do check their population sizes, the protozooplankton often contributes only marginally to the diet of the copepods and, hence, their population growth (e.g. Tiselius, 1989). Thus, it is not a typical mutual predator–prey relationship with feed-back, where the predator population is governed also by the abundance of the prey in a Lotka-Volterra sense.

Several species of copepods have different encounter and hunting strategies for motile ciliates and immobile phytoplankton cells (Jonsson & Tiselius, 1990; Tiselius & Jonsson, 1990; Landry, 1980, 1981; Kiørboe et al., 1996): while phytoplankters are captured by copepods by setting up a feeding current (suspension feeding), motile prey may be perceived and encountered by ambush feeding copepods. The latter feeding behaviour implies that the copepod hangs motionless in the water, it does not produce a feeding current, and hydromechanical disturbances generated by moving prey is perceived by mechanoreceptors on the antennae. Many copepods posses both feeding strategies (e.g. *Acartia* spp., *Centropages* spp., *Calanus finmarchicus*; Tiselius & Jonsson, 1990; Landry, 1980) and are able to switch between the two depending on the relative abundance of the different types of prey in the environment (Kiørboe et al., 1996). This has the implication that predation on ciliates depends on the availability of alternative phytoplankton prey and, conversely, that the predation on phytoplankton depends on the occurrence of ciliates (Landry, 1981; Stoecker & Egloff, 1987; Kiørboe et al., 1996). Specifically, blooms of phytoplankton may provide survival windows for ciliates. Ciliate populations may, therefore, oscillate concurrent with the abundance of phytoplankton, even in environments where they are not resource limited, as demonstrated by Nielsen & Kiørboe (1994).

Mesozooplankton and the vertical flux of particulate material

Grazing and sedimentation – fecal pellets and discarded larvacean houses: Copepod grazing may cause sedimentation of phytoplankton to the sea floor in at least two different ways. One is the traditional, that phytoplankton cells become packed in fecal pellets with high settling velocities. Because the grazing impact of mesozooplankton on the net phytoplankton is typically limited in seasonal environments (cf. above) this often does not represent a significant sink to the phytoplankton. However, the indirect effect of grazing activity of other zooplankters, such as larvaceans, may potentially cause significant sinking of netphytoplankton. Larvaceans feed on bacteria and pico- and nanophytoplankton (Alldredge, 1977) that they capture on mucus filters through which they filter huge volumes of water. For example, a 0.7 mm body length larvacean (typical size of the common *Oikopleura dioca*) may filter 65 ml of water per day (Paffenhöfer, 1975). The filtering structure (the house) of pelagic larvaceans consist of a coarse outer filter that screens larger particles, and a finer meshed inner filter, that retains the food particles. At intervals of a couple of hours the outer filter clogs, and the larvacean discards its house. The discarded house, now a marine snow particle with its attached netphytoplankton cells and other particles from a significant volume of water (Hansen et al., 1996), is rapidly sinking (order 100 m d^{-1}) out of the photic zone. Since larvaceans may occur in tremendous concentrations, up to 10 000 m^{-3} (Seki, 1973), this may at times represent a very significant sink to netphytoplankton populations. The implications to the population dynamics of netphytoplankton and the general significance of the phenomenon, is so far not well understood, but may well be significant.

Coprophagy, flux feeding and retention of material in the euphotic zone

The vertical flux of particulate material in the ocean is believed to be due mainly to the sinking of relatively large particles, such as fecal pellets and marine snow aggregates (Fowler & Knauer, 1986). While the activity of mesozooplankton may enhance the vertical particle flux in the ocean in the form of fecal pellets and discarded mucus feeding webs with attached particles, and even by increasing the aggregation rate of particulate material, it may also have the reverse role, i.e., slowing down the vertical flux (e.g. Smetacek, 1980; Smetacek & Pollehne, 1986). The retardation of vertical flux due to the activity of mesozooplankton is an understudied but potentially very important phenomenon, because it helps conserve matter in the euphotic zone with implications to the overall biomass of pelagic communities. The best known mechanism is likely coprophagy, i.e. feeding on fecal pellets. Many copepods are coprophagous (Paffenhöfer & Knowles, 1979) or may otherwise destroy fecal pellets by their feeding activity (coprohexy) (Noji et al., 1991) and this facilitates their mineralization in the water column and reduces losses from the euphotic zone to vertical flux. This mechanism is believed to be quantitatively important at times. González & Smetacek (1994) even suggested that copepods, particularly of the genus *Oithona*, may constitute a 'coprophagous filter' that significantly reduces the vertical flux of fecal pellets. An other mechanism that would reduce the vertical flux is feeding on (and remineralization of) marine snow aggregates. Marine snow particles are often inhabited by a diverse protozoan fauna, but metazoans are also represented, particularly by copepods of the genus *Oncaea*. *Oncaea* species are abundant in the ocean, they seem not to have a real planktonic life but rather inhabit marine snow aggregates, upon which they feed (Alldredge & Silver, 1988; Paffenhöfer, 1993; Dagg & Green, 1994). Quantitative estimates of their significance in remineralizing sinking aggregates in the upper ocean are lacking.

There are other mechanisms with a similar effect. Some copepods, such as *Neocalanus cristatus*, appear to be adapted to feed directly on sinking phytoplankton aggregates (Dagg, 1993): the animal is oriented horizontally in the water, it remains motionless, and its antennae and caudal furca are supplied with long featherlike plumes, which intercept sinking particles. There are several other copepod species with similar structures on the antennae and/or caudal furca, and these may also be *flux feeders*. Again, quantitative assessments of their significance are lacking. Pteropods are another group of flux feeders. These small snails may have huge mucus feeding webs, up to several cm across (Gilmer & Harbison, 1986), by means of which they collect falling particles (Jackson, 1993). These snails occur abundantly in the ocean. Jackson (1993) compiled abundance data and demonstrated their role in substantially reducing – and at times almost stopping – vertical particle fluxes in the ocean.

Flux feeding and coprophagous mesozooplankters may help conserve material in the upper mixed layer of the ocean. This may be important for the maintenance of plankton biomass in the upper mixed layer during periods of density stratification. And it may be particularly important during blooms of large-sized diatoms. By escaping predator control, blooming phytoplankters may exhaust all inorganic nutrients. Due to subsequent aggregation and sedimentation such blooms might eventually strip the euphotic zone for all elements. This would lead to subsequent plankton communities with very low biomasses. The retardation of vertical material fluxes – and losses – may be the most important role of the mesozooplankton for pelagic communities, and the phenomenon needs much further study. Particularly, quantitative measurements are needed to allow assessments of the significance of the phenomenon.

Summary and conclusions

Whether plankton communities are controlled mainly from below or from above depend strongly on the scale at which the problem is considered. On large spatio-temporal scales, seasonal plankton communities are obviously controlled from below. For example, copepods are most abundant during the summer halfyear in temperate and arctic systems, where the phytoplankton production is highest. However, on smaller scales, copepod population dynamics appear mainly controlled by predation processes: juvenile growth *in situ* is close to the physiological maximum, and although their fecundity appears to be strongly food dependent this seems to have only limited demographic implications. Population maintenance appear to hinge on efficient predator avoidance, and several morphological and behavioral features in copepods suggest strong selection pressure to minimize post-hatching mortality.

The effects of mesozooplankton grazing on prey populations are severalfold. Grazing pressure on the phytoplankton from mesozooplankton is typically low, and the mesozooplankton have generation times that are severalfold longer than those of their netphytoplankton prey. This explains why netphytoplankton blooms in the ocean upon nutrient enrichment of the euphotic zone. And it explains the mere existence of netphytoplankton cells in the oceans, even though they are competitively inferior to nanoplankton cells in terms of nutrient uptake capability, light harvesting efficiency and growth rates: they can outgrow their predators. In contrast, populations of pico- and nanosized phytoplankters are typically controlled by predation. Thus, the size distribution of phytoplankton in the ocean is to a large extent governed by the demography of its zooplankton grazers. The different behaviours used by copepods to feed on motile prey (ciliates) and immobile phytoplankton, allows the copepods to switch to the feeding behaviour that provides the greatest reward; i.e., they feed mainly on phytoplankton *or* ciliates, depending on their relative abundances. Thus, even though copepods may check resident populations of ciliates, prey switching may create survival windows for the ciliates during blooms of phytoplankton and, thus, determine their seasonal distribution.

For the plankton community maybe the most important effect of mesozooplankton activity is that it can modify the vertical flux – and, thus, loss – of elements from the euphotic zone. While the somewhat modest enhancement of the vertical flux due to fecal pellets etc. has been recognized – and emphasized – for a long time, the potentially more important retardation of the flux has not been much studied. This is likely to be particularly important following hydrodynamically mediated injections of nutrients to the surface layer, because the subsequently developing netphytoplankton bloom would remove most elements from the upper mixed layer due to aggregation and sedimentation and lead to plankton communities with much reduced biomass. The retardation of vertical flux due to mesozooplankton activity helps prevent the enervation of plankton communities. This issue is likely to become an important topic for future research.

Acknowledgement

Thanks are due to the organizers of the PELAG symposium for inviting me to present this work. Dr W. T. Peterson critically read the manuscript. The work was supported by a grant from the Danish Natural Science Research Council (9502163).

References

Aksnes, D. L., 1996. Natural mortality, fecundity and development in marine planktonic copepods – implications of behaviour. Mar. Ecol. Prog. Ser. 131: 315–316.

Aksnes, D. L. & T. Magnesen, 1988. A population dynamic approach to the estimation of production of four calanoid copepods in Lindåspollene, western Norway. Mar. Ecol. Prog. Ser. 45: 57–68.

Alldredge, A. L. & M. V. Silver, 1988. Characteristics, dynamics and significance of marine snow. Prog. Oceanogr. 20: 41–82.

Andersson, M., 1996. Regulering af copepodbestande i lavvandede fjorde. Betydning af fødebegrænsning og mortalitet. M.Sc. thesis, University of Copenhagen, 78 pp.

Bakker, C. & P. Van Rijswijk, 1987. Development time and growth rate of the marine copepod *Temora longicornis* as related to food conditions in the Oosterschelde estuary (Southern North Sea). Neth. J. Sea Res. 21: 125–141.

Banse, K., 1995. Zooplankton: Pivotal role in the control of ocean production. ICES J. Mar. Sci. 52: 265–277.

Bechman, B. R. & W. T. Peterson, 1986. Egg production by *Acartia tonsa* in Long Island Sound. J. Plankton Res. 8: 917–925.

Berggreen, U., B. Hansen & T. Kiørboe, 1988. Food size spectra, ingestion and growth of the copepod *Acartia tonsa* during development: implications for determination of copepod production. Mar. Biol. 99: 341–352.

Bollens, S. M & B. W. Frost, 1991. Diel vertical migration in zooplankton: Rapid individual response to predators. J. Plankton Res. 13: 1359–1365.

Bollens, S. M, B. W. Frost & J. R. Cordell, 1994. Chemical, mechanical and visual cues in the vertical migration behaviour of the marine planktonic copepod *Acartia hudsonica*. J. Plankton Res. 16: 555–564.

Burkill, P. H. & Kendall, T. F., 1982. Production of the copepod *Eurytemora affinis* in the Bristol Channel. Mar. Ecol. Prog. Ser. 7: 21–31.

Carrick, H. J. & G. L. Fahnenstiel, 1992. Growth and production of planktonic protozoa in Lake Michigan: In situ versus in vitro comparisons and importance to food web dynamics. Limnol. Oceanogr. 37: 1221–1235.

Checkley, D. M. Jr., 1980. Food limitation of egg production by a marine, planktonic copepod in the sea off southern California. Limnol. Oceanogr. 25: 991–998.

Chrisholm, L. A. & J. C. Roff, 1990a. Size-weight relationships and biomass of tropical neritic copepods off Kingston, Jamaica. Mar. Biol. 106: 71–77.

Chrisholm, L. A. & J. C. Roff, 1990b. Abundances, growth rates, and production of tropical neritic copepods off Kingston, Jamaica. Mar. Biol. 106: 79–89.

Colebrook, J. M., 1979. Continuous plankton records: Seasonal cycles of phytoplankton and copepods in the North Atlantic Ocean and the North Sea. Mar. Biol. 51: 23–32.

Dagg, M., 1978. Estimated, *in situ*, rates of egg production for the copepod *Centropages typicus* (Krøyer) in the New York Bight. J. exp. mar. Biol. Ecol. 34: 183–196.

Dagg, M., 1993. Sinking particles as a possible source of nutrition for the large calanoidcopepod *Neocalanus cristatus* in the subarctic Pacific Ocean. Deep-Sea Res. 40: 1431–1445.

Dagg, M. J. & E. P. Green, 1994. Marine snow in the northern Gulf of Mexico. EOS, Transactions, AGU, 75: 36.

Diel, S. & W. C. M. Klein Breteler, 1986. Growth and development of *Calanus* spp. (Copepoda) during a spring phytoplankton succession in the North Sea. Mar. Biol. 91: 85–92.

Dolan, J. R., 1991. Microphagous ciliates in mesohaline Chesapeake Bay waters: estimates of growth rates and consumption by copepods. Mar. Biol. 111: 303–309.

Durbin, A. G. & E. G. Durbin, 1981. Standing stock and estimated production rates of phytoplankton and zooplankton in Narragansett Bay, Rhode Island. Estuaries 4: 24–41.

Durbin, E. G., A. G. Durban, T. J. Smayda & P. G. Verity, 1983. Food limitation of production by adult *Acartia tonsa* in Narragansett Bay, Rhode Island. Limnol. Oceanogr. 28: 1199–1213.

Eppley, R. W., 1972. Temperature and phytoplankton growth in the sea. Fish. Bull. 70: 1063–1085.

Fenchel, T. & B. J. Finlay, 1983. Respiration rates in heterotrophic, free-living protozoa. Microbiol. Ecol. 9: 99–122.

Fowler, S. W. & G. A. Knauer, 1986. Role of large particles in the transport of elements and organic compounds through the oceanic water column. Prog. Oceanogr. 16: 147–194.

Fransz, H. G. & S. Diel, 1985. Secondary production of *Calanus finmarchicus* (Copepoda:Calanoidea) in a transitional system of the Fladen Ground area (Northern North Sea) during the spring of 1983. In P. E. Gibbs (ed.), Proc. 19th Europ. Mar. Biol. Symp. Cambridge University Press, Cambridge: 123–133.

Frost, B. W., 1988. Variability and possible adaptive significance of diel vertical migration in *Calanus pacificus*, a planktonic marine copepod. Bull. Mar. Sci. 43: 675–694.

Gilmer, R. W. & G. R. Harbison, 1986. Morphology and field behaviour of pteropod molluscs: Feeding methods in the families Cavoliniidae, Limacinidae, and Peraclididae (Gastropoda: Thecosomata). Mar. Biol. 91: 47–57.

Goldman, J. C., 1987. On phytoplankton growth rates and particulate C : N ratios at low light. Limnol. Oceanogr. 31: 1358–1363.

González, H. E. & V. Smetacek, 1994. The possible role of the cyclopoid copepod *Oithona* in retarding vertical flux of zooplankton faecal material. Mar. Ecol. Prog. Ser. 113: 233–246.

Hairston, N. G., F. E. Smith & L. B. Slobodkin, 1960. Community structure, population control, and competition. Am. Nat. 94: 421–425.

Hansen, B. & K. Christoffersen, 1995. Specific growth rates of heterotrophic plankton organisms in a eutrophic lake during a spring bloom. J. Plankton Res. 17: 413–430.

Hansen, J. L. S., T. Kiørboe & A. L. Alldredge, 1996. Marine snow derived from abandoned larvacean houses: sinking rates, particle content and mechanism of aggregate formation. Mar. Ecol. Prog. Ser. 141: 205–215.

Haslund, O. H. & M. Fryd, 1990. *In situ* undersøgelser af juvenile copepoders vækstrater gennem en sæson i Kattegat. M.Sc. thesis, University of Copenhagen, 97 pp.

Huntley, M. & M. D. G. Lopez, 1992. Temperature dependent growth production of marine copepods: a global synthesis. Am. Nat. 140: 201–242.

Hutchings, L., H. M. Verheye, B. A. Mitchell-Innes, W. T. Peterson, J. Huggett & S. Painting, 1995. Copepod production in the Southern Benguela system. ICES J. mar. Sci. 52: 439–455.

Ianora, A. & I. Buttino, 1990. Seasonal cycle in population abundance and egg production in the planktonic copepods *Centropages typicus* and *Acartia clausii*. J. Plankton Res. 12: 473–481.

Jackson, G. A., 1993. Flux feeding as a mechanism for zooplankton grazing and its implications for vertical particle flux. Limnol. Oceanogr. 38: 1328–1331.

Jonsson, P. & P. Tiselius, 1990. Feeding behaviour, prey detection and capture efficiency of the copepod *Acartia tonsa* feeding on planktonic ciliates. Mar. Ecol. Prog. Ser. 60: 35–44.

Kimmerer, W. J., 1983. Direct measurements of the production: biomass ratio of the subtropical calanoid copepod *Acrocalanus inermis*. J. Plankton Res. 5: 1–14.

Kimmerer, W. J., 1991. Predatory influences on prey distributions in coastal waters. Bull. Plankton Soc. Japan, Spec. Vol.: 161–174.

Kimmerer, W. J. & A. D. McKinnon, 1987. Growth, mortality, and secondary production of the copepod *Acartia tranteri* in Westernport Bay, Australia. Limnol. Oceanogr. 32: 14–28.

Kivi, K., S. Kaitala, H. Kuosa, J. Kuparinen, E. Leskinen, R. Lignell, B. Marcussen & T. Tamminen, 1993. Nutrient limitation and grazing control of the Baltic plankton community during annual succession. Limnol. Oceanogr. 38: 893–905.

Kiørboe, T., 1993. Turbulence, phytoplankton cell size, and the structure of pelagic food webs. Adv. mar. Biol. 29: 1–72.

Kiørboe, T., C. Lundsgaard, M. Olesen & J. L. S. Hansen, 1994. Aggregation and sedimentation processes during a spring phytoplankton bloom: A field experiment to test coagulation theory. J. mar. Res. 52: 297–323.

Kiørboe, T. & T. G. Nielsen, 1994. Regulation of zooplankton biomass and production in a temperate, coastal ecosystem. 1. Copepods. Limnol. Oceanogr. 39: 493–507.

Kiørboe, T., F. Møhlenberg & P. Tiselius, 1988. Propagation of planktonic copepods: production and mortality of egg. In G. A. Boxshall & H. K. Schminke (eds), Biology of Copepods. Developments i Hydrobiology 47. Kluwer Academic Press, Dordrecht: 219–225. Reprinted from Hydrobiologia 167/168.

Kiørboe, T. & M. Sabatini, 1994. Reproductive and life cycle strategies in egg-carrying cyclopoid and free-spawning calanoid copepods. J. Plankton Res. 16: 1353–1366.

Kiørboe, T. & M. Sabatini, 1995. Scaling of fecundity, growth and development in marine planktonic copepods. Mar. Ecol. Prog. Ser. 120: 285–298.

Kiørboe, T., E. Saiz & M. Viitasalo, 1996. Prey switching behaviour in the planktonic copepod *Acartia tonsa*. Mar. Ecol. Prog. Ser. 143: 65–75.

Landry, M. R., 1978. Population dynamics and Production of a Planktonic Marine Copepod, *Acartia clausii*, in a Small Temperate Lagoon on San Juan Island, Washington. Int. Revue ges. Hydrobiol. 63: 77–119.

Landry, M. R., 1980. Detection of prey by *Calanus finmarchicus*: implications of the first antennae. Limnol. Oceanogr. 25: 545–549.

Landry, M. R., 1981. Switching between herbivory and carnivory by the planktonic marine copepod *Calanus pacificus*. Mar. Biol. 65: 77–82.

Leakey, R. J. G., P. H. Burkill & M. A. Sleigh, 1994. Ciliate growth rates from Plymouth Sound: comparison of direct and indirect estimates. J. mar. biol. Ass. U.K. 74: 849–861.

Levinsen, H., 1995. Protozooplanktonets betydning i et arktisk Pelagisk fødenet. M.Sc. thesis, Marine Biological Laboratory, University of Copenhagen, 53 pp.

Lonsdale, D. J., E. M. Cosper, W. S. Kim, M. Doall, A. Divadeenam & S. H. Jonasdottir, 1996. Food web interactions in the plankton of Long Island bays, with preliminary observations on brown tide effects. Mar. Ecol. Prog. Ser. 134: 247–263.

Miller, C. B., M. E. Huntley & E. R. Brooks, 1984. Post-collection molting rates of planktonic, marine copepods: Measurement, application, problems. Limnol. Oceanogr. 29: 1274–1289.

Miller, C. B. & R. D. Nielsen, 1988. Development and growth of large, calanid copepods in the ocean Subarctic Pacific, May 1984. Prog. Oceanogr. 20: 275–292.

Mullin, M. M., 1991. Relative variability of reproduction and mortality in two pelagic copepod populations. J. Plankton Res. 13: 1381–1387.

Munk, W. H. & G. A. Riley, 1952. Absorption of nutrients by aquatic plants. J. Mar. Res. 11: 215–240.

Myers, R. A. & J. R. Runge, 1983. Predictions of seasonal natural mortality rates in a copepod population using life history theory. Mar. Ecol. Prog. Ser. 11: 189–194.

Nielsen, T. G. & T. Kiørboe, 1994. Regulation of zooplankton biomass and production in a temperate, coastal ecosystem. 2. Ciliates. Limnol. Oceanogr. 39: 508–519.

Ohman, M. D., 1986. Predator-limited population growth of the copepod *Pseudocalanus* sp. J. Plankton Res. 8: 673–713.

Ohman, M. D., 1988. Behavioral responses of zooplankton to predation. Bull. Mar. Sci. 43: 530–550.

Ohman, M. D., 1990. The demographic benefits of diel vertical migration by zooplankton. Ecol. Monogr. 60: 257–281.

Ohman, M. D. & S. N. Wood, 1995. The inevitability of mortality. ICES J. Mar. Sci. 52: 517–522.

Ohman, M. D. & S. N. Wood, 1996. Mortality estimation for planktonic copepods: *Pseudocalanus newmani* in a temperate fjord. Limnol. Oceanogr. 41: 126–135.

Paffenhöfer, G.-A., 1975. On the biology of appendicularia of the southeastern North Sea. 10th Europ. Symp. Mar. Biol., Ostende, Belgium 2: 437–455.

Paffenhöfer, G. A., 1993. On the ecology of marine cyclopoid copepods (Crustacea, Copepoda, Cyclopoida). J. Plankton Res. 15: 37–55.

Paffenhöfer, G.-A. & S. C. Knowles, 1979. Ecological implications of fecal pellets production and consumption by copepods. J. mar. Res. 37: 35–49.

Peterson, W. T., P. Tiselius & T. Kiørboe, 1991. Copepod egg production, moulting and growth rates, and secondary production, in the Skagerrak in August 1988. J. Plankton Res. 13: 131–154.

Peterson, W. T. & W. J. Kimmerer, 1994. Processes controlling recruitment of the marine calanoid copepod *Temora longicornis* in Long Island Sound: Egg production, egg mortality, and cohort survival rates. Limnol. Oceanogr. 39: 1594–1605.

Pielou, E. C., 1969. An Introduction to Mathematical Ecology. Wiley-Interscience, New York, 286 pp.

Rudstam, L. G., G. Aneer & M. Hildén, 1994. Top-down control in the pelagic Baltic ecosystem. Dana 10: 105–129.

Sabatini, M. & T. Kiørboe, 1994. Egg production, growth and development of the cyclopoid copepod *Oithona similis*. J. Plankton Res. 16: 1329–1351.

Saiz, E. & T. Kiørboe, 1995. Predatory and suspension feeding of the copepod *Acartia tonsa* in turbulent environments. Mar. Ecol. Prog. Ser. 122: 147–158.

Seki, H. Red tide of *Oikopleura* in Saanich Inlet. Lamer Tome 11, No. 3: 153–158.

Smetacek, V., 1980. Zooplankton standing stock, copepod fecal pellets and particulate detritus in Kiel Bight. Estuar. coast. Mar. Sci. 2: 477–490.

Smetacek, V. S., 1984. Growth dynamics of a common Baltic protozooplankter: the ciliategenus *Lohmaniella*. Limnologica (Berlin) 15: 371–376.

Smetacek, V. & F. Pollehne, 1986. Nutrient cycling in pelagic systems: A reappraisal of the conceptual framework. Ophelia 26: 401–428.

Stoecker, D. K. & D. A. Egloff, 1987. Predation by *Acartia tonsa* Dana on planktonic ciliates and rotifers. J. exp. mar. Biol. Ecol. 110: 53–68.

Tiselius, P., 1989. Contribution of aloricate ciliates to the diet of *Acartia clausi* and *Centropages hamatus* in coastal waters. Mar. Ecol. Prog. Ser. 56: 49–56.

Tiselius, P. & P. R. Jonsson, 1990. Foraging behaviour of six calanoid copepods: observations and hydrodynamic analysis. Mar. Ecol. Prog. Ser. 66: 23–33.

Tranter, D. J., 1976. Herbivore production. In D. H. Cushing & J. J. Walsh (eds), The Ecology of the Seas. Blackwell Scientific Publications, Oxford: 186–224.

Tumantseva, N. I. & A. I. Kopylov, 1985. Reproduction and production rates of planktonic infusoria in coastal waters of Peru. Oceanology 25: 390–394.

Uye, S.-I., 1982. Population dynamics and production of *Calanus sinicus* (Copepoda: Calanoida) in inlet waters. J. exp. mar. Biol. Ecol. 57: 55–83.

Verity, P., 1986. Growth rates of natural tintinnid populations in Narragansett Bay. Mar. Ecol. prog. Ser. 29: 117–126.

Verity, P. G. & V. Smetacek, 1996. Organism life cycles, predation, and the structure of marine pelagic ecosystems. Mar. Ecol. Prog. Ser. 130: 277–293.

Vuorinen, I., 1987. Vertical migration of *Eurytemora* (Crustacea, Copepoda): A compromise between the risk of predation and decreased fecundity. J. Plankton Res. 9: 1037–1046.

Walker, D. R. & W. T. Peterson, 1991. Relationships between hydrography, phytoplankton production, biomass, cell size and species composition, and copepod production in the southern Benguela Upwelling system in April 1988. S. Afr. J. mar. Sci. 11: 289–305.

Wiadnyana, N. W. & F. Rassoulzadegan, 1989. Selective feeding of *Acartia clausi* and *Centropages typicus* on microzooplankton. Mar. Ecol. Prog. Ser. 53: 37–45.

Williamson, C. E. & H. A. Vanderploeg, 1988. Predatory suspension-feeding in *Diaptomus*: Prey defense and the avoidance of cannibalism. Bull. Mar. Sci. 43: 561–572.

Yen, J., P. H. Lenz, D. V. Gassie & D. K. Hartline, 1992. Mechanoreceptors in marine copepods: electrophysiological studies on the first antennae. J. Plankton Res. 14: 495–512.

Hydrobiologia **363**: 29–57, 1998.
T. Tamminen & H. Kuosa (eds), Eutrophication in Planktonic Ecosystems: Food Web Dynamics and Elemental Cycling.

Retention versus export food chains: processes controlling sinking loss from marine pelagic systems

Paul Wassmann
Norwegian College of Fishery Science, University of Tromsø, N-9037 Tromsø, Norway

Key words: export and retention food chains, vertical flux, top-down regulation, zooplankton, global carbon flux

Abstract

The role of export and retention food chains for pelagic-benthic coupling is considered by evaluating different food chain scenarios and processes such as aggregation, grazing and zooplankton-mediated fluxes. The consequences of grazing of primary production by different zooplankton for the vertical export of particulate organic matter from the euphotic zone are discussed. Reference is made to existing data and algorithms regarding primary production and vertical export of carbon from the euphotic zone, both on annual and daily time scales. Examples regarding the role of nutrient addition, removal of pelagic carnivores and zooplankton grazing for vertical flux are presented. It is speculated how variable grazing impact of micro- and mesozooplankton, as well as herbivorous, omnivorous and carnivorous feeding strategies of mesozooplankton could compete with aggregation during phytoplankton blooms and influence export fluxes. It is concluded that the transport of particulate organic matter to depth not only depends on bottom-up regulation as determined by physical forcing, but also on the structure and function of the prevailing planktonic food web. Scenarios are presented which indicate that top-down regulation plays a pivotal role for the regulation of vertical flux. This conclusion may have crucial consequences for future biogeochemical programmes investigating pelagic-benthic coupling in the ocean. The endeavours of many research programmes are dominated by lines of thought where straightforward biogeochemistry and bottom-up regulation is the focus. Phyto- and zooplankton as well as process-oriented research activities have to be the focal point of future research if the current comprehension of export from and retention in the upper layers is going to make distinct progress.

Introduction

Retention and export food chains as well as sinking losses from marine pelagic systems have been ill-defined subjects in oceanography for more than 30 years. After the introduction of the food chain and food web concept in the late 1940's, it took two decades before trophic relationships in plankton and benthos became the focus of mainstream research. The concept of new versus regenerated production was introduced by Dugdale & Goering (1967), but it had little influence on biological oceanography until the milestone publication of Eppley & Peterson (1979) (Figure 1). Their scheme suggests that the regulation of vertical flux is basically limited by new production, influenced by planktonic organisms and how these regulate the cycling of nutrient and organic matter (Platt et al., 1988). Retention food chains, which recycle organic matter and nutrients, minimise sinking losses while export food chains, which prevail during new production episodes, maximise them.

Mesozooplankton production, grazing and population dynamics have been intensively studied during the entire period of modern oceanography. For example, in the North Sea model by Steele (1974) all phytoplankton is grazed by mesozooplankton, and export to the benthos is comprised only by faecal pellets. This can be explained by the emphasis which has been given to the grazing food chain, essential for the exploitation of marine pelagic resources (Legendre, 1990), and the concern for the development and future of important fisheries world-wide (Iverson, 1991). The relationship between primary production, recycling, and the classical and microbial food web entered more recently

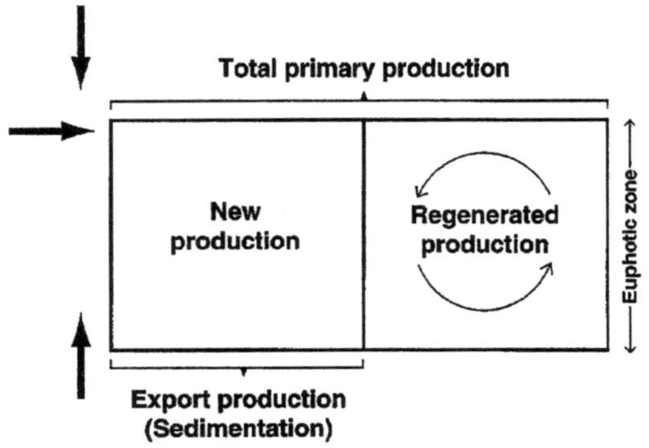

Figure 1. New and regenerated production are based on (a) the supply of the limiting (allochthonous) nutrients from the aphotic zone, by advection, run-off or from the atmosphere (straight arrows) and (b) the recycled (autochthonous) nutrients in the euphotic zone (circular arrows), respectively. New and regenerated production comprise total primary production. Export production is the amount of sinking organic carbon at the bottom of the euphotic zone and reflects new production.

into general consideration (Azam et al., 1983). Azam and coworkers introduced the concept of the microbial loop, by which energy recycled by the microbial community is made available for higher trophic levels of the classical food web through a protozooplankton shunt. Since then, (a) the regeneration of biomass and nutrients by the microbial community, (b) the channelling of microbial biomass and energy into the classical grazing food chain and (c) the role of dissolved organic carbon (Thingstad, 1995) has been a focus of marine research (Figure 2).

The connection between plankton and benthos by means of vertical flux measurements in the ocean was not investigated thoroughly before the 70's (e.g. Smetacek et al., 1984; Graf, 1992). Surprisingly, plankton and benthos were more or less separate entities in the minds of most biological oceanographers until the late 1970's. Vertical flux measurements using sediment traps became a major focus in oceanography in the 1980's (Bloesch, 1996). The current understanding of the regulation of vertical flux is dominated by scientific approaches which perceive vertical flux to be regulated by bottom-up control (Figure 3). The vertical structure of the water column and the dynamics of nutrients, primary production and suspended biomass above the sediment traps are often compared to the resulting vertical flux (e.g. Smetacek et al., 1984; Wassmann, 1991; Bodungen et al., 1995). Indeed, the physical environment determines nutrient availability and hence the particles potentially available for sedimentation (e.g. Wassmann et al., 1991; Kiørboe, 1996). The question arises, however, if the focus on bottom-up control is sufficient to interpret and understand loss of suspended biomass, the vertical flux patterns, and the chemical and biological composition of sinking matter. For a review of top-down control focusing on the role of organism life cycles, predation and the structure of marine pelagic ecosystems, see Verity & Smetacek (1996).

The appearance of automatic sediment traps and long-term moorings in the deep ocean gave rise to substantial data sets of vertical flux with global coverage (e.g. Honjo, 1990; Ittekot et al., 1996). A majority of recent vertical flux investigations such as the Global Ocean Flux Study (GOFS) have been dominated by biogeochemical, bottom-up interpretations. The dominance of biogeochemical approaches and the paucity of simultaneous investigations of the export and retention food chains above the sediment traps have provided information about the magnitude and patterns of biogenic fluxes, but little knowledge about the biological composition of the exported material, let alone the regulating mechanisms of export and retention food chains above the sediment traps, or the regulation of vertical flux.

It is now recognised that the link between biogenic particle production at the surface ocean and export to deeper waters is close (Deuser et al., 1981, Asper et al., 1992). It varies seasonally in dependence on physical and biological processes in the upper water column (Peinert et al., 1989; Kiørboe, 1996). The general succession of pelagic autotrophic communities, from diatom-domination following initial stratification to flagellate-domination under nutrient recycling conditions, has implications on organic matter export from the mixed layer. This is due to the role of diatoms in forming fast-sinking aggregates (e.g. Passow et al., 1994) as well as the evolution of the pelagic heterotrophic community and its role in particle export and recycling (e.g. Smetacek et al., 1984; Frost, 1991). The value of long-term recordings of particle flux using sediment traps, and the ability to conduct a wide range of specific analyses on the samples obtained, has greatly enhanced our understanding of the functioning of the oceanic system with respect to material transport between surface waters and abyssal depths. Today we know that vertical export of organic matter is a most important process which regulates (a) the residence time of phytoplankton, organic matter and nutrients in the upper ocean and (b) determines the amount, quality and temporal variation of organic matter supply to the deep ocean and sediments.

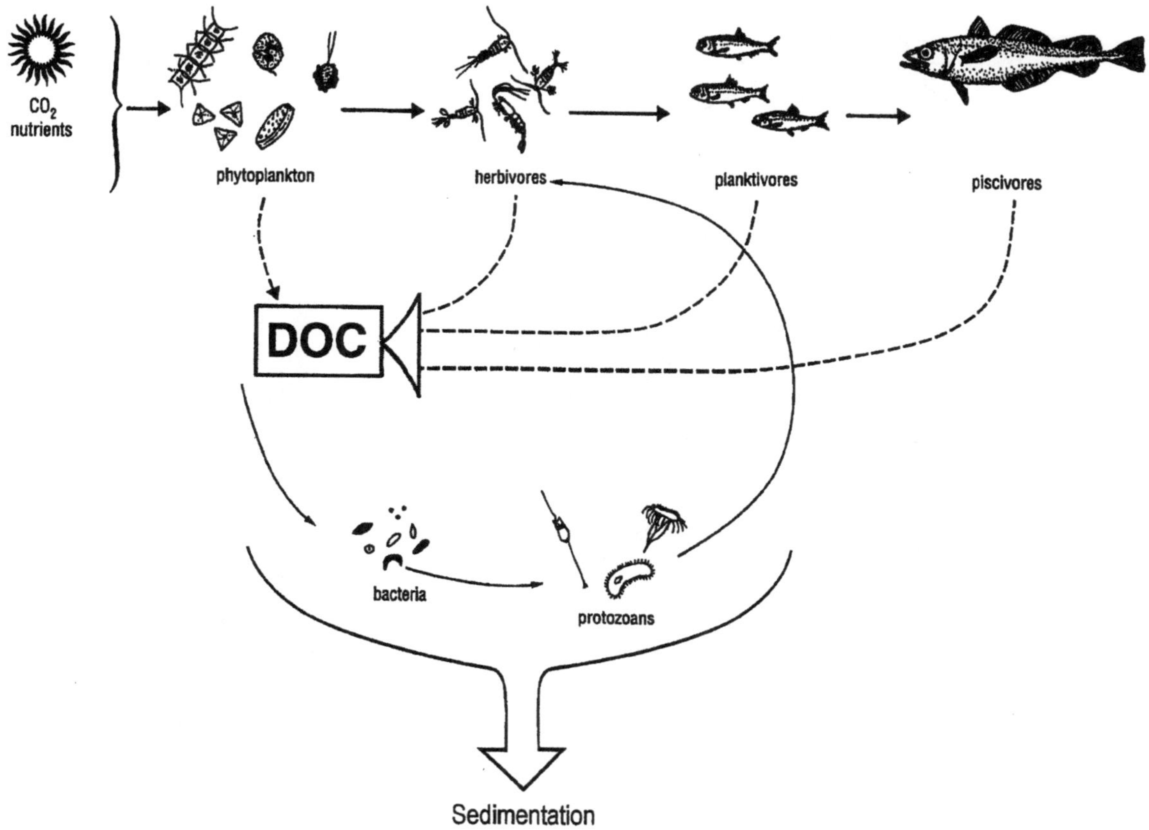

Figure 2. The classical pelagic food web (indicated by the channelling of energy from phytoplankton to pisciviorous fish), the microbial loop (bacteria and protozans) and the release and channelling of the metabolic by-product DOC into the microbial loop and small herbivores. Both the classical and microbial food chain influence the vertical export of organic matter. Redrawn from Lalli & Parson (1993).

The focus here is not a balanced, comprehensive review on the regulation of vertical export derived from a complete analysis of the diverse, abundant and widespread literature. Rather an attempt is made to present the involved processes and discuss their dynamics. Basically, and for no other than practical reasons, many of the examples given are based on research carried out by the sedimentation group at the University of Tromsø in northern Norway.

Processes affecting export and recycling of organic matter

Aggregation. Aggregation strongly promotes export of organic matter. It is the worlds greatest snow storm (Alldredge & Silver, 1988). It is a perpetual storm, but varies in space and time with the greatest intensity during phytoplankton blooms and in upwelling areas. It is rather astonishing that one of the most prominent planktonic features in the ocean has received limited attention until recently. Among phytoplankton species, diatoms seem by far the most important contributors to aggregation. This is due to cell stickiness or the production of particulate and dissolved matter which is sticky (Kiørboe & Hansen, 1993). Physical aggregation mechanisms play a far more important role compared to biological aggregation (mainly by zooplankton activity). Phytoplankton sedimentation can balance the growth of algae at critical concentrations (e.g. Kiørboe et al, 1984). Sedimentation of aggregates can thus regulate phytoplankton population dynamics of species. Different species can also aggregate with each other and the various loss rates are dependent on the stickiness of and collision rate between the various species. There is also the phenomena of mucilaginous matter, not visible until stained. The most important of these type of particles are TEP (transparent exopoly-

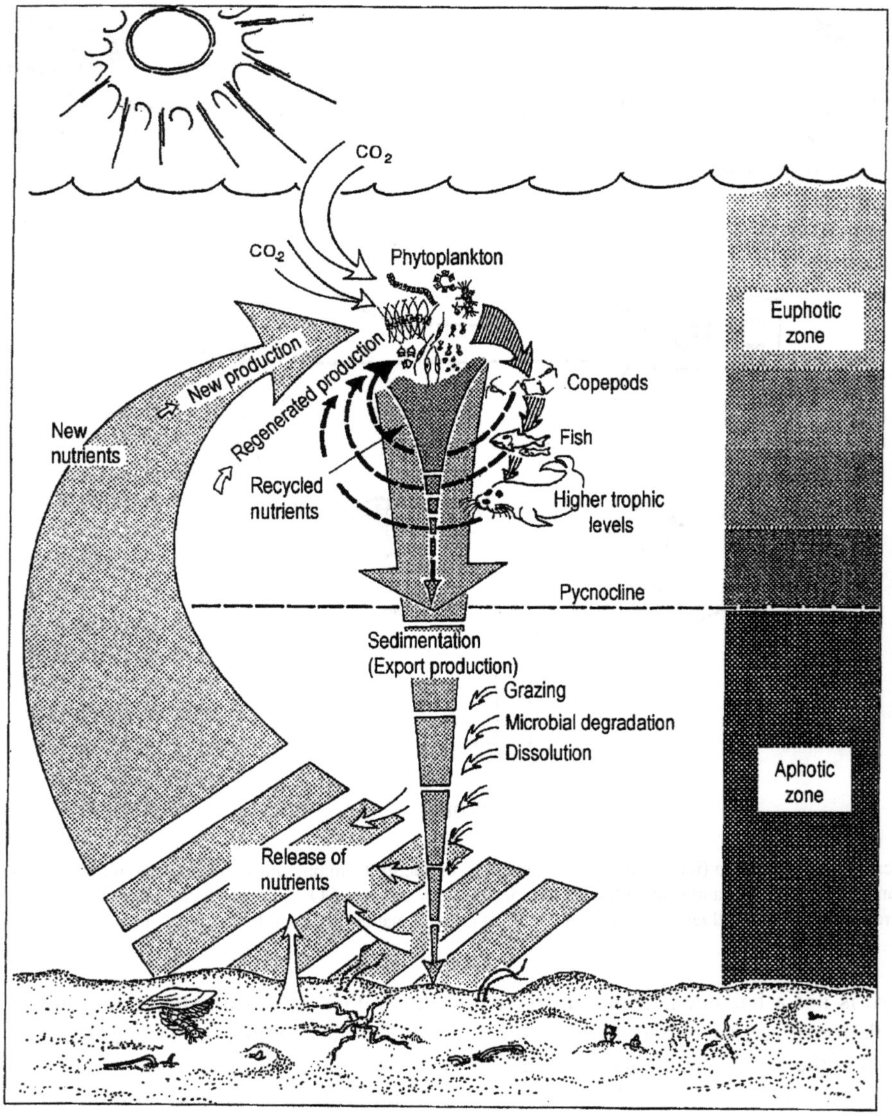

Figure 3. Primary production, vertical flux and regeneration of nutrients in a coastal marine ecosystem. Also shown are some of the involved organisms such as phytoplankton, zooplankton, higher trophic levels and benthic organisms. The massive and narrow vertical arrows indicate scenarios of substantial and insignificant vertical flux. Redrawn from Keck & Wassmann (1994).

mer particles) (Passow et al., 1994). TEP glues cells and other particles together.

The aggregation of even simple diatom blooms is, however, complicated. In a tank experiment simulating a diatom bloom, aggregation was neither linked to nutrient depletion nor to the physiological state of cells (Alldredge & Jackson, 1995). Laboratory experiments by Kiørboe & Hansen (1993) indicated that stickiness could be constant, increase or vary over the length of a bloom. There are even non-sticky diatoms. In summary, aggregation is probably the most important mechanism regulating vertical export of organic matter during major phytoplankton blooms, but unfortunately there are at present no general rules describing stickiness of phytoplankton species. During the last decade some of the basic mechanisms involved have become evident and some first attempts have been made to go from recording to prediction (Figure 4). There is some hope that regulation of vertical flux through aggregation can be incorporated into mathematical models of

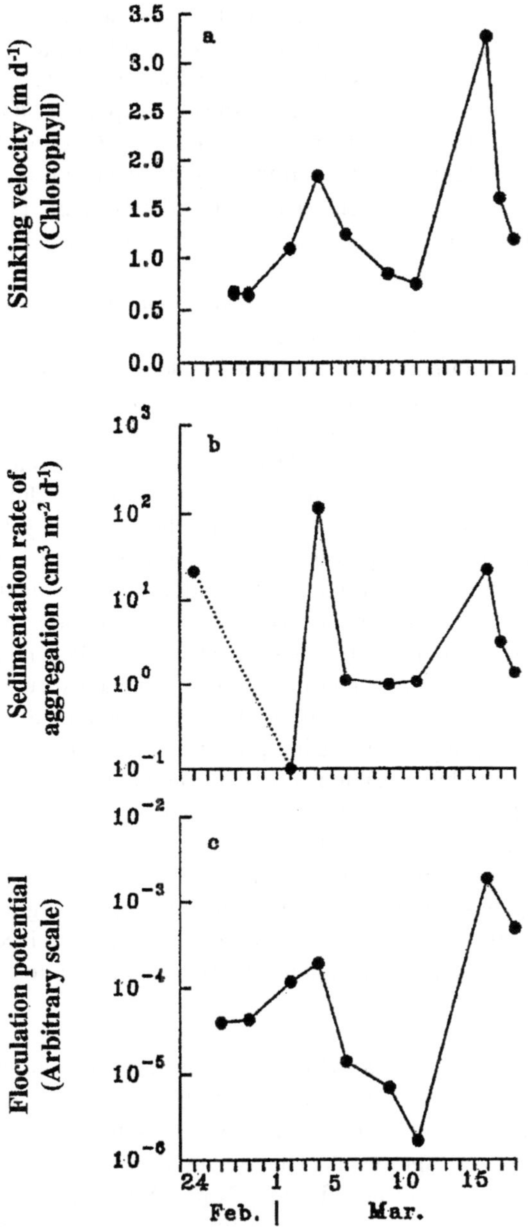

Figure 4. Observed sinking velocity of particulate chlorophyll *a* in sediment traps (a), observed sedimentation rate of aggregates in sediment traps (b) and rate of aggregate formation ('flocculation potential') predicted from coagulation theory (c). Redrawn from Kiørboe et al. (1994).

vertical flux in the future without decreasing their predictability.

Mesozooplankton and fish grazing. Grazing by zooplankton and fish can both promote vertical flux and increase the retention efficiency of the upper layers of the ocean. Large faecal pellets are an important fraction of particles involved in the vertical export of biogenic matter (Angel, 1984; Bodungen et al., 1995). Due to their size (several hundred μm to mm in diameter), their content of relatively fresh biogenic matter and their considerable sinking speed (in the range of about some ten to thousand m d^{-1}), faecal pellets are export vehicles of considerable significance (e.g. Pilskaln & Honjo, 1987; Sasaki et al., 1988). Already in the North Sea model by Steele (1974), supply to the sediment surface was assumed by faecal pellets, but subsequent investigations showed that faecal pellets represented only a limited part of the vertical flux. Usually faecal pellets comprise a few to some ten percent of the vertical POC flux. However, in special cases (for example the presence of krill swarms or large copepod standing stock) a major part of the POC flux is caused by carbon from faecal pellets (Riebesell et al., 1995; Hansen, 1997). Specific organisms such as krill, salps or appendicularians can play a dominant role for vertical export by the rapid and abundant production of large, fast sinking faecal pellets (e.g. Bodungen et al., 1987; Arashkevitch et al., 1994). While krill faeces lack a peritrophic membrane and are loosely packed (and thus are easily broken during their descent), salp and appendicularian pellets are dense and reach in a couple of days depths of several thousand meters (Iseki, 1981; Bathmann, 1988; Arashkevitch et al., 1994). They are thus important for food supply of deep sea organisms. In certain areas fish faecal pellets can play an important role for the vertical export (e.g. Staresinic et al., 1983; Heussner et al., 1987; J. C. Miquel, pers. comm.), but generally they are rarely recorded in traps. This may be due to rapid disintegration of the loosely bound, membrane-lacking fish faeces.

Zooplankton and fish do not only produce large faecal pellets and thus contribute significantly to vertical export. They also reduce the abundance of large-celled phytoplankton and aggregates, and thus contribute to the efficiency of retention food chains. Inevitably, some vertical losses do occur and their production of faeces contributes to these as long as processes counteracting losses, e.g. zooplankton mediated fluxes, are not important (see below).

Mismatch and match between phyto- and zooplankton. Phytoplankton blooms represent scenarios characterised by imbalances between biomass production and mineralization. All phytoplankton blooms are an indication that there is temporary mismatch between producers and consumers. Ecosystems such as shallow coastal areas or shelfs less then 100 m deep are, generally speaking, characterised by mismatch scenarios during spring: there is usually a considerable time lag between the phytoplankton bloom and the increase of major grazers such as copepods (Heinrich, 1962; Fransz & Gieskes, 1984). Export of phytoplankton spring blooms has been recorded in the Baltic Sea (Smetacek et al., 1984; Heiskanen & Kononen, 1994), central North Sea (Davies & Payne, 1984) and shallow or silled, boreal coastal inlets (Hargrave et al., 1985; Laws et al., 1988; Wassmann, 1991). The causes for the mismatch are that the zooplankton community is not able to catch up with the increase in phytoplankton growth, that microzooplankton is preferentially grazed by mesozooplankton (decreasing the grazing pressure on large-celled phytoplankton) and that mesozooplankton does not successfully over-winter at shallow depths (see below: over-wintering and advection of zooplankton). Mismatch will inevitably result in increased suspended biomass and aggregation and consequently export of organic matter.

On the contrary, match between phyto- and zooplankton intensifies recycling. The phytoplankton standing stock is kept low, aggregation of large-celled phytoplankton is limited, and the vertical export is ultimately based on detritus and faecal pellets whose size spectrum depends on the composition of the grazer community. In the tropics with primary production taking place in variable quantities throughout the year, there is usually a good match between phyto- and zooplankton growth and retention food chains are believed to prevail. However, match can also be experienced in boreal and arctic areas, where phytoplankton growth is slow enough to be controlled by heterotrophs. Outside shelf breaks, particularly in sub-polar and polar environments, over-wintering of larger zooplankton forms such as copepods and krill can give rise to excellent match and decreased export (see below: over-wintering and advection of zooplankton).

Microbial loop. The microbial loop increases the recycling of organic matter, promotes the efficiency of retention food chains and counteracts, thus, the loss of biogenic matter from the surface layers to depth. It also makes an important fraction of dissolved organic matter again available for the grazing food chain and limits thus also the vertical export of dissolved organic matter (Figure 2). The microbial loop crops the standing stock of particulate and dissolved biogenic matter, recycles essential nutrients, produces minipellets which do not contribute significantly to vertical flux, and microzooplankton is a prey for larger heterotrophs. A strong microbial loop is thus an excellent way for pelagic ecosystems to omit vertical losses. DOC can, however, also be exported vertically. Export of DOC was first predicted by model studies (e.g. Bacastow & Maier-Reimer, 1991) and recently it has been proved that dissolved biogenic matter can indeed be part of the vertical flux (Carson et al., 1994; Dam et al., 1995).

Over-wintering and advection of zooplankton. Larger zooplankton forms such as copepods or krill can over-winter, in particular at higher latitudes. *Calanus finmarchicus* for example has a life cycle which is spread out over 1 1/2 years in the Norwegian Sea and the Barents Sea (Tande, 1991). *C. finmarchicus* hibernates in the CV stage and migrates, developing into adults, well before the vernal bloom into the surface layers, ready to graze upon the spring bloom and to spawn (e.g. Bathmann et al., 1989b). Adult mesozooplankton populations can be advected across and along the shelf or into fjords (e.g. Aksnes et al., 1989; Pedersen, 1995). By over-wintering, zooplankton can execute a strong grazing pressure on phytoplankton already in early spring. The grazing pressure can be so strong that no phytoplankton spring bloom as characterised by large-celled phytoplankton species (e.g. diatoms) or accumulation of chlorophyll, occurs. Scenarios like these have been supported by data from the north Norwegian shelf (Ratkova et al., 1998). The effect of grazing pressure of over-wintering zooplankton and advection of adult copepod populations is also illustrated in Figure 18.

Zooplankton mediated fluxes. There can be large differences between faecal pellet production and sedimentation, as not all produced faecal pellets of the larger size classes are apt to sink. This is caused by zooplankton mediation of vertical flux (e.g. Noji, 1991; Noji & Rey, 1996). There are at least three different mechanisms by which mesozooplankton reduces vertical loss of faecal pellets:

- Grazing by meso- and macrozooplankton species directly on faecal pellets is called *coprophagy*. Faecal pellets are thus directly removed from suspension and vertical export decreases.
- Manipulation of faecal pellets which results in fragmentation is called *coprorhexy*. Sinking and dissolution rates of faecal matter decrease and increase, respectively.

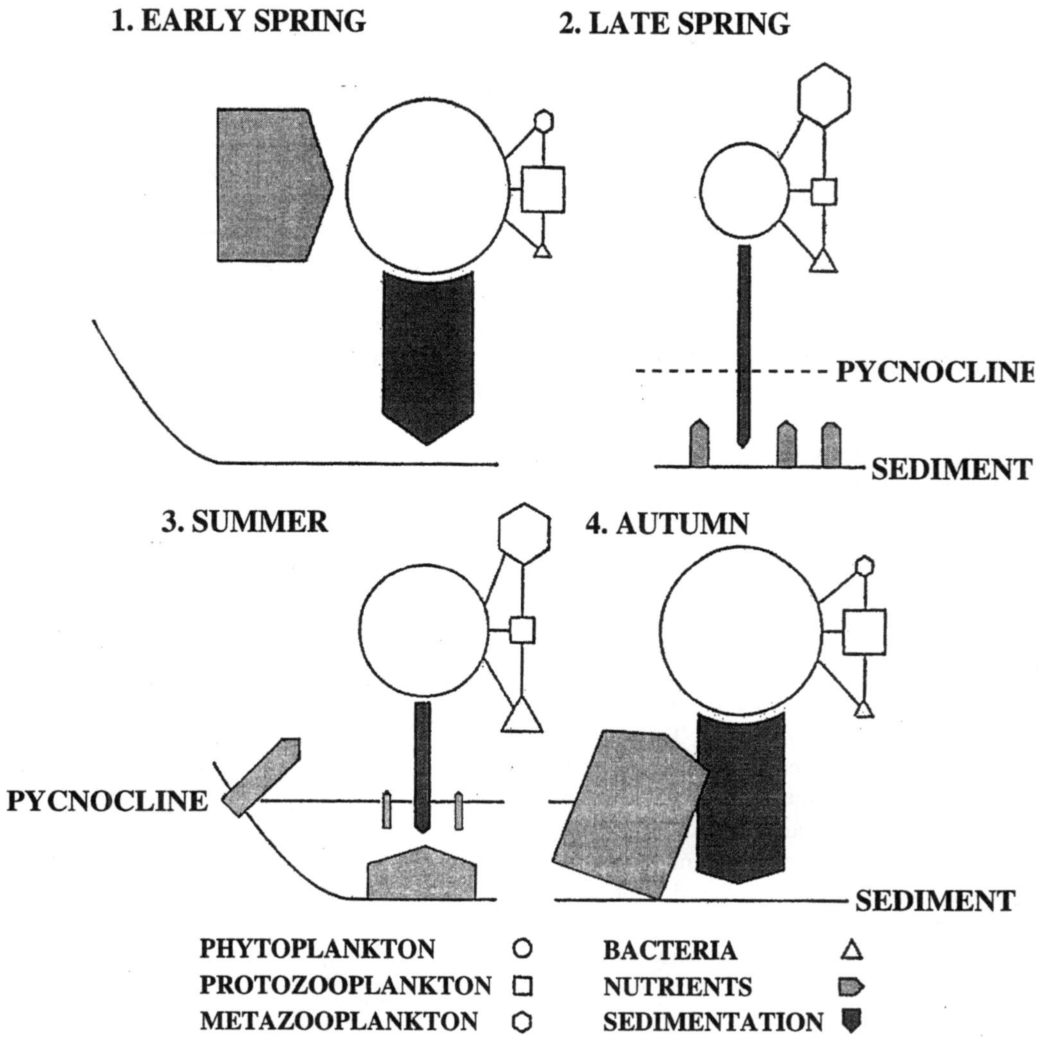

Figure 5. Semiquantitative flow diagrams representing *four stages* of the pelagic systems during the growth season. Arrows indicate input of new nutrients and output of sedimenting matter. Quantitative relationships are denoted by the thickness of the arrows and the size of the symbols representing the four major biological components of the system. Redrawn from Smetacek et al., (1984).

- Colonisation of faecal pellets by bacteria and protists which destroy the periotrophic membrane is called *coprochaly*. The faecal pellet content is diluted by water and dissolved matter. The sinking rate of the faecal matter decreases while dissolution increases. There is also action by bacteria and protists from inside the faecal pellets which derived from the guts of the grazers.

It may be difficult to distinguish between coprophagy and coprorhexy. Furthermore, it is not clear what mechanisms various zooplankton organisms use to take advantage of the rich resources of suspended faeces. Important genera such as *Oncaea*, *Oithona* and *Calanus* are supposed to reduce vertical flux of faecal pellets (Skjoldal & Wassmann, 1986; Noji, 1991; Gonzáles & Smetacek, 1994). They can live on at least 4 sources of suspended particles: phytoplankton, aggregates and detritus, microzooplankton and faecal pellets. Their feeding mechanisms also has implications for the retention of faecal matter in the upper layers. They may prefer microzooplankton, if available (e.g. Kiørboe et al., 1996). In order to feed on microzooplankton, they will use the active *ambush feeding* mode. However, if they select phytoplankton, small aggregates and detritus they will use the passive *suspension feeding* mode, followed by active or passive food item selection by their feeding appendages. These two feeding mechanisms have implications for the fate of faecal pellets in surface layers. When copepods graze upon ciliates, faecal pellets will probably not be recycled and sink to deeper layers, increasing export (A. S. Heiskanen, pers. comm.). However, if suspension feeding predominates, vertical fluxes will be influenced by zooplankton mediation due to coprophagy and coprorhexy.

Successional food web dynamics and vertical export

Modern text books in marine ecology do not always contain information about vertical flux, let alone pelagic-benthic coupling. Even today plankton and benthos are often treated as separate entities, in contradiction to what system ecological research has revealed over the last decades. Our general understanding regarding retention versus export food chains and vertical flux is similar to the evidence presented in Figures 5 and 6. The state of the ecosystem during spring is characterised by *export chains.* The amount of regenerated production increases, as the planktonic system develops and becomes more complex during late spring and summer. Sedimentation of organic material is low and the ecosystem is characterised by *retention chains.* Export chains are based upon new production and represent episodic events on the background of a continuous, whereas seasonally variable regenerated production, which is based on recycled nutrients from retention chains.

Our general understanding of plankton development and vertical flux suggests that vernal and autumn blooms accompanied by phytoplankton biomass accumulation occur in all meso- and eutrophic environments. It also implies that these blooms sink more or less ungrazed to the bottom (Figures 5 and 6). However, this is not always the case. On the contrary, there are numerous eutrophic environments where retention of organic matter in the surface layers can be substantial (Frost, 1991; Wassmann et al., 1997). Obviously our general understanding of retention versus export food chains is rather simplistic. With regard to the processes controlling retention and export from marine pelagic systems, customary comprehension does often not reflect the inherent complexity of pelagic systems, but rather the results of the earliest studies of vertical flux from shallow areas. A more thorough understanding of the dynamics of processes controlling retention and

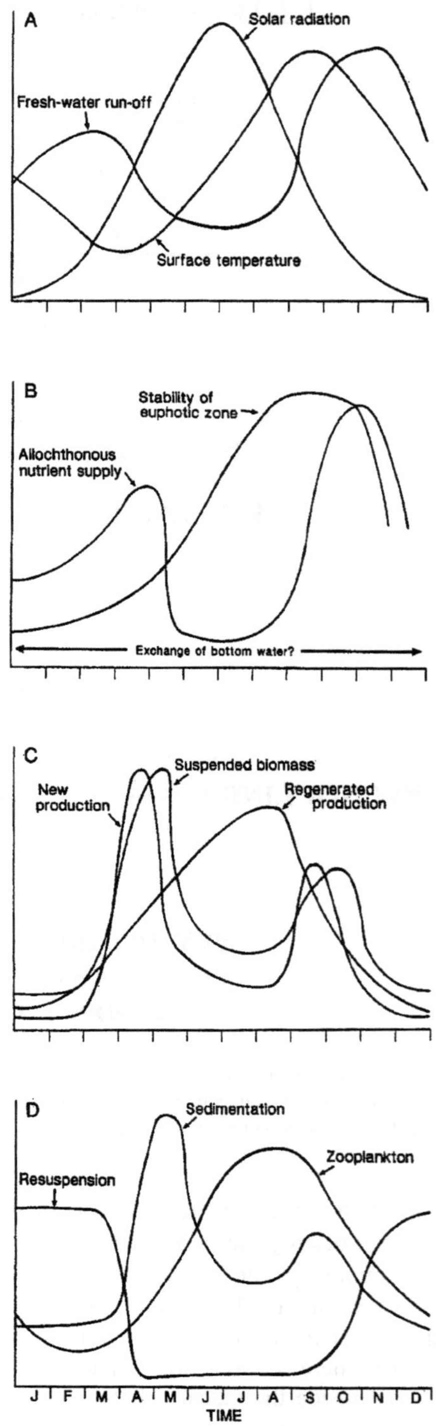

Figure 6. Conceptual diagram illustrating the *annual cycle* of solar radiation, water temperature and freshwater supply in the upper layer (A), allochthonous nutrient supply and stability of the euphotic zone (B), new, regenerated production and suspended biomass (C) and zooplankton, sedimentation and resuspension (D) in an unspecified coastal area of the boreal zone. Redrawn from Wassmann (1991).

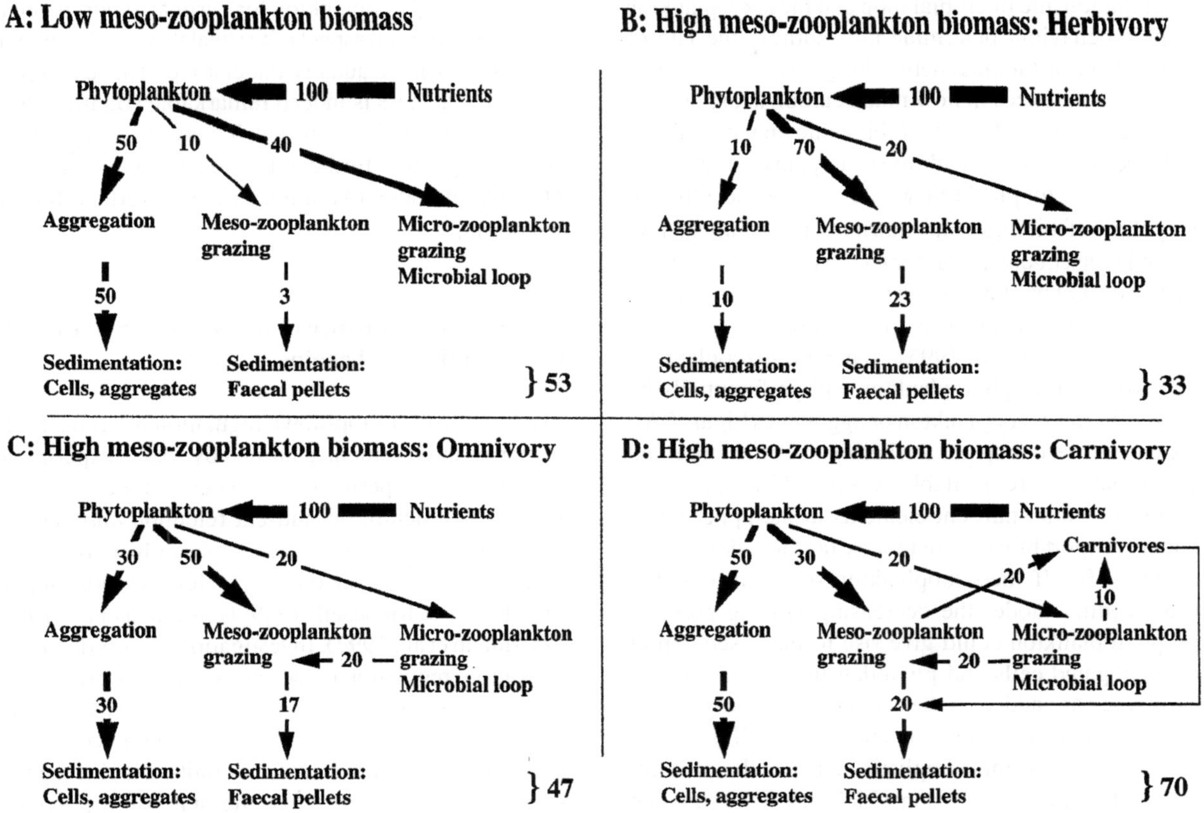

Figure 7. Schematic presentation of the influence of various zooplankton trophic impact on recycling and export of phytoplankton blooms. 100% of nutrients are taken up by phytoplankton. Four scenarios are given: (A) Low mesozooplankton biomass, (B) high zooplankton biomass and assuming an exclusively herbivorous feeding mode, (C) high zooplankton biomass and assuming an omnivorous feeding mode and (D) high zooplankton biomass and assuming omnivory as well as grazing pressure by carnivores. See text for more details.

export of organic matter in marine pelagic systems is obviously needed.

Currently there is a vigorous debate about the relative roles of the meso- and microzooplankton in grazing the phytoplankton (e.g. Banse, 1995), and it would appear that in some situations the microzooplankton exert the dominant grazing pressure (Burkill et al., 1993; Gifford et al., 1995). This enables the grazers to prolong their contact with locations of highest productivity. The zooplankton communities thus play pivotal roles in the upper ocean and the flux of material out of this zone. In order to understand the dynamics of biogenic matter in the upper water layers, it is therefore essential that a clear understanding is obtained about the distribution of the various classes of zooplankton, the feeding pressures they exert, and their productivity.

In order to investigate the dynamic nature of vertical organic matter export as a function of the most prominent mechanisms determining the vertical flux, i.e. aggregation and zooplankton mediation, a simple schematic scenario is presented. The scheme does not claim to be completely realistic. As mentioned previously, most of the vertical flux investigations carried out so far were dominated by ideas of bottom-up regulation. Having in mind this trend in vertical flux research, the following scheme aims at a more dynamic understanding of the regulation of vertical flux, focusing on the potential role of top-down regulation.

In the scheme presented in Figure 7, nutrients incorporated by phytoplankton can either aggregate and sink directly, or be grazed by mesozooplankton and microzooplankton in competition. It is assumed that no organic matter from microzooplankton/microbial loop sinks while 1/3 of the mesozooplankton grazing rate sinks in form of faecal pellets (Figure 7). In order to keep the scheme simple, vertical flux mediation by zooplankton, degradation of aggregates and aggregation of detritus are not taken into consideration.

Aggregation, the microbial loop and mesozooplankton trophic behaviour determine thus entirely the fate of phytoplankton biomass here. This gives, in theory, rise to clear differences in how much carbon is exported.

The export potential of biogenic matter will be reduced by micro- and mesozooplankton grazing. While microzooplankton would add to the efficient recycling of biomass and nutrients in the upper layer, and consequently counteract the export of biogenic matter during phytoplankton blooms, the role of mesozooplankton grazing for vertical export is more complex (e.g. Wassmann, 1993). When mesozooplankton herbivory is low, phytoplankton will preferentially be grazed by microzooplankton or aggregate (Figure 7A). Extensive mesozooplankton herbivory, often observed when diatoms are available, would add to the retention of essential nutrient elements in the upper layer, but also export biogenic matter by faecal pellet sinking (Figure 7B). If mesozooplankton select an omnivorous feeding mode, the decreasing grazing pressure on phytoplankton could give rise to increased export of aggregated cells and phytodetritus (Figure 7C). If carnivores are added to the scheme, additional phytoplankton derived matter may sink (Figure 7D).

According to this simplistic scheme, the percentage of phytoplankton carbon which leaves the euphotic zone could vary widely between 33 to 70%, depending on the trophic behaviour of various types of zooplankton and the structure of the zooplankton community (Figure 7). On the one hand, in order to promote recycling of organic matter, herbivory by meso- and in particular microzooplankton must be efficient and compete with aggregation. Omnivory and carnivory, on the other hand, may promote export of organic matter. The trophic behaviour of zooplankton and the competition between zooplankton grazing and aggregation thus mediate vertical fluxes. In addition, not only grazing, but also the survival of zooplankton offspring is important to understand the successional dynamics retention and export chains. A growing literature has documented the inhibitory effect of a diatom diet on egg spawning, hatching success and nauplii survival (Ianora & Poulet, 1993; Ianora et al., 1995; Uye, 1996, but see also Jonasdottir & Kiørboe, 1996). Mixotrophic flagellates are apparently superior to diatoms in copepod diets (see Uye, 1996 and references therein). The involvement of mesozooplankton in retention and export food chains is thus not only influenced by processes such as match/mismatch, advection and over-wintering, but also by dietary circumstances.

The role of zooplankton for vertical export is obviously an important aspect of vertical flux studies, which at present is inadequately understood. This conceptual inadequateness is indeed remarkable (Banse, 1994) and supports the notion that it may be essential for future progress to further elucidate the potentially pivotal role of zooplankton mediation of vertical flux in the ocean.

Experimental evidence of the regulation of export and retention food chains

The knowledge of top-down regulation of vertical flux in marine environment is limited by sparse experimental evidence. Experiments investigating the impact of food web structure on nutrient retention *and* vertical losses of biogenic matter in the ocean have been carried out recently and only by a few research teams (e.g. Keller & Riebesell, 1989; Heiskanen et al., 1996; Wassmann et al., 1997). Investigations of top-down and bottom-up regulation in the eastern Baltic were carried out in mesocosms with additions of nutrients and carnivorous fish. They revealed the emergence of two basically different systems: (a) units which received nutrients immediately after the start of the experiment subsequently developed towards more regenerating systems and (b) units which received nutrients after a 5-day lag period, showed a new production response. In all units, vertical export of biogenic matter followed closely the development of the autotrophs in the euphotic zone. The cascading effect of top-down manipulation influenced the plankton community and resulted in different functional response in the various treatments. The results indicated that during the process of eutrophication, the food web structure, timing of fertilisation and alternative grazing/predation strategies of the planktonic heterotrophs have a crucial impact on the retention and loss of nutrients from the pelagic zone (Heiskanen et al., 1996).

Addition of nutrients which are important for new production and which influence phytoplankton species composition, e.g. dissolved silicate (Dugdale et al., 1995), has a pronounced effect on primary production, species succession (Egge, 1993) and vertical carbon export. An experiment in two sea enclosures with nitrate and phosphate (NP), and nitrate, phosphate and dissolved silicate (NPS) (Wassmann et al., 1997) revealed that primary production was 31% higher in the NPS enclosure as compared to the NP enclosure over the experimental period of 27 days (Table 1).

Table 1. Sedimentation, carbon fixation and nutrient consumption during the experimental period (27 days). Data from Wassmann et al. (1997)

		NP-enclosure	NPS-enclosure	NPS-NP
C-fixation	($g\ C\ m^{-2}$)	13.1	19.0	5.9
DSi-consumption	($g\ DSi\ m^{-2}$)	0.66	1.51	0.85
N-consumption	($g\ N\ m^{-2}$)	1.49	1.35	–0.14
P-consumption	($g\ P\ m^{-2}$)	0.23	0.24	0.01
C-sedimentation	($g\ C\ m^{-2}$)	13.7	16.8	3.1
N-sedimentation	($g\ N\ m^{-2}$)	2.0	2.5	0.5
Chl *a* sedimentation	(mg Chl *a* m^{-2})	28.7	53.4	24.7

Table 2. Results from a study in Lake Haugatjern which had a large population of whitefish (*Coregonus lavaretus*) and perch (*Perca fluviatilis*). Vertical flux was measured at 8 m depth from 1979 to 1982. Ice-cover: 200 d y^{-1}; primary production starts in late May. Vertical flux was calculated for a productive season of 130 d y^{-1}. Rotenone treatment killed the fish, cladoceran and copepod population in September 1980

Year	Time period	# days	POC vertical flux ($g\ C\ m^{-2}\ 130\ d^{-1}$)
1979	20/6-5/10	110	101
1980	29/5-2/9	92	115
1981	24/5-14/9	109	203
1982	4/6-18/10	131	82

Increased phytoplankton growth was mainly caused by mass development of diatoms in the NPS enclosure. Enhanced growth was accompanied by an increased vertical flux of organic matter (about 16% in terms of particulate nitrogen and particulate carbon) which was dominated by diatoms. For each g of dissolved silicate added vertical flux was enhanced by 3.6 g C, implying that the ratio of dissolved silicate added/carbon exported was close to the Redfield ratio. Thus the presence of dissolved silicate appears to decrease the nutrient turnover time in the euphotic zone by increasing vertical export. As expected, bottom-up control of vertical flux is significant when the impact of the planktonic heterotrophs is limited.

To the best of my knowledge, the only experimental investigations regarding the consequences of addition or removal of fish on plankton dynamics and vertical flux in marine environments have been conducted in the Baltic Sea (see above, Heiskanen et al., 1996). In lakes and rivers, on the other hand, numerous inquiries have been carried out (e.g. Langeland, 1990), but vertical flux studies were too infrequent to generalise. In order to illustrate the dynamics which may be expected with regard to vertical flux when fish is removed from the upper layers, some unpublished results from a study in Lake Haugatjern (Reinertsen & Olsen, 1984) are presented. This lake had a large population of whitefish (*Coregonus lavaretus*) and perch (*Perca fluviatilis*). Rotenone treatment killed the fish, cladoceran and copepod populations in September 1980. The vertical flux of organic carbon during the years prior to rotenone treatment was alike (Table 2). After the treatment, vertical carbon export doubled, probably as a function of the elimination of both fish and zooplankton. Due to the lack of grazers, a majority of phytoplankton sank ungrazed into an anoxic bottom layer. In 1981, however, vertical carbon export was lower compared to the pre-rotenone years. The decreased vertical export in 1981 as compared to 1979/80 seems to reflect re-establishment of the zooplankton grazer community in the absence of predation by carnivorous fish. This is clearly seen in the low sedimentation rates in summer 1981 which follow after the regular spring bloom deposition of diatoms (data not shown).

The perception from limnetic environments is far more complicated than the evidence of processes controlling sinking loss from marine pelagic systems. For example, increased zooplankton grazing pressure has been observed to promote large, grazer-resistant cells (e.g. Hessen et al., 1986; Mazumder et al., 1988) which together with the faecal pellet flux gives rise to an *increase* of vertical flux rather than the expected *decrease*. Visually feeding, planktivorous fish can reduce the abundance of large herbivores (e.g. Hessen et al., 1996; Drenner et al., 1989), which can either give rise to *increased* vertical flux due to a lack of grazing pressure on phytoplankton, or *decreased* vertical flux due to increased growth of small zooplankton. It may well be that our present understanding of the marine environments is based on assumptions and oversimplifications which will have to be revised in years to

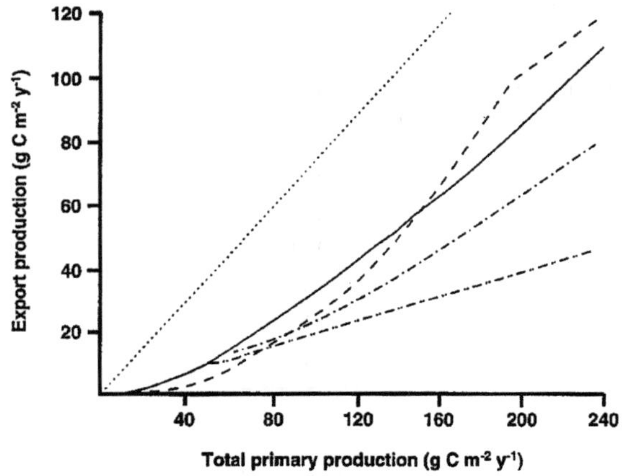

Figure 8. Export production as a function of total primary production on an annual scale in marine ecosystems. Algorithms from various publications are presented. Suess (1980) (..........), Eppley & Petersen (1979) (- - - - -), Betzer et al. (1984) (–·–·–·–), Pace et al. (1987) (–··–··–··–) and Wassmann (1990) (——). Redrawn from Wassmann (1990).

come, and that we can expect as complex responses of retention and export food chains in marine as in limnetic environments.

Primary and export production: Processes, dynamics and problems in quantifying and predicting vertical export

Several international research initiatives such as the Global Ocean Flux Study (GOFS) attempt to comprehend carbon cycles of the ocean. These programmes in part used satellites for determination of ocean colour, from which, using complicated computations, phytoplankton stock, primary production and finally export production is calculated (e.g. Sathyendranath & Platt, 1993). Having in mind the rather limited knowledge of the regulation of export production, the question arises whether these estimates of export production are precise. As will be seen below, no universal algorithm exists which predicts export as a function of total primary production with any precision. To my mind the only feasible way of approaching this challenge is by attempting to quantify pelagic-benthic coupling regionally and to characterise functional ecosystem types on a global scale.

Algorithms of primary production versus vertical carbon export on an annual scale

In the following section, total primary and export production will be abbreviated PT and PE, respectively. An overview on algorithms predicting export production in marine environments on the basis of total primary production on an annual scale has been presented by Wassmann (1990, 1993) (Figure 8). Some of the algorithms are based on comparisons of daily rates of PT and PE and are thus of limited value when data were consecutively extrapolated to an annual scale. Only data based on investigations of continuously integrated PE (close to the bottom of the euphotic zone) over time intervals of an year or the productive season and PT (at the same site, the same time interval and the same year) should be accepted. These criteria were only met by two data sets from the boreal, coastal zone of the North Atlantic and subalpine lakes (Figure 9A). Significant variability with regard to the PE versus PT relationship was detected. What algorithm should be selected for a global carbon flux model? Obviously, there is no universal algorithm. Does the variability of the algorithms reflect real difference in the PE vs. PT relationships in the various ecosystems from which they were derived (Figure 8)? If so, then different algorithms should be applied in different regions.

The results of the model of Aksnes & Wassmann (1993) indicate that domination by copepods in the marine environments and cladocerans in lakes can give rise to very different relationships between primary versus export production (Figure 9A). Mesozooplankton species composition obviously influence the pelagic-benthic coupling: for example, copepods and cladocerans have different reproductive strategies (hence different grazing pressure), and cladocerans do not produce distinct faecal pellets. A comparison of retention and export food chains, and vertical flux in lakes dominated by copepods (e.g. Lake Baikal) or marine environments strongly influenced by cladocerans (e.g. the eastern Baltic Sea), would be advantageous to analyse in greater detail the contrasting scenarios of copepod and cladoceran dominance for pelagic-benthic coupling.

In case the algorithms depicted in Figure 8 truly predict annual PE on the basis of PT, why are there significant differences? In the case of subalpine lakes and boreal coastal areas, we have already recognised that differences in the zooplankton community species composition result in the observed variance. The question can be raised if the results presented in Figures 8

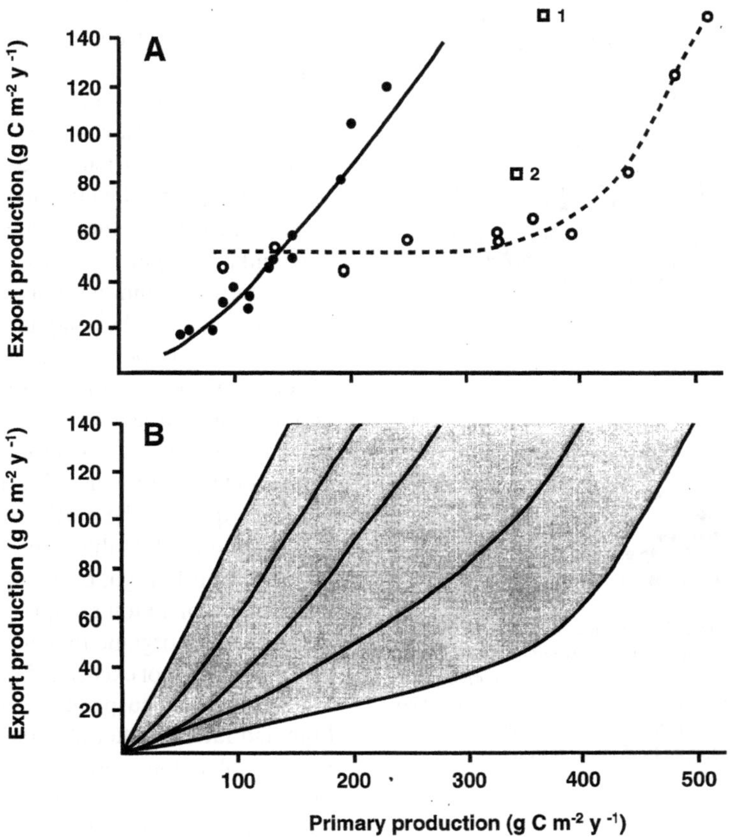

Figure 9. (A) Export production as a function of total primary production from the North Atlantic, boreal coast (Wassmann, 1990; full line) and subalpine lakes (Aksnes & Wassmann, 1993; broken line) on an annual scale. The zooplankton of the former ecosystems is often dominated by copepods, the latter one by cladocerans. Also shown are two data points from Dabob Bay, a boreal, North Pacific fjord, and a tropical lagoon, Kaneohe Bay on Hawaii (open squares, 1 and 2, respectively). Redrawn from Wassmann (1993). (B) Schematic diagram on the conceivable relationship between annual export production and total primary production in miscellaneous ecosystems with different production, recycling and export regimes. The functional lines of the various ecosystems could be spread in the shaded area. The relationships could fall onto a suite of lines contrasting between maximum export (steep angle, straight relationship) and high retention (flat angle, strong curvature) efficiencies.

and 9A suggest that various types of top-down regulation are the basis for the observed variability? The few data which do exist from non-boreal environments outside the North Atlantic suggest that coastal areas and tropical bays in the North Pacific Ocean experience more efficient retention in the upper layers and less vertical export (Figure 9A). This interpretation is consistent with the notion that tropical environments are characterised by effective retention food chains. This may also be true for the North Pacific Ocean where at least the open ocean is characterised by extensive microzooplankton grazing, which prevents major accumulation of phytoplankton biomass (Frost, 1991; Dagg, 1993). PE as a function of PT in miscellaneous ecosystems with different production, recycling and export regimes could fall onto a suite of lines falling between maximum export (steep angle, straight line) and high retention (flat angle, curved line) efficiencies (Figure 9B). In order to test this hypothesis, annual investigations of plankton dynamics and vertical flux have to be carried out in different biogeographical zones with variable productivity rates and f-ratios.

General trends in production and vertical flux variability in north-east Atlantic ecosystems

The patterns of production and sedimentation presented by Smetacek et al. (1984) and Wassmann (1991) find representations in many shallow, boreal coastal areas (e.g. Hargrave et al., 1985; Peinert et al., 1982; Laws et al., 1988; Heiskanen & Kononen, 1994). However, quite different patterns can be found in other areas. In

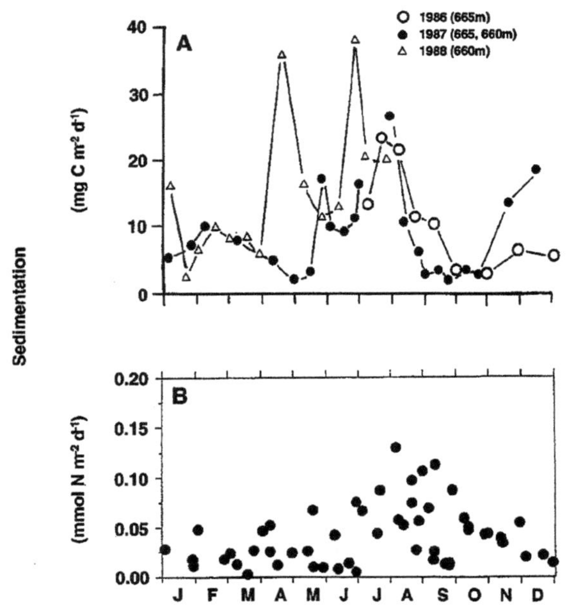

Figure 10. Vertical export of particulate organic carbon on the Vøring Plateau from 1986 to 1988 and the central Norwegian Sea. Redrawn from Wassmann et al., (1996). Data covering 8 consecutive years of investigation from the central Norwegian Sea are projected into graph B. Observe the large scatter in seasonal rates. Redrawn from Wassmann et al. (1991) and Haupt (1995), respectively.

various parts of the Norwegian Sea under the influence of the North Atlantic Current, export of vernal bloom material in general does not give rise to increased vertical flux at depth (Figure 10). Increases in sedimentation during spring are minor and more part of a general trend towards increased vertical export during the productive part of the year. Maxima of export flux at trap depths ranging between 660 to 1034 m (average 807 m) on the Vøring Plateau were generally recorded between July–August (Figure 10A) and in the central Norwegian Sea between August–October (Figure 10B). The reasons why (a) the extensive new production during spring and early summer does not give rise to increased vertical fluxes which penetrate down to several hundred to thousand meters depths and (b) the seasonal vertical flux maximum is in late summer/autumn, are not self evident. Simultaneous, seasonal and detailed investigations of the plankton dynamics in those remote oceanic areas are lacking.

Wassmann et al. (1991) suggested that the retention efficiency of the planktonic community in the Norwegian Sea is rather strong, with calanoid copepods playing one of the principal roles. CV stages of *Calanus finmarchicus*, for example, descend to over-wintering depths, exceeding 500 m in midsummer, cease feeding and remain at depth until next spring when the population ascends to surface waters as adults, ready to graze extensively on the spring bloom (Bathmann et al., 1990b). Mesozooplankton apparently have the ability to retain nutrients and matter from the spring bloom in the upper layers, thereby minimising vertical losses. The role of other pelagic forms such as microzooplankton, larger crustacea, pteropods, salps and fish with regard to retention versus export in Norwegian Sea is not well known, but may be essential (Bathmann et al., 1990a, b). Export losses increase after the decline of retention food chain efficiency in late summer and autumn, and ontogenetic vertical migration may be a reason for the increased vertical export in autumn.

Recent investigations of the trophic interactions and vertical export on the north Norwegian shelf revealed that despite the fact that annual new production must have been fairly high, probably > 75 g C m^{-2}, large-celled phytoplankton comprised only a small part of the suspended organic matter, and chlorophyll *a* concentrations did not exceed 2.5 μg l^{-1} (Wassmann et al., 1998). However, suspended POC concentrations were high and faecal matter abundant (Hansen, 1997). The mesozooplankton biomass at the shelf edge was also high, about 2.5 g C m^{-2} throughout the productive season (Nordby & Tande, 1998). Grazing impact of microzooplankton, especially heterotrophic flagellates, was apparently substantial (Ratkova et al., 1998), and vertical export of POC was moderate to high, on average about 170 mg C m^{-2} d^{-1} (Andreassen et al., 1998). Obviously meso- and microzooplankton did play an important role for the retention of organic matter, as they were able to prevent the accumulation of phytoplankton biomass which could aggregate and sink. As a consequence, faecal pellets by mesozooplankton contribute significantly to the export of carbon (Hansen, 1997). Although indication for zooplankton vertical flux mediation were recorded (Urban-Rich et al., 1998), the retention food chain at the north Norwegian shelf is obviously not efficient enough to reduce export losses to a minimum. However, it is an ecosystem where the retention food chain plays a far more significant role compared to west-Norwegian fjords or the Baltic Sea. Similar scenarios of strong top-down control have been observed in the central North Pacific Ocean (Frost, 1991; Odate, 1994). An example for the variation in export fluxes where differences in physical forcing (i.e. bottom-up regulation) are the main cause for the development of the patterns of production and sedimentation is given in Figure 11. In the Barents

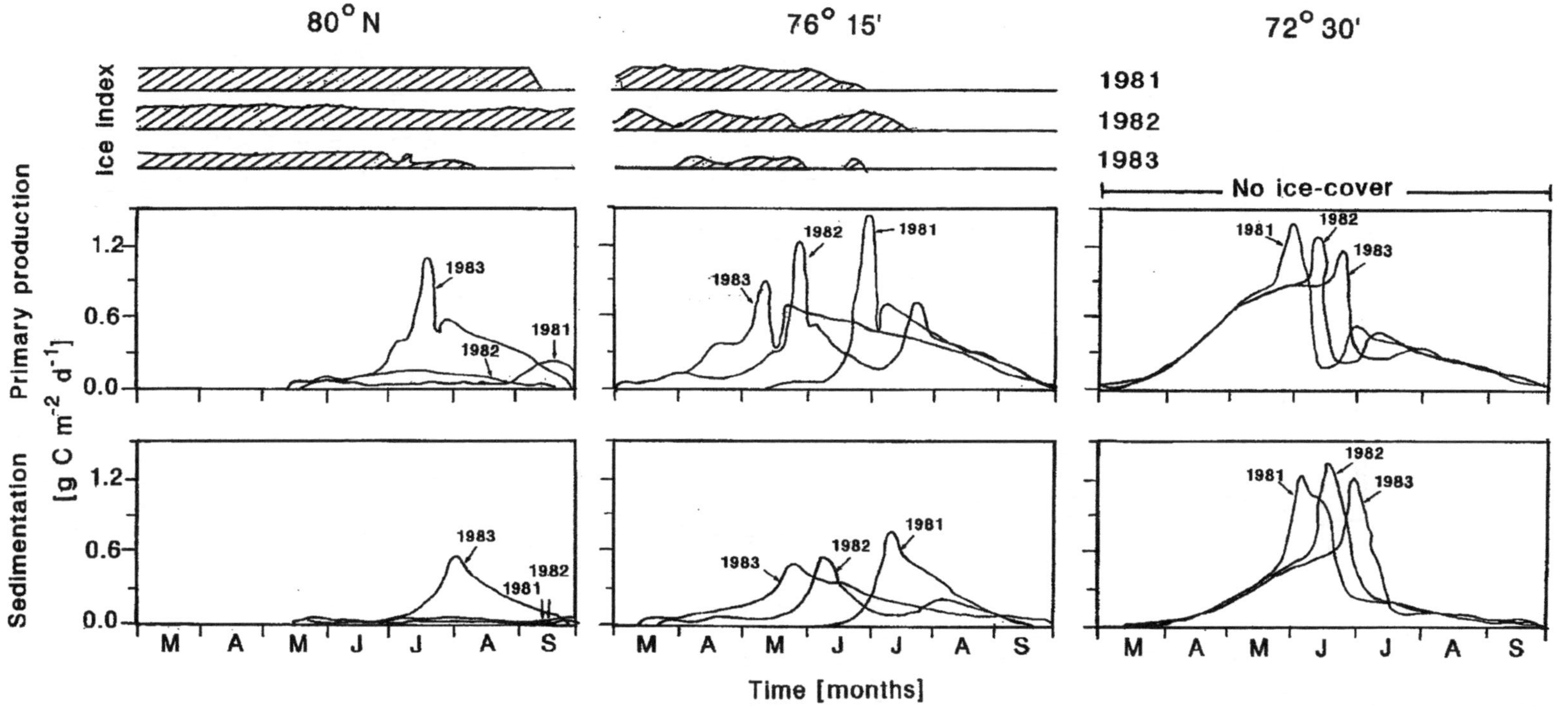

Figure 11. Ice index, primary production and vertical carbon flux at different latitudes in the central Barents Sea for the years 1981–1983. The results are replotted according to a physical-biological coupled carbon model presented by Wassmann & Slagstad (1993). redrawn from Wassmann et al. (1996).

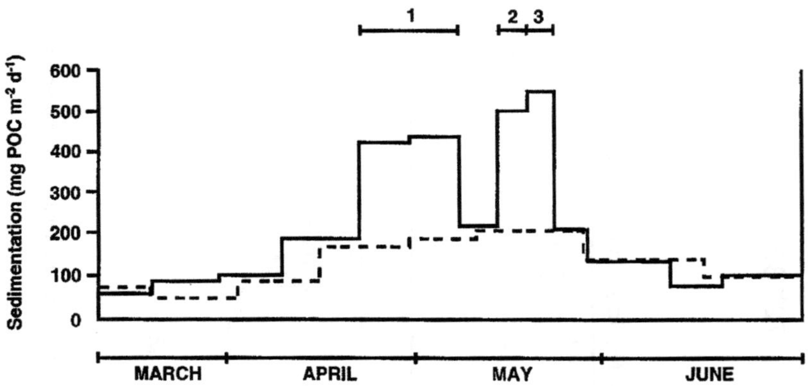

Figure 12. Vertical export of particulate organic carbon at 50–60 m in 1982 (broken line) and 1992 (full line) in Balsfjord, northern Norway. The horizontal bars indicate three episodes when bottom water was exchanged in 1992. Redrawn from Reigstad & Wassmann (1996).

Sea, the development of stratification and variable ice-coverage exerts a strong control on the export of organic matter. Over a distance of about 500 km, primary and export production decrease (south to north) from about 73 to 18 and 48 to 9 g C m^{-2} y^{-1}, respectively (Wassmann & Slagstad, 1993). Sea ice cover has a dominant impact on plankton dynamics in the central and northern parts of the Barents Sea (Figure 11). Although top-down regulation of vertical export contributes to seasonal and interannual variability (Wassmann et al., 1998), the physical environment sets up the basic structure, dominating biological production and recycling in this arctic ecosystem. In areas where the impact of physical forcing is strong, large differences in the patterns due to top-down regulation may be less prominent.

Annual variability in production and vertical flux in north-east Atlantic ecosystems

Figures 10 and 11 also reveal indications of inter-annual variations in vertical export. These may reflect inter-annual variations in the efficiency of retention and export food chains and/or climate. Examples given in Figure 10A from the Vøring Plateau indicate clearly that the year 1988 gave rise to major variations in carbon export while 1986–1987 revealed more consistency. Support for the assumption that changes in the pelagic food chain structure on the Vøring Plateau were the cause for the inter-annual differences in vertical flux was provided by microscopical analysis of the sedimented matter (Bathmann et al., 1990a). Such inter-annual variability over 8 years was also observed in the central Norwegian Sea (Figure 10B) as the variance of the monthly vertical export rates is significant. It is not known if interannual variation in the pelagic food chain takes place on a basin-wide scale in the Norwegian Sea.

The interannual variations depicted in Figure 11 are more or less related to physical forcing as changing pelagic food chains were not the focus of this part of the model study. Changes in climate, insolation, stratification, wind mixing etc. can all give rise to significant modifications in the patterns of production and sedimentation. This is indicated in the southern Barents Sea where despite no ice cover, substantial differences in the timing of the spring bloom were depicted by the model. During the warmer year (1983), the spring bloom developed one month later compared to the cold year (1981). This is caused by the extended periods of mixing and the changing weather patterns during warm years which delays the intensifying of stratification and, as a consequence, the maximum of the vernal bloom.

A final example regarding inter-annual variation in export fluxes comes from Balsfjord, northern Norway (Figure 12). Significant changes in vertical carbon flux were found between the two years 1982 and 1992. These differences seem to coincide with advective events. While no major advection of shelf water could be detected in 1982, 3 episodes of water exchange were recorded in 1992. These episodes introduced new nutrients, suspended biomass, as well as zooplankton populations. This gave rise to two 'additional' maxima in vertical carbon flux in 1992 while the remaining sedimentation rates were surprisingly similar. The first maximum in vertical carbon flux in 1992 was caused by the introduction and export of a krill population whose faecal pellets comprised close to 100% of the vertical carbon flux (Riebesell et al., 1995). Obviously, annually and inter-annually changing zooplankton populations can have a strong impact on the relationship between retention and export of nutrients and organic matter.

Variability of vertical export in the pelagic zone

All investigations of export of biogenic matter indicate that the export flux decreases more or less exponentially with depth in the upper part of the ocean, with minor decreases below 200–500 m depth [for algorithms predicting the depth variation of vertical carbon flux see Berger et al. (1989)]. Resuspension and protrusion of advective, particle-rich layers may alter this general feature of vertical flux. The degradation rate of organic matter in the water column and, in particular, of fast sinking particles, is of pivotal importance for the quantitative regulation of pelagic-benthic coupling. Depending on this rate, the absolute vertical export of organic matter at a certain depth could be small or large, irrespective the size of the new production from which it derives. Comparing primary production with vertical carbon export without quantifying the pelagic degradation rate and food web dynamics, involves the risk that an adequate understanding of pelagic-benthic coupling is not within reach.

The vertical decrease of more biologically reactive, sinking material is greater compared to less reactive matter, as shown by the variability of 'half life' depths which decrease from 1700 m for POC to 700 m for amino acids (Figure 13). However, there is a substantial variability in vertical flux in the upper layers, predominantly the upper 500 m (Figure 13), which is caused by seasonal changes in vertical flux and patchiness. The wide range of biogenic sedimentation rates in the upper layers could also reflect analytical problems such as variable degradation, leakage, contamination by swimmers, or it could be caused by differences in downward fluxes caused by migrating zooplankton (e.g. Longhurst & Harrison, 1989). The scatter is so substantial that attempts to make a regression analysis through the rates in the upper layer is inappropriate.

Food concentrations within the euphotic zone are frequently too low to support growth of the suspension-feeding organisms that consume phytoplankton and other forms of particulate matter. How, then, are particle-consuming zooplankton living in the aphotic zone able to exist in their impoverished milieu? Jackson (1993) suggested that it is not by filtering

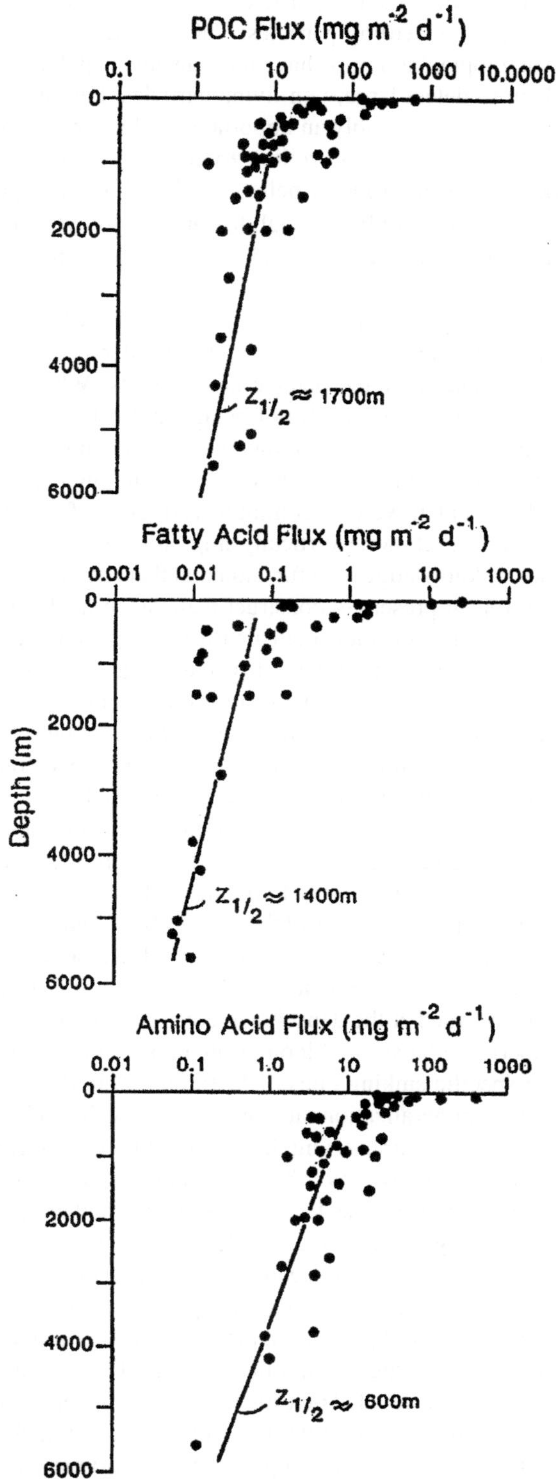

Figure 13. Vertical flux of POC, fatty acids and amino acids at several locations in the ocean. Note the extensive scatter in vertical flux estimates in the upper 1000 m and the depth $z_{1/2}$ where vertical flux is reduced by 50%. Redrawn from Lee & Wakeham (1991).

water to strain the few particles available, but rather by intercepting settling particles. This implies, however, that zooplankton must have mechanisms which allow them to detect large, non-living particles. Faecal pellet export is probably an important mechanism for the subsistence of deep-dwelling zooplankton. They may be dependent on faecal pellets produced in the upper layers which are big enough to sink fast to overcome degradation in the warm and more densely populated surface waters.

It is argued here that the general form of the vertical flux curvature as well as the scatter is strongly influenced by the pelagic heterotrophs. While we know that all size spectra of pelagic heterotrophs add to the degradation of sinking matter, we are much less aware of what relation exists between sinking particle cohorts and the vertical community structure of the heterotrophs, let alone vertically migrating species. Our lack of knowledge is so fundamental that it may not be possible at present to construct realistic vertical algorithms where the biota realistically reduces the vertical flux (Silver et al., 1991). Together with the lack of adequate algorithms for export production out of the euphotic zone (Figure 8), the inadequacy of our current understanding continues when particles penetrate deeper into the ocean on their way to the sediment.

The strength of changes in vertical flux can change over a couple of meters as depicted in Figure 14. Vertical flux was measured at 14 depths in the upper 200 m of the marginal ice zone of the Barents Sea (Andreassen & Wassmann, 1998). A conspicuous decrease in vertical POC flux of about 70% between 20 and 50 m was recorded at the densely ice-covered station which experienced a spring bloom at its maximum. Assuming specific sinking rates of 5, 10 and 30 m d^{-1}, the pelagic mineralization rate ranges between 70-800 mg C m^{-2} d^{-1}, emphasising the impact of particle degradation in a narrow layer below the euphotic zone. The vertical decrease was less evident in the southern part of the marginal ice zone with less ice-cover, but even here vertical POC flux decreased by about 50% over a 40 m depth interval. Much of the quantitatively important processes of pelagic-benthic coupling occur in the upper parts of the ocean, often over short vertical distances. More detailed investigations in the upper layers are a prerequisite in order to comprehend and mathematically describe the coupling between primary production, suspended matter and vertical flux.

It is suggested that the dynamics of retention and export food chains determine these extensive changes in quality and quantity of the vertical flux of biogenic matter. Accordingly, there is an apparent need to investigate more profoundly the 'pelagic mill', i. e. the upper 500 m of the ocean with its food chains, mineralization and biogenic fluxes (Figure 15). Assuming a vertical distribution of pelagic heterotrophs which decreases exponentially, an equivalent decrease in vertical flux would be the result. The maximum vertical flux is equivalent to the export production, i.e. the amount of biogenic carbon leaving the euphotic zone. However, vertically migrating zooplankton (on a daily, seasonal or ontogenetic scale) and uneven distribution of zooplankton on the vertical scale may give rise to deviations from an exponential decrease. Depth intervals with large and minor decreases would supposedly be found (Figure 15). Repackaging of faecal pellets could give rise to intermediate increases in vertical flux. In summary, the short-term vertical flux profile of biogenic matter will be rather complex, reflecting a diverse group of organisms and mechanisms which 'grind' the sinking matter to small entities or produce a few bigger ones. The efficiency of the 'pelagic mill' will vary in space and time, and only a few types of sinking particles will escape from it and reach the sea floor. On shelfs, the efficiency of the 'pelagic mill' is supposedly not as high as in the deep sea because of high new production and the limited vertical extension for the mill. More detailed investigations of the 'pelagic mill' in a majority of biogeographical areas seem to be inevitable to comprehend better the export of biogenic matter from the surface.

Dynamics of primary and export production on a daily basis

The regulation of PE as a function of PT is not easily understood by long-term evaluations such as the algorithms presented in Figure 8. The regulation of processes controlling sinking loss from marine pelagic systems rather have to be studied on a short-term basis. However, few studies exist, to the best of my knowledge, which would have the necessary detail in order to investigate the PE vs. PT relationship for longer periods of time on a daily basis. Such investigations are necessary to prepare algorithms which have the necessary complexity and precision in order to answer the question how PT in the ocean can give rise to PE. To prepare this challenging task, I present below some theoretical evaluations of the concepts and scan through some already existing examples to verify the theoretical evaluations.

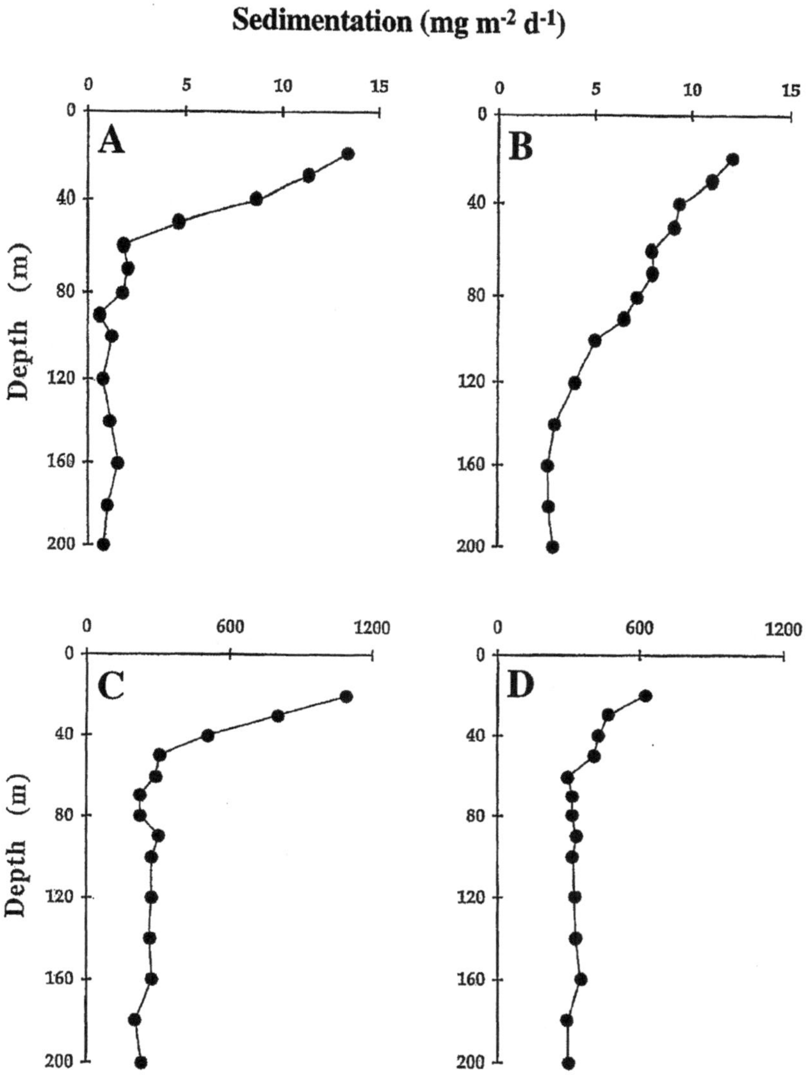

Figure 14. Vertical flux of chlorophyll *a* (A and B) and POC (C and D) in the upper 200 m of the marginal ice zone of the Barents Sea at the end of May 1993 (mg C m^{-2} d^{-1}). Two stations with 60% (left) and 20% (right) sea ice coverage are presented. Data from Andreassen & Wassmann (1998).

Theoretical evaluations

Some basic relationships between daily PT and PE are presented in Figure 16. If aggregate formation, sinking rate and grazing are constant or exponential as compared to PT, a linear or curvilinear relationship between daily estimates of PT and PE is expected (Figure 16A). Depending on efficiency of the retention food chain, high and low cases could be recorded. Such scenarios are unlikely to exist. In Figure 16B it is assumed that aggregate formation, sinking rate and grazing are variable and that the relationship between PT and PE is not proportional, but greatly variable in time. This should give rise to 'loops' of variable size depending on the closeness of the coupling between PT and PE over time (Figure 16B). While PT increases during spring and gives rise to an accumulation of biomass, PE will initially remain constant, but after some time a high (critical) suspended concentration will be attained and aggregation as well as faecal pellet production will increase PE. While PT decreases at the cessation of the bloom, PE can still increase, produce a loop, the 'export loop', but will sooner or later fall back to a line, the 'retention line' (Figure 16). The angle and curvature of the retention line and the circumference,

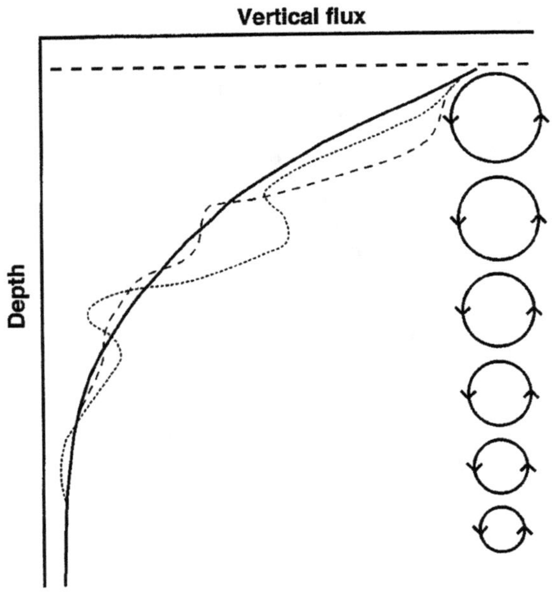

Figure 15. Schematic presentation of the 'pelagic mill' in the upper part of the ocean and its regulation of biogenic vertical flux. The *full line* assumes a continuous mineralisation of export production, giving rise to a decline in flux which follows a power function. The *broken line* indicates a step-wise decrease in vertical flux caused by extensive grazing at certain depth horizons. The *stippled line* indicates that vertical flux can increase intermittently due to repackaging. The recycling by the zooplankton community is schematically indicated to the right.

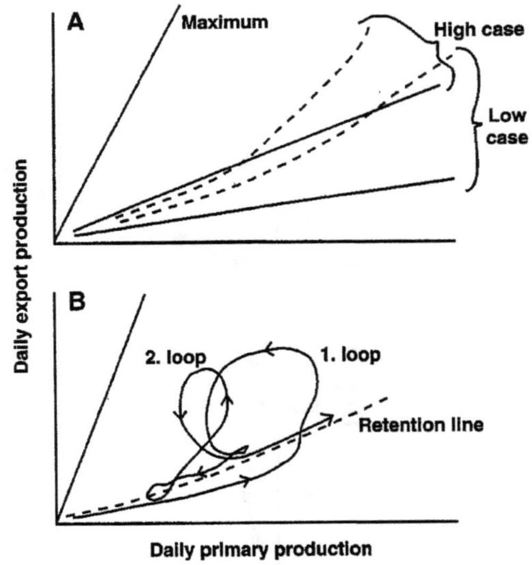

Figure 16. Schematic presentation of the potential configuration of the relationship between daily total primary (PT) and export production (PE). (A) is based on the assumption that aggregate formation, sinking rate and grazing are constant fraction of PT or follow a power function. There may be low or high cases of export production. (B) indicates the most probably development of the PT vs. PE relationship. A 'retention line' (stippled) may appear and 'loops' (full line) of variable size depending on the closeness of the coupling between PT and PE over time may take place. See text for details.

form and area of the export loop are *measures for the retention efficiency* in aquatic ecosystems.

Any disturbance or stochastic event in the pelagic ecosystem, may it be changes in light, turbulence, nutrients, advection of new algae etc., may give rise to deviations from the retention line of the PT vs. PE relationship. Disturbances are probably reflected in looping behaviour, followed by periods of zooming back onto a retention line or a retention point. It is hypothesised that over successional, long-term and evolutionary times the pelagic system develops towards (a) a flattening of the loops, (b) a minimisation of the loop area and (c) a low angle of the retention line. The retention efficiency is thus reflected in the angle of the retention line, the size of the loop area and the time period involved to complete an eventual loop of the PT vs. PE relationship. There may be different retention lines and loops throughout the year since the line and loops are a function of (a) the predominant phytoplankton types, their aggregation potential and the behaviour, composition and the dynamics of the zooplankton community, but also (b) the type, frequency and length of episodic events in physical forcing. Is there any systematic pattern in the PT vs. PE short-term relationship, a pattern which could be described mathematically and used for predictive and classification purposes?

Examples from mathematical models

There are almost no data sets where PT and PE were measured frequently enough to study the details of the PT vs. PE relationship. Therefore results from a mathematical model on planktonic production and vertical flux from the southern Barents Sea, dominated by Atlantic water and the adjacent marginal ice zone to the north (Slagstad & Wassmann, 1993) were investigated. The vertical carbon flux in this model is a function of: (a) sinking diatoms, whose sinking rate is a function of nitrate concentration (an attempt to mimic aggregation), (b) *Phaeocystis pouchetii* colonies, (c) detritus from flagellates and (d) faecal pellets from mesozooplankton grazing (*Calanus finmarchicus* or *C. glacialis*). Three measurements per week, i.e. about 70 data points per seasonal cycle, were selected for the climatic different years 1981–1983.

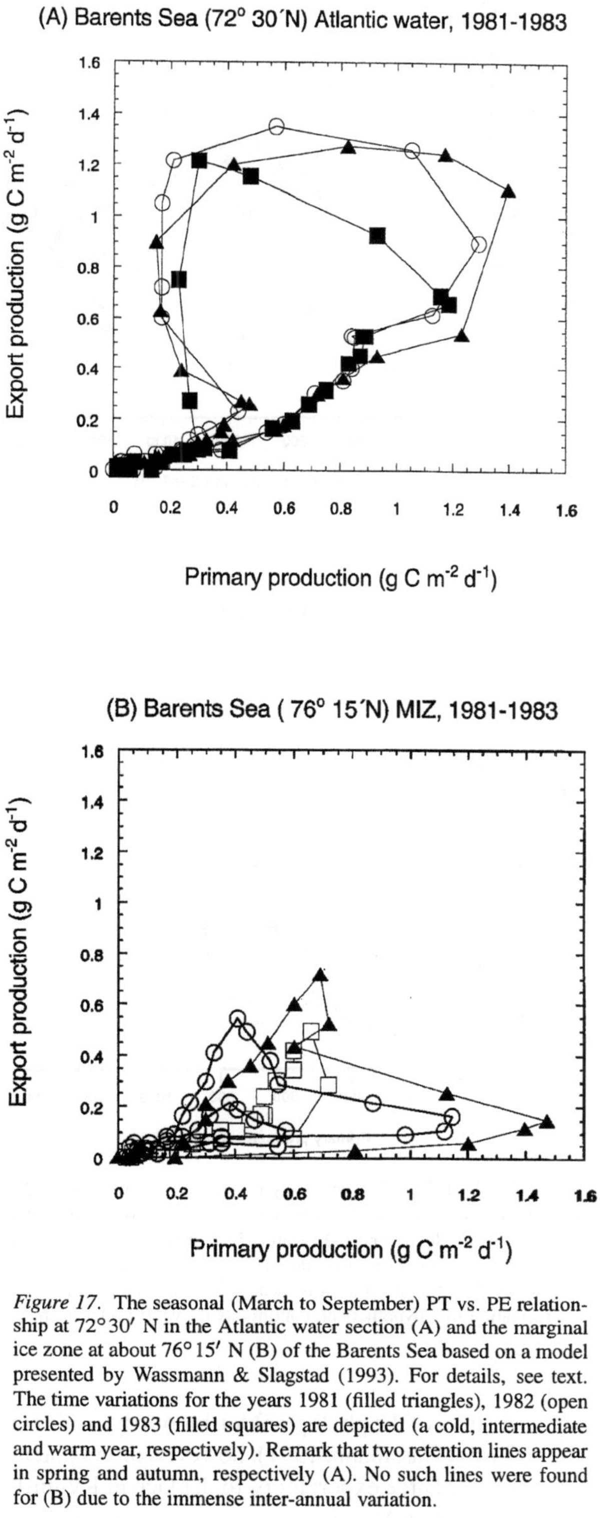

Figure 17. The seasonal (March to September) PT vs. PE relationship at 72°30′ N in the Atlantic water section (A) and the marginal ice zone at about 76°15′ N (B) of the Barents Sea based on a model presented by Wassmann & Slagstad (1993). For details, see text. The time variations for the years 1981 (filled triangles), 1982 (open circles) and 1983 (filled squares) are depicted (a cold, intermediate and warm year, respectively). Remark that two retention lines appear in spring and autumn, respectively (A). No such lines were found for (B) due to the immense inter-annual variation.

The seasonal variation in the PT vs. PE relationship from the Atlantic water, which experiences almost no stratification during spring and develops a weak density gradient later on, indicates clearly that this ecosystem develops slowly along a retention line during spring and early summer until a rapid export loop takes off from this line (Figure 17A). The loop lasts for only 3 weeks after which the system falls back onto a second retention line which is similar to the first one. Although some interannual variability appears, the general impression is that these variations are confined less to the size, but first of all to the timing of the loop episodes.

The situation in the marginal ice zone is different. Stratification of the surface water is strong due to melting sea ice, and ice-coverage varies greatly between years (see Figure 11). Thus 1981 was a cold year where PT started first in the middle of May. 1982 was a medium-cold year where the ice cover was more variable. The penetration of radiation allowed the spring bloom to take place in May, but the sea-ice coverage increased again in thickness thereafter, before it disappeared completely in July. This is nicely reflected in Figure 17B which shows two loops of the PE vs. PT relationship, a larger one to begin with and a subsequent smaller one. The lack of significant ice coverage in 1983 gave rise to a scenario were the ecosystem developed more slowly with regard to PT and PE rates and tendencies of a retention line were found. General retention lines were not discernible due to the large variability during the other years.

The differences between the two stations is apparent by comparing Figure 17A with 17B. Stratification of surface water results in far less nutrient supply in the marginal ice zone compared to the more turbulent Atlantic water stations. The dynamics of a marginal ice zone ecosystem with its more limited nutrient availability is reflected in the significant interannual variability in size, form and development of the export loops. The PE rates are larger and smaller and the export loops have a larger and smaller size in Atlantic water and the marginal ice zone, respectively.

Grazing by mesozooplankton can give rise to conspicuous changes in the PT vs. PE relationship (Figure 18). When no grazing is involved, large amounts of phytoplankton-derived matter are exported to the aphotic zone, as reflected by the largest export loop. Increased grazing decreases PE as well as the size of the loop. With very large amounts of over-wintering mesozooplankton present in March and their consecutive grazing and growth, the export loop almost disappears

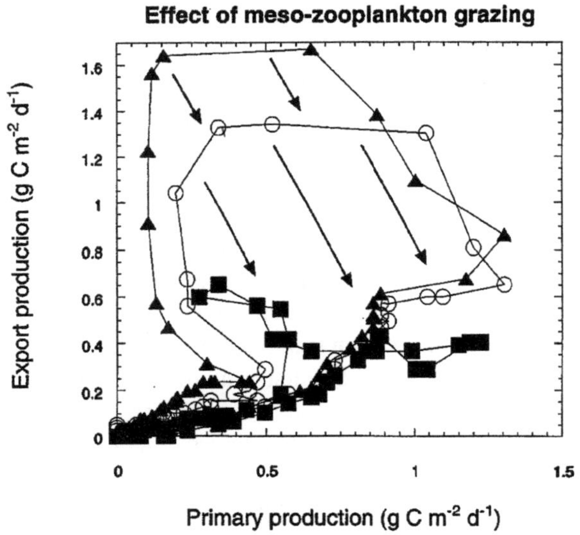

Figure 18. The seasonal (March to September) PT vs. PE relationship at 72°30′ N in the Atlantic water section of the Barents Sea as a function of variable over-wintering mesozooplankton abundance in March, i.e. variable grazing. The arrows indicate how the export loop caused when grazing does not take place (filled triangles) is reduced by increased grazing (2,000 and 10,000 individuals of over-wintering CV stages of *Calanus finmarchicus* (open circles and filled squares, respectively)). The figure is based on a model presented by Wassmann & Slagstad (1993). There is an obvious retention line during spring.

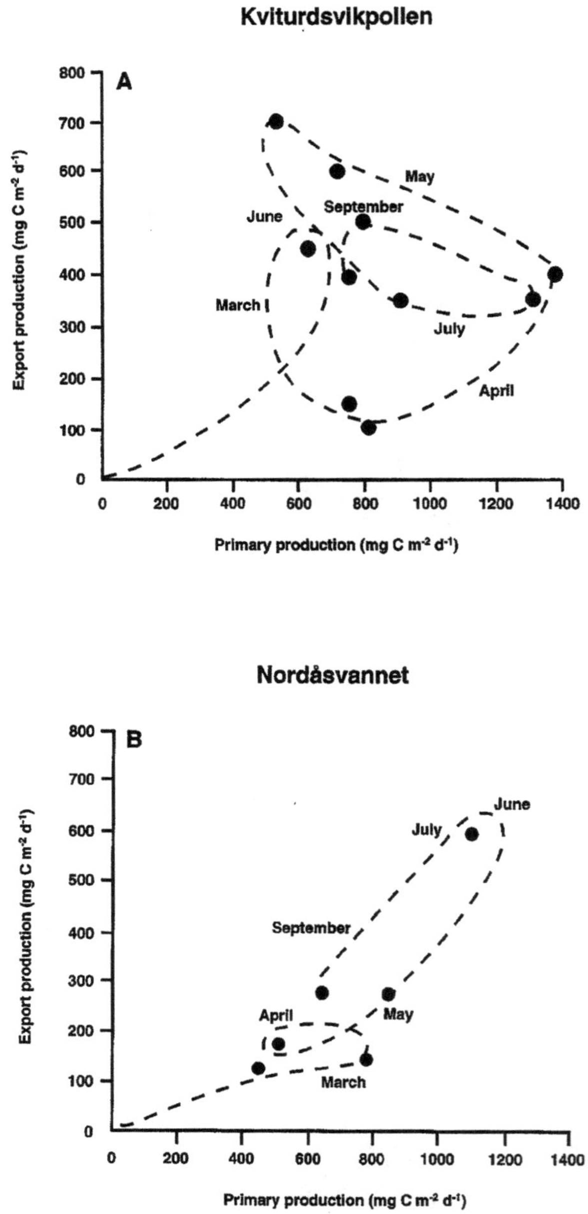

Figure 19. The PT vs. PE relationship in two west-Norwegian fjords, Kviturdvikpollen (B) and Nordåsvannet (B). Data from Wassmann (1991). The productive period (March–September) was covered in 1982. PT represents the average rate during the time period when PE just below the euphotic zone was measured. The stippled line is an interpretation of a hypothetical, continuous PT vs. PE relationship.

and most of the PE and PT data points fall back onto the retention line. Obviously mesozooplankton has the capacity to increase retention in the upper layers to an extent that major vertical loss can be minimised. Adequate abundance of mesozooplankton results in that excursions of PE are 'zoomed' back onto a retention line.

Examples from coastal areas

The lack of data with an adequate time resolution on primary and export production makes the presentation of 'real' data problematic. Selected data from two west Norwegian fjords (Wassmann, 1991), comprising 10–12 data points over the productive season, were selected. These data points have to be interpreted with regard to the size and form of the loops as they are time integrals of 2–3 weeks.

PT and PE data from the shallow fjord Kviturdvikpollen are presented in Figure 19A. The dilemma of studying the time variations of the PT vs. PE relationship on a daily basis with the majority of existing data gets immediately apparent: there is an extensive scatter in data points with no obvious relationship in sight. A study of the phytoplankton bloom dynamics in these fjords (Wassmann, 1991), however, offers potential insights. Three export loops took place in Kviturdvikpollen: one in March, one in May and a final small one in August/September. A similar interpretation of data points from the strongly eutrophied land-locked fjord Nordåsvannet revealed two export loops: a small one in March/April and a second one in summer (Figure 19B). These lines are seriously speculative. The daily dynamics of the PT vs. PE relationship are not really recoverable by the time integrals which are the basis of Figure 19. This is, among other facts, reflected in the PE rates which are far smaller than in Figure 18. The strong variability of short-term PE rates are by no means reflected in the averages supported by sediment traps exposed for longer time periods. The details of the dynamics of retention and export food chains probably can not be studied in greater detail on the basis of existing vertical flux data sets.

Examples from an experimental mesocosm study
More adequate data sets of PT and PE for the purpose of daily investigations of pelagic-benthic coupling are available from mesocosm studies. Data from an experiment where the effect of dissolved silicate on PT and PE was tested (see above, Wassmann et al., 1997) are shown in Figure 20. In the stratified water of the mesocosms and with abundant nutrient concentrations, PT increased rapidly and decreased after little more than a week. PE increased first after PT had declined. Both PT and in particular PE were greater in the NPS (Figure 20) compared to the NP mesocosm (Figure 22A). Two loops developed. In the NP mesocosm they were small and focused upon a retention point with PT $\approx$ 300 mg C m^{-2} d^{-1}, while in the NPS case the first export loop was larger, and the second one was in the same range as for the NP mesocosm. After intense excursions of the PT vs. PE relationship, caused by the episodic supply of abundant nutrients, both mesocosms developed towards one retention point. The disturbance of the ecosystem prior to the start of the experiment gave rise to a stimulation of the export food web (e.g. growth of diatoms and aggregation), until nutrient limitation and the retention food web reduced vertical loss and introduced a new 'status quo' which was comparable for both mesocosms. Strong stratification probably gives rise to this type of PT vs. PE relationship (compare Figure 20 with Figure 17B).

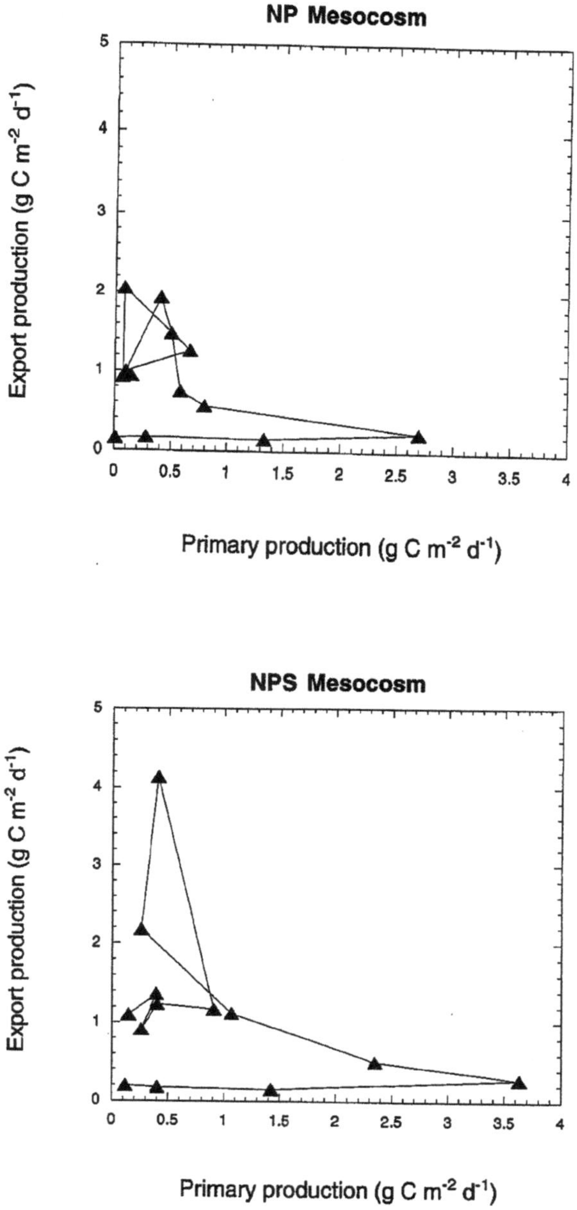

Figure 20. The PT vs. PE relationship from a 27 day long mesocosm experiment where N (nitrate) and P (phosphate) and N, P and DSi (dissolved silicate) were added to the mixed upper layers of the sea enclosure (NP mesocosm and NPS mesocosm, respectively). PT was measured every second day, PE continuously over two days periods. Data from Wassmann et al., (1997).

Systematic patterns in the relationship between primary versus export production: Is prediction and ecosystem classification conceivable?

One-pulse and *multi-pulse systems* can be distinguished. One major 'disturbance' (e.g. light, strati-

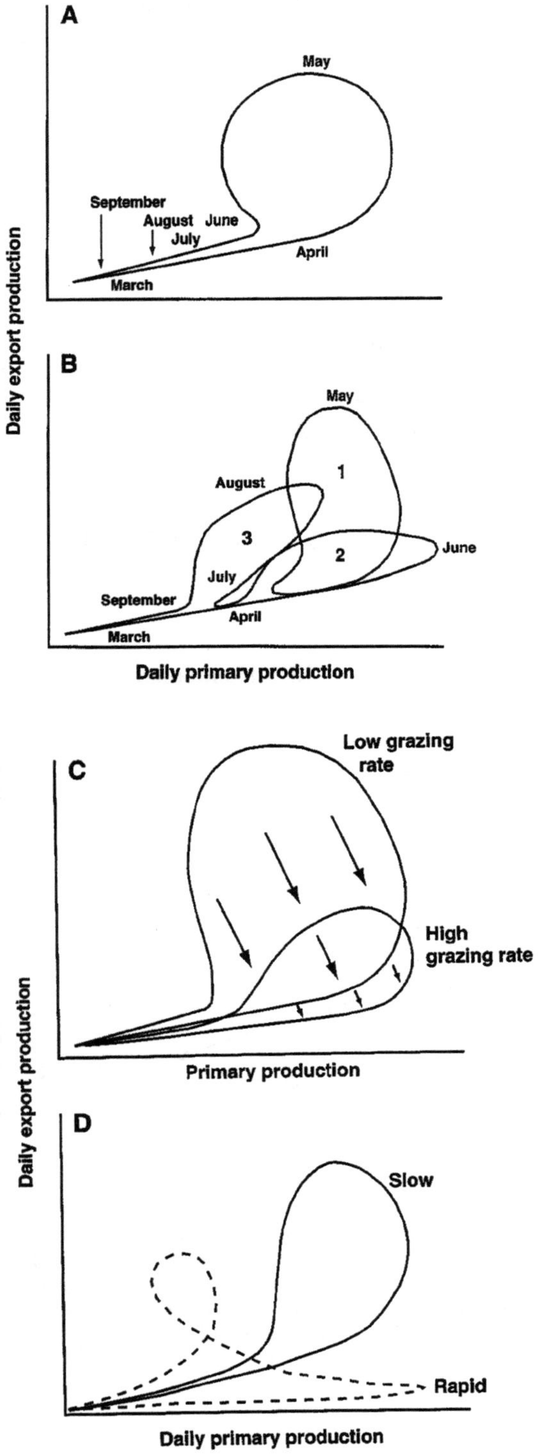

Figure 21. Schematic presentation of central elements of the PT vs. PE relationship. The time variation of one and multi-pulse systems, (A) and (B), respectively. The influence of zooplankton grazing pressure on the PT vs. PE relationship (C). The influence of rapid and slow increases of PT on PE (D).

fication, nutrients, advection, upwelling) gives rise to 1–2 retention lines and *one loop*. The size and form of the loop is dependent on the type of phytoplankton present, the types of grazers and grazing efficiency (Figure 21A). Several 'disturbances' may give rise to *several loops*. Again, the size of the loops is dependent on the type of phytoplankton present, the grazer types, the grazing efficiency as well as the time between the disturbances (Figure 21B). Zooplankton grazing will lower the angle of the retention line and lower the excursion of the export loop (Figure 21C). Also the rapidity of the temporal development will play a role for the functional relationship between PT and PE (Figure 21D). When PT increases rapidly (for example due to strong stratification), a flat line will appear at the beginning, followed by a small loop at low PE rates. A more gradual development of both PT and PE (for example due to increased turbulence) will give rise to a more pronounced loop and smoother development over time.

The schematic features exemplified in Figure 21 could probably be depicted and quantified in the ocean, should adequate data be available. The features contain the potential to be expressed mathematically and, thus, be used as tools for quantifying the export of organic carbon from the surface ocean in order to build up a more sound basis for future global carbon models and ecosystem characterisation. For different biogeographic scenarios in the world ocean, algorithms as those illustrated in Figure 9B could be developed. However, achieving such a goal would represent a great endeavour, as detailed investigations of principal rates and basic components of retention and export food chains have to be conducted on a daily basis over lengthy periods of time and for principal regions of the world ocean. From a global point of view, our prospects of a strongly improved, quantitative understanding of export and retention food chain and their principal rates, constants and mechanisms are not encouraging. However, focused, systemecological investigations of contrasting ecosystems, and scenarios which include the necessary detail with regard to components and time resolution, may pave the road to significant improvements of our understanding of the dynamics of the export and retention food chains. If these investigations are carried out in a systematic manner and incorporated into detailed mathematical models, extrapolations over larger biogeographic zones and contrasting ecological scenarios may lead to more valid generalisations.

Summary and epilogue

Pelagic systems are potentially capable of retaining and recycling all autochthonous organic material, although some losses due to sinking particles inevitably occur. However, in coastal areas and on shelfs not subjected to extensive grazing, a major proportion of the phytoplankton production is channelled into the benthos (Smetacek et al., 1984; Wassmann, 1991). Relating processes in the surface layers quantitatively to vertical particle flux is difficult in the open ocean, because only a small percentage of the total production (e.g. large phytoplankton cells, various kinds of aggregates and faecal pellets) is lost annually via sinking. The 'pelagic mill', grinding suspended and sinking particles in the mesopelagic zone of the ocean, is usually effective, but rather poorly understood. It deserves immediate attention in order to improve our understanding of the regulation of vertical flux. The physical environment determines nutrient availability, influences the aggregation of sticky particles and hence the particles potentially available for sedimentation. Thus there is a strong impact of bottom-up regulation of vertical flux. However, there is also an obvious, but in marine ecosystems so far less considered, top-down regulation. It is suggested that the strength and variability of top-down control is an important driving force for the fate of organic matter which is reflected in observed geographic and interannual variability. Differences in patterns and magnitude of export production in various pelagic ecosystems may remain unexplained when exclusively regarded from a bottom-up perspective.

The relationship between phyto- and zooplankton governs vertical flux seasonality, and zooplankters with different life cycles and feeding strategies further modify the principal patterns of export production. The relationship between various zooplankton communities and life patterns of major forms has not been adequately studied in order to understand the regulation of export of organic matter from the upper layer (Verity & Smetacek, 1996). For example, herbivores with life-cycle strategies which involve over-wintering of large biomass and predictable seasonal appearance (copepods, euphausiids) will have a different impact than opportunistic organisms with very low overwintering biomass (e.g. salps, pteropods). In some areas, predictable seasonal appearance by over-wintering zooplankton can be reduced, amplified or be stochastic due to inter-annual variations in advection. The role of omnivory and carnivory, as well as the coupling between the microbial loop, mesozooplankton and suspended biomass on vertical flux is poorly known. Also, the role of different zooplankton functional groups to promote (e.g. grazing and production of large faecal pellets) or remove sinking matter (e.g. grazing on aggregates and processes such as coprophagy) influence the export and retention efficiency of various pelagic ecosystems.

In general we assume that large, ungrazed cells, aggregates and some faecal pellets dominate vertical export, in particular at the end of major phytoplankton blooms. This opinion is based on the fact that so far a majority of vertical flux studies have been carried out in areas where mesozooplankton is excluded from over-wintering, or where other processes result in a mismatch between phyto- and zooplankton production. The assumption of bottom-up regulation of vertical flux gains also support by the fact that many of the biogeochemically oriented vertical flux studies with long-term deployed multitraps were rarely accompanied by extensive planktonic research; i.e. the vertical export of carbon was estimated, but less is known about the composition of the sedimented matter, let alone the plankton dynamics above the sediment traps. To understand the impact of top-down regulation on vertical export of biogenic matter implies that greater emphasis on research focused on planktonic heterotrophs is needed in basically biogeochemically oriented approaches. This research may focus on species-specific approaches or investigations based on major functional types of plankton.

Careful consideration of pelagic investigations and vertical flux studies in the open ocean, in particular in the North Pacific (Takahashi, 1986; Karl et al., 1996), Antarctica (Bodungen, 1986; Fischer et al., 1988), the Polar Ocean (Andreassen et al., 1996) and to a lesser extend the North Atlantic (Wassmann et al., 1991; Bodungen et al., 1995), support the assumption that our general understanding of export production and regulation is the *exception* rather than the rule. No global algorithm describing export as a function of primary production seems to be valid. Should we be tempted to look for global rules in plankton ecology in general and vertical export of biogenic matter in particular, these algorithms should be derived from 'old', open ocean ecosystems with circular circulation pattern, such as the Pacific Ocean and Antarctica, not from a 'young', advective systems such as the North Atlantic where most of our expertise derives from. Shelfs are persistently subjected to episodic events, and they therefore teach us more about deviations (e.g. episodes of new production and advection of grazers) than letting us understand the basic patterns. As a consequence, shelfs generally favour export rather than retention food chains. The general rule that pelagic systems are potentially capable of retaining and recycling autochthonous organic material has so far not been adequately substantiated in the literature due to the fact that a majority of investigations have been carried out in the 'wrong' areas. Adequate generalisations are thus difficult and those which are attempted may easily lead into misconception.

Integrated research of plankton ecology and vertical flux has been inadequate to date. In this respect the emphasis on long-term measurements of vertical flux has not improved our understanding on the *regulation* of vertical flux, but taught us what orders of magnitude and what kind of seasonal patterns we can expect. This implies that the conspicuous amount of vertical flux data from all over the oceans are interpreted on the basis of a rather vague and insufficient theory of vertical flux regulation at present. Future investigations aiming at an improvement of our understanding of vertical flux and its regulation have to focus on (a) concerted actions of detailed planktonic investigations combined with short-term studies of vertical flux in the upper 500 m, (b) experimental studies in mesocosms and small, enclosed lagoons and (c) sensitivity analyses of the pelagic-benthic coupling using mathematical models.

Acknowledgements

The invitation to the PELAG symposium in August 1996 is gratefully acknowledged. Over the years the co-operation with the PELAG group has been a steady inspiration for a joint interest in planktonic cycling and vertical flux. Thanks to the members of the sedimentation group at the Norwegian College of Fishery Science, University of Tromsø, who by active co-operation in the laboratory and the field, interpretations of the results, vivid discussions and patient listening to the development of the ideas presented here contributed significantly to this manuscript. The comments, recommendations and corrections by K. Kivi and P. Verity are gratefully acknowledged. I am also indebted to Y. Olsen and H. Reinertsen for letting me use unpublished vertical flux data from Lake Haugatjern. I am grateful to the ambience of Plage du Bestouan with its view of Cap Canaille which provided me with the idea of the structure and dynamics of the relationship between daily primary and export production. The manuscript was written during a stay at the Station Zoologique, Ecologie du Plankton Marin, Villefranche-sur-Mer. Thanks to Fereidoun Rassoulzadegan for his hospitality. The work was supported by the Norwegian Research Council (NFR) and the Nordic Council of Ministers Programme for environmental co-operation in the Baltic.

References

Aksnes, D. L., J. Aure, S. Kaartvedt, T. Magnesen & J. Richard, 1989. Significance of advection for the carrying capacities of fjord populations. Mar. Ecol. Prog. Ser. 50: 263–274.

Aksnes, D. L. & P. Wassmann, 1993. Modelling the significance of zooplankton grazing for export production. Limnol. Oceanogr. 38: 978–985.

Alldredge, A. & G. A. Jackson, 1995. Aggregation in marine systems. Deep-Sea Res. 42: 1–273.

Alldredge, A. L. & M. W. Silver, 1988. Characteristics, dynamics and significance of marine snow. Prog. Oceanogr. 20: 41–82.

Arashkevitch, E. G., A. V. Drits, T. N. Semenova & V. P. Shevchenko, 1994. Contents, production and sinking rate of faecal pellets of salps and pyrosomas in the south-western part of the Atlantic ocean. Russian J. aquat. Ecol. 3: 143–153.

Andreassen, I., E.-M. Nöthig & P. Wassmann, 1996. Sedimentation of particulate matter on the shelf of northern Spitzbergen. Mar. Ecol. Prog. Ser. 137: 215–228.

Andreassen, I. J. & P. Wassmann, 1998. Vertical flux of biogenic matter in the marginal ice zone of the Barents Sea in May. Mar. Ecol. Prog. Ser. (in press).

Angel, M. V., 1984. Detrital organic fluxes through pelagic ecosystems. In M. J. R. Fasham (eds), Flows of Energy and Materials in Marine Ecosystems: 475–516.

Asper, V. L., W. G. Deuser, G. A. Knauer & S. E. Lohrenz, 1992. Rapid coupling of sinking particle fluxes between surface and deep ocean waters. Nature 357: 670–672.

Azam F, T. Fenchel, J. G. Field, J. S. Gray, L. A. Meyer-Reil & F. Thingstad, 1983. The ecological role of water-column microbes in the sea. Mar. Ecol. Prog. Ser. 10: 257–263.

Bacastow, R. & E. Maier-Reimer, 1991. Dissolved organic carbon in modelling the oceanic new production. Global Biogeochem. Cycles 5: 71–85.

Banse, K., 1994. Grazing and zooplankton production as key controls of phytoplankton production in the upper ocean. Oceanography 7: 13–20.

Banse, K., 1995. Zooplankton: Pivotal role in the control of ocean production. ICES J. mar. Sci. 52: 265–277.

Bathmann, U., 1988. Mass occurrence of *Salpa fusiformis* in the spring of 1984 off Ireland: implication for sedimentation processes. Mar. Biol. 97: 127–135.

Bathmann, U., T. Noji & B. von Bodungen, 1990b. Copepod grazing potential in late winter in the Norwegian Sea – a factor in the control of spring phytoplankton growth? Mar. Ecol. Prog. Ser. 60: 225–233.

Bathmann, U., R. Peinert & B. von Bodungen, 1990a. Pelagic origin and fate of sedimenting particles in the Norwegian Sea. Prog. Oceanogr. 24: 117–125.

Berger, W., V. Smetacek & G. Wefer, 1989. Ocean productivity and paleoproductivity. In W. Berger, V. Smetacek, & G. Wefer (eds),

Productivity of the Ocean: Present and Past, John Wiley & Sons, New York: 1–34.

Betzer, P. R., W. J. Showers, E. D. Laws, C. D. Winn, G. R. DiTullio & P. M. Kroopnick, 1984. Primary productivity and particle fluxes on a transect of the equator at 153° W in the Pacific Ocean. Deep Sea Res. 31: 1–11.

Bloesch, J., 1996. Towards a new regeneration of sediment traps and a better measurement/understanding of settling particle flux in lakes and oceans: A hydrodynamical protocol. Aquat. Sci.: 58: 283–296.

Bodungen, B. von, 1986. Phytoplankton growth and krill grazing during spring in the Bransfield Strait, Antarctica. Implications for sediment trap collections. Polar Biol. 6: 153–160.

Bodungen, B. von, A. Antia, E. Bauerfeind, O. Haupt, I. Peeken, R. Peinert, S. Reitmeier, C. Thomsen, M. Voss, M. Wunsch, U. Zeller & B. Zeitzschel, 1995. Pelagic processes and vertical flux of particles: an overview over a long-term comparative study in the Norwegian Sea and Greenland Sea. Geologische Rundschau 84: 28–48.

Bodungen B. von, G. Fischer, E.-M. Nöthig & G. Wefer, 1987. Sedimentation of krill faeces during spring development of phytoplankton in Bransfield Strait, Antarctica. SCOPE/UNEP 62: 243–257.

Burkill, P. H., E. S. Edwards, A. W. G. John & M. A. Sleigh, 1993. Micro-zooplankton and their herbivorous activity in the northeast Atlantic Ocean. Deep Sea Res. 40: 479–494.

Carson, C. A., H. W. Ducklow & A. F. Michael, 1994. Annual flux of dissolved organic carbon form the euphotic zone in the northwestern Sargasso Sea. Nature 371: 405–408.

Dagg, M., 1993. Grazing by the copepod community does not control phytoplankton production in the Subarctic Pacific Ocean. Prog. Oceanogr. 32: 163–183.

Dam, H. G., M. R. Roman & M. I. Youngbluth, 1995. Downward export of respiratory carbon and dissolved inorganic nitrogen by diel-migrating mesozooplankton at the JGOFS Bermuda time-series station. Deep Sea Res. 42: 1187–1197.

Davies, J. M. & R. Payne, 1984. Supply of organic matter to the sediment in the northern North Sea during a spring phytoplankton bloom. Mar. Biol. 78: 315–324.

Deuser W. G., E. H. Ross & R. F. Anderson, 1981. Seasonality in the supply of sediment to the deep Sargasso Sea and implications for the rapid transfer of matter to the deep ocean. Deep Sea Res. 28: 495–505.

Drenner, R. W., S. T. Threlkeld, J. D. Smith, J. R. Mummert & P. A. Cantrall, 1989. Interdependence of phosphorous, fish and site effects on phytoplankton biomass and zooplankton. Limnol. Oceanogr. 34: 1315–1321.

Dugdale R. C. & J. J. Goering, 1967. Uptake of new and regenerated forms of nitrogen in primary productivity. Limnol. Oceanogr. 12: 196–206.

Dugdale, R. C., F. P. Wilkerson & H. J. Minas, 1995. The role of a silicate pump in driving new production. Deep Sea Res. 42: 697–719.

Egge, J. K., 1993. Nutrient control of phytoplankton growth: Effects of macro nutrient composition (N, P, Si) on species succession. Dr. Scient. thesis, Univ. Bergen, 104 pp.

Elser, J. J., D. K. Foster & R. E. Hecky, 1995. Effects of zooplankton on sedimentation in pelagic ecosystems: Theory and test in two lakes of the Canadian shield. Biogeochemistry 30: 143–170.

Eppley, R. & B. J. Peterson, 1979. Particulate organic flux and planktonic new production in the deep ocean. Nature 282: 677–680.

Fischer, G., D. Fütterer, R. Gersonde, S. Honjo, D. Ostermann & G. Wefer, 1988. Seasonal variability of particle flux in the Weddell Sea and its relationship to ice cover. Nature 335: 426–428.

Fransz, H. G. & W. W. C. Gieskes, 1984. The unbalance of phytoplankton and copepods in the North Sea. Rapp. P.-v. Réun. Cons. int. Explor. Mer 183: 218–225.

Frost, 1991. The role of grazing in nutrient-rich areas of the open sea. Limnol. Oceanogr. 36: 1616–1630.

Gifford, D. J., L. M. Fessenden, P. G. R. Garrahan & E. Martin, 1996. Grazing by microzooplankton and mesozooplankton in the high-latitude North Atlantic Ocean: Spring versus summer dynamics. J. Geophys. Res. 100: 6665–6675.

González H. & V. Smetacek, 1994. The possible role of the cyclopoide copepod *Oithona* in retarding vertical flux of the zooplankton faecal material. Mar. Ecol. Prog. Ser. 113: 233–246.

Graf, G., 1992. Pelagic-benthic coupling: a benthic perspective. Mar. Biol. Annu. Rev. 30: 149–190.

Hansen, L., 1997. Suspenderte og sedimenterte fekalier langs et transekt over Nordvestbanken, Nord-Norge, i 1994. Cand. Scient. thesis, University of Tromsø, Norway (in Norwegian).

Hargrave, B. T., G. C. Harding, K. F. Drinkwater, T. C. Lambert & W. G. Harrison, 1985. Dynamics of the pelagic food web in St. Georges Bay, southern Gulf of St. Lawrence. Mar. Ecol. Prog. Ser. 20: 221–240.

Haupt, O., 1995. Modellstudien zum pelagischen Stoffumsatz und vertikalen Partickelfluß in der Norwegensee. Ber. Sonderforschungsbereich 313, Univ. Kiel, 60: 1–140.

Hedges, J. I., W. A. Clarke & G. L. Cowie, 1988a. Organic matter sources to the water column and surficial sediments of a marine bay. Limnol. Oceanogr. 33: 1116–1136.

Hedges, J. I., W. A. Clarke & G. L. Cowie, 1988b. Fluxes and reactivities of organic matter in a coastal marine bay. Limnol. Oceanogr. 33: 1137–1152.

Heinrich, A. K., 1962. The life histories of plankton animals and seasonal cycles of plankton communities in the oceans. J. Cons. perm. int. Explor. Mer 27: 15–24.

Heiskanen, A.-S. & K. J. Kononen, 1994. Sedimentation of vernal and late summer phytoplankton communities in the coastal Baltic Sea. Arch. Hydrobiol. 131: 175–198.

Heiskanen, A.-S., T. Tamminen & K. Gundersen, 1996. The impact of planktonic food web structure on nutrient retention and loss from a late summer pelagic system in the coastal northern Baltic Sea. Mar. Ecol. Prog. Ser. 145: 195–208.

Heussner, S., A. Monaco & S. W. Fowler, 1987. Characterisation and vertical transport of settling biogenic particles in the northwestern Mediterranean. In E. T. Degens, E. I. Izdar & S. Honjo (eds), Particle Flux in the Ocean, Mitt. Geol.-Paläont. Inst., Univ. Hamburg, SCOPE/UNEP Sonderband 62: 127–147.

Hessen, D. O., J. P. Nilsen & T. O. Eriksen, 1986. Food size spectra and species replacement within herbivorous zooplankton. Int. Revue ges. Hydrobiol. 71: 1–10.

Honjo, S., 1990. Particle fluxes and modern sedimentation in polar oceans. In W. O. Smith, (ed.), Polar Oceanography, Academic Press, New York: 687–739.

Ianora, A. & S. Poulet, 1993. Egg viability in the copepod *Temora stylifera*. Limnol. Oceanogr. 38: 1615–1626.

Ianora, A., S. Poulet & A. Miralto, 1995. A comparative study of the inhibitory effect of diatoms on the reproductive biology of the copepod Temora stylifera. Mar. Biol. 121: 533–539.

Iseki, K., 1981. Particulate organic matter transport to the deep sea by salp faecal pellets. Mar. Ecol. Prog. Ser. 5: 55–60.

Ittekot, V., P. Schäfer, S. Honjo & P. J. Depetris, 1996. Particle Flux in the Ocean. John Wiley & Sons Ltd., 372 pp.

Iverson, R., 1991. Control of fish production. Limnol. Oceanogr. 35: 1593–1604.
Jackson, G. A., 1993. Flux feeding as a mechanism for zooplankton grazing and its implications for vertical particulate flux. Limnol. Oceanogr. 38: 1328–1331.
Jonasdottir, S. H. & T. Kiørboe, 1996. Copepod recruitment and food composition: do diatoms affect hatching success? Mar. Biol. 125: 743–750.
Karl, D. M., J. R. Christian, J. E. Dore, D. V. Hebel, R. M. Letelier, L. M. Tupas & C. D. Winn, 1996. Seasonal and interannual variability in primary production and particle flux at station ALOHA. Deep-Sea Res. 43: 539–568.
Keller, A. A. & U. Riebesell, 1989. Phytoplankton carbon dynamics during a winter-spring diatom bloom in an enclosed marine ecosystem: primary production, biomass and loss rates. Mar. Biol. 103: 131–142.
Keck, A. & P. Wassmann, 1993. Den sibirske kontinentalsokkel og Polhavet. II. Betydning for den globale karbonkretsløp? Naturen 6: 264–272.
Kiørboe, T., 1993. Turbulence, phytoplankton cell size and the structure of pelagic food webs. Adv. mar. Biol. 29: 1–72.
Kiørboe, T., 1996. Material fluxes in the water column. In B. B. Jørgensen & K. Richardson, (eds), Coastal and Estuarine Studies 52, American Geophysical Union, Washington D. C.: 67–94.
Kiørboe, H. & J.-L. Hansen, 1993. Phytoplankton aggregate formation: observations of patterns and mechanisms of cell sticking and the significance of exopolymeric material. J. Plankton Res. 15: 993–1018.
Kiørboe, T., C. Lundsgaard, M. Olesen & J. L. Hansen, 1994. Aggregation and sedimentation processes during a spring phytoplankton bloom: A field experiment to test coagulation theory. J. mar. Res. 52: 297–323.
Lalli, C. M. & T. R. Parson (1993). Biological Oceanography: An Introduction. Pergamon Press, Oxford, 301 pp.
Lampitt, R. S., T. Noji & B. von Bodungen, 1990. What happens to zooplankton fecal pellets? Implications for material flux. Mar. Biol. 104: 15–23.
Langeland, A., 1990. Biomanipulation development in Norway. Hydrobiologia 200/201: 535–540.
Laws, E. A., P. K. Bienfang, A. D. Ziemann & L. D. Conquest, 1988. Phytoplankton population dynamics and the fate of production during the spring bloom in Auke Bay, Alaska. Limnol. Oceanogr. 33: 57–65.
Lee, C. & S. Wakeham, 1991. Production, transport and alteration of particulate organic matter in sea water. In P. Wassmann, A.-S. Heiskanen & O. Lindahl (eds), Sediment Trap Studies in the Nordic Countries 2. Proceedings. NurmiPrint OY, Nurmijärvi, 61–75.
Legendre, L., 1990. The significance of microalgal blooms for fisheries and for the export of particulate organic carbon in the ocean. J. Plankton Res. 12: 681–699.
Longhurst, A. R. & W. G. Harison, 1989. Vertical nitrogen flux from the oceanic photic zone by diel migrant zooplankton and nekton. Deep Sea Res. 35: 881–889.
Mazumder, A., D. J. McQueen, W. D. Taylor & D. R. S. Lean, 1988. Effects of fertilisation and planktivorous fish (yellow perch) predation on size distribution of particulate phosphorus and assimilated phosphate: Large enclosure experiments. Limnol. Oceanogr. 33: 421–430.
Noji, T. T., 1991. The influence of macrozooplankton on vertical particulate flux. Sarsia 76: 1–9.
Noji, T. T. & F. Rey, 1996. Old and new perspectives on zooplankton and vertical particulate flux. ICES report cm 1996/O:10, 14 pp.
Nordby, E. & K. S. Tande, 1998. Zooplankton biomass distribution and plankton transport at Nordvestbanken. Sarsia (in prep).
Odate, T., 1994. Plankton abundance and size structure in the northern North Pacific Ocean in early summer. Fish. Oceanogr. 3: 267–278.
Olesen, M. & C. Lundsgaard, 1995. Seasonal sedimentation of autochthonous material from the euphotic zone of a coastal system. Estuar. coast. Shelf-Sci. 41: 475–490.
Pace, M. L., G. D. Knauer, D. M. Karl & J. H. Martin, 1987. Primary production, new production and vertical flux in the eastern Pacific Ocean. Nature 325: 803–804.
Passow, U., A. Alldredge & B. Logan, 1994. The role of particulate carbohydrate exudates in the flocculation of diatom blooms. Deep Sea Res. 41: 335–357.
Pedersen, G., 1995. Factors influencing the size and distribution of the copepod community in the Barents Sea with special emphasis on *Calanus finmarchicus* (Gunnerus). Dr Scient. thesis, University of Tromsø, Norway.
Peinert, R., B. von Bodungen & V. Smetacek, 1989. Food web structure and loss rates. In W. Berger, V. Smetacek & G. Wefer (eds), Productivity of the Ocean: Present and Past, John Wiley & Sons, New York: 34–48.
Peinert, R., A. Saure, P. Stegmann, C. Stienen, H. Haardt & V. Smetacek, 1982. Dynamics of primary production and sedimentation in a coastal ecosystem. Neth. J. Sea Res. 16: 276–289.
Pilskaln C. H. & S. Honjo, 1987. The fecal pellet fraction of biogeochemical particle fluxes to the deep sea. Global Biogeochem. Cycles 1: 31–48.
Platt, T., W. G. Harrison, M. L. Lewis, W. K. W. Li, S. Sathyendranath, R. Smith & A. F. Vezina, 1988. Biological production and the oceans: the case for a consensus. Mar. Ecol. Prog. Ser. 52: 77–88.
Ratkova, T., I. Andreassen & P. Wassmann, 1998. Phytoplankton and protozoa abundance and biomass along a transect across the Nordvestbank, north Norwegian shelf, in 1994. Sarsia (in prep).
Reigstad, M. & P. Wassmann, 1995. The importance of advection for the pelagic-benthic coupling in north Norwegian fjords. Sarsia 80: 245–257.
Reinertsen, H. & Y. Olsen, 1984. Effects of fish elimination an the phytoplankton community of a eutrophic lake. Verh. Int. Ver. Limnol. 22: 649–657.
Riebesell, U., M. Reigstad, P. Wassmann, U. Passow & T. Noji, 1995. On the trophic fate of Phaeocystis pouchetii. VI. Significance of Phaeocystis-derived mucus for vertical flux. Neth. J. Sea. Res. 33: 193–203.
Sasaki, H., H. Hattori & S. Nishizawa, 1988. Downward flux of particulate matter and vertical distribution of calanoid copepods in the Oyashio water in summer. Deep Sea Res. 35: 505–515.
Silver, M. W., C. H. Pilskaln & D. Steinberg, 1991. The biologists' view of sediment trap collections: problems of marine snow and living organisms. In P. Wassmann, A.-S. Heiskanen & O. Lindahl (eds). Sediment Trap Studies in the Nordic countries 2. Proceedings. NurmiPrint OY, Nurmijärvi: 76–93.
Sathyendranath, S. & T. Platt, 1993. Remote sensing of water-column primary production. In W. K. W. Li & S. Y. Maestrini (eds), Measurements of Primary Production from the Molecular to the Global Scales. ICES Marine Science Symposia, Vol. 197: 236–243.
Skjoldal, H. R. & P. Wassmann, 1986. Sedimentation of particulate organic matter and silicium during spring and summer in Lindåspollene, western Norway. Mar. Ecol. Prog. Ser. 30: 49–63.
Smetacek, V., B. von Bodungen, B. Knoppers, R. Peinert, F. Pollehne, P. Stegmann & B. Zeitzschel, 1984. Seasonal stages

characterising the annual cycle of an inshore pelagic system. Rapp. P.-v. J. Cons. int. Explor. Mer 183: 126–135.

Staresinic, N., J. Farrington, R. B. Gagosian, C. H. Clifford & E. M. Hulburt, 1983. Downward transport of particulate matter in the Peru coastal upwelling: Role of anchoveta, *Engraulis ringens*. In E. Suess & J. Thiede (eds), Coastal Upwelling, its Sediment Record. NATO Conference Series IV: 225–240.

Steele, J., 1974. The Structure of Marine Ecosystems. Harvard University Press, Cambridge, Massachusetts.

Suess, E., 1980. Particulate organic carbon flux in the oceans: surface productivity and oxygen utilization. Nature 288: 260–263.

Taguchi, S., 1982. Sedimentation of newly produced particulate organic matter in a subtropical inlet, Kaneohe Bay, Hawaii. Estuar. coast. Shelf Sci. 14: 533–544.

Tande, K. S., 1991. *Calanus* in high latitudes. Polar. Res. 10: 389–407.

Takahashi, K., 1986. Seasonal fluxes of pelagic diatoms in the subarctic Pacific 1982–1983. Deep Sea Res. 33: 1225–1251.

Thingstad, T. F., 1995. Feedback mechanisms between degradation and primary production in the pelagic environment. In M. Beran (ed.), Carbon Sequestration in the Biosphere. Processes and prospects. NATO ASI Series, series I. Global environmental change, 33: 113–128.

Uye, S.-I., 1996. Induction of reproductive failure in the planktonic copepod *Calanus pacificus* by diatoms. Mar. Ecol. Prog. Ser. 133: 89–97.

Verity, P. & V. Smetacek, 1996. Organism life cycle, predation and the structure of marine pelagic ecosystems. Mar. Ecol. Prog. Ser. 130: 277–293.

Wassmann, P., 1990. Relationship between primary and export production in the boreal, coastal zone of the North Atlantic. Limnol. Oceanogr. 35: 464–471.

Wassmann, P., 1991. Dynamics of primary production and sedimentation in shallow fjords and polls of western Norway. Oceanogr. Mar. Biol. annu. Rev. 29: 87–154.

Wassmann, P., 1993. Regulation of vertical export of particulate organic matter from the euphotic zone by planktonic heterotrophs in eutrophicated aquatic environments. Mar. Pollut. Bull. 26: 636–643.

Wassmann, P., I. Andreassen, M. Reigstad & D. Slagstad, 1996. Pelagic-benthic coupling in the Nordic Seas: The role of episodic events. P.S.Z.N. I: Mar. Ecol. 17: 447–471.

Wassmann, P., J. K. Egge, M. Reigstad & D. L. Aksnes, 1997. Influence of dissolved silicate on vertical flux of particulate biogenic matter. Mar. Pollut. Bull. 33: 10–21.

Wassmann, P., I. Andreassen & F. Rey, 1998. Seasonal variation of nutrients and suspended biomass along a transect across the Nordvestbank, north Norwegian shelf, in 1994. Sarsia (in prep).

Wassmann, P., R. Peinert & V. Smetacek, 1991. Patterns of production and sedimentation in the boreal and polar Northeast Atlantic. Polar Research 10: 209–228.

Wassmann, P. & D. Slagstad, 1993. Seasonal and interannual dynamics of carbon flux in the Barents Sea: a model approach. Polar Res. 13: 363–372.

Hydrobiologia **363:** 59–72, 1998.
T. Tamminen & H. Kuosa (eds), Eutrophication in Planktonic Ecosystems: Food Web Dynamics and Elemental Cycling.

A theoretical approach to structuring mechanisms in the pelagic food web

T. Frede Thingstad
Dept. of Microbiology, University of Bergen, Jahnebakken 5, N-5020 Bergen, Norway

Abstract

In the literature there is a commonly used idealized concept of the food web structure in the pelagic photic zone food web, based to a large extent on size dependent relationships. An outline is here given of how the elementary size-related physical laws of diffusion and sinking, combined with the assumption of predators being size selective in their choice of prey, give a theoretical foundation for this type of structure. It is shown how such a theoretical fundament makes it possible to relate a broad specter of phenomena within one generic and consistent framework. Phenomena such as Hutchinson's and Goldman's paradoxes, the influence of nutrients and water column stability on the balance between microbial and classical food webs, bacterial carbon consumption, new production and export of DOC and POC to the aphotic zone, eutrophication and diversity, can all be approached from this perspective. By including host-specific viruses, this approach gives a hierarchical structure to the control of diversity with nutrient content controlling the maximum size of the photic zone community, size selectivity of predators regulating how the nutrient is distributed between size-groups of osmotrophic and phagotrophic organisms, and viral host specificity regulating how the nutrients within a size group is distributed between host groups. I also briefly discuss how some biological strategies may be successful by *not* conforming to the normal rules of such a framework. Analyzing the behavior of these idealized systems is thus claimed to facilitate our understanding of the behavior of complex natural food webs.

Introduction

The pelagic food web is characteristically a complex and dynamic system, and although our knowledge contains a lot of detail in the fields of taxonomy, physiology, chemistry and physics, it is a far from straight forward matter to extract from all this knowledge the essential features and understand the behavior of such ecosystems. The problem lies not only in system complexity, but also in the poor ability of human intuition to serve as a tool for understanding the consequences of cause-effect relationships which are circular rather than linear. Such circularity is a typical feature of the mature successional stage of photic zone food webs where production is based largely on recycled rather than on 'new' nutrients. Typical of this is the loop circulating elements from the free mineral form, via uptake into phytoplankton, by grazing into zooplankton, and from there a remineralization back to the free dissolved form. Understanding of how rates and biomasses are controlled in such loops is greatly facilitated when some kind of model is used, particularly when this is given a form available to mathematical analysis. Seen in this perspective, modeling becomes the search for a tool allowing us to structure, explain, and analyze existing knowledge. The art of the game is then not to make a model bundling together as much as possible of the massive amount of existing knowledge about the pelagic ecosystem, but rather to extract what is believed to be essential in a form as compact and as comprehensible as possible. In a sense, the strive is towards a maximization of the *ratio* between what loosely could be called explaining power, and complexity. The best model according to such a philosophy would be one which from as small a set of assumptions as possible (small denominator), can explain as large a range of phenomena and properties of the system as possible (large numerator).

With variations over a common theme, many authors have formulated ideas based on a size-structured pelagic food web (Sheldon, 1972; Azam et al., 1983; Fenchel, 1987; Rodriguez & Mullin, 1986; Thingstad & Sakshaug, 1990; Moloney & Field, 1991; Moloney, 1992; Gaedke, 1993; Armstrong, 1994).

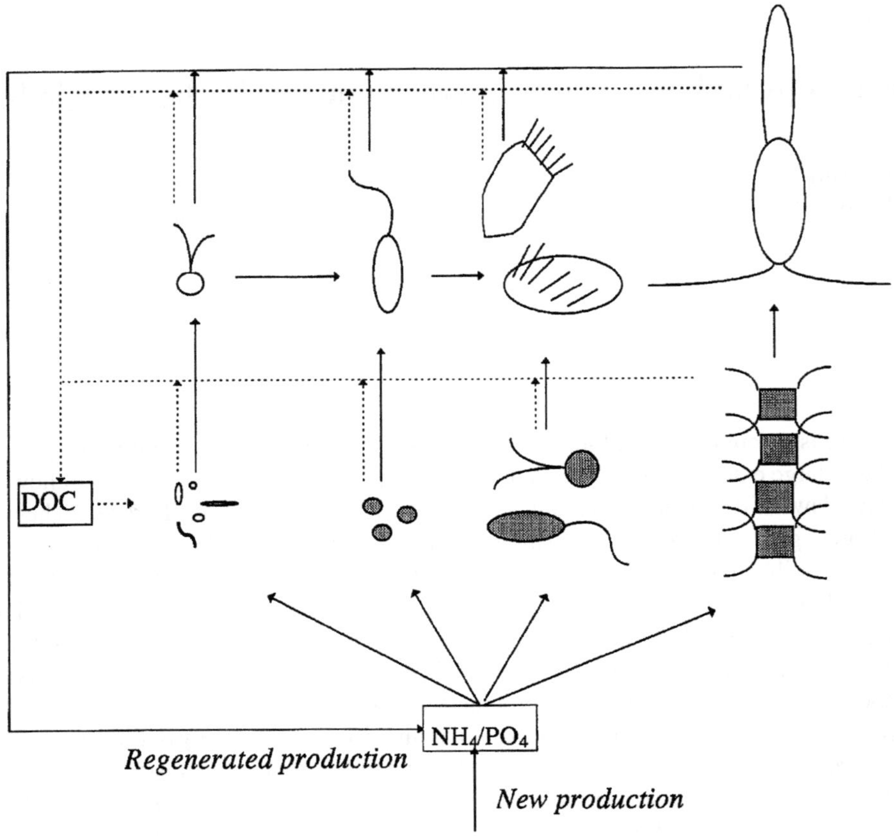

Figure 1. Idealized food web model for the photic zone. Osmotrophic organisms of increasing size from left to right in the lower line are fed upon by size-selective predators in the upper line which also feed on the smaller size class of predators. All phytoplankton growth rates are limited by the same mineral nutrient, while heterotrophic bacteria to the left in the line of osmotrophs may be limited either by dissolved organic carbon (DOC) or by mineral nutrients. Solid lines represent flows of mineral nutrient, dotted lines represent flows of organic carbon. DOC released at various levels in the food web can be reincorporated into particulate form via uptake in heterotrophic bacteria. Photosynthetic organisms shaded.

These may contain different size classes of phytoplankton and heterotrophic bacteria as osmotrophic organisms, and grazing on these by size-selective predators. The food web outlined in Figure 1 is one version summarizing some of these ideas. If the food web in Figure 1 really represents some kind of valid generalization of the photic zone food web, it seems logical to assume that the formation of such structures must be rooted in very basic and generic principles, and behavior of this system should be deducible from those basic principles. I here attempt to explore what such a minimum set of assumptions could be, and outline how such a theory may provide a common framework for discussion of many central features of the food web. I readily acknowledge that the real world photic zone food web is much more complex than what is outlined in Figure 1. The suggestion is, however, that in the search for a way to understand the complexity of the real world system it is a worthwhile, maybe even a necessary, step to explore the mechanisms leading to the formation of this idealized structure.

The outline of such a theory presented here is a synthesis of articles previously published on different of its aspects: Thingstad & Sakshaug (1990) analyzed the steady state of models of this type, but did not include bacteria. Thingstad & Pengerud (1985) and Pengerud et al. (1987) analyzed theoretically and experimentally a simple three-member food web of bacteria, their predators (protozoa), and their competitors for mineral nutrients (phytoplankton), and Thingstad et al. (1997) have discussed the consequences of such models for the oceanic carbon cycle. Combining these ideas, Thingstad & Rassoulzadegan (1995) discussed the relationships between phosphorus limitation and food web structure in the Mediterranean, while Thingstad et al. (in press) have discussed how microbial diversity can be seen as an integrated aspect of such biogeochemical flux models. Thingstad & Lignell (1997) have focused on particular aspects of different types of control of bacterial biomass and growth rate.

Food chain structure as a function of nutrient content

The advantage of small cell size for osmotrophs

The minimum concentration (N_C) of limiting substrate at which a spherical osmotrophic cell can balance sinking loss by growth, can be derived from the physical laws of diffusion and sinking (Frame 1). When other factors such as cell buoyancy (ρ) and volume-specific nutrient content (σ) are assumed constant, this theoretical minimum substrate concentration increases with the 4th power of cell radius (Frame 1, Equation 6). To escape from such a strong penalty, large cells would have to compensate by active swimming, or by mechanisms allowing them to lower the values of ρ or σ.

Central to our further discussion is what I will term nutrient content (N_T) of the photic zone. N_T is defined as the sum of the limiting element contained in all biologically available pools, dissolved as well as particulate, in living or dead material. The concentration N_T thus represents the maximum amount of the limiting nutrient available for sharing among the organisms of the food web. A measure for the success of an organism at steady state of the system is thus the fraction of N_T sequestered in its biomass. N_T also represents a measure of the total carrying capacity of the system. If one organism is successful in sequestering a large fraction of N_T, mass balance necessarily means that there are other populations left with a only a low fraction. High diversity is in this terminology equivalent to a wide distribution of N_T among many populations, each only being able to take hold of a small fraction, i.e. in a sense a society of many unsuccessful organisms.

The discussion will be focused around the mental experiment of increasing the nutrient content N_T of the photic zone in small steps, starting from zero. This way we can explore how the steady state structure of the food web changes as we go from extreme oligotrophy (N_T close to zero) towards eutrophy.

For sufficiently low N_T, the substrate concentration would be too low for any organism to establish, and all of N_T would be in the form of free mineral nutrients. Assuming no allochthonous import of organic carbon, the first organism to establish in such a system would have to be a primary producer. In our hypothetical world of strict size-regulation, it would be the phytoplankton species which is best at stacking all the necessary tools: photosynthetic apparatus, DNA, enzymes, etc., into the smallest cell volume. The concentration of N_T corresponding to the establishment of this first phytoplankton species would thus be given by Equation 6 for the smallest achievable radius r. If the phytoplankton, when alone in the ecosystem, has an additional loss rate due to i.e. autolysis, a higher nutrient content would be required before establishment could take place.

This relationship between phytoplankton cell size and nutrient content could be used as an argument for the observed dominance of the phytoplankton community by unicellular cyanobacteria and prochlorophytes in oligotrophic environments (Campbell & Vaulot, 1993).

Coexistence of competitors and the effect of someone 'killing the winner'

As long as our single phytoplankton species is alone without predators, its loss rate can be assumed to remain constant. When the system is in equilibrium (growth balances loss), the phytoplankton growth rate thus also remains independent of a further increase in nutrient content. An unchanging growth rate requires an unchanging concentration of free, dissolved, nutrients N (Equation 4). As long as there is no increase in N, no competing phytoplankton species with a higher N_C can establish. With only two pools, N and A_1, of which N is constant, mass balance means that the increase in nutrient content will be channeled into increasing abundance (A_1) of the established phytoplankter. Such an increase in abundance is creating a new niche in the system: that of a predator (P_1) which can establish once prey concentration is high enough (Figure 2). We then enter a new situation where predator growth must balance loss, or:

$$Y_{P1}\alpha_{P1}A_1P_1 = \delta_{P1}P_1, \quad (7)$$

where δ_{P1}, Y_{P1}, and α_{P1} are specific loss rate, yield coefficient, and clearance rate, respectively, of the predator on phytoplankton species number 1. Equation 7 can be solved for phytoplankton abundance:

$$A_1 = \delta_{P1}/Y_{P1}\alpha_{P1}, \quad (8)$$

DIFFUSIVE TRANSPORT:
Maximum diffusive transport J of a substance towards a spherical cell of radius r is given by (see e.g. Fenchel, 1987):

$$J = 4\pi DrN, \qquad 1.$$

where D is the diffusion coefficient, and N is the bulk substrate concentration. For these organisms, maximum specific population growth rate under diffusion limitation is given by:

$$\mu = YJ = Y4\pi DrN, \qquad 2.$$

where Y is cell yield. For non-respired substrates like phosphorus or nitrogen compounds, Y is the inverse of cell content of the element and can be expressed as

$$Y = ((4\pi/3)r^3\sigma)^{-1}, \qquad 3.$$

where σ is the volume-specific content of the element (in e.g. mol-P/cm^3). Insertion of Eqn.3 into Eqn.2 gives:

$$\mu = 3D\sigma^{-1}r^{-2}N. \qquad 4.$$

SINKING LOSS:
In the upper ocean, sinking may lead to transport out of the upper mixed layer and thus out of the photic zone. According to Stokes' law, sinking rate v is proportional to the square of cell radius (see e.g. Walsby and Reynolds, 1980):

$$v = (2/9)(g/\eta)(\rho-\rho_w)\, r^2, \qquad 5.$$

where g is the gravity constant, η is the viscosity of water, and ρ and ρ_w are the specific densities of the cell and the water, respectively.

With a depth h of the photic zone, the fraction of cells lost per time unit is v/h.

SINKING LOSS BALANCED BY DIFFUSION LIMITED GROWTH
Assuming sinking to be the only loss factor, the minimum substrate concentration N_C where a non-motile cell can balance its sinking loss (μ = v/h) is obtained from combining Eqns. 4 and 5 :

$$N_C=(2g(\rho-\rho_w)\sigma/27h\eta D)r^4, \qquad 6.$$

i.e. N_C is increasing as the 4. power of cell radius r.

Frame 1. Derivation of the minimum substrate concentration at which a non-motile spherical cell can maintain itself in the photic zone.

We have thus entered a new state where further increases in nutrient content no longer leads to increased abundance A_1 of our established phytoplankton species. A_1 remains constant at the level where it allows the predator to grow at a rate balancing its loss. An increase in nutrient content will now give *both* increased abundance P_1 of the predator *and* increased concentration N of free mineral nutrients. The increase in N is linked to the increase in P_1: With increasing loss rate due to predation, the phytoplankton will have to grow faster. To allow such faster growth, the concentration N of free mineral nutrients will now have to increase if A_1 is to remain in the system. In this situation *two* new potential niches are created: either (1)the predator abundance reaches a high enough level for a larger, 2.order predator (P_2) to establish, or (2) the mineral nutrient

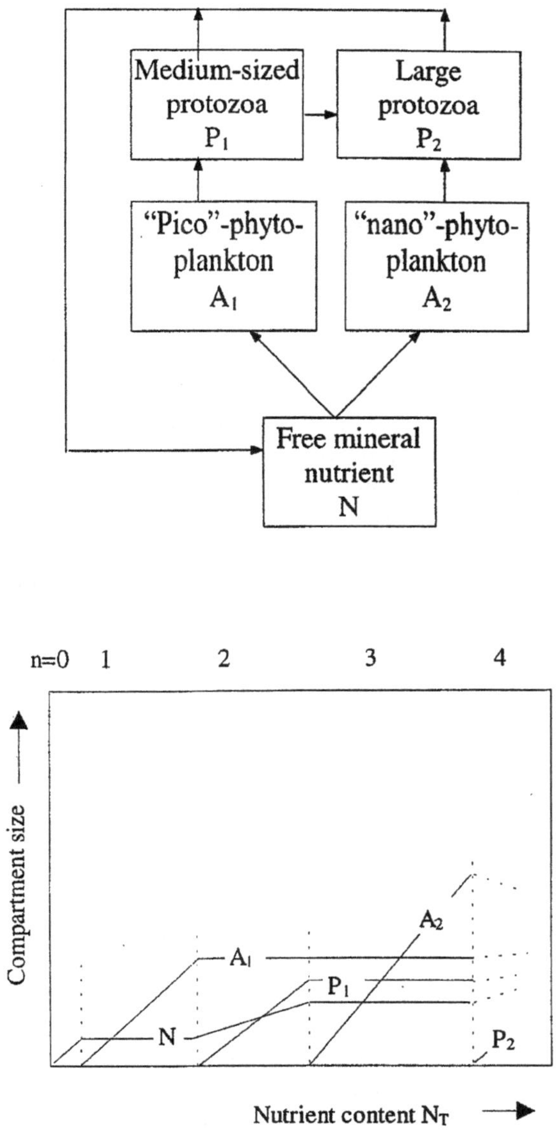

Figure 2. Upper: Hypothetical 4-membered food web of two phytoplankton species of different size, and two protozoan species of different size and with different size-selectivity. Lower: Change in population abundance as a function of system nutrient content N_T. Number of populations established change with nutrient content as denoted along upper side. Phytoplankton species A_2 assumed to establish at a lower N_T than predator P_2 preying on both P_1 and A_2.

concentration reaches a level corresponding to N_C for a larger phytoplankton (A_2) with a size outside the prey size-spectrum of P_1. The latter situation is interesting because we by this line of reasoning have constructed a situation where two species competing for one limiting resource can coexist. This suggests a simple solution to Hutchinson's paradox of why there are so many phytoplankton species in such an apparently homogenous environment (Hutchinson, 1961): Stable coexistence of two populations competing for one resource in a homogenous environment can be obtained if there is a mechanism which selectively 'kills the winner'. In our case, this was a predator selective for the superior competitor for the free mineral nutrient.

If we pursue our mental exercise and increase nutrient content further, we get a system with repeated transitions into new states, either as there is a lengthening of the food chain when higher predators, selective for larger prey, establish, or there is a widening of the base of the food web with larger phytoplankton species. Figure 2 shows a simple 4-membered food chain of two phytoplankton and two prey size classes, and the shift in steady state structure of this as a function of N_T. For the first three members this is simple. It becomes more complicated with 4 or more members when e.g. A_2 is consumed by a predator feeding on the predator of its competitor. The generalized picture is one where the number n of established populations becomes a stepwise function of N_T as long as additional larger species exist. Lengthening of the food chain by adding larger predators will as a general trend tend to increase the concentration N of free mineral nutrients, and thus in turn allow for a widening by establishment of larger phytoplankton species. Establishment of new, large phytoplankton species, unedible by the zooplankton so far present, will tend to keep the free nutrient concentration from increasing with N_T, but create a niche for larger zooplankton, able to feed on this species. This type of relationship between nutrient content and food web structure was discussed by Thingstad & Sakshaug (1990) and later by Armstrong (1994). The theory of top-down and bottom-up effects in linear food chains have been extensively analyzed in both terrestrial and aquatic ecology (see e.g. Power, 1992; Strong, 1992). Models of the type discussed here add the general feature that strong top-down (predatory) control opens for the establishment of a competitor. Once established, the competitor may keep the nutrient level low, and these models thus contains a mechanism allowing for situations combining simultaneous strong top-down and bottom-up control of a population (low growth rate and low biomass). If Goldman's paradox of phytoplankton growing close to their maximum growth rates in oligotrophic oceans (Goldman et al., 1979) is really true, it could be explained in this model with a sufficiently long food chain (many levels of predators) being established, even at the nutrient contents of oligotrophic oceans. Recent evidence of nutrient lim-

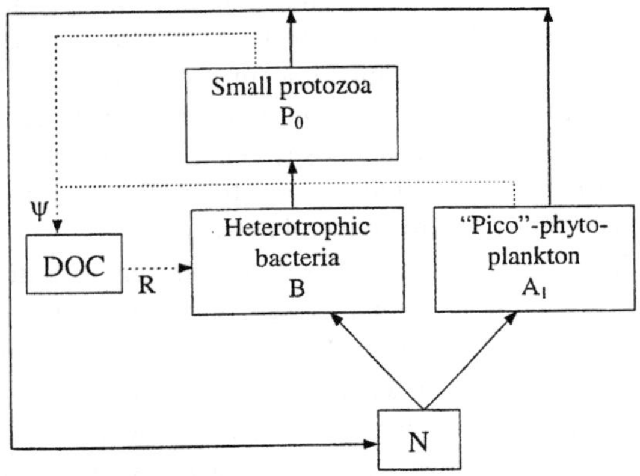

Figure 3. Hypothetical 3-membered food web where heterotrophic bacteria are locked to a low growth rate due to mineral nutrient competition with phytoplankton, and a low biomass due to predation. As discussed in the text, this system may be at steady state in the mineral cycle (solid lines) while DOC is produced at a higher rate (ψ) than it is consumed (R) (carbon flows dotted).

ited picoplankton growth rates in oligotrophic environments (Vaulot et al., 1996), do, however, suggest that Goldman's paradox may not adequately reflect the state in nature.

Adding the heterotrophic bacteria

In the food web of Figure 1, there are organisms at all levels which may produce dissolved organic carbon (DOC) by processes such as excretion, egestion, sloppy feeding, viral lysis etc. As a slightly modernized description of the 'microbial loop' (Azam et al. ,1983), the ensemble of these processes divert organic material from the upwards flux towards larger particles in the food web, to be released into the pool of dissolved organic carbon. From the dissolved pool, the organic material may be reintroduced into the particulate form by uptake into heterotrophic bacteria. At the extreme oligotrophic end of our gradient, when bacteria become first established, they have to be C-limited. By analogy to the phytoplankton-phytoplankton competition situation in Figure 2, the 3-membered food web in Figure 3 only allows phytoplankton-bacteria coexistence on the same limiting mineral nutrient if there is a predator on the superior competitor. Until establishment of the bacterial predator, bacteria thus have to be C-limited. Bacteria will in this situation consume all the mineral nutrient they need to degrade whatever is produced of DOC, the rest of mineral nutrients is left for phytoplankton. A more detailed analysis of different types of carbon and mineral nutrient limitation of bacteria is given in Thingstad & Lignell (1997), and more detailed analyses of the algal-bacterial-bacterial predator situation can be found in Thingstad & Pengerud (1985), and Thingstad et al. (1997).

In the case of mineral nutrient limited bacteria, degradable organic material must be produced at a rate faster than it is consumed (or we would have C-limited bacteria). The excess production would in this case accumulate in the system and may be transported to the aphotic zone by diffusion or down-welling. The bacterial carbon demand (R) of the system in Figure 3 is derived in Frame 2, Equation12. Carbon demand of the bacteria is proportional to the product $\delta_{A1}\delta_{P0}$ of the loss rates of their phytoplankton competitors (A_1) and the protozoan predators (P_0). If we assume loss of P_0 and A_1 to be dominated by predation, and our predator P_1 (Figure 2) to have a size-specificity covering both A_1 and the bacterial predator P_0 with the same clearance rate α_{P1}, we can replace both loss rates in Equation 14 (Frame 2) with $\alpha_{P1}P_1$. This gives:

$$R = Y_{BN}(Y_{A1}Y_{BC}Y_{P0})^{-1}\,(\alpha_B/\alpha_{A1}\alpha_{P0})\,\alpha_{P1}^2 P_1^2. \quad (12)$$

We thus get a bacterial carbon demand which increases as the square of the abundance of predator P_1. Since P_1 increases with nutrient content, so will bacterial carbon demand R. Depending upon whether production or consumption increase most with N_T, the system may or may not change to carbon limitation of the bacteria as we go to high nutrient contents. While all models of this kind will have carbon limited bacteria in the oligotrophic end of the N_T-gradient, it is possible to construct models where mineral nutrient limitation of heterotrophic bacteria may either not occur at all, occur only for intermediate values of N_T, or for all N_T above some given value. Observations describing mineral nutrient limited bacteria in natural surface waters have been reported (Pomeroy et al., 1995) and suggest that the two latter cases should be investigated more closely.

Combining Figures 2 and 3, and adding one more level to the right of the ladder containing the classical food chain from net-plankton to copepods, we have obtained the structure shown in Figure 4 which is analogous to the one in Figure 1.

New and export production

As in the classical analysis of Dugdale & Goering (1967), the equilibrium condition of the photic zone

Using the equilibrium argument of growth ($Y\alpha_{A1}NA_1$) balancing loss ($\delta_{A1}A_1$), the stable existence of the phytoplankton in the food web of Fig. 3 gives the concentration of free nutrients:

$$N = \delta_{A1}/Y_{A1}\alpha_{A1}, \qquad 10.$$

where δ_{A1} is the sum of sedimentation and other phytoplankton loss rates, and α_{A1} is phytoplankton affinity for the limiting nutrient N. Assuming N to be so low that bacterial growth rate μ_B is also well below saturation, we get (from Eqn. 10) the growth rate of bacteria when limited by the same mineral nutrient as the phytoplankton:

$$\mu_B = Y_{BN}\alpha_B N = (Y_{BN}/Y_{A1})\,(\alpha_B/\alpha_{A1})\,\delta_{A1}, \qquad 11.$$

where Y_{BN} is bacterial yield on the mineral nutrient.

Applying our equilibrium condition to the bacterial predator (in a manner analogous to Eqn. 7 and 8), gives bacterial abundance B:

$$B=\delta_{P0}/Y_{P0}\alpha_{P0} \qquad 12.$$

Bacterial carbon demand R is given by the product:

$$R=Y_{BC}^{-1}\mu_B B, \qquad 13.$$

where Y_{BC} is the bacterial yield on DOC. Insertion of the expressions for bacterial growth rate and biomass in Eqns. 11 and 12 gives:

$$R= Y_{BN}(Y_{A1}Y_{BC}Y_{P0})^{-1}\,(\alpha_B/\alpha_{A1}\alpha_{P0})\,\delta_{A1}\,\delta_{P0} \qquad 14.$$

Frame 2. Derivation of carbon demand R by mineral nutrient limited bacteria in the 3-membered food web of Figure 3.

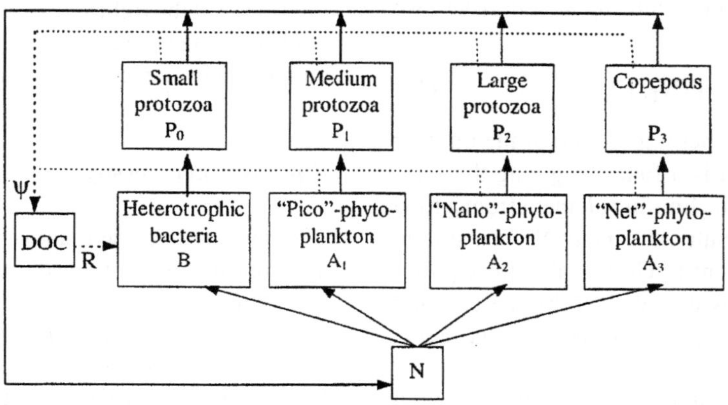

Figure 4. Joining of the food webs in Figures 2 and 3, and extension of the food web to the right with a classical food chain to give an idealized structure analogous to Figure 1.

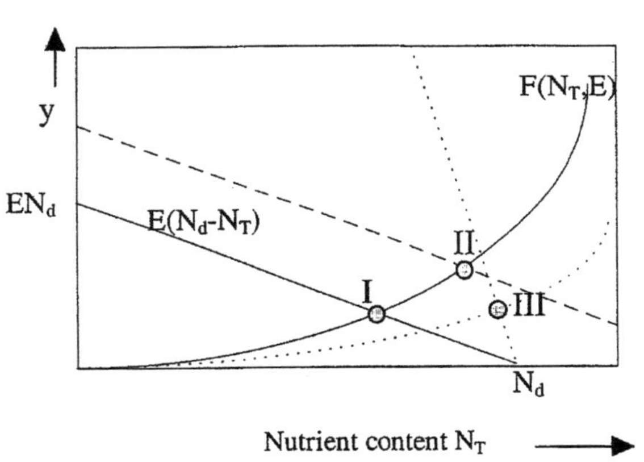

Figure 5. Graphical analysis of the effects of increased nutrient import rate to the photic zone. The equilibrium point (I) is determined by the requirement that net nutrient import $E(N_d-N_T)$ due to water exchange, must balance sedimentary loss $F(N_T,E)$. Changes in the topology of this figure with N_d and E give the new equilibria corresponding to either an increase in N_d (II) or increased E (III). See explanation in Frame 3.

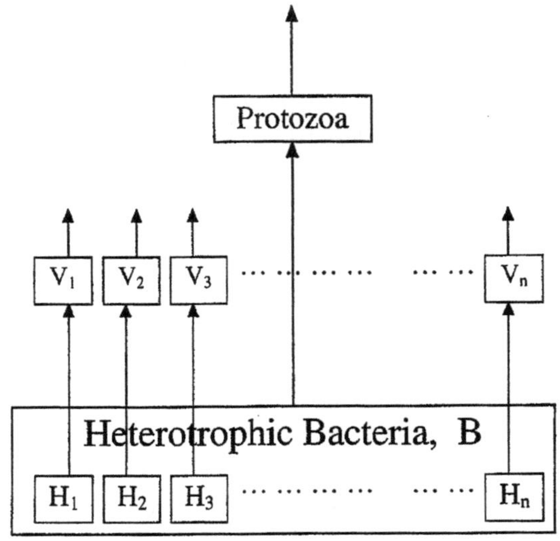

Figure 6. Hypothetical model for diversity control in the bacterial community. Total abundance B of bacteria at equilibrium is controlled by the requirement that protozoan loss rate must equal growth. A similar argument for stable existence of viruses give the abundance in each host-population H_i. Assuming all H_i to be of comparable size, the number of simultaneously dominating host populations is $n = B/H$.

must be that the loss rate of the limiting nutrient must balance the rate at which it is imported. In our terminology, this means that the photic zone nutrient content N_T becomes a dependent variable: N_T will adjust to a value corresponding to a width (number of parallel food chains) and length (number of trophic levels) of the food web where the limiting element is lost from the photic zone at the same rate as it is imported. There are two, not equivalent, factors influencing import of nutrients to the photic zone: the fraction E of water in the photic zone exchanged with the aphotic zone per time unit, and nutrient concentration N_d in the aphotic zone. Assuming no active swimming by any group of organisms between the photic an aphotic zone, we have:

$$dN_T/dt = E(N_d-N_T)-F(N_T, E), \qquad (15)$$

where $E(N_d-N_T)$ is the net import rate of the limiting element, and $F(N_T, E)$ is a function giving the sedimentation loss rate of the limiting element from a food web in equilibrium at nutrient content N_T in a system with exchange coefficient E. Graphically, the equilibrium nutrient content is thus defined by the crossing point between the line $y = E(N_d-N_T)$ and the curve $y = F(N_T,E)$ (Figure 5).

A graphical analysis of the changes in Figure 5 with changing N_d and E is given in Frame 3. The conclusion is that increased nutrient concentration in the deep water will give an equilibrium in the photic zone with higher nutrient content and a food web more dominated by the classical side producing sedimenting particles. Increased exchange coefficient will also give a higher nutrient content, but may give either more or less dominance of the classical side, and correspondingly more or less sedimentation.

Ryther's classical work demonstrating short food chains in nutrient rich and long food chains in oligotrophic areas (Ryther, 1969) could in this framework be re-interpreted as a dominance of the short classical food chain at the right side of the food web in Figure 1, while the long food chain in oligotrophic would be one dominated by the left, microbial side of Figure 1.

Equilibrium diversity

The model discussed above included mechanisms allowing diversity at steady state in an environment homogenous in time and space. By a mechanism selectively 'killing the winner', the predators prevented any single size class of organisms from successfully monopolizing a large fraction of the photic zone nutrient content. This principle can be extended to the action of viruses within each of these size classes. Viruses have recently been shown to be abundant

Shape of the sedimentation function $y = F(N_T,E)$:

F as a function of nutrient content N_T: Increasing N_T is argued to lead to establishment of larger organisms generally producing more sedimenting material. F would therefore be expected to be a generally increasing function of N_T. (It is possible to make special models where F is not monotonously increasing by introducing mechanisms such as e.g. higher sedimentation rates for nutrient limited diatoms.)

F as a function of exchange coefficient E: For a fixed nutrient content, an increase in exchange coefficient E will lead to an increased loss rate for all organisms. To remain in balance, their growth rate must therefore increase, meaning that all food concentrations must increase. This is another way of stating that more of N_T must be stored in free mineral nutrients, and in the left microbial part of the food web in Fig. 1. This means less large organisms, and less sedimentary loss at a given value of N_T. F must therefore decrease with increasing E (dotted curve in Fig. 5).

Changes in the line $y = E(N_d - N_T)$

Effect of a change in aphotic zone nutrient concentration N_d: Increasing N_d will give a parallel displacement of the line to the right (broken line in Fig. 5).

Effect of a change in exchange coefficient E: Increasing the exchange coefficient will give a clockwise rotation of the line around its crossing point with the x-axis (dotted line in Fig. 5).

Net effect of increasing N_d:

Inspection of Fig. 5 shows that an increase in aphotic nutrient content always will shift the crossing point to right, and, as long as F is monotonously increasing, upwards (from I to II in Fig. 5.). This corresponds to a new equilibrium with higher photic zone nutrient content and higher sedimentary loss.

Net effect of increased exchange coefficient E:

As can be seen from Fig. 5, increasing E will also always shift the crossing point to the right (from I to III in Fig. 5) i.e. to a higher N_T. However, depending upon the magnitude of the effect of E on F, point II may be above or below I, corresponding to a food web with more or less sedimentation loss, i.e. more extended towards the classical side to the right, or with more of the nutrients stored in the microbial side to the left in Fig. 1.

Frame 3. Analysis of how aphotic zone nutrient concentration (N_d) and exchange coefficient (E) between the aphotic and photic zones influence equilibrium nutrient content (N_T) of the photic zone.

(Bergh et al., 1989) in the pelagic environment. At least for bacteria, their activity seem largely to be through lytic, rather than lysogenic cycles (Wilcox & Fuhrman, 1994). In a simple model where the protozoan predator is completely unselective for bacteria and the viruses completely selective for one host (Figure 6), applying similar equilibrium arguments as used previously (Equation 8), but now for equilibrium of protozoa and for each type of viruses, will give a total abundance B of bacteria controlled by the protozoan predator, and an abundance H_i of bacteria in a host population controlled by its host-specific virus. The number of types of dominant bacterial hosts can then be estimated as $n = B/H$ (Thingstad et al., in press), where H is the average abundance of bacteria of each host-type.

The line of reasoning used here thus creates a hierarchical structure for the control of diversity: Nutrient content determines the carrying capacity of the whole food web, the size selectivity of predators determines how the nutrient content is distributed between size classes of osmotrophs and phagotrophs, and the host specificity of viruses determine how the nutrient content of one size class is distributed between host populations.

A somewhat disturbing prediction from this theory in its simplest form is that increasing eutrophication (here defined as increasing nutrient content) may seem to increase diversity. This appears to contradict the usual paradigm that oligotrophic systems characteristically have high diversity and that diversity decreases with eutrophication (Margalef, 1969; Horowitz et al., 1983). Thingstad et al. (in press) suggested that a solution to this apparent contradiction between observation and theory could be found in a finite maximum width and length of the food web; i.e. that, in the real world, there is a finite size to large phytoplankton and to large predators. At sufficiently high nutrient content, there are no longer any existing new species to add to the right side of Figure 1, and increasing nutrient content will be channeled into increase in biomass of the final organisms to the right. The biomass of the classical side will then dominate over the microbial part to the left, and the evenness component of diversity will decrease. More mechanisms for decreased diversity at high nutrient content levels emerges, however, if we relax our simplifying assumption of each osmotroph only being eaten by one predator. In a model of what Leibold (1996) called keystone predators, one predator eating several species of prey competing for one resource was shown to be able to maintain diversity among the competitors. These models allowed some prey species to be replaced by others as resources increase. Diversity in the prey population was a uni-modal function of productivity with superior resource exploiters dominating at low productivity, and predator resistant forms at high productivity. One would expect from this that models where the predators consume several size classes of prey with different clearance rates, could predict small species of phytoplankton to be replaced by larger ones at higher nutrient content levels.

Some possible 'dirty tricks' of the food web game

The mere existence of an orderly society creates opportunities for those not sticking to the rules. Whether such behavior is classified as innovation and creativity or as criminal activity is of less relevance in the present context. In the hypothetical case where the only allowed evolutionary adjustment is a change in size, the success of one species would only mean that it could replace another. The number of niches, the structure of the food chain, and the explaining principles, would not really change. In a slightly more complicated world where the size-selectivity of predators is allowed to evolve, high selectivity could lead to a more finely divided distribution of the nutrient content and thus a high diversity. A virus able to evolve high resistance to decay will in these models lower the diversity inside a size class, but the underlying principles would still be the same. In the complexity of the real world organisms are free, not only to evolve the parameters of the above model within the physically possible limits, but also too seek for alternative strategies. An excellent overview of strategies in the pelagic food web as seen from such a perspective is given in the recent article by Verity & Smetacek (1996). If such strategies are moderately successful, they would add complexity to the idealized picture. If really successful, however, such strategies would become the dominating norm themselves (meaning that our size-ordered structure would be of minor relevance). In the photic zone food web, there is a multitude of strategies which could be seen as attempts to escape from the limitations imposed by the mechanisms of the size-structured model above.

The 'apparent size' trick

If increase in size of osmotrophic organism also gives more flexibility for control of buoyancy (ρ) and/or volume specific nutrient content (σ), large organisms could to some extent have an additional means to swimming to escape the severe 4th-power size penalty in minimum substrate requirement (N_c, Equation 6). The large vacuole in diatoms may serve as a tool to regulate ρ (Taylor, 1980), and could also presumably serve as a means to lower the volume specific nutrient content σ. If, however, the price for a large vacuole is the need for a silicate wall, it means that there is a silicate switch in the system. Import of silicate will allow the establishment of organisms which in a sense are smaller (in terms of ρ and σ) than they look (in terms of r). Upwelling of nutrient rich water may therefore be a more complex matter than was discussed above in terms of exchange coefficient E and nutrient concentration N_d only. If the exchanged water in addition contains silicate, the effect may not only be an extension of food web width and length. The silicate may allow diatoms with a low N_C (but large r) to replace already established species and give the food chain a completely new size structure. Silicate may thus in theory act as a switch between two modes, although the basic regulating principles would be the same for both modes. This picture is somewhat different from the simple idea conveyed by Figure 1 that large diatoms follows automatically as the group of phytoplankton larger than phytoflagellates. The question of phytoplankton succession following nutrient enrichment is a classical theme of phytoplankton ecology, and generalizations in terms of diatom cell size and diatoms versus flagellate abundance has been attempted (Smayda, 1980). If a model like the one outlined here is adopted, the equilibrium size structures of the diatom and flagellate communities, as well as the balance between these two modes, depends on how fast the added nutrients are lost from the photic zone (decay in N_T), as well as the ratio between silicate and other nutrients in the enrichment and in the rates at which the elements are lost. The observation that successional patterns may be different at different locations (Smayda, 1980) is therefore not astonishing.

The 'baleen whale' trick

In an ocean dominated by the left microbial part of the food web, food for a phagotrophic organism occurs mostly as small particles. A large predator able to feed directly on such prey therefore experience a much higher food concentration than one who sticks to the normal rule of eating food at a size $\approx 1 : 10$ of its own body size (Fenchel, 1987). A traditional example of this in the microbial world is the appendicularians (Alldredge, 1981). This strategy obviously complicates the orderly structure of Figure 1 by transferring parts of the material directly to larger size classes without the respiration loss of carbon and the recycling of mineral nutrients occurring at the intermediate levels. It would also mean that the predators would compete for prey, a very relevant situation not treated here.

The 'eating your competitor' trick

A mineral nutrient limited photosynthetic organism with the ability to feed on nutrient competitors of smaller size would obtain two things, not only would it gain the nutrients already taken up by the smaller and superior competitor, it would also remove a part of this competitor population. This may the explanation for the importance of mixotrophic flagellates in some nutrient limited environments (Havskum & Riemann, 1996). Theoretically, the success of such a strategy depends upon the price paid by the organism for simultaneously harboring the required tools for a photosynthetic/osmotrophic, and a heterotrophic/phagotrophic strategy (Thingstad et al., 1996). This mixed strategy would complicate the food web structure by introducing organisms which combine the roles of the organisms in the upper and lower rows of Figure 1.

The 'inedibility' trick

Any prey organism able to lower either the yield or the clearance rate of its predator would increase its own steady state biomass (Equation 8). In the extreme case of a completely inedible phytoplankton, the establishment of this species would cut off any further widening or lengthening of the food chain. All increases in nutrient content will only result in increased biomass of the inedible species. We would have a case where there is no one to kill the winner. The probable importance of such strategies is reflected in observations like the takeover of long, filamentous bacteria when mixed culture bacterial chemostats are inoculated with heterotrophic flagellates (Havskum et al., pers. comm.).

Discussion

Our discussion essentially used two basic principles: (1) that osmotrophic organisms caught between the physical laws of diffusion and sinking will be able to survive at lower ambient concentration of the limiting nutrient, and that (2) there are predators selectively exploiting the organisms otherwise successful in monopolizing the nutrients available for sharing in the food web. These are of course not new and revolutionary assumptions concerning the way pelagic food webs function. Indeed the main value of this exercise hinges on the degree to which the approach succeeds in explaining features at one level of system organization (the food web) from principles at a lower level (cell physiology and physical/chemical properties of the environment), and whether these explanations differ from present concepts. I feel that the idea of a relatively small set of common principles possibly governing main features of food web structure, diversity, and biogeochemical cycles, is an idea of great conceptual and practical value. In the search for a prac-

Table 1. Symbols used

ABUNDANCES AND RELATED VARIABLES	
A_i	Abundance of phytoplankton population i
P_i	Abundance of predator i grazing on A_i and on P_{i-1}
P_0	Abundance of predator on heterotrophic bacteria
B	Total abundance of heterotrophic bacteria
H_i	Abundance of bacterial hosts for virus population i
H	Average bacterial abundance in host-groups
n	Number of host-groups in the bacterial community
N	Concentration of free mineral nutrient
N_C	Minimum concentration of limiting nutrient where a given phytoplankton population can balance loss with growth
N_T	System nutrient content
SPECIFIC LOSS AND GROWTH RATES	
δ_{Ai}	Specific loss rate of A_i
δ_{Pi}	Specific loss rate of P_i
μ_B	Specific bacterial growth rate
AFFINITIES AND CLEARANCE RATES	
α_{A1}	Affinity for mineral nutrients by phytoplankton A_1
α_B	Affinity for mineral nutrients by heterotrophic bacteria
α_{P0}	Predator P_0's clearance rate for heterotrophic bacteria
α_{P1}	Predator P_1's clearance rate for A_1 and for P_0
YIELD COEFFICIENTS	
Y_{A1}	Yield of A_1 on mineral nutrients
Y_{BN}	Yield of B on mineral nutrients
Y_{BC}	Yield of bacteria on DOC
Y_{P0}	Yield of P_0 on heterotrophic bacteria
Y_{P1}	Yield of P_1 on A_1
OTHERS	
ρ	Density of organism
ρ_W	Density of seawater
σ	Volume-specific content of limiting nutrient
η	Viscosity of water
D	Diffusion rate of dissolved mineral nutrient
r	Cell radius
h	depth of photic zone
g	gravity constant
ψ	System production rate of dissolved organic carbon
R	Bacterial carbon demand

tical tool for design of experiments and interpretation of field observations, such an approach seems much more fruitful than focusing on the beautiful, but bewildering complexity of the real world. Numerical models based on very similar assumptions are already available and have been shown to be able to reproduce important aspects of field observations (Moloney, 1992) and mesocosm experiments (Baretta-Bekker et al., 1994). Such simulation models also link assumptions at one level (physics and physiology) to model output at a higher level (population changes) and thus also offer an explanation for observed behavior, even at a much more quantitative level than attempted here. The more analytical approach used here may however provide a somewhat different type of understanding since it allows a better comprehension of the mechanisms linking assumptions and model output.

To understand the behavior of an ecosystem with a high degree of regenerated production, there is a need for a tool by which mechanisms of growth and loss, of nutrient uptake and nutrient recycling, of bottom-up and top-down control, can be integrated into one picture. In the present analysis this aspect is included through the steady state requirement that growth must equal sum of losses for all components in the mineral nutrient cycle. The assumption may appear trivial and of little information content, but analyzing the consequences of this assumption reveals patterns not easily grasped by pure intuition, even in systems of fairly low complexity. Since no food web is probably ever in a true steady state, models of this type should of course be treated with caution. The organisms with a generation time in the order of hours and days in the left, microbial part of Figure 1 would be expected to fairly rapidly form a sub-system fluctuating around an equilibrium driven either by external forces such as copepod growth and migration, diel light cycles, and nutrient input events, or by internal mechanisms caused by the many predator-prey relationships in the system. However, in many physically stable environments fluctuations in phytoplankton biomass over time scales much longer than cell generation times are small (Malone, 1980), implying that growth nearly balances loss. The equilibrium condition may thus be a reasonable approximation to reality, at least in the left, microbial, side of Figure 1. Understanding the properties of these equilibria, even if they are not ever really attained, is therefore an important step.

One very important feature of the above framework is that the number of state variables is in itself a variable, the number of size classes of osmotrophic organisms, the number of predator levels and the number of hosts groups inside the bacterial community will vary with nutrient content. This property removes part of a serious problem with traditional models where a fixed food web structure seriously limits the possible responses of the model. In the framework suggested here, the degree of resolution into separate boxes is to a much larger extent a function emerging from lower level assumptions such as predator selectivity and system nutrient content. This is the property of the suggested approach which makes it of potential value for analysis of mechanisms controlling diversity. Very much of the classical thinking around niches and diversity was focused on resources and different organisms exploiting different resources. In models of the type suggested here, this bottom-up type of diversity control could in one sense explain the coexistence of predators since each size-selective predator in Figure 1 is feeding on its own resource. On the other hand, here the predator selectivity is also in itself the main mechanism creating diversity in resources by 'killing the winner' and thus allowing coexisting size classes of osmotrophs. Together with lytic viruses, such top-down mechanism allow a diverse community of osmotrophs to coexist on one common limiting resource (the free mineral nutrient). The top-down and bottom-up effects on diversity are therefore quite intimately linked in such a model. Tett & Barton (1995) pointed out that, while Hutchinson's paradox of an unexpectedly large phytoplankton diversity seems true at a local scale, the opposite is true on a global scale where there has been estimated to be only ≈ 5000 different species. The theory presented here seems to a large extent to resolve the local problem. The intriguing low number of species on a global scale may seem to support the idea of fairly universal niche-forming mechanisms in water as suggested here.

Any researcher in this field could probably extend our list of 'dirty tricks' by examples spanning from nitrogen fixation, via bacteria preying on bacteria (*bdellovibrios*), to diel vertical migrations, and justifiably claim that such mechanisms have to be incorporated to get the full picture. The increase obtained in explanatory power will however usually have to be paid for in terms of loss in model clarity or transparency and may not necessarily improve the model. The approach taken here was to start from as simple assumptions as possible, and then add complexity in small steps, hoping to increase explanatory power in a satisfactory proportion to the complexity added. I believe the analysis given here shows that this is both a possible and a worthwhile path to explore in the search for understanding the pelagic food web.

Acknowledgement

This work was financed through contract MAS3-CT95-0016 'MEDEA' of the EU MAST-III program.

References

Alldredge, A. L., 1981. The impact of appendicularian grazing on natural food concentrations *in situ*. Limnol. Oceanogr. 26: 247–257.

Armstrong, R. A., 1994. Grazing limitation and nutrient limitation in marine ecosystems: Steady state solution of an ecosystem model with multiple food chains. Limnol. Oceanogr. 39: 597–608.

Azam, F., T. Fenchel, J. G. Field, J. S. Gray, L. A. Meyer-Reil & T. F. Thingstad, 1983. The ecological role of water-column microbes in the sea. Mar. Ecol. Prog. Ser. 10: 257–263.

Baretta-Bekker, J. G., B. Riemann, J. W. Baretta & E. K. Rasmussen, 1994. Testing the microbial loop concept by comparing mesocosm data with results from a dynamical simulation model. Mar. Ecol. Prog. Ser. 106: 187–198.

Bergh, Ø., K. Y. Børsheim, G. Bratbak & M. Heldal, 1989. High abundance of viruses found in aquatic environments. Nature 340: 467–468.

Campbell, L. & D. Vaulot, 1993. Photosynthetic picoplankton community structure in the subtropical North Pacific Ocean near Hawaii (Station ALOHA). Deep Sea Res. 40: 2043–2060.

Dugdale, R. C. & J. J. Goering, 1967. Uptake of new and regenerated forms of nitrogen in primary productivity. Limnol. Oceanogr. 12: 196-206.

Fenchel, T., 1987. Ecology- Potentials and limitations. In Kinne, O. (ed.), Excellence in Ecology. Ecology Institute, Oldendorf/Luhe.

Gaedke, U., 1993. Ecosystem analysis based on biomass size distributions: A case study of a plankton community in a large lake. Limnol.Oceanogr. 38: 112–127.

Goldman, J. C., J. J. McCarthy, & D. G. Peavey, 1979. Growth rate influence on the chemical composition of phytoplankton in oceanic waters. Nature 279: 210–215.

Havskum, H. & B. Riemann, 1996. Ecological importance of bacterivorous, pigmented flagellates (mixotrophs) in the Bay of Aarhus, Denmark. Mar. Ecol. Prog. Ser. 137: 251–263.

Horowitz, A., M. I. Krichevsky, & R. M. Atlas, 1983. Characteristics and diversity of subarctic marine oligotrophic, stenoheterotrophic, and euryheterotrophic bacterial populations. Can. J. Microbiol. 29: 527–535.

Hutchinson, G. E., 1961. The paradox of the plankton. Am. Nat. 95: 137–145.

Leibold, M. A., 1996. A graphical model of keystone predators in food webs: trophic regulation of abundance, incidence, and diversity patterns in communities. Am. Nat. 147: 784–812.

Malone, T. C., 1980. Algal size. In Morris, I. (ed.), The physiological ecology of phytoplankton. Studies in Ecology. Vol. 7. pp. 433–463. Blackwell, Oxford.

Margalef, R., 1969. Diversity and stability: A practical proposal and a model of interdependence. In Diversity and stability in ecological systems. Brookhaven Symposia in Biology. Vol. 22. Brookhaven National Laboratory, Upton,. New York.

Moloney, C. L. & J. G. Field, 1991. The size-based dynamics of plankton food webs. 1. A simulation model of carbon and nitrogen flows. J. Plankt. Res. 13: 1003–1038.

Moloney, C. L., 1992. Simulation studies of trophic flows and nutrient cycles in Benguela upwelling foodwebs. In Payne, A. I. L., K. H. Brink, K. H. Mann, & R. Hilborn (eds), Benguela Trophic Functioning. S. afr. J. mar. Sci. 12: 457–476.

Pengerud, B., E. F. Skjoldal, & T. F. Thingstad, 1987. The reciprocal interaction between degradation of glucose and ecosystem structure. Studies in mixed chemostat cultures of marine bacteria, algae, and bacterivorous nanoflagellates. Mar. Ecol. Prog. Ser. 35: 111–117.

Pomeroy, L. R., J. E. Sheldon, W. M. Sheldon Jr. & F. Peters, 1995. Limits to growth and respiration of bacterioplankton in the Gulf of Mexico. Mar. Ecol. Prog. Ser. 117: 259–268.

Power, M. E., 1992. Top-down and bottom-up forces in food webs: Do plants have primacy. Ecology 73: 733–746.

Rodriguez, J. & M. M. Mullin, 1986. Relation between biomass and body weight of plankton in a steady state oceanic system. Limnol. Oceanogr. 31: 361–370.

Ryther, J., 1969. Photosynthesis and fish production in the sea. The production of organic matter and its conversion to higher forms of life throughout the world ocean. Science 166: 72–76.

Sheldon, R. W., A. Prakash, A. & W. H. Sutcliffe, 1972. The size distribution of particles in the ocean. Limnol. Oceanogr. 17: 327–340.

Smayda, T. S., 1980. Phytoplankton species succession. In Morris, I. (ed.), The physiological ecology of phytoplankton. Stud. Ecol. 7: 493–570. Blackwell Scientific Publications, Oxford.

Strong, D. R., 1992. Are trophic cascades all wet? Differentiation and donor-control in speciose ecosystems. Ecology 73: 747-754.

Taylor, F. J. R., 1980. Basic biological features of phytoplankton cells. In Morris, I. (ed.) The physiological ecology of phytoplankton. Stud. Ecol. 7: pp. 3–55. Blackwell Scientific Publications, Oxford.

Tett, P. & E. D. Barton, 1995. Why are there about 5000 species of phytoplankton in the sea? J. Plankt. Res. 17: 1693–1704.

Thingstad, T. F. & B. Pengerud, 1985. Fate and effect of allochthonous organic material in aquatic microbial ecosystems. An analysis based on chemostat theory. Mar. Ecol. Prog. Ser. 21: 47–62.

Thingstad, T. F. & E. Sakshaug, 1990. Control of phytoplankton growth in nutrient recycling ecosystems. Theory and terminology. Mar. Ecol. Prog. Ser. 63: 261–272.

Thingstad, T. F. & F. Rassoulzadegan, 1995. Nutrient limitations, microbial food webs, and 'biological C-pumps': suggested interactions in a P-limited Mediterranean. Mar. Ecol. Prog. Ser. 117: 299–306.

Thingstad, T. F., H. Havskum, K. Garde, & B. Riemann, 1996. On the strategy of 'eating your competitor'. A mathematical analysis of algal mixotrophy. Ecology 77: 2108–2118.

Thingstad, T. F., G. Bratbak, M. Heldal & I. Dundas, (in press). Trophic interactions controlling diversity in pelagic microbial food webs. Proceedings of the 7th International Symposium in Microbial Ecology (ISME-7) Santos Brazil, 27.08–01.09 1995.

Thingstad, T. F., Å. Hagstrøm & F. Rassoulzadegan, 1997. Export of degradable DOC from oligotrophic surface waters: caused by a malfunctioning microbial loop? Limnol. Oceanogr. 42: 398–404.

Thingstad, T. F. & R. Lignell, 1997. Theoretical models for the control of bacterial growth rate, abundance, diversity and carbon demand. Aquat. Microbiol. Ecol. 13: 19–27.

Vaulot, D., N. LeBot, D. Marie & E. Fukai, 1996. Effect of phosphorus on *Synechococcus* cell cycle in surface Mediterranean waters during summer. Appl. Envir. Microbiol. 62: 2527–2533.

Verity, P. & V. Smetacek, 1996. Organism life cycles, predation, and the structure of marine pelagic ecosystems. Mar. Ecol. Prog. ser. 130: 277–293.

Walsby, A. E. & C. S. Reynolds, 1980. Sinking and floating. In Morris, I. (ed.), The Physiological Ecology of Phytoplankton. Stud. Ecol. 7: 371–412. Blackwell Scientific Publishers, Oxford.

Wilcox, R. M. & J. A. Fuhrman, 1994. Bacterial viruses in coastal seawater: Lytic rather than lysogenic production. Mar. Ecol. Progr. Ser. 114: 35–45.

Hydrobiologia **363:** 73–79, 1998.
T. Tamminen & H. Kuosa (eds), Eutrophication in Planktonic Ecosystems: Food Web Dynamics and Elemental Cycling.

30 years' eutrophication in shallow brackish waters – lessons to be learned

U. Schiewer
Universität Rostock, Institut für Ökologie, Freiligrathstr. 7/8, D-18051 Rostock, Germany

Key words: brackish waters, eutrophication, pelagic community, regulation

Abstract

The anthropogenically induced eutrophication of the Darss-Zingst bodden chain, a tideless shallow brackish water, has been followed over a period of more than 30 years. The whole process was dominated by increasing nutrient loads and irregular exchange processes with the Baltic Sea. Dramatic changes started between 1964/72 in the western parts of the system, which is more influenced by nutrient loads and less connected to the exchange with the Baltic Sea. The first signs of increased anthropogenic loading were a massive reduction in submerged macrophytes. 10 to 15 years later the same processes occurred in a slightly modified form in the eastern parts of the bodden chain. The whole eutrophication process could be followed in more detail in the Barther Bodden. Since 1968 there have been: – no changes in soluble reactive phosphate

– increases in the spring values of nitrate

– stepwise changes in the community composition leading to a remarkable increase of the importance of the microbial food web.

The influences and consequences of eutrophication on the phytoplankton, bacterioplankton, protozooplankton and metazooplankton are generalized. The first signs for remesotrophication have been observed in the last two years.

Introduction

The German coast of the Baltic Sea is characterized by several Boddens and Haffs, inner coastal tideless estuaries (Figure 1). They are important buffers and filters for the coastal waters and the Baltic Proper. During the last 40 years they have gradually become eutrophic, mainly owing to agricultural activities in the discharge area. Figure 2 shows the primary productivity in the various water basins of the Bodden chain at the end of the eighties.

The Department of Biology at the Rostock University has been studying the different processes connected with eutrophication in the Darss-Zingst Bodden chain for more than 30 years (e.g. Arndt 1989, Nausch & Schlungbaum 1991, Schiewer 1991, Wasmund & Kell 1991, Winkler 1991).

The results obtained during the investigation period reveal many common features in shallow tideless estuaries which might also apply to other estuarine ecosystems receiving high nutrient loads.

The methods used for these investigations are typical of coastal research (see Schiewer et al., 1986). The research strategy was strongly oriented towards an experimental ecological approach using lab experiments, lab microcosms, field mesocosms and field investigations (Schiewer, 1990). This approach proven very useful in connection with the microbial food web which dominates the pelagic community.

Selected results

The first important sign of the ongoing eutrophication in shallow waters was the decline of the submersal macrophytes. This started in the more severely loaded western parts of the bodden chain during the seventies (Lindner, 1978; Behrens, 1982).

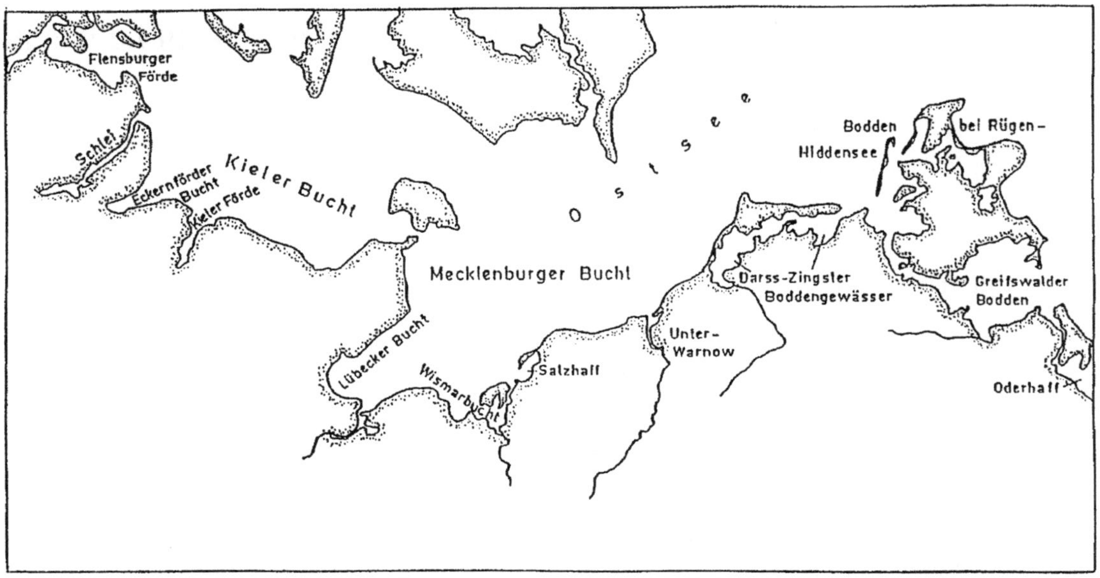

Figure 1. The German Baltic coast. Fjords, Boddens, Haffs – typical inner coastal waters.

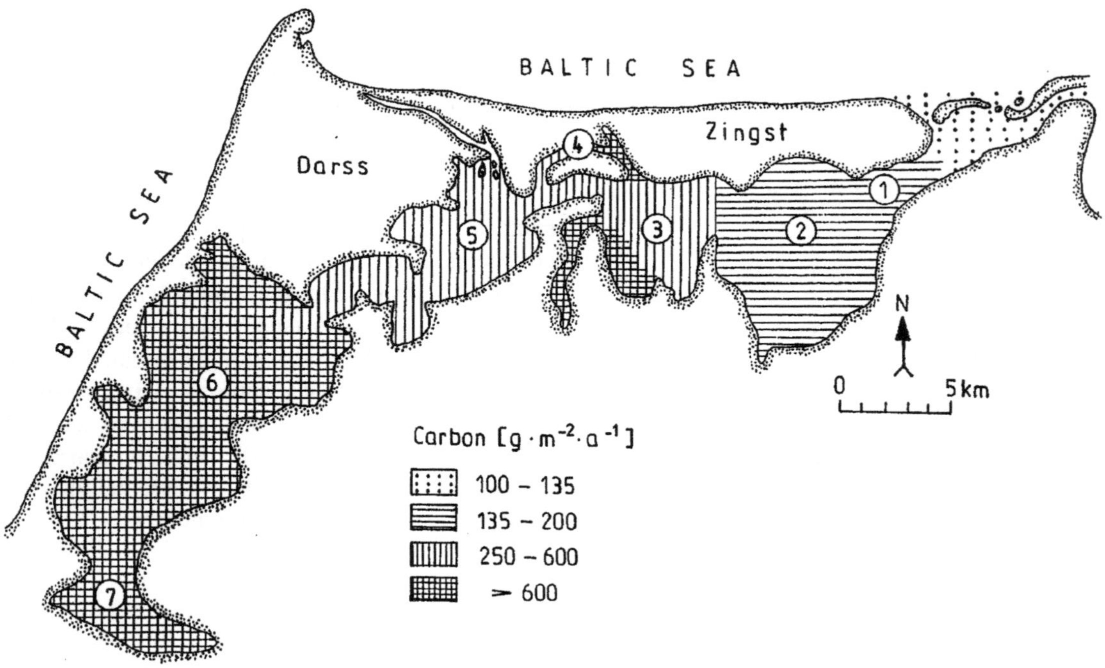

Figure 2. The Darss-Zingst Bodden Chain (DZBC). Annual productivity in the various water basins. Decreasing salinity from east to west, increasing eutrophication from east to west.

In the eastern parts, e.g. the Barther Bodden, we are able to follow this process from the beginning of the eighties. The phytoplankton was nitrogen limited in spring, mainly due to the high nitrogen uptake of the phytoplankton and high denitrification rates at water temperatures higher than 6 °C (S. Dahlke, pers. comm.).

In 1981, a massive decline of submerse macrophytes took place in the eastern parts of the Bodden chain (Figure 3). The immediate reasons were a rainy spring season with low water salinities in the Boddens, high nutrient loads and high water turbidity. The absence of macrophytes increased sediment mobility and water turbidity. As the system became dominated by phytoplankton, its productivity was enhanced but not as much as expected (Barther Bodden: Hübel 1972/73: 265 g C $\times$ m^{-2} $\times$ a^{-1}; Börner 1982/83: 287 g C $\times$ m^{-2} $\times$ a^{-1}; see Börner, 1984). At the same time the microbial food web became the most important route for energy flow and cycling of matter owing to a decrease in copepod (see Figure 5) and rotifer biomass (not shown here) and an increase in protozooplankton biomass and activity (Schiewer et al., 1993).

Mesocosm experiments showed that increasing nutrient loads induce a change in the phytoplankton community leading to dominance by green algae (Schiewer et al., 1986; Schiewer et al., 1988).

Phytoplankton monitoring using the traditional Utermöhl-technique showed good agreement with this prediction during the subsequent few years (Wasmund & Börner, 1992).The green alga *Tetrastrum triangulare* (Chodat) Komarek flourished in particular. However, fluorescence microscopic studies undertaken in 1991 showed that the phytoplankton abundance estimations obtained by the traditional Utermöhl-method were incorrect (Schumann, 1993). Indeed, the cyanobacterial community has changed from one of microplanktonic species to one of nano- and picoplanktonic species (Figure 4). The percentage distribution had changed in favour of the cyanobacteria. During the seventies and up to 1982, Wasmund & Börner (1992) observed a mean annual cyanobacterioplankton biomass of 50% with a reduction to 10% in the winter season. Schumann (1993) estimated that cyanobacterioplankton accounted for 60% of the biomass during the vegetation period and 43% during the winter season.

More detailed investigations underscored the increased importance of the picophytoplankton. The annual mean percentage of picophytoplankton was

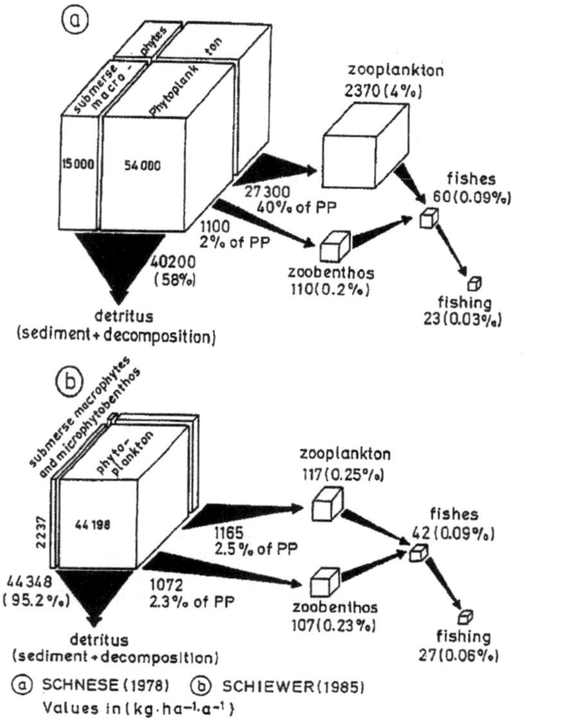

Figure 3. Changes in submersal macrophytes in the Barther Bodden 1981, DZBC. Food web calculation for the Barther Bodden based on measurements of its biomasses: a. calculation for the period before 1981, b. calculation for the period since 1981 (Schiewer, 1995). PP = primary production. Note the dramatic decline of submerse macrophytes! The main changes took place in 1981.

10%, with maxima up to 19% in May and November. They were well adapted to low and high light intensities (Schubert, 1996), had high growth rates and were well grazed by protozoans (Schumann & Schiewer, 1994; Klinkenberg & Schumann, 1995). *Aphanothece clathrata* W. et G. S. West is now the most important cyanobacterium. Normally a colony forming species, it appears here mainly as single cells, obviously owing to the high nutrient content of the ecosystem.

At the same time, the heterotrophic component of the microbial community had changed. There was a strong and stepwise reduction (Heerkloss & Schnese, 1995) in the copepod *Eurytemora affinis* (Poppe), (Figure 5) and of the larger rotifers (not shown). These changes were accompanied by growth of the protozoan community composed mainly of heterotrophic flagellates and ciliates. The high grazing pressure of the protozoans reduced the bacterioplankton and picoplankton abundances as shown by Klinkenberg & Schumann (1995), but simultaneously enhanced the activity of both groups.

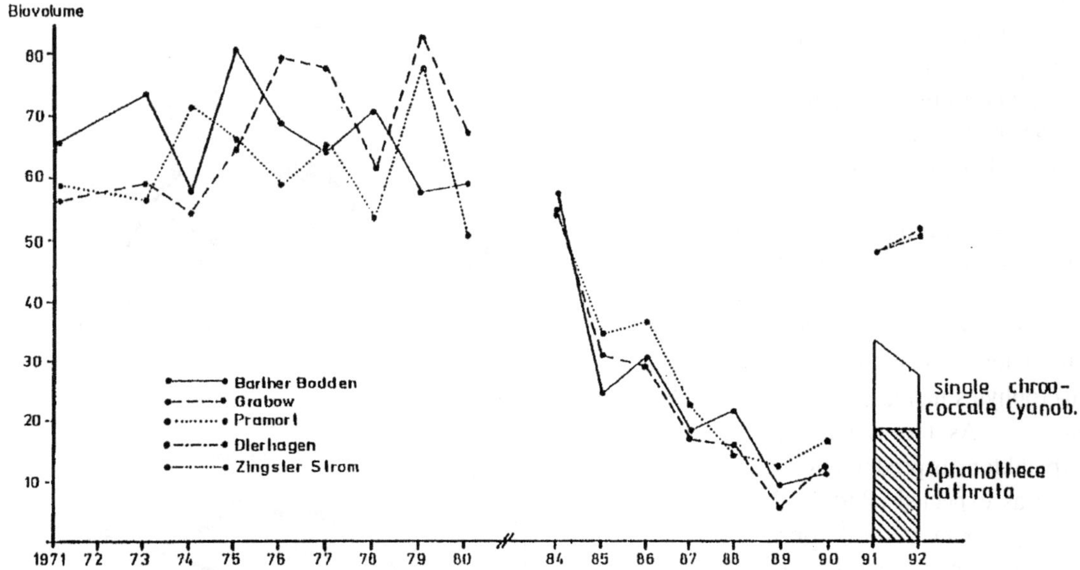

Figure 4. Phytoplankton biovolume development in the DZBC. 1971 – 90: monitoring using the traditional Utermöhl-technique. 1991–92: fluorescence microscope (Image) analysis. Columns: some main forms; lines: total percentage of cyanobacteria. y-axis: cyanobacteria biovolume as % of total phytoplankton biovolume (= 100%).

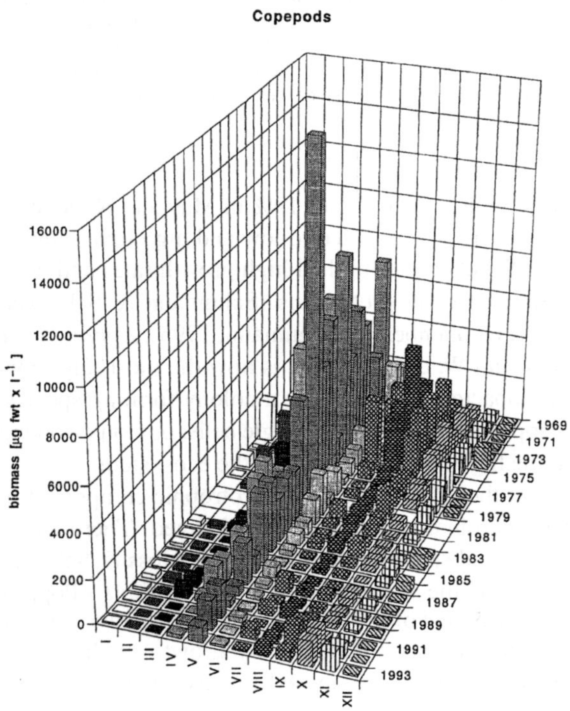

Figure 5. Biomass of copepods: Long-term changes in the Barther Bodden (Heerkloss & Schnese, 1995). Note the stepwise decline!

Finally, all changes together led to an increase in the dominance and importance of the microbial food web in the pelagic community (Figure 6). Its main part II is responsible for more than 90% of the carbon turnover in the pelagic ecosystem. The internal loop of the ciliates short-circuits the cycle and increases its activity. This enhances the respiration of organic substances. Mesocosm experiments showed that even very intensive primary production (e.g. primary productivity on June 18–19, 1990: 944 μg C $\times$ $l^{-1} \times d^{-1}$ in the control enclosure A versus 3020 μg C $\times$ $l^{-1} \times d^{-1}$ in the loaded enclosure B) over a period of 3 weeks did not lead to a higher seston content than in the nutrient limited control mesocosm (Figure 7).

On the other hand, the increased flow of organic matter was associated with high remineralization rates for nitrogen and phosphorus (Fig. 8). Compared to the external annual load of 1.3 g P $\times$ $m^{-2} \times a^{-1}$ and 14 g N $\times$ $m^{-2} \times a^{-1}$, these remineralization rates are important for the nutrient cycle balance. Remineralization eliminated the nutrient limitation for the dominant nano- and picophytoplankton and stabilized the nutrient supply to the ecosystem.

Such an ecosystem has a high self-eutrophication potential. External nutrient inputs become less important than the internal load under such circumstances. When this stage is reached, the function of the coastal

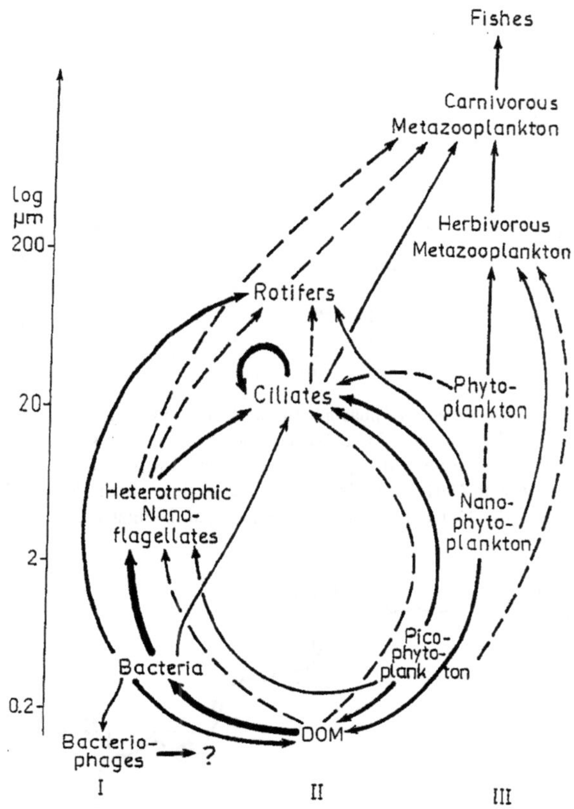

Figure 6. Microbial food web. DZBC, late spring/early summer situation. I = side chain through bacteriophages (proven abundances of virus sized particles: $10^8 \times ml^{-1}$).
II = microbial food web. Main pathway for carbon turnover, mainly by pico- and nanoplankton. Internal loop in the ciliate community (arrow) can involve up to 3 additional trophic levels.
III = 'classical' pelagic food web. The typical components are net plankton and fishes. Minor role only.
DOM = Dissolved Organic Matter.

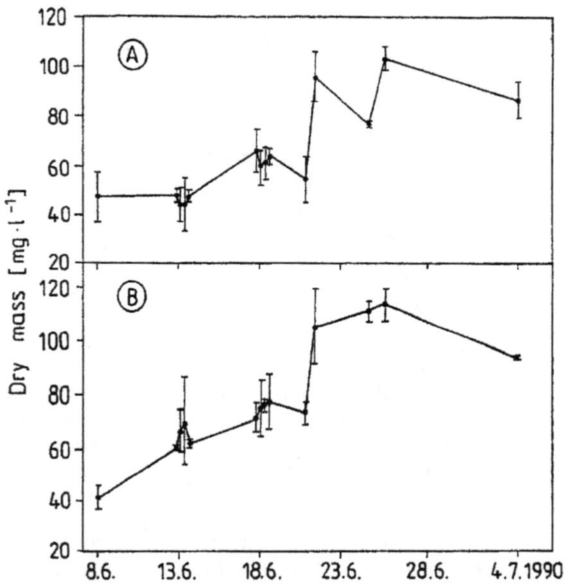

Figure 7. Dry mass development in mesocosms. Summer situation in the Zingster Strom, DZBC. For setup of the mesocosm experiments see Schiewer et al., 1986, 1997. A = Control enclosure: polytrophic pelagic community. Increasing N-limitation of phytoplankton during the 4-week experiment led to development of N_2-fixing cyanobacteria. B = (N + P)-loaded mesocosm. Very high phytoplankton productivity for 3 weeks; Further increase in productivity and biomass was limited by light.

ecosystem for the Baltic Sea changes: it no longer acts as an important filter and buffer, but as an increasing load source! Light and temperature are the most important abiotic factors regulating the phytoplankton communities at the present stage of polytrophication. More recent experiments indicate (Schubert, 1996; Schubert, Forster & Sagert, 1995) that unexpectedly high light stress is also important for the outcome of competition between green algae and cyanobacteria. The shallow bodden ecosystem is characterized by frequent Langmuire spirals which, on average, can transport the phytoplankton between high light intensities at the surface and nearly zero light at the bottom every 10 to 20 minutes. Stress by high light intensities can occur in the upper 6 cm of the water column by direct and/or reflected light. Reflected light can have up to five times the light intensity of direct surface light (Figure 9). Over short term intervals of around 90 minutes, cyanobacteria are much better able to cope with such light stresses than green algae. This can be measured by the phaeophytin production (Figure 10). The reason for the better light tolerance of cyanobacteria lies in their partly coupled photosynthesis and respiration mechanisms (Shyam et al., 1993; Schubert et al., 1995a; Schubert et al., 1995b). This gives the cyanobacteria an additional competitive advantage under such circumstances.

Finally, Figure 11 shows a more general overview of the eutrophication process in the Darss-Zingst Bodden chain since the middle ages, and especially for the last 35 years. It shows that eutrophication took place in steps consisting of more or less stable phases (marked by o) and critical states. The signs of the transitions include the reduced importance of pelagic diatoms in 1969 and the dramatic decline of the submerged macrophytes in 1981. Since 1985, a relatively stable polytrophic situation has evolved. It is based on the more pronounced development of the microbial food web. The microbial food web now established allows the pelagic ecosystems to react very quickly to perturbations. On the other hand, self-eutrophication has stabilized the trophic situation reached. Consequently,

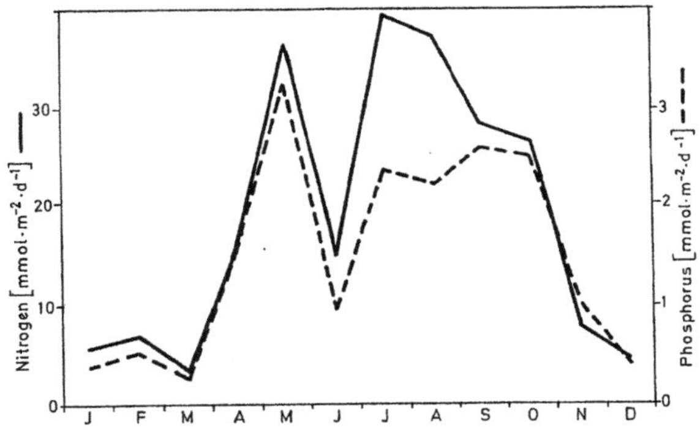

Figure 8. Nitrogen and phosphorus remineralization rates. Monthly means calculated from the respiration of pelagic phagotrophic consumers (HNF 1–3 μm; 3–5 μm; 5–10 μm; rotifers, copepods and larvae of polychaetes) using the Redfield ratio 106:16:1 for C:N:P for phytoplankton. Basis are the monthly means of biomasses and the temperature depence of the P/B-ratios. Netto productivity efficiency P/A = 0.4. Water depth 2 m. Daily P/B-ratios for 20 °C: HNF 1–3 μm = 8.7; HNF 3–5 = 5.5; HNF 5–10μm = 3.6; ciliates = 3.0; rotifers = 0.42; calanids = 0.32; larvae of polychaetes = 0.32 (for further details see Schiewer et al., 1994).

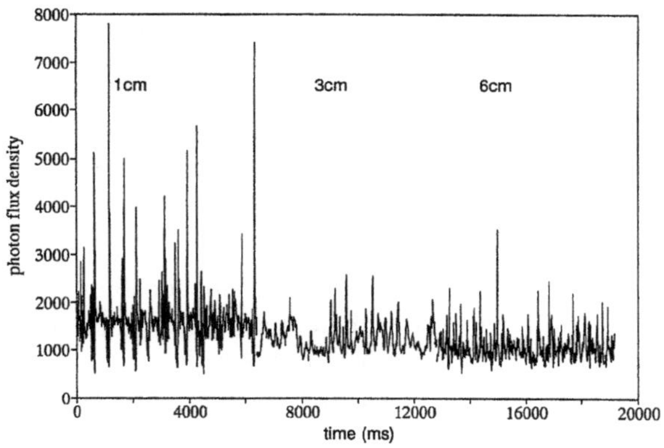

Figure 9. Short-time underwater light measurements in the Zingster Storm (Schubert 1996). Photonflux density (μmol photons × cm^{-2} × s^{-1}) at 1, 3 and 6 cm depth. X-axis: measured time in ms; measurement interval 10 ms, spectral resolution 400–710 nm.

the buffer capacity for increased loads is considerable, but load reductions will not lead to the fast restoration of the former ecosystem.

Concluding remarks

Based on our investigations in the Darss-Zingst Bodden Chain we have developed a 5 step classification system for inner coastal shallow waters. It allows the classification of our coastal ecosystem (Schlungbaum, Schiewer & Arndt, 1994) from oligotrophic (class 1) to hypertrophic (class 5). Figure 12 shows the re-

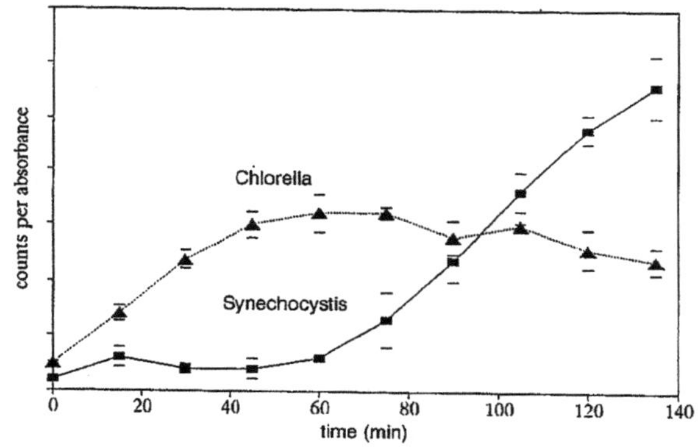

Figure 10. Kinetics of phaeophytin production after enhanced photon flux density from 25 to 260 μmol photons × cm^{-2} × s^{-1} (Schubert, 1996). Radioactive phaeophytin counts (algae preincubated for 6 h with NaH^{14}CO3) by absorbance of phaeophytin. Standard deviations of 4 parallels after the beginning of the light stress.

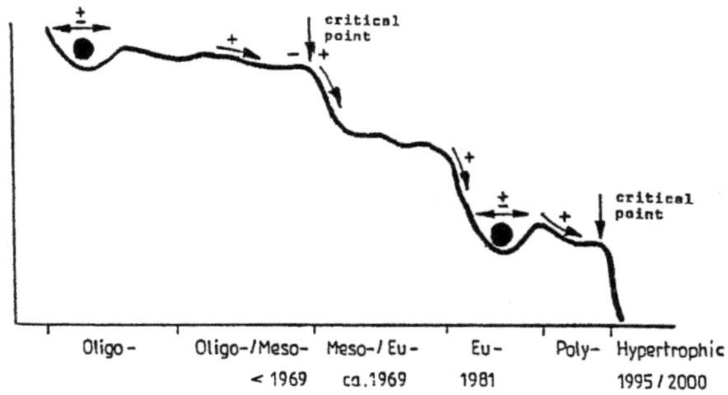

Figure 11. Hypothetical eutrophication model of the Barther Bodden, DZBC (Schiewer, 1994). Various eutrophication stages can be distinguished during the development of the DZBC . During the past 30 years a stepwise development from meso-/eutrophic to polytrophic status has taken place. The prognosis of the hypertrophic status was based on the nutrient input changes during the period 1981/85. o = stabilized situation.

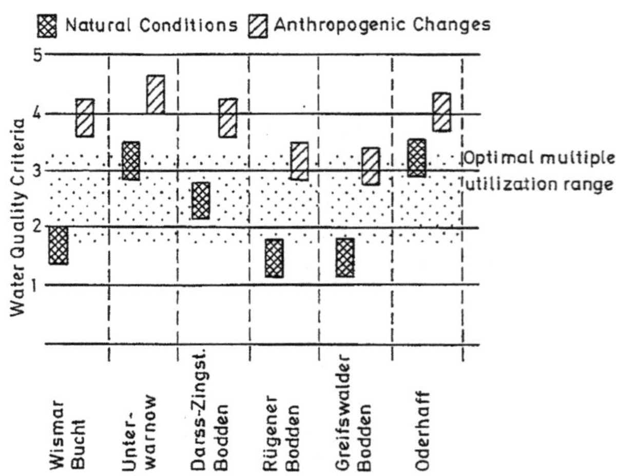

Figure 12. Water Quality of different German coastal waters (Schlungbaum et al., 1994).

sults of such a classification. In addition we have calculated the achievable normal natural potentials of the systems if anthropogenic loads are more or less eliminated. Concerning a multivalent use of the inner coastal waters as the best management strategy, we can define an optimal classification range for each of our coastal ecosystems. Reducing eutrophication to this level would lead to the maintenance of adequate biodiversity and functions of the ecosystems. It would also reduce the costs of restoration and regulation.

In the last 5 years, anthropogenic loads have decreased by over 40%. First signs of an ongoing remesotrophication have been observed in the eastern parts of the Darss-Zingst Bodden Chain during the past two years. Investigation of this remesotrophication will be of great scientific and practical value at least for the southern and eastern coastal region of the Baltic Sea.

References

Arndt, E. A., 1989. Ecological, physiological and historical aspects of brackish water fauna distribution. In J. S. Ryland & P. A. Tyler (eds), Proc. 23rd Eur. Mar. Biol. Symp. Swansea UK, Olsen & Olsen, Fredensborg: 327–338.

Behrens, J., 1982. Soziologische und produktionsbiologische Untersuchungen an den submersen Pflanzengesellschaften der Darß-Zingster Boddengewässer. Diss. A Univ. Rostock.

Börner, R., 1984. Produktionsbiologisch-ökologische Untersuchungen am Phytoplankton des Zingster Stromes. Diss. A Univ. Rostock.

Heerkloss, R. & W. Schnese, 1995. Long term monitoring in the bodden waters south of Darss-Zingst. In J. Köhn & U. Schiewer (eds), The Future of the Baltic Sea. Metropolis-Verlag, Marburg: 55–62.

Klinkenberg, G. & R. Schumann, 1995. Abundance changes of autotrophic and heterotrophic picoplankton in the Zingster Strom, a shallow, tideless estuary south of the Darss-Zingst peninsula (Southern Baltic). Arch. Hydrobiol. 134: 359–377.

Lindner, A., 1978. Soziologisch-ökologische Untersuchungen an der submersen Vegetation in der Boddenkette südlich des Darß und des Zingst (südliche Ostsee). Limnologica 11: 229–305.

Nausch, G. & G. Schlungbaum, 1991. Eutrophication and restoration measures in the Darß-Zingst Bodden Chain. Int. Rev. ges. Hydrobiol. 76: 451–463.

Schiewer, U., H. Arndt, G. Baader, G. Ballin, R. Börner, F.-K. Evert, F. Georgi, R. Heerkloss, G. Jost, V. Kell, B. Krüger & Th. Walter, 1986. The bounds and potential effects of NH_4 (loading) on the pelagic system of a Baltic estuary. Limnologica (Berlin) 17: 7–28.

Schiewer, U., R. Börner & N. Wasmund, 1988. Deterministic and stochastic influence of nutrients on phytoplankton function and structure in coastal waters. Kieler Meeresforsch, Sonderh. 6: 173–183.

Schiewer, U. , 1990. Werner Schnese and the development of coastal waters ecology in Rostock, GDR. Int. Rev. ges. Hydrobiol. 75: 1–13.

Schiewer, U. & G. Jost, 1991. The microbial food web in eutrophic shallow estuaries of the Baltic Sea. Int. Rev. ges. Hydrobiol. 76: 339–350.

Schiewer, U., R. Heerkloss, K. Gocke, G. Jost, H.-P. Spittler & R. Schumann, 1993. Experimental bottom-up influences on microbial food webs in eutrophic shallow waters of the Baltic Sea. Verh. int. Ver. Limnol. 25: 991–994.

Schiewer, U., 1994. Regulationsmechanismen und Wechselwirkungen zwischen Pelagial und Benthal. Rostock. Meeresbiolog. Beitr. 2: 179–189.

Schiewer, U., 1995. Inner coastal waters of the Baltic Sea – Ecological status and links between ecology and economic uses. In J. Köhn & U. Schiewer (eds), The Future of the Baltic Sea. Metropolis-Verlag, Marburg: 63–74.

Schiewer, U., G. Jost, K. Gocke, R. Schumann, P. Spittler & R. Heerkloss, 1997. Daily pattern of microbial communities in mesocosms. In E. Ojaveer (ed.), Proc. 14th Baltic Marine Biologists Symp., Pärnu/Estonia, 248–259.

Schlungbaum, G., U. Schiewer & E. A. Arndt, 1994. Beschaffenheitsentwicklung und Klassifizierung der Darß-Zingster Boddengewässer mit ausgewählten Vergleichen zu anderen Bodden und Haffen. Rostock. Meeresbiolog. Beitr. 2: 191–202.

Schubert, H., H. C. P. Matthijs, L. R. Mur & U. Schiewer, 1995a. Blooming of cyanobacteria in turbulent water with steep light gradients: the effect of intermittent light and dark periods on the oxygen evolution capacity of *Synechocystis* sp. PCC 6803. FEMS Microbiol. Ecol. 18: 237–245.

Schubert, H., R. M. Forster & S. Sagert, 1995b. In situ measurement of state transition in cyanobacterial blooms: kinetics and extent of the state change in relation to underwater light and vertical mixing. Mar. Ecol. Prog. Ser. 128: 99–108.

Schubert, H., 1996. Starklichtanpassung – Strategien von Grünalgen und Cyanobakterien. Nova Acta Leopoldina Suppl. 14: 109–124.

Schumann, R., 1993. Zur Rolle des Pico- und Nanophytoplanktons im mikrobiellen Nahrungsgefüge der Darß-Zingster Boddenkette. Diss. A Univ. Rostock.

Schumann, R. & U. Schiewer, 1994. Influence of abiotic induced phytoplankton changes on protozoan communities from the Darss-Zingst Bodden Chain (Germany). Mar. Microb. Food Webs 8: 265–282.

Shyam, R., A. S. Raghavendra & P. V. Sane, 1993. The role of dark respiration in photoinhibition of photosynthesis and its reactivation in a cyanobacterium *Anacystis nidulans*. Physiol. Pl. 88: 446–452.

Wasmund, N. & V. Kell, 1991. Characterization of brackish coastal waters of different trophic levels by means of phytoplankton biomass and primary production. Int. Rev. ges. Hydrobiol. 76: 361–370.

Wasmund, N. & R. Börner, 1992. Phytoplanktonentwicklung in der Darß-Zingster Boddenkette. Wasser und Boden 44: 643–647.

Winkler, H. M., 1991. Changes of structure and stock in exploited fish communities in estuaries of the Southern Baltic Coast (Mecklenburg-Vorpommern, Germany). Int. Rev. ges. Hydrobiol. 76: 413–422.

Hydrobiologia **363**: 81–95, 1998.
T. Tamminen & H. Kuosa (eds), Eutrophication in Planktonic Ecosystems: Food Web Dynamics and Elemental Cycling.

Relations between planktivorous fish abundance, zooplankton and phytoplankton in three lakes of differing productivity

Jouko Sarvala[1], Harri Helminen[2], Vesa Saarikari[1], Seppo Salonen[3] & Kristiina Vuorio[1]
[1] *University of Turku, Department of Biology, FIN-20014 Turku, Finland*
[2] *Southwest Finland Regional Environment Centre, Inkilänkatu 4, FIN-20300 Turku, Finland*
[3] *University of Turku, Satakunta Environmental Research Centre, Konttorikatu 1, FIN-28900 Pori, Finland*

Key words: phytoplankton, zooplankton, planktivorous fish, phosphorus, biomanipulation, trophic interactions

Abstract

Water chemistry, phytoplankton, zooplankton and fish populations were studied over several years in three shallow, non-stratified lakes with differing nutrient loadings and fish communities in southwest Finland. Lake Pyhäjärvi was weakly mesotrophic in 1980–1996, Lake Köyliönjärvi was highly eutrophic in 1991–1996, and Lake Littoistenjärvi was mesotrophic in 1993–1996 and eutrophic in 1992. In Lake Pyhäjärvi, natural year-class fluctuations of vendace and smelt (range of combined biomass 5–28 kg ha^{-1}) caused significant variation in planktivory. The very dense fish stocks of Lake Köyliönjärvi (mainly roach, bream and smelt) were decimated from > 175 kg ha^{-1} in 1991 to about 50 kg ha^{-1} in 1996 by removal fishing. The roach stock of Lake Littoistenjärvi declined from about 71 kg ha^{-1} to about 28 kg ha^{-1} during 1993–1996. In Lake Pyhäjärvi, strong stocks of planktivorous fish were accompanied with depressed crustacean zooplankton biomass, reduced role of calanoids and cladocerans, a low proportion of larger cladocerans (length > 0.5 mm), and a high chlorophyll level. In the lakes Littoistenjärvi and Köyliönjärvi, zooplankton was dependent on both fish and phytoplankton: in spite of dense fish stocks, a high crustacean biomass developed in a phytoplankton peak year, but it was dominated by very small cladocerans. In Lake Pyhäjärvi, late summer chlorophyll concentration was predictable from total phosphorus in water and cladoceran biomass ($r^2 = 0.68$), both factors explaining roughly similar fraction of total variation. In combined data from all three lakes, chlorophyll was almost solely dependent on total phosphorus, while the cladocerans were regulated both from below by productivity and from above by fish. Our data from Pyhäjärvi lend support to consumer regulation of late summer phytoplankton; low chlorophyll values prevailed when planktivorous fish biomass was below 15 kg ha^{-1}. In large eutrophic lakes it may be difficult to reduce fish stocks to such a low level: in Lake Köyliönjärvi, after six years of removal fishing, fish biomass still remained higher, and changes in plankton were accordingly small. Unexpectedly, in 1993–1996, phytoplankton biomass in Littoistenjärvi remained low in spite of low crustacean zooplankton biomass; submerged macrophytes probably regulated the water quality.

Introduction

Phytoplankton chlorophyll in lakes is predictable from external nutrient loading, but only to the correct order of magnitude (e.g. Jones & Lee, 1986). The remaining variation can be attributed to food web interactions and physical factors (Mazumder, 1994). According to the trophic cascade hypothesis, predation by zooplanktivorous fish determines the abundance of herbivorous zooplankton, which in turn regulates the level of phytoplankton biomass (Carpenter et al., 1985). However, the proposed role of planktivorous fish in the regulation of lower trophic levels in freshwater ecosystems has been debated for a long period, and several alternative interaction mechanisms have been suggested (e.g. DeMelo et al., 1992; Carpenter & Kitchell, 1992, 1993; Reynolds, 1994). A recent meta-analysis of published studies (Brett & Goldman, 1996) confirmed that changes in the abundance of planktivorous fish do affect both the zooplankton and the phytoplankton biomasses, as predicted by the trophic cascade hypothesis (Carpenter et al., 1985). However, most of the available information comes from experimental enclosures and ponds, and much less is known about trophic interactions in whole lakes (Brett & Goldman, 1996).

Interactions between planktivorous fish, zooplankton and phytoplankton have been studied over several years in three shallow, non-stratified lakes in SW Finland. In one of these lakes, Lake Pyhäjärvi, the late summer phytoplankton chlorophyll *a* in 1979–1992 was correlated with both the total phosphorus concentration in water and the total biomass of crustacean zooplankton, and the latter was negatively correlated with planktivorous fish abundance (Helminen & Sarvala, 1997). In the present paper, we expand our earlier analysis on Pyhäjärvi, adding new data for 1993–1996 and new details for all years. We also examine the relationships between fish, zooplankton and phytoplankton in two other lakes with different nutrient levels and fish communities. We focus here on the late summer situation, because it is then that the strongest biotic interactions are expected (Helminen & Sarvala, 1997).

Study lakes, material and methods

All three study lakes are located in southwestern Finland, in an area of generally fertile soils, flat topography and few lakes. Lake Pyhäjärvi is the largest of these lakes (Table 1), but yet relatively shallow. It is mesotrophic, but known for its clear water (Secchi depth usually 3–4 m) and productive fishery on the coregonids vendace (*Coregonus albula*) and whitefish (*C. lavaretus*) (Sarvala et al., 1984; Sarvala & Jumppanen, 1988; Helminen & Sarvala, 1997). During the last decades, Pyhäjärvi has been slowly eutrophicated, mainly due to diffuse loading from agriculture. Recently a specific Lake Pyhäjärvi Protection Fund has been organized to promote reduction of the external loading.

The nutrient and chlorophyll concentrations in Pyhäjärvi were monitored during 1980–1992 by the Water Protection Association of SW Finland and in 1993–1996 by the Southwest Finland Regional Environment Centre, taking vertical profiles of usually 0–8 m with a Ruttner sampler (in 1993–1996 with a 0.5-m high Limnos sampler) in the central deep of the lake, 6–8 times during the open water period in 1980-1991, and at two-week intervals in more recent years. Owing to the openness of the lake, there is no permanent stratification during the summer, and the water is both horizontally and vertically homogeneous with regard to nutrient and plankton concentrations (Sarvala & Jumppanen, 1988). Phytoplankton was sampled along with nutrients, and counted with an inverted microscope. Zooplankton was sampled throughout the summer months in 1980 (Vuorinen & Nevalainen, 1981), in July-August 1982, in June and August 1986, and through the open water season in 1984 and 1987–1996, usually at weekly intervals (in 1984 partly daily). From 1984 onwards, samples were taken from surface to bottom with a 1-m high tube sampler (volume 6.8 l) at ten locations, selected with a stratified random design. Samples were concentrated with a 25 or 50 μm mesh net, and combined in the laboratory to form one composite sample each date. Using an inverted microscope, crustacean zooplankton was identified and counted from subsamples until 50–200 individuals of each dominant species had been measured. Length measurements were converted to carbon biomass using carbon to length regressions (as in Sarvala et al., 1988). Because of different sampling design and lacking length measurements, zooplankton data for 1980 and 1982 were not included in the statistical analyses but are shown in Figure 3.

The declining catch per unit effort in the seine net fishery during each winter enabled estimates of the year-class strength and population biomass of the major planktivore, vendace, for the period 1970–1996 (DeLury method, Helminen et al., 1993b and unpublished). In 1980-1996, the 0+ vendace accounted for most of the total vendace biomass and food consumption in late summer (e.g. Helminen et al., 1990; Helminen & Sarvala, 1994b); therefore, we used the 0+ autumn biomass here. Judging from the annual catches recorded by the local fisheries management association, the smelt (*Osmerus eperlanus*) stock was strong in the early 1980s (peak year 1983; Sarvala et al., 1994), but exact population data are lacking. In 1986–1992 the smelt stock was sparse, but the year-class 1993 was very strong (Karjalainen et al., 1997), and its development could be followed up to the winter 1996–1997 from winter seine catch samples (n = 10 in 1994–1995, n = 28 in 1995–1996). Because smelt catches in winter 1996–1997 were negligible, a reasonable estimate for smelt biomass in autumn 1995 was obtained directly from the total smelt catch in the winter 1995–1996. Smelt biomass in autumn 1994 was similarly obtained from the catch in winter 1994-1995; the back-calculated biomass of smelt caught in 1995–1996 was added, assuming a 40% mortality from autumn 1994 to autumn 1995. Smelt biomass in 1993

Table 1. Some morphometric, hydrological and chemical characteristics of the study lakes (for the period 1980–1996 in Pyhäjärvi, for 1992–1996 in Köyliönjärvi, and for 1983–1996 in Littoistenjärvi)

		Pyhä-järvi	Köyliön-järvi	Littoisten-järvi
Drainage area (excluding the lake)	km^2	461	129	3.0
Lake area	km^2	154	12.5	1.5
Volume	10^6 m^3	840	38.7	3.25
Mean depth	m	5.4	3.1	2.2
Maximum depth	m	25	13	3.0
Mean retention time of water	a	3.2	1.0	1.8
Total phosphorus, range of annual means for the open water season	mg m^{-3}	11–21	79–96	20–53
Total nitrogen, range of annual means for the open water season	mg m^{-3}	330–560	1090–1270	370–1040
Chlorophyll *a*, range of annual means for the open water season	mg m^{-3}	4–9	42–57	4-39

was then calculated backwards from the cohort numbers in autumn 1994, assuming an 81% annual mortality. The mortality rates used correspond to those reported for smelt from Lake Sjamozero in Russian Karelia, closely resembling Lake Pyhäjärvi in many respects (Reshetnikov et al., 1982), and to those found for similar-sized vendace juveniles in Lake Pyhäjärvi in the 1990s (Helminen & Sarvala, unpublished manuscript). Relative strength of perch (*Perca fluviatilis*) year-classes was assessed for the years 1986–1993 from the catch samples (Sarvala & Helminen, 1996). Whitefish year-class variation is currently being evaluated, but the numbers were much lower than those of vendace. Due to the inadequacies of the data from the early 1980s, statistical analyses of the fish-zooplankton relationships were restricted to the years 1984 and 1986–1996.

Lake Köyliönjärvi is highly eutrophic (Table 1; late summer maxima of total phosphorus up to 170 mg m^{-3} and of chlorophyll *a* up to 180 mg m^{-3}), mainly because of diffuse nutrient runoff from agriculture. Extensive cyanobacterial blooms occur in late summer, and transparency is poor (late summer Secchi depth 0.3–0.5 m; Sarvala et al., 1995). A restoration project has been established to cut down the external loading. To achieve a more rapid improvement of water quality, food web manipulation has also been initiated, and for this purpose the fish stocks (mainly roach (*Rutilus rutilus*), bream (*Abramis brama*) and smelt) have been reduced through removal fishery since 1992 (Hirvonen & Salonen, 1995; Salonen et al., 1996).

Nutrients, chlorophyll, phytoplankton and zooplankton in Köyliönjärvi were monitored in 1991–1996, mostly from weekly samples (twice a week in 1992). Three water columns from the surface to bottom were taken with a 6.8-litre tube sampler at three sites both in the southern and northern basin of the lake and combined into a single composite sample for each basin and date. Nutrient and chlorophyll analyses were made in the laboratory of the Southwest Finland Regional Environment Centre. Plankton analyses were performed in the University of Turku with similar methods as for Pyhäjärvi.

The fish community of Köyliönjärvi was studied from two-stage catch samples (procedure described in Salonen et al., 1996) taken from 227 (74.9%) of the total of 303 seine net hauls during 1992-1996; the samples covered 93.8% of the total winter catch of 319900 kg. During the winter fishing seasons of 1993 and 1996, there was a significant decrease in the catch per seine net haul, allowing estimates of the total catchable fish stock with the removal (DeLury) method (for details of the method, see Helminen et al., 1993a). From the age and size distribution of roach during the first fishing winter, a production to biomass ratio (P/B) of 0.24 a^{-1} was calculated. It was further assumed that the P/B ratio increased with the reduction of the fish stock by 0.05 units each year, until a P/B of 0.50 was achieved. This change of P/B corresponds to observations on a heavily exploited roach stock in Lake Syrjänalunen in the southern Finland (Sarvala et al., 1992 and unpublished). Using the estimated fish stock in autumn 1995, the mentioned P/B ratios, and the removal fishery catches, the development of the fish stock was reconstructed backwards until the autumn 1991. The estimated fish stock obtained in this way for autumn 1992 (201096 kg) was in good agreement with the estimate obtained with the DeLury method from the winter fishery in 1993 (200000 kg). Another estimate for the fish stock development was obtained by regressing the cumulative catch from the southern basin against the log-transformed number of seine hauls each winter, and calculating the expected catch after 31 hauls (the smallest number of seine hauls in any year from the southern basin). This method effectively smoothes the random variation of the catches; all regressions were highly significant.

Lake Littoistenjärvi is small and shallow. The lake oscillates with a 5–6 year period between a short turbid phase lasting one summer, and a longer clear-water phase (Sarvala & Perttula, 1994; Mäkinen et al., unpubl. manuscr.). In the clear-water years the lake is weakly mesotrophic (total P 15–30 mg m^{-3}, total N 300–500 mg m^{-3}, chlorophyll *a* 1–9 mg m^{-3}), but in the turbid phase it is clearly eutrophic (total P 25–75, total N 500–2100 mg m^{-3}, chlorophyll *a* 13–120 mg m^{-3}). In the turbid phase, transparency of water is < 1 m, in the clear phase > 2.5 m (often down to the bottom in the whole lake). The most abundant fish species are roach and perch, and also pike (*Esox lucius*).

Since 1970, Littoistenjärvi has been used as a municipal water source, and since the early 1980s, practically all of the outflow has been used for tap water production. In the 1960s and 1970s, chemical water quality data on the lake were collected at varying intervals by the water treatment plant and the regional water administration (now Southwest Finland Regional Environment Centre). In the 1980s, the number of analyses increased, and since 1992, samples were regularly collected weekly or twice a month throughout each open water season. In 1983, phytoplankton, chlorophyll and zooplankton were sampled weekly or twice a week (Sainio, 1985). Plankton samples were also available from 1988 and 1989, and beginning from 1992, sampling covered the open water season usually at weekly intervals (twice a week in 1992). Each sample consisted of two or three water columns taken with a 1-m or 0.5-m high tube sampler (6.8 or 3.5 l) from the surface to bottom in different parts of the central open area, combined to form one composite plankton sample for each date. Plankton samples were analysed in the University of Turku with similar methods as for Pyhäjärvi.

The stock sizes of roach and perch in Littoistenjärvi were estimated in 1993 and 1994 with mark-recapture methods (Kurkilahti & Rask, 1996, and M. Kurkilahti, unpublished report: in 1993 about 8000 roach and 1000 perch were marked and released), with parallel test fishing using so-called Nordic survey nets (height 1.5 m, length 30 m, each net contains 12 mesh sizes from 5 to 55 mm, knot to knot). Similar test fishing, using a stratified random design, was repeated in later years, and the average catch per net compared to the year 1994 was used to derive fish stock estimates for the years 1995–1996.

Our study lakes are normally covered by ice from early November to early May or for about half of the year. Because of the long and dark winter, the plankton development starts every spring from very low biomass levels irrespective of the trophic level of the lake (Heinonen, 1982), and differences develop gradually during the summer. Due to the low zooplankton biomass, phytoplankton growth and succession in spring are more affected by variable weather conditions, while by late summer the biotic interactions become more prominent. Helminen & Sarvala (1997) showed that in Lake Pyhäjärvi the average chlorophyll-phosphorus relationship was similar for the whole open-water period and for August alone, but the chlorophyll variation was much wider in August, improving the resolution of statistical analyses. Therefore, and because planktivorous fish predation on zooplankton in our study lakes is maximal in late summer (Helminen & Sarvala, 1997), we use here mean values for the late summer (26 July – 15 September) for the nutrient and plankton variables, and annual estimates of fish year-class strength or biomass. This seven-week period normally wholly covers the late summer peaks of both phytoplankton and zooplankton in our study lakes, and makes our analysis less sensitive to variation in the timing of samples and plankton succession in different years. More detailed analyses covering the whole seasonal development will be dealt with in a later paper.

Results

In Lake Köyliönjärvi, the late summer chlorophyll levels relative to total phosphorus concentration corresponded to predictions from Mazumder (1994) for non-stratified lakes lacking large herbivorous zooplankton (Figure 1). Phytoplankton was very diverse ($>$ 650 taxa identified), diatoms and cyanobacteria being the most important groups. In Lake Pyhäjärvi,

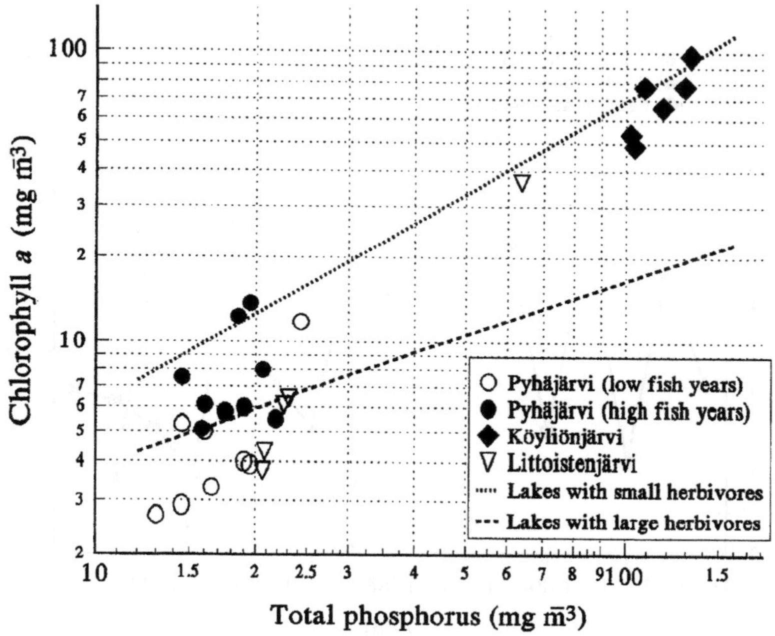

Figure 1. Late summer chlorophyll *a vs.* total phosphorus in the lakes Pyhäjärvi (1986–1996), Köyliönjärvi (1991–1996) and Littoistenjärvi (1992–1996). Each symbol corresponds to an annual average for the period 26 July – 15 September. For Pyhäjärvi, years with lower than average planktivorous fish biomass are distinguished from those with higher fish biomass. Broken lines denote the regressions from Mazumder (1994) for non-stratified lakes with large or small herbivores dominant in zooplankton.

chlorophyll levels relative to total phosphorus varied from low to high, low values appearing in years with small stocks of planktivorous fish. Characteristic phytoplankton groups were diatoms, chrysophyceans, cryptophyceans and, especially during the 1990s, cyanobacteria. Lake Littoistenjärvi showed two alternative states: in 1993–1996 water was clear and chlorophyll levels low relative to total phosphorus, but high nutrient and chlorophyll levels and a high chlorophyll to total phosphorus ratio prevailed in 1992. The turbid water state coincided with a collapse of submerged vegetation, which was moderately abundant in the clear-water years (Mäkinen et al., unpubl. manuscript). Phytoplankton composition was very variable from year to year, the main groups being chrysophyceans, diatoms, chlorophyceans, cryptophyceans, and in some years cyanobacteria.

In Lake Pyhäjärvi, planktivorous fish abundance seemed to control crustacean zooplankton biomass and the size distribution of cladocerans, although an intensive fishery on vendace and whitefish kept the fish stocks relatively sparse (vendace stock biomass variation in 1979–1996 about 5–25 kg ha^{-1}). Major fluctuations of year-class strength were caused by variable weather conditions (vendace: Helminen & Sarvala, 1994a and unpublished manuscript; perch: Sarvala & Helminen, 1996). Vendace biomass was higher than average in 1981, 1983, 1985, 1986, 1988, 1989 and 1992 (Figure 2). In 1993 and 1994, when the vendace stock was weak, the biomass of the strong 1993 smelt year-class was high; its rapid decline was due to a subsidized fishery focused on smelt and other less valuable fish in the winters 1994–1995 – 1996–1997. In 1986–1992 and 1996 the smelt stock was so weak that its biomass can be ignored here. Perch year-classes of 1986, 1988 and 1992 were strong (Sarvala & Helminen, 1996), accentuating thus the planktivory variation due to vendace. However, even in these years perch biomass in the planktivorous juvenile stage seemed to remain small compared to vendace (some kg ha^{-1}) and was not included in the statistical analyses.

In the years with abundant planktivorous fish (Figure 2), the total biomass of crustacean zooplankton was depressed (Figure 3), and most of the cladocerans were less than 0.5 mm long (Figure 4). Consequently, the grazing pressure by zooplankton diminished, making possible an increase in hytoplankton chlorophyll (Figure 3). In Pyhäjärvi, cladoceran biomass differences between years explained most of the variation in the chlorophyll-phosphorus relationship in Figure 1. Late summer phytoplankton chlorophyll level showed significant negative correlation with the biomass of

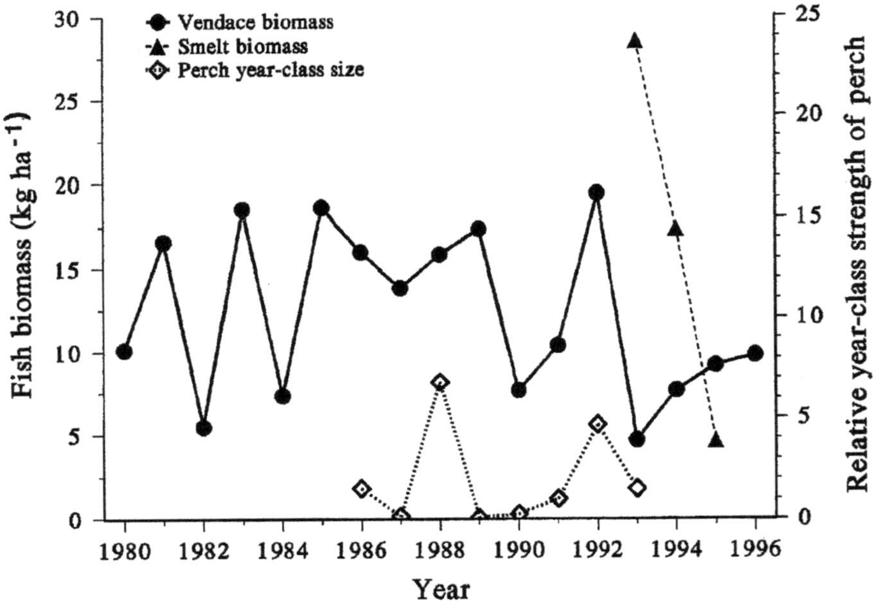

Figure 2. The autumn biomass of the main planktivorous fish in Lake Pyhäjärvi. Vendace biomass for 1980–1996 based on year-class estimates from Helminen et al. (1993b and unpublished); perch year-class strength for 1986–1993 from Sarvala & Helminen (1996). For smelt biomass, see text; smelt was also abundant around the year 1983.

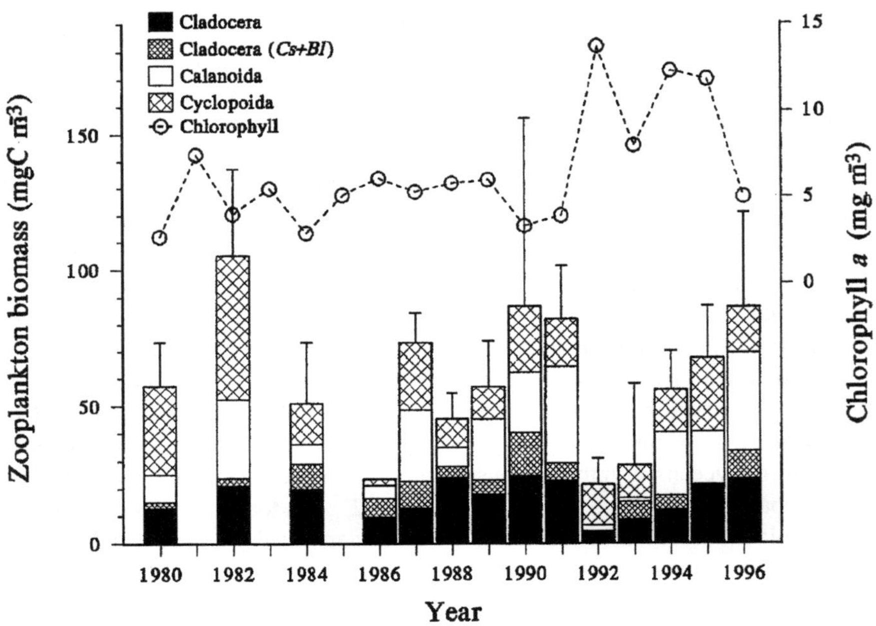

Figure 3. Average crustacean zooplankton biomass (Cladocera, Calanoida and Cyclopoida) and phytoplankton chlorophyll *a* for the period 26 July – 15 September in Lake Pyhäjärvi in 1980–1996 (no zooplankton samples for 1981, 1983 and 1985). Within Cladocera, the contribution of *Chydorus sphaericus* and *Bosmina longirostris* (*Cs + Bl*) is shown separately. Vertical bars denote one standard deviation of the total crustacean biomass.

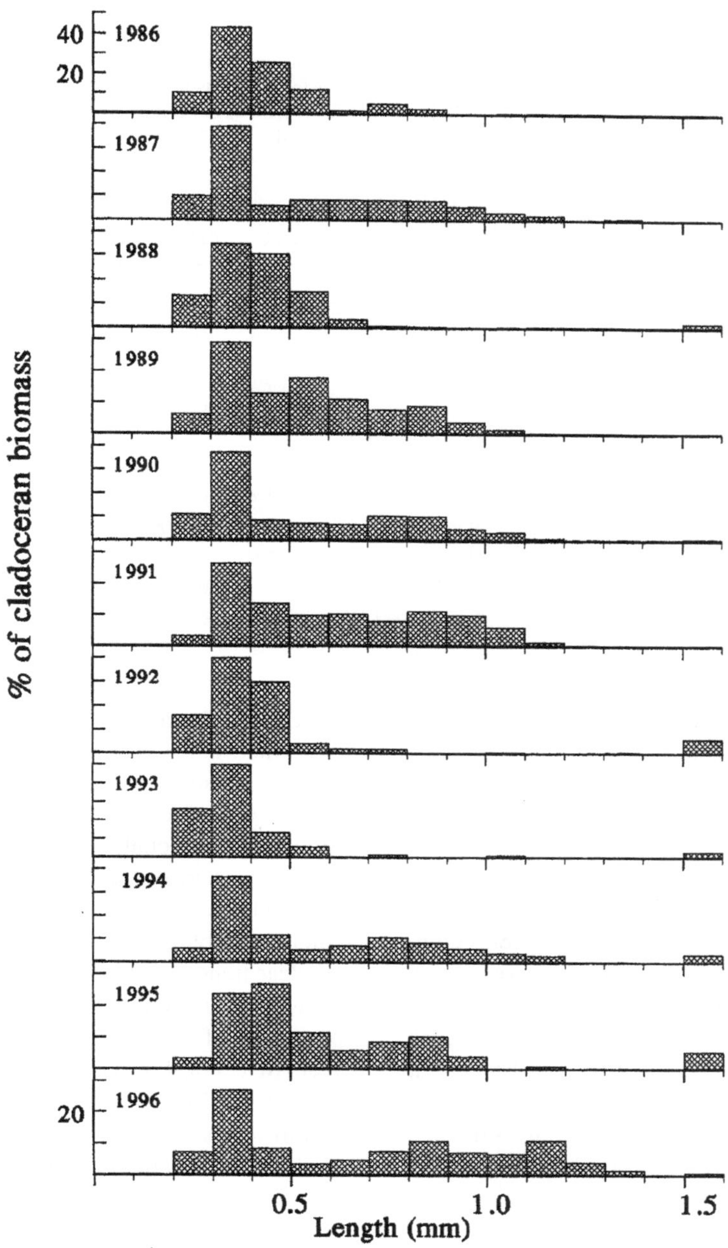

Figure 4. Size frequency distribution of late summer cladoceran zooplankton in Lake Pyhäjärvi in 1986–1996 (averages for the period 26 July – 15 September).

total cladocerans, larger cladocerans, small cladocerans and herbivorous crustaceans (Table 2). Correlation between chlorophyll and fish biomass was positive; the positive correlation with total phosphorus and the negative correlation with total crustacean zooplankton biomass did not quite reach the 0.05 significance level in this time series. However, multiple regression analyses clarified that phytoplankton chlorophyll was predictable from total phosphorus in water and cladoceran biomass (log(chlorophyll) = −0.42 + 1.54* log(total phosphorus) − 0.54* log(cladoceran biomass); adjusted $r^2 = 0.68$, df = 9, $P = 0.0024$; both independent variables significant), both factors explaining roughly similar fractions of the total variation (36.8 and 31.4%, respectively). Further, the negative relationship between small cladocerans and chlorophyll disap-

Table 2. Simple correlations between planktivorous fish biomass, zooplankton variables, temperature, total phosphorus and phytoplankton chlorophyll in Lake Pyhäjärvi in late summer (26 July – 15 September) in 1984–1996 (for CQ2 and CQ3 in 1986–1996). All values were $\log_{10}$-transformed to stabilize variance and to account for the non-linearity of the relationships. FISH = planktivorous fish biomass (B); ZPL = total crustacean B; CLA = total cladoceran B; LCL = herbivorous cladoceran B without SCL; SCL = small cladoceran (*Chydorus sphaericus* + *Bosmina longirostris*) B; CAL = calanoid B; CYC = cyclopoid B; HER = herbivorous crustacean B (Cladocera + the calanoid *Eudiaptomus*); CQ2 = biomass-weighted median of cladoceran length distribution; CQ3 = biomass-weighted third quartile of cladoceran length distribution; TEMP = mean water temperature for June-August; TOTP = total phosphorus concentration; CHL = chlorophyll *a* concentration ($^{1} = P < 0.05$; $^{2} = P < 0.01$)

	FISH	ZPL	CLA	LCL	SCL	CAL	CYC	HER	CQ2	CQ3	TEMP	TOTP
ZPL	-0.718^{1}											
CLA	-0.651^{1}	0.808^{2}										
LCL	-0.675^{1}	0.830^{2}	0.947^{2}									
SCL	−0.189	0.249	0.507	0.254								
CAL	-0.683^{1}	0.918^{2}	0.708^{1}	0.742^{2}	0.233							
CYC	−0.340	0.675^{1}	0.228	0.311	−0.204	0.461						
HER	-0.675^{1}	0.932^{2}	0.935^{2}	0.911^{2}	0.431	0.905^{2}	0.377					
CQ2	−0.550	0.691^{1}	0.482	0.587	0.085	0.787^{2}	0.296	0.675^{1}				
CQ3	-0.675^{1}	0.904^{2}	0.632^{1}	0.661^{1}	0.188	0.961^{2}	0.527	0.850^{2}	0.847^{2}			
TEMP	0.003	−0.303	−0.152	0.047	−0.357	−0.190	−0.202	−0.229	0.156	−0.256		
TOTP	0.316	−0.143	−0.279	−0.066	-0.752^{2}	−0.232	0.231	−0.261	−0.043	−0.180	−0.195	
CHL	0.695^{1}	−0.528	-0.758^{2}	-0.625^{1}	-0.728^{1}	−0.447	−0.005	-0.650^{1}	−0.281	−0.357	0.142	0.564

peared when phosphorus was included in the regression. The zooplankton-chlorophyll correlations were mainly due to the larger Cladocera.

On the other hand, cladoceran biomass was predictable from the biomass of planktivorous fish (cladoceran biomass = 40.5–1.13*(planktivorous fish biomass); adjusted $r^2 = 0.51$, df = 10, $P = 0.006$). Fish abundance also influenced zooplankton species composition and especially size structure; however, cladoceran size was not a good predictor of chlorophyll (Table 2). The size distribution of the total crustacean zooplankton was even less informative, because it was dependent on the abundance of cyclopoids and calanoids, which were weakly or not at all correlated with water quality and fish (Table 2). The biomass-weighted third quartile of the cladoceran length distribution was sensitive to the abundance of the larger (> 0.5 mm long) cladocerans, and it was also more tightly correlated with fish biomass than the median. All correlations with summer temperature were very weak.

The fish stocks of Lake Köyliönjärvi were very dense in autumn 1991 (> 175 kg ha^{-1}; mainly roach, bream and smelt), but were by 1996 decimated to about 50 kg ha^{-1} by removal fishing; both estimates of the decline were in good agreement (Figure 5). Decreasing fish biomass in Köyliönjärvi led to a slight increase of the biomass of larger cladocerans (Figure 6); the high peak in 1993 was mainly due to *Chydorus sphaericus* which is a very small species and less affected by fish predation than larger cladocerans. The cladocerans were clearly smaller than in Lake Pyhäjärvi (Figure 7), although the last two years showed an increasing share of *Daphnia cucullata* > 0.5 mm long. Small cladoceran size was consistent with the high chlorophyll-phosphorus ratios (Figure 1). There were no clear trends in chlorophyll values (Figure 6) or in phytoplankton composition, except a decrease in the proportion of cyanobacteria from the premanipulation years (data not shown here).

In Lake Littoistenjärvi, in spite of higher phosphorus levels, the late summer crustacean zooplankton biomass was much lower than in Lake Pyhäjärvi; even the peak values in 1992 with exceptionally high phytoplankton abundance were only somewhat higher (Figure 8). The roach stock in Lake Littoistenjärvi was moderately strong (71 kg ha^{-1}) after the high abundance of submerged macrophytes in the early 1990s, declining to about 28 kg ha^{-1} by 1996 (Figure 8). Perch biomass in Littoistenjärvi was notable, but consisted mainly of large (> 20 cm) individuals that were probably piscivorous. In 1994–1996, the total fish biomass did not decrease, but there was a clear shift in the size distribution towards larger fish; especially small roach were almost absent in 1995–1996. This size shift probably led to a lowered predation

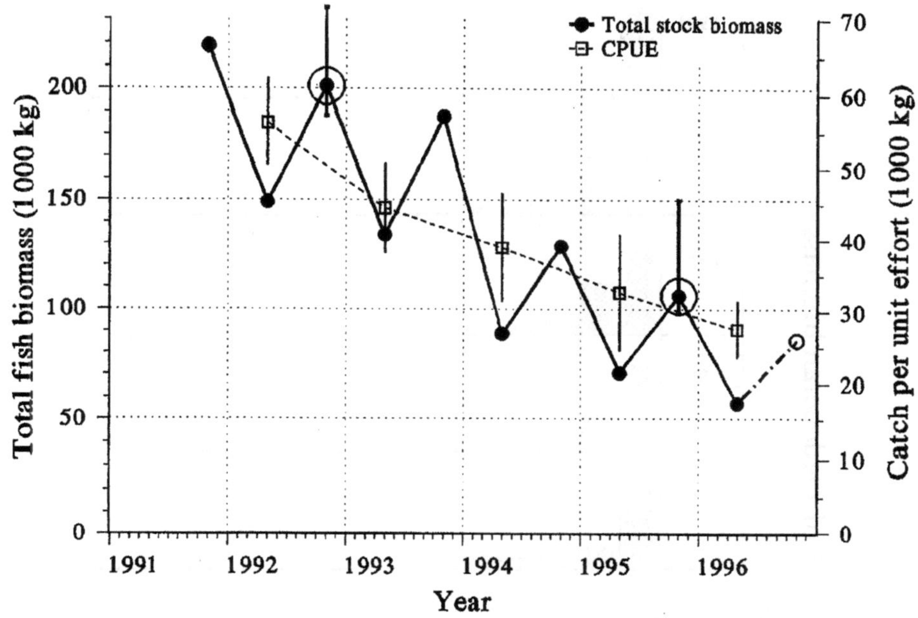

Figure 5. Estimated development of the total fish stock biomass in Lake Köyliönjärvi from autumn 1991 to spring 1996 (dots and solid line; left scale); vertical bars show the 95% confidence limits of the direct De Lury estimates for the autumns 1992 and 1995 (see text). Winter declines show the total catches of the removal fishery, and summer increases correspond to estimated production. Cumulative catch per standardized fishing effort (31 seine net hauls) is also shown for each winter (open squares and dashed line, vertical bars give the standard error; right scale).

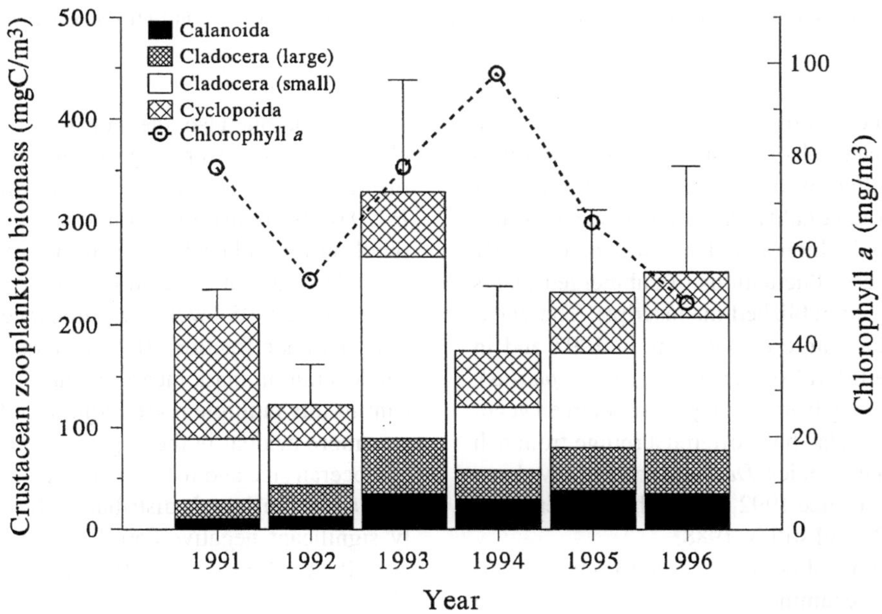

Figure 6. Average crustacean zooplankton biomass (Cladocera, Calanoida and Cyclopoida) and phytoplankton chlorophyll *a* for the period 26 July – 15 September in Lake Köyliönjärvi in 1991–1996. Other explanations as in Figure 3.

pressure on zooplankton, and the cladoceran biomass showed some increase in 1994–1996. The size distributions of cladocerans were very variable (Figure 9). In 1992, with very abundant phytoplankton, cladocerans were very small (like those in Köyliönjärvi); however, small cladocerans also prevailed in 1994–

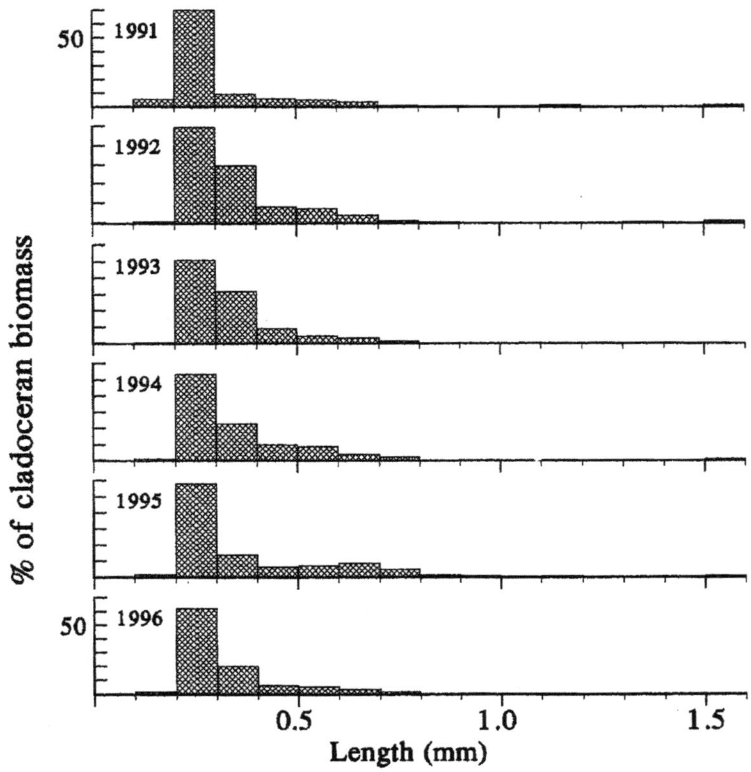

Figure 7. Size frequency distribution of late summer cladoceran zooplankton in Lake Köyliönjärvi in 1991–1996 (averages for the period 26 July–15 September).

1995, but yet the chlorophyll level remained low. In Lake Littoistenjärvi, cladoceran zooplankton was thus probably controlled by fish predation as in the other two lakes, but water quality was not dependent on zooplankton. Phytoplankton abundance was more tightly linked to long-term fluctuations of submerged plants (Mäkinen et al., unpublished manuscript). The unexpectedly abundant large cladocerans in 1993 and in later years were littoral species (*Sida*, *Simocephalus*), probably originating from among the submerged vegetation, which may also have offered a refuge from fish predation for other species. *Daphnia* species have been practically absent since 1992, although they were very abundant in 1983 and still in 1988.

The interactions of bottom-up and top-down regulation were also examined using the combined data from all three lakes. All simple correlations (log-transformed data) between the zooplankton biomass variables and fish, chlorophyll and phosphorus were positive and significant (except that between fish and calanoid copepods), showing the importance of bottom-up regulation. Chlorophyll was tightly correlated with total phosphorus ($r = 0.95$, df $= 20$, $P < 0.001$), but also the negative correlations with cladoceran size were significant ($r = -0.76$ for the median, $r = -0.68$ for the third quartile, df $= 19$, $P < 0.01$ for both). In multiple regressions of chlorophyll on total phosphorus and one of the cladoceran variables, both of the cladoceran size measures were close to the 0.05 level of significance, but explained only a minor fraction (0.9%) of the total variation. Cladoceran biomass measures did not appear significant in this multiple regression. Top-down regulation was more evident in the negative correlation between cladoceran size and fish: the median and third quartile of cladoceran length distribution both showed highly significant negative correlation with fish biomass ($r = -0.52$, df $= 19$, $P < 0.02$, and $r = -0.57$, df $= 19$, $P < 0.01$). In a multiple regression of cladoceran biomass on fish biomass and chlorophyll, the fish entered with a negative coefficient, which was significant for the regression of large cladocerans and almost significant for total cladoceran biomass. The use of chlorophyll as a measure of food availability in this analysis is problematic, however, because zooplankton itself regulates the chlorophyll levels, as shown above for Lake

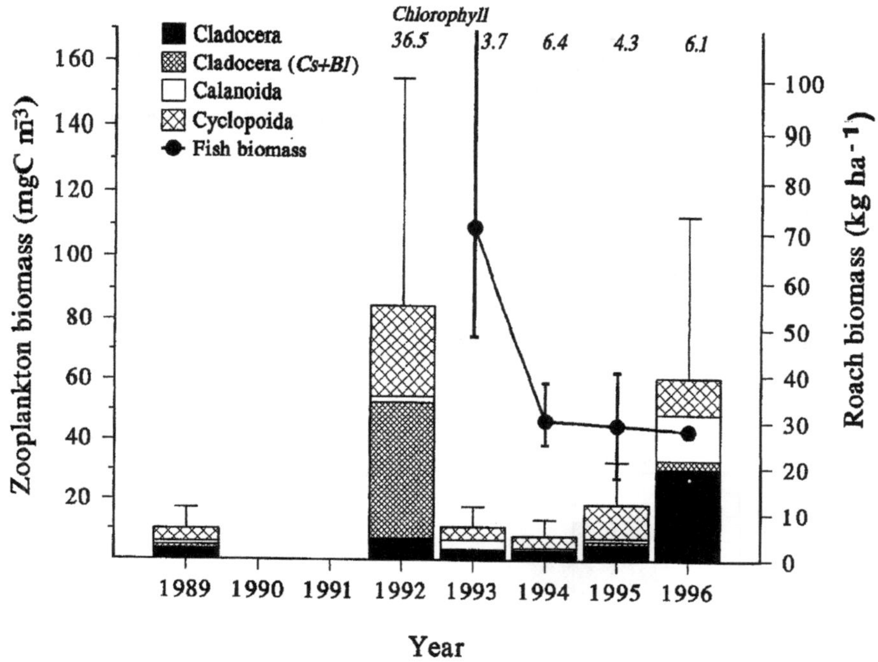

Figure 8. Average crustacean zooplankton biomass (Cladocera, Calanoida and Cyclopoida) for the period 26 July–15 September in Lake Littoistenjärvi in 1989–1996 (no samples for 1990 and 1991). Roach biomass estimates are shown for 1993–1996; vertical bars denote the 95% confidence limits for the estimate (in 1993–1994 from mark-recapture analysis by M. Kurkilahti (Dept. Biol., Univ. Turku), in 1995 from catch-per-unit-effort in experimental gill net catches). Phytoplankton chlorophyll *a* for 1992–1996 shown above each year; no chlorophyll data for late summer in 1989, but the transparency of water was equal to that in 1993. Other explanations as in Figure 3.

Pyhäjärvi. Therefore, it may be better to use total phosphorus concentration as an indicator of productivity; this is justified by the tight general relationship between phytoplankton chlorophyll and total phosphorus in lakes (e.g. Mazumder, 1994). A multiple regression of cladoceran biomass on fish biomass and total phosphorus was highly significant: log(cladoceran biomass) = –0.23 + 0.91* log(total phosphorus) – 0.43 * log(fish biomass); adjusted $r^2 = 0.709$, df = 21, $P = 0.0000$; both independent variables highly significant). The bottom-up regulation was stronger here: the impact of fish was slightly less than one third of that of the total phosphorus. We must remember, however, that our data did not include any case with a high phosphorus level combined with low fish biomass (an unlikely situation except in biomanipulated lakes). Although the exact figures are thus certainly sensitive to further data points, the analysis confirms that both bottom-up and top-down forces must be considered when trying to understand zooplankton biomass variation.

Discussion

In Lake Pyhäjärvi, fish abundance markedly influenced zooplankton biomass, species composition and especially size structure. Strong stocks of planktivorous fish were usually accompanied with depressed zooplankton biomass, reduced role of calanoids and cladocerans, and a low proportion of larger cladocerans (length >0.5 mm). These observation agree well with paleolimnological data from Lake Pyhäjärvi, which show a reduction in the mean size of the cladoceran *Bosmina* following the introduction of planktivorous coregonids into the lake in the early 1900s (Salo et al., 1989; Räsänen et al., 1992). However, in the lakes Littoistenjärvi and Köyliönjärvi also bottom-up effects on zooplankton were seen in years of extremely high chlorophyll levels, when a high crustacean biomass developed in spite of dense fish stocks (in the years 1992 and 1993, respectively). In these cases the top-down regulation was apparent in the scarcity of the medium-sized cladocerans. The significance of both bottom-up and top-down forces in regulating zooplankton was also evident in a statistical analysis of the combined data from all three lakes.

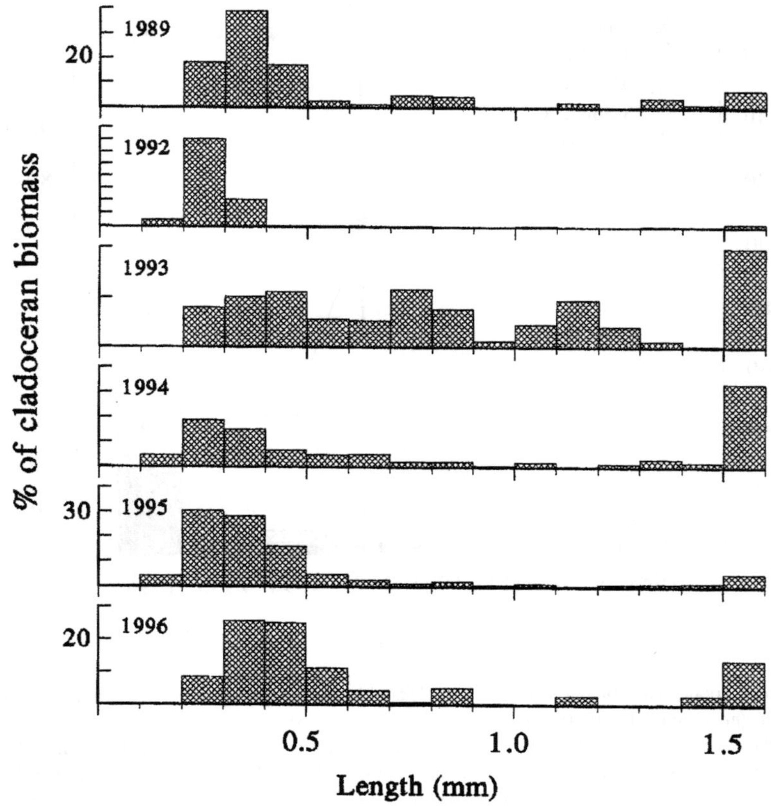

Figure 9. Size frequency distribution of late summer cladoceran zooplankton in Lake Littoistenjärvi in 1989–1996 (averages for the period 26 July – 15 September).

In Lake Pyhäjärvi, where the time series was long enough for a statistical analysis, late summer phytoplankton chlorophyll showed positive correlation with both total phosphorus in water and planktivorous fish abundance, and negative correlation with cladoceran biomass. Thus the chlorophyll-to-phosphorus relationship varied according to the abundance of cladoceran zooplankton, and a similar share of the between-year variation in chlorophyll was explained by cladoceran biomass differences as by total phosphorus. The significant influence of crustacean zooplankton on chlorophyll, found earlier in Pyhäjärvi (Helminen & Sarvala, 1997), was now shown to be mainly due to the cladocerans. On the other hand, in our data set the cladoceran size had much weaker connection to phytoplankton abundance than usually reported (e.g. Mazumder, 1994; Carpenter et al., 1996), the total biomass being more decisive. This is in line with the fact that in all of our study lakes even the larger cladocerans were small (i.e. < 1.2 mm long) according to usual standards (e.g. Mazumder, 1994). Yet the cladoceran size covaried with cladoceran biomass, and thus a significant correlation with phytoplankton would be expected in a larger data set.

Enclosure and pond experiments have earlier shown that planktivorous fish abundance regulates crustacean zooplankton abundance and size distribution (Brett & Goldman, 1996), and there is also evidence that natural fish stock variation in lakes similarly affects zooplankton (Mills & Schiavone, 1982; Gophen et al., 1992; Karjalainen & Viljanen, 1993). However, usually it has not been possible to link the long-term fish stock changes in lakes with phytoplankton abundance variation (McQueen et al., 1992; Ramcharan et al., 1995). In Lake Pyhäjärvi, in contrast, the effects of natural year-class fluctuations of fish cascaded down the food-web and were detectable at the phytoplankton level in late summer.

Temperature is known to be one important factor regulating plankton communities (e.g. Adrian & Deneke, 1996). However, in our data set from Lake Pyhäjärvi, summer temperature was not correlated with any of the plankton variables studied. This is not surprising considering that we examined only mean values from late summer, when the between-year variation of temperature in our study lakes is minimal; the largest temperature differences are found in spring (own unpublished data). However, from earlier studies we know that both spring and summer temperature conditions in several ways affect the year-class variation of vendace and perch (Helminen & Sarvala, 1994a; Sarvala & Helminen, 1996), and hence indirectly zooplankton and phytoplankton. Climatic fluctuations thus ultimately control the food web structure, and, through the top-down regulation, water quality in Lake Pyhäjärvi.

In Pyhäjärvi, the calculated rate of phosphorus recycling by zooplankton changed relatively little between years, but the calculated grazing rates were sufficient to control phytoplankton (Helminen & Sarvala, 1997). Thus, the planktivorous fish regulated phytoplankton primarily through cladoceran grazing. Similar mechanisms seemed to apply to Köyliönjärvi, although there the changes induced by the removal fishing were still small.

Our data from Pyhäjärvi thus lend support to consumer regulation of productivity; low chlorophyll values prevailed when planktivorous fish biomass was below 15 kg ha^{-1}. It is noteworthy that so clear top-down effects were observed at such a low fish biomass level. The major planktivore in Pyhäjärvi, vendace, is the most specialized and apparently very efficient zooplankton feeder in northern Europe. Moreover, as hypothesized by Sarvala et al. (1984), in shallow, unstratified lakes, such as our study lakes, zooplankton have fewer possibilities to escape fish predation than in deeper, stratified lakes. Mills & Forney (1983) also reported that 20 kg ha^{-1} of young yellow perch were enough to decimate the *Daphnia* population in Lake Oneida. It may be difficult – but not impossible – to reduce fish populations below this level in larger eutrophic lakes. Our experiences from Lake Köyliönjärvi also show that the biomass level of 50 kg ha^{-1} was still too high for notable water quality changes. Lake managers contemplating biomanipulation through fish removal in large eutrophic lakes should be aware from the beginning that a very intense fishing effort will be required. Fish community composition and species characteristics are likely to be important in the trophic interactions, and similar considerations probably apply to the lower trophic levels, as suggested for zooplankton by Hessen et al. (1995).

In Littoistenjärvi, in contrast, the zooplankton changes were not reflected to the phytoplankton level: in 1993–1996, phytoplankton biomass remained low in spite of very low crustacean zooplankton biomass during most of the summer. The most likely cause of the suppressed development of phytoplankton in Littoistenjärvi was a moderately high abundance of submerged macrophytes (Mäkinen et al., unpubl. manuscript). Similar dominant role of submerged plants has been earlier reported in some other lake ecosystems (Blindow et al., 1993; Schriver et al., 1995).

In Lake Pyhäjärvi, we could only detect top-down regulation of the cladoceran biomass, while in the combined data of our three study lakes, both a strong bottom-up regulation and weaker top-down effects were discernible. In contrast, in Lake Pyhäjärvi, the bottom-up and top-down effects on chlorophyll were roughly equal, while over all three lakes the bottom-up regulation was dominant. Our perception of, and ability to evaluate the relative importance of the top-down and bottom-up mechanisms are clearly dependent on the observed range of the study variables. From chlorophyll-phosphorus plots, such as published by Jones & Lee (1986), or Figure 1 of the present paper, it is easy to see that over the possible range of nutrient input levels, the bottom-up mechanisms are quantitatively more important than the top-down control potential. This was also recently the conclusion of Carpenter et al. (1996). However, even this limited potential is large enough to predict important improvements of water quality through food web manipulation even in eutrophic lakes.

Acknowledgements

Studies on the lakes Pyhäjärvi and Köyliönjärvi were mainly financed by the Academy of Finland (research contract 09/109, research grants 09/074, 1071004, 1071149, 1071255, 1071292, 4158 and 35619), with additional funding from the Foundation for the Study of Natural Resources in Finland and local sources. Studies on Lake Littoistenjärvi were funded by the company Littoistenjärven säännöstely-yhtiö as well as the municipalities of Kaarina and Lieto. Long-term fruitful cooperation with the Southwest Finland Regional Environment Centre, Southwest Finland Water Protection Association, Southwest Finland Fisheries District, Pyhäjärvi Fisheries Management Association, Pyhäjärvi Institute and the Lake Pyhäjärvi Protection Fund, and the Köyliönjärvi Restoration Project is gratefully acknowledged. Particular thanks go to our long-term collaborator Arto Hirvonen, and many other persons who contributed to data collection. Comments by two anonymous referees helped to improve the presentation.

References

Adrian, R. & R. Deneke, 1996. Possible impact of mild winters on zooplankton succession in eutrophic lakes of the Atlantic European area. Freshwat. Biol. 36: 757–770.

Blindow, I., G. Andersson, A. Hargeby & S. Johansson, 1993. Long-term pattern of alternative stable states in two shallow eutrophic lakes. Freshwat. Biol. 30: 159–167.

Brett, M. T. & C. R. Goldman, 1996. A meta-analysis of the freshwater trophic cascade. Proc. natn. Acad. Sci. USA 93: 7723–7726.

Carpenter, S. R. & J. F. Kitchell, 1992. Trophic cascade and biomanipulation: Interface of research and management – A reply to the comment by DeMelo et al.. Limnol. Oceanogr. 37: 208–213.

Carpenter, S. R. & J. F. Kitchell (eds), 1993. The Trophic Cascade in Lakes. Cambridge University Press, Cambridge, ISBN 0-521-43145-X, 385 pp.

Carpenter, S. R., J. F. Kitchell & J. R. Hodgson. 1985. Cascading trophic interactions and lake productivity. BioScience 35: 634-639.

Carpenter, S. R., J. F. Kitchell, K. L. Cottingham, D. E. Schindler, D. L. Christensen, D. M. Post & N. Voichick, 1996. Chlorophyll variability, nutrient input, and grazing: evidence from whole-lake experiments. Ecology 77: 725–735.

DeMelo, R., R. France & D. J. McQueen, 1992. Biomanipulation: Hit or myth? Limnol. Oceanogr. 37: 192–207.

Gophen, M., S. Serruya & P. Spataru, 1990. Zooplankton community changes in Lake Kinneret (Israel) during 1969–1985. Hydrobiologia 191: 39–46.

Heinonen, P., 1982. On the annual variation of phytoplankton biomass in Finnish inland waters. Hydrobiologia 86: 29–31.

Helminen, H. & J. Sarvala, 1994a. Population regulation of vendace (*Coregonus albula*) in Lake Pyhäjärvi, southwest Finland. J. Fish Biol. 45: 387–400.

Helminen, H. & J. Sarvala, 1994b. Changes in zooplanktivory by vendace (*Coregonus albula*) in Lake Pyhäjärvi (SW Finland) due to variable recruitment. Verh. int. Ver. Limnol. 25: 2128–2131.

Helminen, H. & J. Sarvala, 1997. Responses of Lake Pyhäjärvi (SW Finland) to variable recruitment of the major planktivorous fish, vendace (*Coregonus albula*). Can. J. Fish. aquat. Sci. 54: 32–40.

Helminen, H., J. Sarvala & A. Hirvonen, 1990. Growth and food consumption of vendace (*Coregonus albula* (L.)) in Lake Pyhäjärvi, SW Finland: a bioenergetics modeling analysis. Hydrobiologia 200/201: 511–522.

Helminen, H., K. Ennola, A. Hirvonen & J. Sarvala, 1993a. Fish stock assessment in lakes based on mass removal. J. Fish Biol. 42: 255–263.

Helminen, H., H. Auvinen, A. Hirvonen, J. Sarvala & J. Toivonen, 1993b. Year-class fluctuations of vendace (*Coregonus albula*) in Lake Pyhäjärvi, southwest Finland, during 1971-90. Can. J. Fish. aquat. Sci. 50: 925–931.

Hessen, D. O., B. A. Faafeng & T. Andersen, 1995. Replacement of herbivore zooplankton species along gradients of ecosystem productivity and fish predation pressure. Can. J. Fish. aquat. Sci. 52: 733–742.

Hirvonen, A. & S. Salonen, 1995. Ravintoketjukunnostuksen alkutaival Köyliönjärvellä (The first stage in restoring Lake Köyliönjärvi by fish removal). Vesitalous 36(3): 11–14; 39 (in Finnish, with English summary).

Jones, R. A. & G. F. Lee. 1986. Eutrophication modeling for water quality management: an update of the Vollenweider-OECD model. WHO Water Quality Bulletin 11: 67–74, 118.

Karjalainen, J. & M. Viljanen, 1993. Changes in the zooplankton community of Lake Puruvesi, Finland, in relation to the stock of vendace (*Coregonus albula* (L.)). Verh. int. Ver. Limnol. 25: 563–566.

Karjalainen, J., T. Turunen, H. Helminen, J. Sarvala & H. Huuskonen, 1997. Food selection and food consumption of 0+ smelt (*Osmerus eperlanus* (L.)) and vendace (*Coregonus albula* (L.)) in the pelagial zone of Finnish lakes. Arch. Hydrobiol. spec. Issues Advanc. Limnol. 49: 37–49.

Kurkilahti, M. & M. Rask, 1996. A comparative study of the usefulness and catchability of multimesh gill nets and gill net series in sampling of perch (*Perca fluviatilis* L.) and roach (*Rutilus rutilus* L.). Fish. Res. 27: 243–260.

Mazumder, A., 1994. Phosphorus-chlorophyll relationships under contrasting herbivory and thermal stratification: predictions and patterns. Can. J. Fish. aquat. Sci. 51: 390–400.

McQueen, D. J., E. L. Mills, J. L. Forney, M. R. S. Johannes & J. R. Post, 1992. Trophic level relationships in pelagic food webs: comparisons derived from long-term data sets for Oneida Lake, New York (USA), and Lake St. George, Ontario (Canada). Can. J. Fish. aquat. Sci. 49: 1588–1596.

Mills, E. L. & J. L. Forney, 1983. Impact on *Daphnia pulex* of predation by yellow perch in Oneida Lake, New York. Trans. am. Fish. Soc. 112: 154–161.

Mills, E. L. & A. Schiavone Jr., 1982. Evaluation of fish communities through assessment of zooplankton populations and measures of lake productivity. North am. J. Fish. Mgmt. 2: 14–27.

Ramcharan, C. W., D. J. McQueen, E. Demers, S. A. Popiel, A. M. Rocchi, N. D. Yan, A. H. Wong & K. D. Hughes, 1995. A comparative approach to determining the role of fish predation in structuring limnetic ecosystems. Arch. Hydrobiol. 133: 389–416.

Räsänen, M., V.-P. Salonen, J. Salo, M. Walls & J. Sarvala, 1992. Recent history of sedimentation and biotic communities in Lake Pyhäjärvi, SW Finland. J. Paleolimn. 7: 107–126.

Reshetnikov, Yu. S., O. A. Popova, O. P. Sterligova, V. F. Titova, L. G. Bushman, E. P. Ieshko, N. P. Makarova, R. P. Malakhova, I. V. Pomazovskaya & Yu. A. Smirnov, 1982. Izmenenie struktury rybnogo naseleniya evtrofiruemogo vodoema. Izd. Nauka, Moskva, 248 pp. (in Russian).

Reynolds, C. S., 1994. The ecological basis for the successful biomanipulation of aquatic communities. Arch. Hydrobiol. 130: 1–33.

Sainio, J., 1985. Eräiden limnoplanktisten vesikirppujen populaatiodynamiikkaa Littoistenjärvessä. M.Sc. thesis, Department of Biology, University of Turku, Finland. 64 pp. In Finnish.

Salo, J., M. Walls, M. Rajasilta, J. Sarvala, M. Räsänen & V.-P. Salonen, 1989. Fish predation and reduction in body size in a cladoceran population: palaeoecological evidence. Freshwat. Biol. 21: 217–221.

Salonen, S., H. Helminen & J. Sarvala, 1996. Feasibility of controlling coarse fish populations through pikeperch (*Stizostedion lucioperca*) stocking in Lake Köyliönjärvi, SW Finland. Ann. zool. fenn. 33: 451–457.

Sarvala, J. & H. Helminen, 1996. Year-class fluctuations of perch (*Perca fluviatilis*) in Lake Pyhäjärvi, southwest Finland. Ann. zool. fenn. 33: 389–396.

Sarvala, J. & K. Jumppanen, 1988. Nutrients and planktivorous fish as regulators of productivity in Lake Pyhäjärvi, SW Finland. Aqua fenn. 18: 137–155.

Sarvala, J. & H. Perttula, 1994. Littoistenjärvi. Littoistenjärvityöryhmä, Kaarinan kaupunki & Liedon kunta, Kaarina, ISBN 951-97062-0-8, 80 pp. (in Finnish).

Sarvala, J., H. Helminen & A. Hirvonen, 1994. The effect of intensive fishing on fish populations in Lake Pyhäjärvi, south-west Finland. In I. G. Cowx (ed.), Rehabilitation of freshwater fisheries. Fishing News Books, Oxford: 77–89.

Sarvala, J., H. Helminen, A. Hirvonen & S. Salonen, 1995. Köyliönjärven veden laatu ja siihen vaikuttavat mekanismit (Effects of cyprinid fish reduction on water quality of Lake Köyliönjärvi). Vesitalous 36(3): 15–17, 39 (in Finnish, with English summary).

Sarvala, J., H. Helminen, J.-P. Ripatti, V. Pruuki & J. Ruuhijärvi, 1992. Tehokalastuksen vaikutus Evon Syrjänalusen särkikantaan (Impact of intensive fishing on the roach population in Lake Syrjänalunen, southern Finland). Suomen Kalatalous 60: 191–205 (in Finnish, with English summary).

Sarvala, J., K. Aulio, H. Mölsä, M. Rajasilta, J. Salo & I. Vuorinen, 1984. Factors behind the exceptionally high fish yield in the lake Pyhäjärvi, southwestern Finland – hypotheses on the biological regulation of fish production. Aqua fenn. 14: 49–57.

Sarvala, J., M. Rajasilta, C. Hangelin, A. Hirvonen, M. Kiiskilä & V. Saarikari, 1988. Spring abundance, growth and food of 0+ vendace (*Coregonus albula* L.) and whitefish (*C. lavaretus* L. s.l.) in Lake Pyhäjärvi, SW Finland. Finn. Fish. Res. 9: 221–233.

Schriver, P., J. Bøgestrand, E. Jeppesen & M. Søndergaard, 1995. Impact of submerged macrophytes on fish-zooplankton-phytoplankton interactions: large-scale enclosure experiments in a shallow eutrophic lake. Freshwat. Biol. 33: 255–270.

Vuorinen, I. & J. Nevalainen, 1981. Säkylän Pyhäjärven eläinplanktontutkimus 1980. Lounais-Suomen Vesiensuojeluyhdistys Julkaisu 47: 89–117 (in Finnish).

Hydrobiologia **363:** 97–105, 1998.
T. Tamminen & H. Kuosa (eds), Eutrophication in Planktonic Ecosystems: Food Web Dynamics and Elemental Cycling.

Planktonic food web in marine mesocosms in the Eastern Mediterranean: bottom-up or top-down regulation?

Paraskevi Pitta[1], Antonia Giannakourou[2], Pascal Divanach[1] & Maroudio Kentouri[3]
[1] *Institute of Marine Biology of Crete, P.O. Box 2214, 71003 Heraklion, Crete, Greece*
[2] *National Center for Marine Research, 16604 Hellinikon, Athens, Greece*
[3] *Biology Department, University of Crete, P.O. Box 1470, 71110 Heraklion, Crete, Greece*

Key words: Mesocosms, nutrients, bacteria, phytoplankton, zooplankton, fish larvae, microbial food web, E Mediterranean

Abstract

A mesocosm experiment was conducted in order to study the structure of the planktonic food web. The dynamics of pico-, nano- and microplankton populations were followed during 40 days in four large (40 m^3) enclosures. In three tanks a gradient of added nutrients (nitrogen and phosphorus) was applied, while a fourth tank was used as a control. On day 14, the top predator (sea bream *Sparus aurata* larvae) was introduced into the tanks and part of the water column in each tank was isolated in a plastic bag without fish larvae, to act as a control for predation. Physical parameters, chlorophyll *a* and nutrient concentrations, as well as plankton concentrations were monitored. A diatom bloom was observed in all four tanks, in the first phase ending with silicate depletion. Flagellate and dinoflagellate abundance subsequently increased, these organisms being limited by zooplankton grazing. The zooplankton populations were controlled by both resources (mostly flagellates) and predation (by fish larvae) as indicated by the results of the control experiments.

Introduction

Two main concepts have been proposed regarding the mechanisms regulating the biomass and abundance of organisms in the food webs. Resource availability regulates the biomass on a trophic level from the bottom (bottom-up force) while the predation affects it from the top (top-down force). There has been an antagonism between the two theories, in both the terrestrial and the aquatic literature, as to which is the principal regulating force.

For a long period of time, resource availability (bottom-up force) was considered to be a more important factor than predation (top-down force, reviewed by Northcote, 1988; Verity & Smetacek, 1996) for the regulation of structure and function of freshwater pelagic ecosystems. However, top-down regulation has started to attract more attention in recent years. Other theories have resulted from this concept, such as the cascading trophic interactions model (Carpenter et al., 1985), the trophic biomanipulation theory (Shapiro & Wright, 1984) and the theory of simultaneous action of both forces, the 'bottom-up: top-down' model (McQueen et al., 1986).

In the marine environment, only a small amount of evidence exists for trophic cascade including more than three trophic levels (Verity & Smetacek, 1996). Some of this evidence is based on field observations (Roff et al., 1988) while others are based on large enclosure experiments (Gamble et al., 1977; Riemann et al., 1990). However it is not yet clear to what degree a trophic cascade links microbial and classic food webs. One reason for the lack of a clear trophic cascade from metazooplankton to bacteria may be the omnivory of many aquatic species as well as the mixotrophy of protists (Laval-Peuto & Rassoulzadegan, 1988; Sanders, 1991).

A considerable research effort has been devoted to the study of the carbon flux in the food web through the bacteria and the flagellates (Azam et al., 1983;

Bjornsen et al., 1988; Malone & Ducklow, 1990). The role of ciliates has been studied less thoroughly (Rassoulzadegan & Sheldon, 1986), while the link to the higher trophic levels of the metazoan food chain rather less adequately (Horsted et al., 1988). In other words considerably more emphasis has been given to sink than to link processes (Sherr et al., 1987). Studies on the regulation at the ecosystem level are very rare and even in the recent studies in mesocosms, data on multiple trophic levels are hard to find.

The aim of this paper was to investigate the regulation that shapes the pelagic ecosystem in the coastal zone of the Eastern Mediterranean, combining the study of groups involved in both the microbial and the classical food web. The use of large mesocosms made it possible to cover the entire food web from bacteria to the copepods and the top predator, the fish. The study comprised also a nutrient gradient in order to assess the impact of the addition of nutrients on the structure of the food web and consequently on the regulation mechanisms.

Materials and methods

The experiment was carried out in four 40 m^3 mesocosms at the Institute of Marine Biology of Crete during summer 1993 (1st June to 9th July). The tanks were filled with sea water filtered through 300 μm net in order to exclude large predators. While the tanks were being filled, increasing quantities of nitrogen and phosphorus were added into three of the tanks (M1: low addition (0.5 g N m^{-3} and 0.047 g P m^{-3}), M2: medium addition (1 g N m^{-3} and 0.093 g P m^{-3}) and M3: high addition (2 g N m^{-3} and 0.187 g P m^{-3}), while in the fourth control-tank (M0) no nutrients were added. Nitrogen was added in the form of nitrate and ammonium (molar ratio 1 : 1), while phosphorus in the form of phosphate.

On day 14, the top predator was introduced into the mesocosms (20 000 sea bream larvae in each tank at the moment of the first exogenous feeding). Just before the introduction of the fish larvae into the tanks, 2 m^3 of each mesocosm were isolated in a plastic bag placed in each tank; these bags were used as predation-controls in order to assess the influence of the larvae on the planktonic food web. Towards the end of the experiment there was a limited water renewal (5% volume per day) in order to assure good water quality for the fish larvae.

Temperature, pH, salinity and illumination were measured every second day in the sea, in the mesocosms and in the plastic bags throughout the entire experimentation period. A long tube-sampler was used in order to sample the entire column (2 m depth). Samples for nutrients determinations (Strickland & Parsons, 1972) and chlorophyll *a* concentrations (Yentsch & Menzel, 1963) were taken every second day. Bacterial and cyanobacterial counts were made every fourth day in formalin (2% v/v) preserved samples, using a fluorescence microscope after a DAPI staining and filtration on 0.2 μm polycarbonate membranes (Porter & Feig, 1980). Nano- and microplankton were observed by means of an inverted microscope in samples taken every fourth day and preserved with acid Lugols' solution. After the sedimentation (Utermöhl, 1958), organisms of different planktonic groups (diatoms, flagellates, dinoflagellates, ciliates, rotifers, copepods) were identified to species or genus level (with the exception of flagellates) and enumerated. Flagellates and ciliates were grouped into size-classes. The above mentioned method results in an underestimation of the abundance of nanoflagellates; however, it can be used in order to compare concentrations among samples (Estrada, 1991). Analysis of variance was used in order to compare concentrations of various parameters among tanks. In these analyses experimentation day was used as block in order to remove from the experimental error the variation related to the temporal evolution of parameters, thereby increasing the precision of the analysis (Krebs, 1989).

Results

Inorganic nutrients

Nutrients (nitrate, phosphate, silicate) were rapidly consumed in all the enclosures at the beginning of the experiment (6–8 days) irrespective of the nutrient additions (Figure 1). After a few days, a depletion of nutrients (mainly silicate and ammonium) was observed, this phenomenon being less pronounced in the enclosure with high nutrient additions. During a later phase, coinciding with the water renewal period, silicate concentrations increased again due to the passage of water through a sand filter.

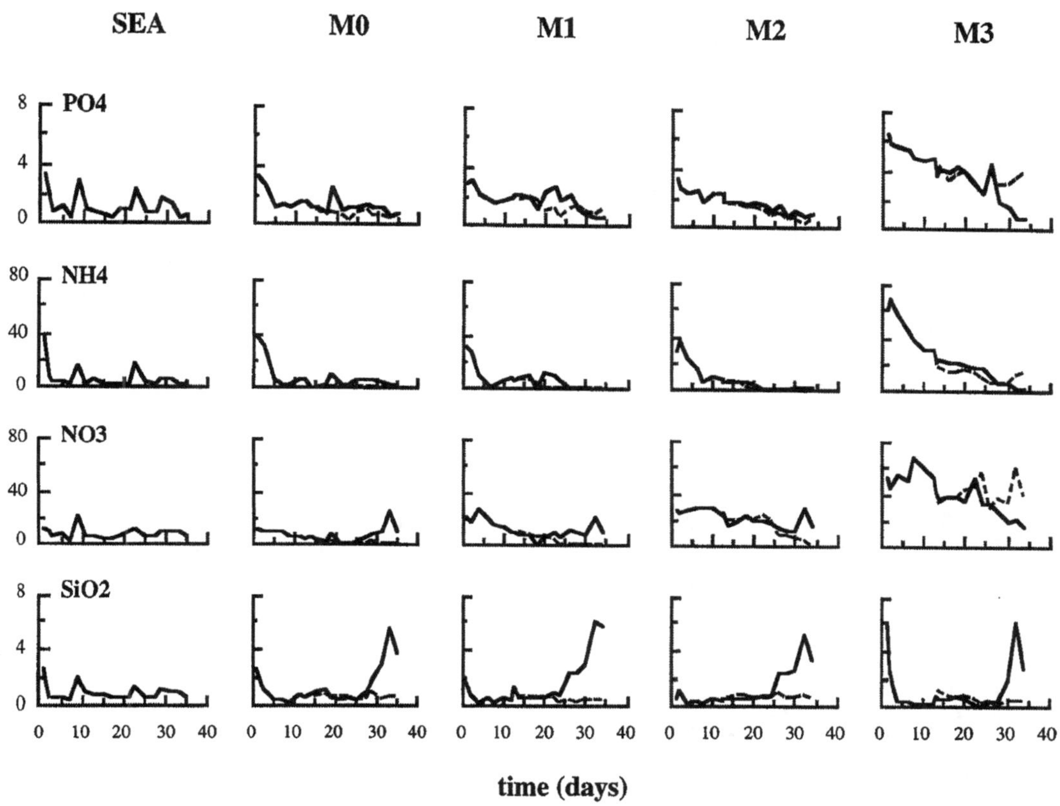

Figure 1. Concentrations of nutrients (μgat l^{-1}) at the sea, the enclosures (solid line) and the bags-controls of predation (dashed line). M0: zero, M1: low, M2: medium and M3: high nutrients' addition.

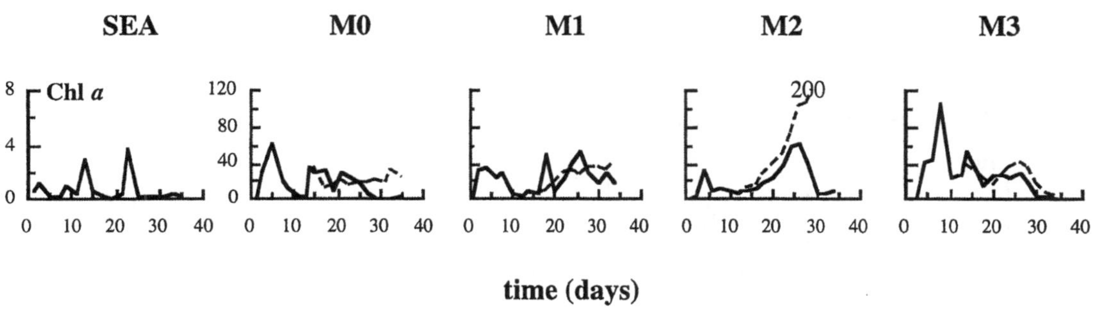

Figure 2. Chlorophyll *a* concentrations (μg l^{-1}) at the sea, the enclosures (solid line) and the bags-controls of predation (dashed line). M0: zero, M1: low, M2: medium and M3: high nutrients' addition.

Chlorophyll a concentrations

Chlorophyll *a* concentration showed maximal values soon after the beginning of the experiment in all tanks, but not in the sea (Figure 2). These phytoplankton blooms were also evident in the control-mesocosm M0, where no nutrients were added and therefore the nutrient levels were identical to those of the sea, at least initially. After the first, other phytoplankton blooms followed. No significant differences were detected among the tanks in respect of chlorophyll *a* concentrations (ANOVA, df = 3, $F = 1.104$, $p = 0.356$).

Succession of planktonic groups

The succession of different planktonic groups was apparent from the phyto- and zooplankton analysis (Figure 3). The succession pattern did not differ among

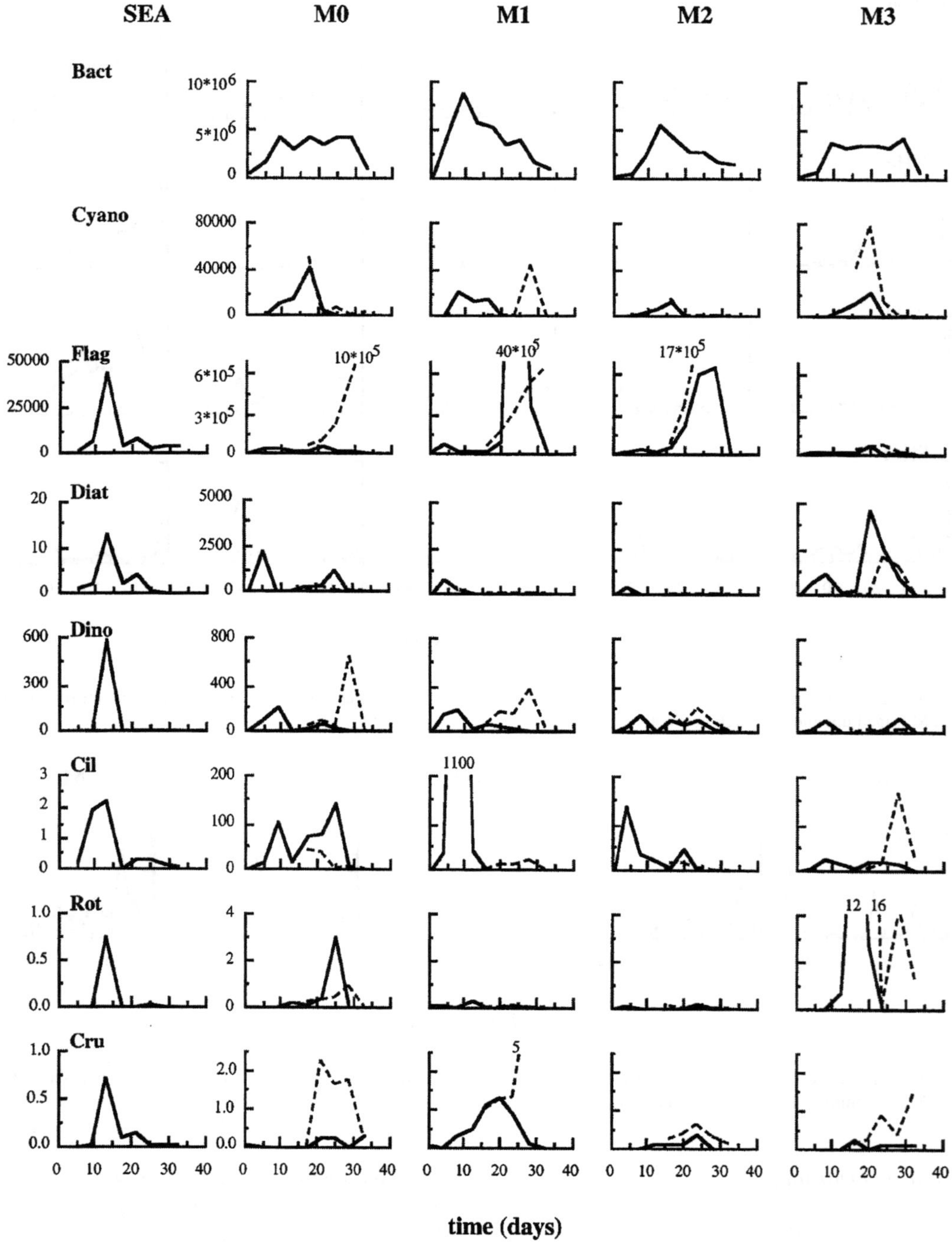

Figure 3. Variation in time of the abundance (ind. ml^{-1}) of different plankton groups at the sea, the enclosures (solid line) and the bags-controls of predation (dashed line). M0: zero, M1: low, M2: medium and M3: high nutrients' addition. Abbreviations used in labels: Bact: Bacteria, Cyano: Cyanobacteria, Flag: Flagellates, Diat: Diatoms, Dino: Dinoflagellates, Cil: Ciliates, Rot: Rotifers, Cru: Crustaceans.

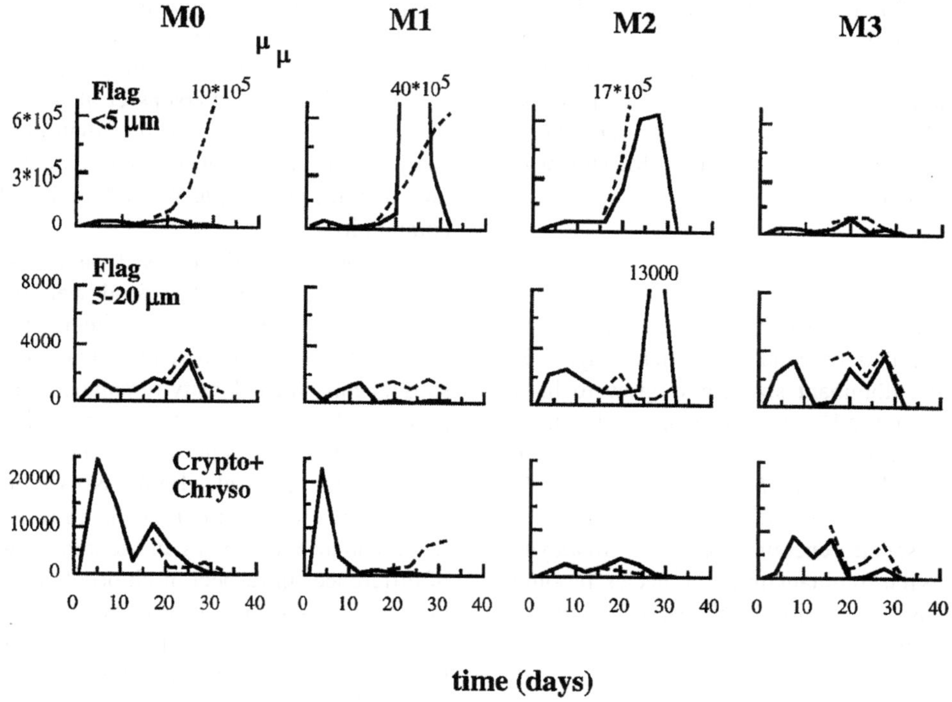

Figure 4. Variation in time of the abundance (ind. ml^{-1}) of the different size classes of flagellates in the enclosures (solid line) and the bags-controls of predation (dashed line). M0: zero, M1: low, M2: medium and M3: high nutrients' addition. Abbreviations used in labels: Flag: Flagellates, Crypto + Chryso: Cryptophyceae and Chrysophyceae.

tanks. Diatoms were the first organisms to develop in the tanks after the water inclusion. After a few days silicate was depleted and the diatom growth ended, these organisms sedimenting on the bottom of the tanks. Flagellates increased in numbers a little after or simultaneously with diatoms. Large cells (Cryptophyceae and Chrysophyceae) grew first (Figure 4) perhaps competing with diatoms for nutrients, while smaller cells grew after the decline of the populations of the other flagellates and diatoms. Dinoflagellates increased in abundance following the diatoms with longer or shorter time lags. Ciliate populations either grew simultaneously with the dinoflagellates or slightly prior to them; they usually followed the flagellate populations. Rotifers grew at the same time as ciliates did or followed them since their diet is also based on flagellates. Crustaceans (mainly copepods) always were the last organisms to grow in the tanks as a result of their longer life cycles. All three zooplanktonic groups (ciliates, rotifers, crustaceans) were preyed on by the fish larvae as has also been found in other experiments (Pitta, 1996).

Picoplankton

The heterotrophic bacteria grew in all the tanks immediately after the first phytoplankton bloom and subsequently remained at high concentrations (3-9 × 10^6 cells ml^{-1}) until the end of the experiment (Figure 3). The addition of nutrients did not produce significant differences in bacterial concentrations among the tanks (ANOVA, df = 3, $F = 0.325$, $p = 0.808$). Picocyanobacteria grew in all the tanks from the initial phase of the experiment, with their concentration displaying a peak by day 15 and then a decrease. The comparison of their abundance among the tanks did not reveal statistically significant differences (ANOVA, fd = 3, $F = 1.549$, $p = 0.228$).

Nanoflagellates

In all the tanks nanoflagellates grew after the diatoms and small cell-size species were numerically dominant in abundance (Figure 4). In the case of tanks M0 and M3, the abundance of nanoflagellates in the control bags was higher than that in the respective mesocosms. The abundance of nanoflagellates decreased with increasing abundance of ciliates and peaked again after the decline of the ciliates' bloom. Clear predator-prey oscillations were also observed between the concentration curves of Cryptophyceae-Chrysophyceae and those of the rotifers in three out of four experiments (M0, M1, M3) while in the fourth (M2) this was only partly verified.

Protozooplankton

Oligotrich species were the dominant component within the ciliate populations. Small oligotrichs (Figure 5) which were the dominant size class, grew to high abundance (10^7 ind. ml^{-1}) as a response to the small flagellates' increased abundance. Tintinnids also grew up to 20 ind. ml^{-1}. *Strombidium sphaericum, S. parvum, Lohmaniella oviformis* were the most abundant oligotrich species encountered, while *Metacylis jorgensenii* was the dominant tintinnid species. In some cases, ciliates oscillated with flagellates several times, with a time lag of 2–4 days.

Metazooplankton

Rotifers attained 12 ind. ml^{-1} and crustaceans 1 ind. ml^{-1}. The addition of the sea bream larvae influenced the abundance of the rotifers and the crustaceans (mainly nauplii of pelagic copepods and some cladocerans). The fish larvae introduced into the tanks survived to the end of the experiment and presented considerable growth. This is a strong indication that predatory pressure was exerted on the zooplankton community. As can be observed in Figure 3, in all cases the crustacean abundances were much higher in the bags than in the respective mesocosms, this observation indicating the predation of the fish larvae on these populations (paired t-test, df = 1, $F = 8.772$, $p = 0.006$).

Discussion

The force determining the ecosystem structure was not the same at all trophic levels. Either resource availability or predation seemed to operate at different trophic levels, or in some cases both were in operation in different stages of the succession.

Diatoms seemed to be controlled by the resources since their growth stopped with the depletion of silicate. The role of silicate is particularly evident in the case of the M3 tank where diatoms ceased to grow although the other nutrients (N, P) were present in adequate quantities. It is well known that diatoms are a group entering and leaving the plankton ecosystem according to their strict physiological demands (Verity & Smetacek, 1996). Parsons et al. (1978) modelled the conditions for the growth of diatoms; when these conditions (nutrient concentration, water mixing) are met, diatoms grow rapidly and dominate the plankton community.

The present study focused on the simultaneous examination of the entire food web from bacteria to copepods. During the experiment, the growth of bacteria followed the first phytoplankton bloom and subsequently their density remained almost constant, probably due to predation or even to viral activity as demonstrated in recent studies (Fuhrman & Suttle, 1993; Maranger et al., 1994). No significant differences were revealed among tanks in respect of bacterial abundance, this being in accordance with the lack of differences in respect of chlorophyll *a*. Bottom-up limitation of bacterial growth has been stated (Tranvik & Sieburth, 1989). Top-down control of bacteria has been supported by experimental work (Lignell et al., 1992) as well as by field observations (Sherr et al., 1986). Kuuppo-Leinikki et al. (1994) have also suggested that bacteria shifted from bottom-up regulation in the initial phase to top-down control during the succession in a mesocosm experiment.

The results of the present study suggest that flagellates were controlled by predation of ciliates and rotifers, since they presented the typical predator-prey oscillations with these two groups. Although autotrophic flagellates are better competitors than diatoms for nutrients at low concentrations (Parsons et al., 1978), they can be resource-limited when nutrients are extremely scarce. However this was not the case in the present experiment since after a first collapse, their populations raised again after the decline of their predators. As regards heterotrophic flagellates there is also no reason to assume resource limitation since there was abundance of bacteria. The control of flagellates through predation by ciliates (Weisse, 1991), rotifers (Dolan & Gallegos, 1991) as well as cladocerans and copepods (Stoecker & Capuzzo, 1990) has also been demonstrated. In mesocosm experiments involving addition of nutrients, Kuuppo-Leinikki et al. (1994) observed predator-prey-like coupled oscillations of ciliates and heterotrophic flagellates. After the incubation of filtered sea water from the coastal environment, Rassoulzadegan & Sheldon (1986) suggested that the development of picoflagellates is controlled by nanociliate predation. Laboratory experiments by

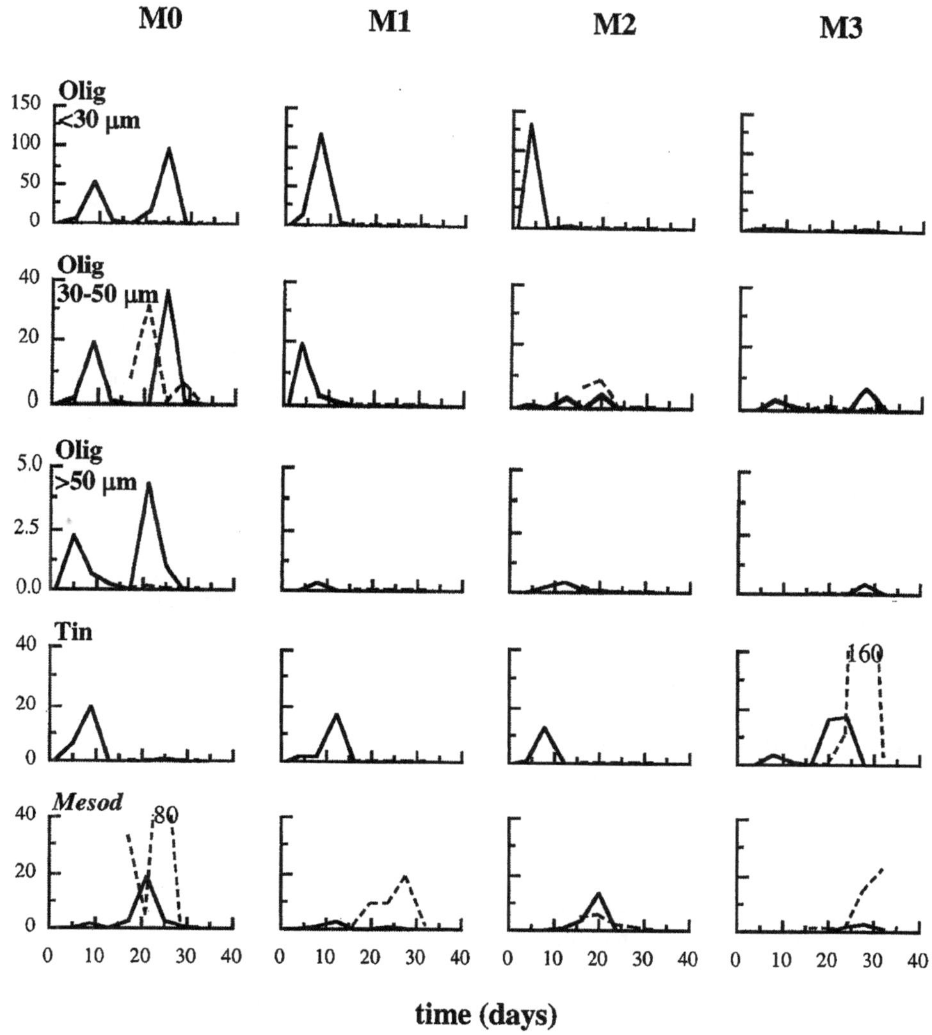

Figure 5. Variation in time of the abundance (ind. ml^{-1}) of the different size classes of ciliates in the enclosures (solid line) and the bags-controls of predation (dashed line). M0: zero, M1: low, M2: medium and M3: high nutrients' addition. Abbreviations used in labels: Olig: Oligotrich ciliates, Tin: Tintinnids, Mesod: *Mesodinium* spp.

Kivi et al. (1996) showed that the total phytoplankton biomass depends on both nutrients and predation by metazoan or protozoan-dominated communities.

In contrast to what occurs with the phytoplankton (control by one only mechanism per group, i.e. diatoms, flagellates), in most zooplankton groups, a double control has been observed during the present study: as can be inferred from the control-bags, the growth of the ciliate, rotifer and crustacean populations was controlled by resource availability (flagellates), but also by the predation from the top predator (fish larvae). In all the bags (free of fish) the densities of the crustaceans increased in comparison to the respective mesocosms, this fact implying that predation by the fish larvae controlled the growth of crustaceans. Rotifers and ciliates also increased in most of the control-bags. This increase indicates that these groups were used as food by the fish larvae but also that they contributed to the crustacean diet. However, in later phases their densities (the densities of all three groups) declined in the control-bags as a result of resource availability. The importance of protists in the copepod diet has been demonstrated by several authors. In laboratory experiments it was found that diets rich in ciliates and rotifers increase the egg production of copepods (Stoecker & Egloff, 1987). After field and laboratory research Kleppel et al. (1991) concluded that when other food recources are limited, diatoms may become critical in sustaining copepod populations. According to Kivi et al. (1993) the growth of protozooplankton in microcosms was restricted due to competition for food and predation by metazooplankton.

The predation of fish on copepods and cladocerans has been demonstrated in mesocosm studies (Horsted et al., 1988) where a decrease of large holoplanktonic species (*Acartia tonsa* and *Pleopis polyphemoides*) was found. The importance of ciliates and rotifers in the diet of coastal fish larvae, such as *Sparus aurata*, has been illustrated in mesocosm studies (Kentouri & Divanach, 1983, 1986).

The secondary effect of planktivorous fish on phytoplankton, i.e. the increase in phytoplankton due to decreased zooplankton grazing, has been shown (Horsted et al., 1988). This effect was only partly confirmed by our experiments, due to the fact that prey selection by fish changes progressively (Pierce & Turner, 1992) towards larger prey size as fish grow, for instance from ciliates and rotifers to copepods (Pitta, 1996); therefore the fish exert predatory pressure on different parts of the food web during the different larval stages. This shift in food preferences favors or suppresses different groups, and this factor may at any time mask the results obtained. The impact of fish on the structure of plankton community has been extensively studied in freshwater lakes and ponds (Drenner et al., 1986; Vanni, 1987). After the first approach: physical and chemical factors-phytoplankton-zooplankton-fish, emphasis was placed on the top-down impact of fish on the food-web.

The results of the present study showed that the effect of the introduction of planktivorous fish larvae in the food web was readily detectable at the metazooplankton level, as well as on ciliates and rotifers which consist their prey, but it was less apparent (attenuating) at the lower trophic levels. It seems that the trophic cascade grows weaker with successive links due to the ability of the microbial food web to modify the rates of transferring material through trophic levels as well as to the omnivory of many aquatic species.

Acknowledgements

Thanks are due to the support staff of IMBC for technical assistance during the experiment. Mrs S. Zivanovic and Mrs H. Dafnomili kindly helped with chemical analyses. The authors are grateful to Prof. A. Eleftheriou, the Editor-in-chief and an anonymous reviewer for their critical comments on the manuscript and to Mrs M. Eleftheriou for linguistic improvement.

References

Azam, F., T. Fenchel, J. G. Field, J. S. Gray, L. A. Meyer-Reil & F. Thingstad, 1983. The ecological role of water-column microbes in the sea. Mar. Ecol. Prog. Ser. 10: 257–263.

Bjornsen, P. K., B. Riemann, S. J. Horsted, T. G. Nielsen & J. Pock-Sten, 1988. Trophic interactions between heterotrophic nanoflagellates and bacterioplankton in manipulated seawater enclosures. Limnol. Oceanogr. 33: 409–420.

Carpenter, S. R., J. F. Kitchell & J. R. Hodgson, 1985. Cascading trophic interactions and lake productivity. Fish predation and herbivory can regulate lake ecosystems. Bioscience 35: 634–639.

Dolan, J. R. & C. L. Gallegos, 1991. Trophic coupling of rotifers, microflagellates and bacteria during fall months in the Rhode River Estuary. Mar. Ecol. Prog. Ser. 77: 147–156.

Drenner, R. W., S. T. Threlkeld & M. D. McRacken, 1986. Experimental analysis of the direct and indirect effects of an omnivorous filter-feeding clupeid on plankton community structure. Can. J. Fish. aquat. Sci. 43: 1935–1945.

Estrada, M., 1991. Phytoplankton assemblages across a NW Mediterranean front: Changes from winter mixing to spring stratification. In J. D. Ros & N. Prat (eds), Homage to Ramon Margalef; or, Why there is such pleasure in studying nature. Oecol. aquat. 10: 157–185.

Fuhrman, J. A. & C. A. Suttle, 1993. Viruses in marine planktonic systems. Oceanography 6: 51–63.

Gamble, J. C., J. M. Davies & J. H. Steele, 1977. Loch Ewe bag experiment, 1974. Bull. mar. Sci. 27: 146–175.

Horsted, S. J., T. G. Nielsen, B. Riemann, J. Pock-Steen & P. K. Bjornsen, 1988. Regulation of zooplankton by suspension-feeding bivalves and fish in estuarine enclosures. Mar. Ecol. Prog. Ser. 48: 217–224.

Kentouri, M. & P. Divanach, 1983. Contribution à la connaissance du comportement et de la biologie des larves de marbré *Lithognathus mormyrus* (Sparides) en élevage. Ann. Zootech. 32: 135–152.

Kentouri, M. & P. Divanach, 1986. Sur l'importance des ciliés pélagiques dans l'alimentation des stades larvaires de poissons. Ann. Biol. 25: 307–318.

Kivi, K., H. Kuosa & S. Tanskanen, 1996. An experimental study on the role of crustacean and microprotozoan grazers in the planktonic food web. Mar. Ecol. Prog. Ser. 136: 59–68.

Kivi, K., S. Kaitala, H. Kuosa, J. Kuparinen, E. Leskinen, R. Lignell, B. Marcussen & T. Tamminen, 1993. Nutrient limitation and grazing control of the Baltic plankton community during annual succession. Limnol. Oceanogr. 38: 893–905.

Kleppel, G. S., D. V. Holliday & R. E. Pieper, 1991. Trophic interactions between copepods and microplankton: A question about the role of diatoms. Limnol. Oceanogr. 36: 172–178.

Krebs, C. J., 1989. Ecological methodology. Harper & Row, New York, 654 pp.

Kuuppo-Leinikki, P., R. Autio, S. Hallfors, H. Kuosa, J. Kuparinen & R. Pajuniemi, 1994. Trophic interactions and carbon flow between picoplankton and protozoa in pelagic enclosures manipulated with nutrients and a top predator. Mar. Ecol. Prog. Ser. 107: 89–102.

Laval-Peuto, M. & F. Rassoulzadegan, 1988. Autofluorescence of marine planktonic Oligotrichina and other ciliates. Hydrobiologia 159: 99–110.

Lignell, R., S. Kaitala & H. Kuosa, 1992. Factors controlling phyto- and bacterioplankton in late spring on a salinity gradient in the northern Baltic. Mar. Ecol. Prog. Ser. 84: 121–131.

Malone, T. C. & H. W. Ducklow, 1990. Microbial biomass in the coastal plume of Chesapeake Bay: phytoplankton-bacterioplankton relationships. Limnol. Oceanogr. 35: 296–312.

Maranger, R., D. F. Bird & S. K. Juniper, 1994. Viral and bacterial dynamics in Arctic sea ice during the spring algal bloom near Resolute, N.W.T., Canada. Mar. Ecol. Prog. Ser. 111: 121–127.

McQueen, D. J., J. R. Post & E. L. Mills, 1986. Trophic relationships in freshwater pelagic ecosystems. Can. J. Fish. aquat. Sci. 43: 1571–1581.

Northcote, T. G., 1988. Fish in the structure and function of freshwater ecosystems: A 'top-down' view. Can. J. Fish. aquat. Sci. 45: 361–379.

Parsons, T. R., P. J. Harrison & R. Waters, 1978. An experimental simulation of changes in diatom and flagellate blooms. J. exp. mar. Biol. Ecol. 32: 285–294.

Pierce, R. W. & J. T. Turner, 1992. Ecology of planktonic ciliates in marine food webs. Rev. Aquat. Sci. 6: 139–181.

Pitta, P., 1996. Dynamics of the plankton community in sea bream (*Sparus aurata*) rearing mesocosms. Ph. D. thesis, University of Crete, Greece, 229 pp.

Porter, K. G. & Y. S. Feig, 1980. The use of DAPI for identifying and counting aquatic microflora. Limnol. Oceanogr. 25: 943–948.

Rassoulzadegan, F. & R. W. Sheldon, 1986. Predator-prey interactions of nanozooplankton and bacteria in an oligotrophic marine environment. Limnol. Oceanogr. 31: 1010–1021.

Riemann, B., H. M. Sorensen, P. K. Bjornsen, S. J. Horsted, L. M. Jensen, T. G. Nielsen & M. Sondergaard, 1990. Carbon budgets of the microbial food web in estuarine enclosures. Mar. Ecol. Prog. Ser. 65: 159–170.

Roff, J. C., K. Middlebrook & F. Evans, 1988. Long-term variability in North Sea zooplankton off the Northumberland coast: productivity of small copepods and analysis of trophic interactions. J. mar. biol. Ass. U.K. 68: 143–164.

Sanders, R. W., 1991. Mixotrophic protists in marine and freshwater ecosystems. J. Protozool. 38: 76–81.

Shapiro, J. & D. I. Wright, 1984. Lake restoration by biomanipulation: Round Lake, Minnesota, the first two years. Freshwat. Biol. 14: 371–383.

Sherr, E. B., B. F. Sherr & L. J. Albright, 1987. Bacteria: Link or sink? Science 235: 88–89.

Sherr, E. B., B. F. Sherr, R. D. Fallon & S. Y. NewellL, 1986. Small, aloricate ciliates as a major component of the marine heterotrophic nanoplankton. Limnol. Oceanogr. 31: 177–183.

Stoecker, D. & J. M. Capuzzo, 1990. Predation on Protozoa: its importance to zooplankton. J. Plankton Res. 12: 891–908.

Stoecker, D. K. & D. A. Egloff, 1987. Predation by *Acartia tonsa* Dana on planktonic ciliates and Rotifers. J. exp. mar. Biol. Ecol. 110: 53–68.

Strickland, J. D. H. & T. R. Parsons, 1972. A practical handbook of seawater analysis. Can. J. Fish. aquat. Sci. 167: 1–310.

Tranvik, L. J. & J. McN. Sieburth, 1989. Effects of flocculated humic matter on free and attached pelagic microorganisms. Limnol. Oceanogr. 34: 688–699.

Utermöhl, H., 1958. Zur Vervollkommnung der quantitativen Phytoplankton-methodik.. Mitt. int. Ver. Limnol. 9: 323–332.

Vanni, M. J., 1987. Effects of food availability and fish predation on a zooplankton community. Ecol. Monogr. 57: 61–88.

Verity, P. G. & V. Smetacek, 1996. Organism life cycles, predation, and the structure of marine pelagic ecosystems. Mar. Ecol. Prog. Ser. 130: 277–193.

Weisse, T., 1991. The annual cycle of heterotrophic freshwater nanoflagellates: Role of bottom-up versus top-down control. J. Plankton Res. 13: 167–185.

Yentsch, C. S. & D. W. Menzel, 1963. A method for the determination of phytoplankton chlorophyll and phaeophytin by fluorescence. Deep Sea Res. 10: 221–231.

Hydrobiologia **363:** 107–115, 1998.
T. Tamminen & H. Kuosa (eds), Eutrophication in Planktonic Ecosystems: Food Web Dynamics and Elemental Cycling.

Coupling of autotrophic and heterotrophic processes in a Baltic estuarine mixing gradient (Pomeranian Bight)

Günter Jost & Falk Pollehne
Baltic Sea Research Institute, Dept. Biological Oceanography, Seestraße 15, D-18119 Rostock, Germany

Key words: Estuarine mixing, salinity gradient, phytoplankton, primary production, community respiration, bacterial production

Abstract

Primary production and decompositional processes were measured within the mixing gradient of lagoonal and coastal water of the Pomeranian Bight during summer/autumn on four cruises between 1993 and 1995. Although different sampling strategies were applied, the results fitted well in a general pattern. Nearly all measured variables (e.g. POC, chlorophyll *a*) appear to be conservatively mixed along the salinity gradient between 2 and 8 PSU which is typical for the southern Baltic area. That pattern is, however, not due to a conservative behaviour of the components but to a balanced state of auto- and heterotrophic processes with a continuous, closely coupled recycling of matter. This is particularly evident in periods of nutrient limitation. Within the mixing gradient, changes were restricted to structural components (species composition) whereas the functional equilibrium was maintained.

Introduction

The mixing between freshwater runoff and seawater and its effects on biological activities has attracted a lot of investigations, especially in tidally influenced estuaries and river plumes in oceanic areas (e.g. Ducklow, 1982; Ducklow & Kirchman, 1983; Fuhrman et al., 1980; Jacobsen et al., 1983; Shiah & Ducklow, 1995). Production and respiration during mixing events are generally thought to be very complex. There is evidence that on annual time scales and larger spatial extensions bacterial biomass and production correlate well with the respective phytoplankton parameters (Cole et al., 1988; White et al., 1991). On smaller scales, different patterns of interactions have been reported (Ducklow & Kirchman, 1983; Fuhrman et al., 1980; Riemann & Søndergaard, 1984; Shiah & Ducklow, 1994). Therefore, Shiah & Ducklow (1995) emphasized the need for investigations on both small temporal and spatial scales. Baltic estuarine mixing patterns are different from oceanic conditions as the salinity difference of the endmembers is only in the range of 5 to 10 PSU. The functioning of the pelagic community in the open Baltic is, however, quite comparable to open ocean conditions (von Bodungen & Zeitzschel, 1996).

Before starting the study in the mixing area of the river Oder, we expected the input of 'new' nutrients by the river to start a temporal and spatial succession of algal, bacterial and metazoan growth that would lead to distinct functional belts around the river mouth. We also expected a small zone of heterotrophy at the outlet, where bacteria would respire introduced organic compounds, and that the water would be too turbid for algal growth. Further out a larger area was envisaged where algae could make use of increasing light penetration to turn the introduced nutrients into biomass and would switch the system output to net autotrophy. With the depletion of nutrients and increased mixing depths in the outer bight, production and respiration were thought to get in balance again. At a certain range of the mixing gradient we expected a sudden shift in species composition due to the fallout of limnetic species stressed by salinity and a concomitant minimum in metabolism of both auto- and heterotrophs. From this scheme the following two hypotheses were derived:

1. The high nutrient load of the river Oder increases autotrophic and heterotrophic activity and produces a 'new' production system in the Pomeranian Bight whose decay leads to different spatial and temporal patterns of auto-heterotrophic relationships.

2. In the salinity gradient a large part of the limnetic population of both auto- and heterotroph species are stressed and vanish. The replacement by a population typical for the open Baltic Sea takes some time and leads to a visible shift in species composition and a detectable gap in metabolic activity of both components.

These hypotheses were tested on several cruises during the growing seasons 1993–1995. As the feeding rate of metazoans in this area was estimated to be in the range of only a few percent of the primary production (Postel et al., 1995), we restricted our rate measurements to the interaction between those auto- and heterotrophic organisms, the activity of which could be determined in water samples of equal size and comparable incubation time (algae, bacteria). To detect the functional gap we concentrated our attention on primary production and community respiration as the main auto- and heterotrophic functions. Bacterial production was measured because it is believed to be the main heterotrophic process (Cole et al., 1988; Ducklow & Carlson, 1992). The results of this study, however, reflect a pattern of processes that is (as not unusual in ecology) very different from what we expected.

Research area and methods

The river Oder is one of the largest tributaries of the Baltic Sea (annual water outflow of about 18 km^3) and also one of the largest sources of nutrients (about 55 000 t N and 6000 t P y^{-1} between 1989 and 1994) (Pastuszak et al., 1996) and organic load (about 88400 t BOD5 y^{-1} in 1988 and 1989)(Rybinski et al., 1992) with seasonal variations. The main outflow occurs via the Swina mouth (see Figure 1).The mean residence time in the bight is about 20 days, while the main mixing process of the water from the inner lagoonal system (between 2 and 4 PSU) into the water of the Pomeranian Bight (about 7 to 8 PSU) usually takes less than one to two days (Sattler, pers. comm.). The riverine water enters the Pomeranian Bight stochastically and pulsewise from the Szczecin Lagoon (Lass, pers. comm.) forming distinct water bodies of different sizes. These water bodies mix with the bight water at different locations, depending on the current wind regime (Siegel et al., 1996).

Between autumn 1993 and autumn 1995 we investigated mixing processes of riverine water from the Szczecin lagoon into the Pomeranian Bight four times over the growing season. Water samples were taken in September 1993, June/July 1994, June/July and September/October 1995 employing two different strategies. One was based on consecutive discrete bottle sampling from a water body marked by a drifting buoy. Starting points of these drift experiments were located by ship-based thermosalinograph measurements usually near the mouth of the Swina. The other method was based on sampling of surface water from a rubber dinghy along a horizontal salinity gradient. This sampling procedure was used in autumn 1995 along salinity gradients near the Peene mouth. Water samples were processed within 30 minutes after arrival on board of the R/V 'Professor Albrecht Penck'.

Salinity and water temperature were measured by CTD (OM-87, FRG, or Seabird Inc.,USA) or from surface water samples by a hand-held TS-probe (WTW, FRG). Underwater light intensities were recorded at 0.5 m depth intervals using a submersible PAR-probe (LI-COR, Ltd., USA). For estimation of particulate organic carbon and nitrogen (POC, PON) and chlorophyll *a*, water samples of varying volume between 100 ml and 500 ml were filtered through precombusted (500 °C) glass-fibre filters (Whatman, GF/F, 25 mm), rinsed with prefiltered sea water and stored at –20 °C until analysis. Chlorophyll *a* was measured fluorometrically after extraction in 90% acetone. POC and PON content were determined using a Carlo Erba CHN-analyzer (CE Instruments, Italy).

Rate measurements were determined immediately after the collection of samples at *in situ* temperatures ($\pm$1 °C). Primary production was measured following JGOFS-protocols (UNESCO, 1994). Samples were incubated in 275 ml polycarbonate bottles with 740 kBq $NaH^{14}CO_3$ *in situ*, or at simulated *in situ* conditions by using neutral screens (3 to 4 different depths according to *in situ* light measurements). After 3 to 4 h of incubation, aliquots were filtered through Nuclepore membrane filters of 0.2 μm, 0.8 μm, 2.0 μm, 3 μm and 5 μm pore size and glass-fibre filters (Whatman, GF/F). Filters were fumed with concentrated HCl for 15 min, and radioactivity was counted in a Tri-Carb 2560 TR/X liquid scintillation counter (Packard, USA) using UltimaGold XR (Packard) as scintillation cocktail. Surface primary production in the salinity gradient was measured additionally on one occasion in June

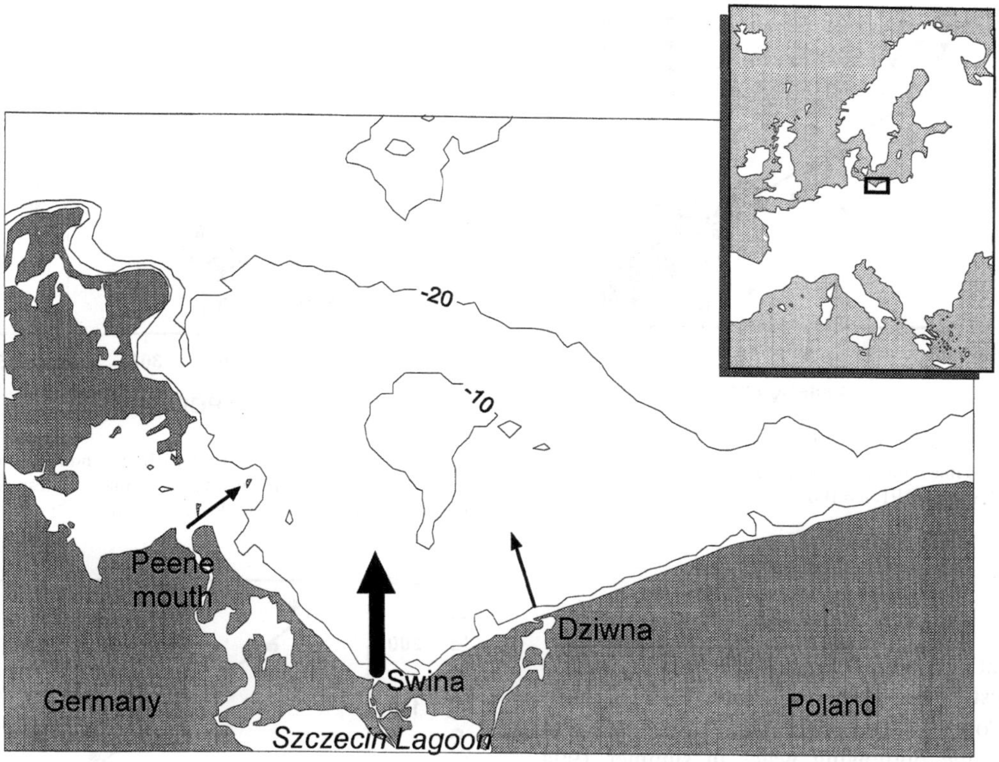

Figure 1. Map of the Pomeranian Bight area. The approximate location in central Europa is also shown. Arrows indicate the relative water discharge along the different outlets.

1996. In order to avoid effects of different insolation on production rates, surface water was sampled on a transect between 2 and 8 PSU salinity within 1 hour and kept in the dark at in situ temperature. Then, the samples were incubated under the same light conditions (20% light reduction neutral screen) for a 2 hour period over noon.

Bacterial production was estimated by thymidine and leucine incorporation with a dual- label method as in Jonas et al. (1988). We incubated triplicate 10-ml aliquots with ^{3}H-methyl-thymidine (925 GBq mmol^{-1}, final concentration 10 nmol l^{-1}) and ^{14}C-U-L-leucine (10.8 GBq mmol^{-1}, final concentration 100 nmol l^{-1}) for 1 h. Incubation was stopped by adding formalin (final concentration 0,35%). The samples plus prekilled blanks were filtered through 0.2 μm nitrocellulose filters, then rinsed ten times with 1 ml ice-cold 5% trichloroacetate. For estimates of size fractionated bacterial production, Nuclepore membrane filters of 0.2 μm and 2.0 μm pore size were used instead of nitrocellulose filters. Turnover rates of glucose and aspartic acid were also estimated by a dual-labelling technique. We incubated triplicate 10 ml aliquots with ^{3}H-2,3-L-aspartic acid (777 GBq mmol^{-1}, final concentration 0.2 nmol l^{-1}) and ^{14}C-U-D-glucose (1.85 GBq mmol^{-1}, final concentration 80 nmol l^{-1}) for 1 h. The incubation was stopped by adding formalin and followed by filtration. Rinsing of filters was done by using prefiltered sea water. In addition to radioassaying the filters, also 100 μl aliquots of several blanks were prepared to estimate total concentrations of added substrates. Radioactivity of air-dried filters was counted at least 3 days after adding 4 ml scintillation fluid by liquid scintillation counting.

Community respiration was determined as difference in oxygen concentration measured by Winkler titration at $t = 0$ and after 24 hours of incubation. Incubation was performed in 100 ml oxygen bottles at *in situ* temperatures in the dark. Bacterial production was calculated as mean of leucine incorporation according to Simon & Azam (1989) and thymidine incorporation by using 1.1*10^{18} bacteria (Mol thymidine)$^{-1}$ (Riemann et al., 1987), 0.08 μm^{3} (bacterium)$^{-1}$ and 350 fg C μm^{-3} (Bjørnsen, 1986) as conversion factors.

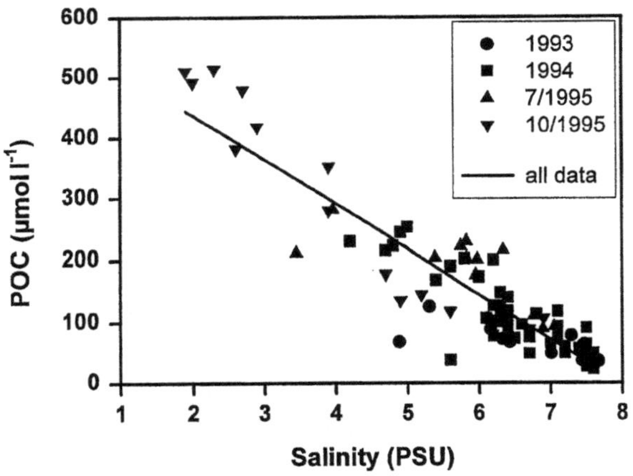

Figure 2. Correlation of POC and salinity for all data of the four cruises. The solid line represents linear regression including all data: POC = 584–73 Sal; $r = 0.91$; $n = 100$.

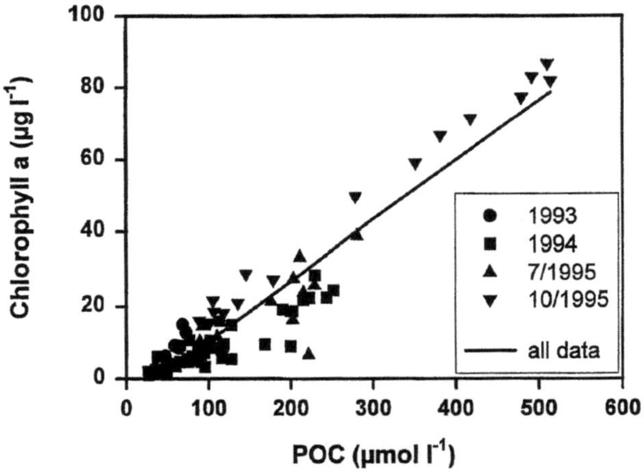

Figure 3. Correlation between chlorophyll *a* and POC for all data of the four cruises. The solid line represents linear regression including all data: Chl. $a = -6.2 + 0.165$ POC; $r = 0.94$; $n = 81$.

Results

The nutrient concentrations in the water leaving the lagoonal system during the vegetation period are usually low. Comparisons between the central Szczecin lagoon and the outflowing water in summer 1994 (Dahlke & Kerstan, pers. comm.) showed that values at the outlet were considerably lower (0.8 μm DIN (dissolved inorganic nitrogen); 0.1 μm DIP (dissolved inorganic phosphorus)) than in the inner lagoon (34 μm DIN; 0.8 μm DIP).

The infrequent and irregular outflow from the lagoonal system into the Pomeranian Bight resulted in a high temporal and horizontal variability of discrete mixing events and did not allow to allocate distinct areas or a fixed timescale to certain ranges of the mixing gradient. Therefore, salinity was directly used as the independent variable for mixing, and data from different cruises and types of experiments were pooled. Although all sets of variables were measured under different experimental conditions (small and large scale gradients, drift experiments) the results are remarkably consistent, when normalized to salinity. A strong linear decrease of POC with increasing salinity (shown in Figure 2) was observed during all cruises. Linear regression analysis of all values showed that the variability in POC content could be predicted by salinity to more than 80% ($r^2 = 0.81$). The same basic pattern was found for salinity and chlorophyll *a* ($r^2 = 0.83$).

The slope of the linear regression between POC and chlorophyll *a* (see Figure 3) yields a POC/chlorophyll *a* ratio of 72 on a μg/μg basis. This suggests that phytoplankton made up a large and relatively constant proportion of POC over the whole mixing gradient. Figure 4 shows the linear relationship between salinity and surface primary production rates under the same simulated in situ light conditions. However, as the integrated water column production is affected by the vertical (turbidity dependend) light distribution on each station, its development in the mixing area is not linear over the whole range. Therefore, imbalances in the production/respiration budget were particularly found close to the Swina outlet. The mean depth-integrated primary production value over the salinity gradient in 1994 (4 to 8 PSU) was found to be 1590 mgC m^{-2} d^{-1}, whereas mean total respiration amounted to 1330 mgC m^{-2}

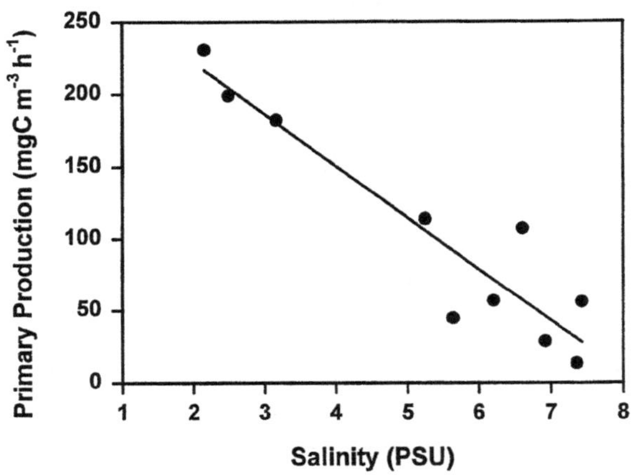

Figure 4. Correlation of surface primary production and salinity along a selected salinity gradient in June 1996. The solid line represents linear regression: PP = 294–35,8 Sal; $r = 0{,}93$; $n = 10$.

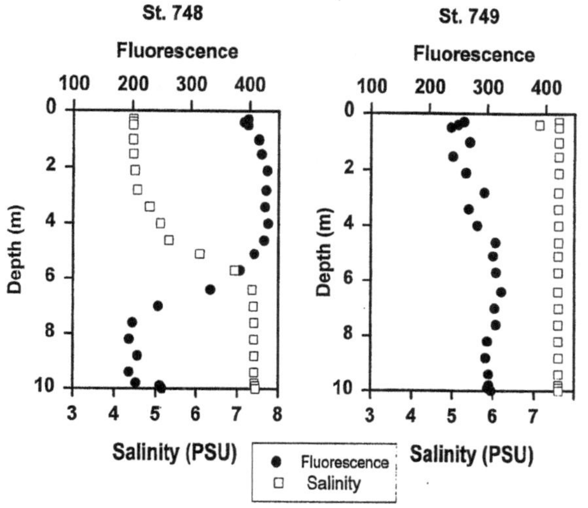

Figure 5a.

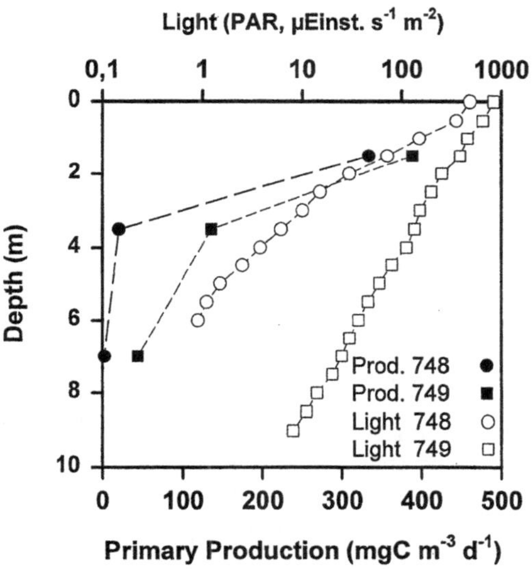

Figure 5b.

Figure 5. Vertical profiles of salinity, Chl *a* -fluorescence (5a), light (note logarithmic scale) and primary production (5b) at drift stations 748 and 749 in September 1993.

d^{-1}. In this frame the net carbon balance ranged from negative values of –1200 mgC m^{-2} d^{-1} close to the Swina mouth to positive ones of around +140 mgC m^{-2} d^{-1} in the whole outer mixing area. The transition in salinity and fluorescence from a still stratified to a well mixed state of the water column is depicted in Figure 5a. In this example, the mixing occured within 24 hours and a range of 11 nautical miles from the Swina mouth. Primary production profiles (Figure 5b) in the mixing zone generally reflected the vertical light distribution. While light intensity (PAR) at a given depth between these stations differed over several orders of magnitude, the integrated primary production rates were not that far apart.

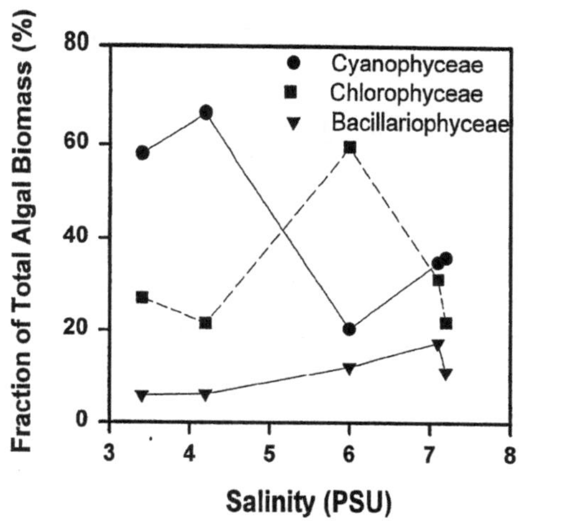

Figure 6. Structural changes in the autotrophic community along a selected salinity gradient in September 1995 indicated by proportions of different algal groups.

There was a distinct shift from an autotrophic community dominated by colonial cyanobacteria to green algae and diatoms at higher salinities (Figure 6). This shift was also reflected in the size spectrum of production. A shift from larger to smaller algae is indicated in the vertical profile of Figure 7, where production in the top layer with a lower salinity was dominated by algae > 5 μm (colonial cyanobacteria) and in the saline bottom layer by a mixture of these aggregates with small flagellates in the size ranges between 0.4 and 3 μm.

We used community respiration as an integrative measure for the sum of heterotrophic processes in the water (Figure 8). A strong negative correlation between community respiration and salinity was evident. In the course of our investigation, seasonal differences in water temperature ranging from 10.7 to 19.2 °C were observed. Normalisation of the respiration values from these temperatures to 15 °C, assuming a Q_{10} of 2.5 (see Valiela, 1995), increased the predictability of the values by salinity from 48% to 61%. During the mixing process, the community respiration was reduced nearly by a factor of 4 in a gradient from 2 to 7.5 PSU. A less distinct decrease was observed in bacterial production. Here, the difference of the initial and final values of the mixing process was about 3.3-fold (see Figure 9). Temperature normalisation to 15 °C increased the predictability by salinity from 38% to 59%, which is very

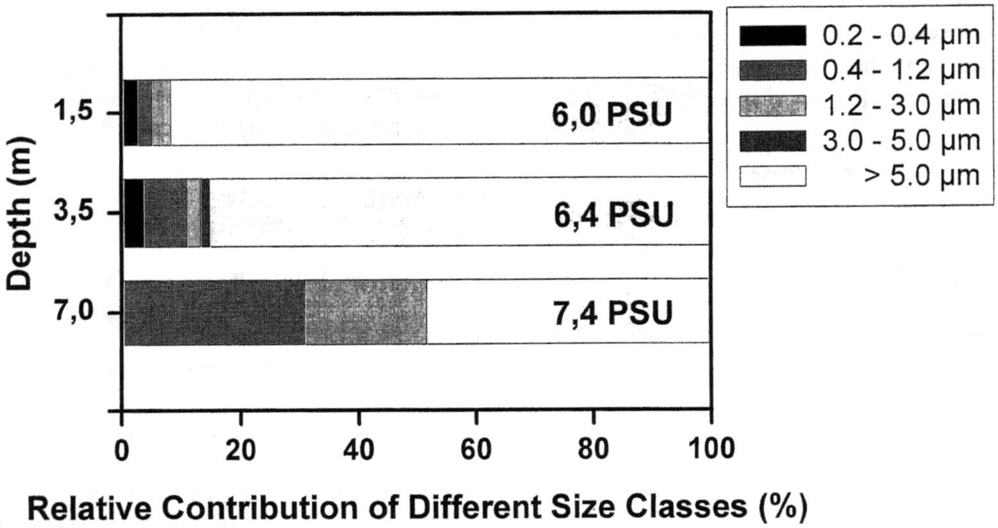

Figure 7. Relative proportions of size fractions of primary production at different depths in a still stratified state (st. 775, September 1993, 5 nautical miles off the Swina mouth). Salinity of the selected depths is indicated.

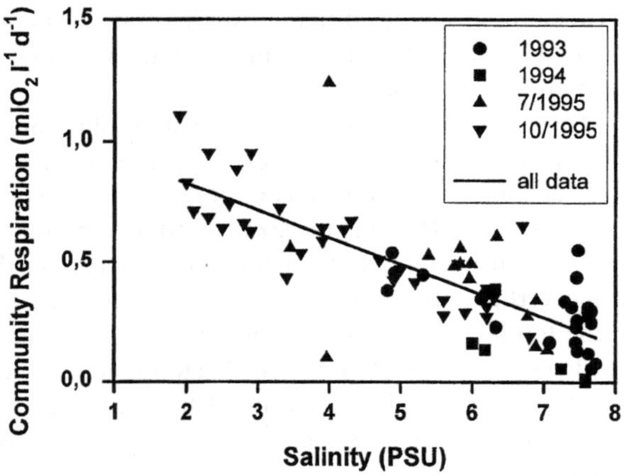

Figure 8. Correlation of community respiration (normalized to 15 °C; see text) and salinity for all data of the four cruises. The solid line represents linear regression including all data: Com.Resp. = 1.05–0.11 Sal.; $r = 0.78$; $n = 139$.

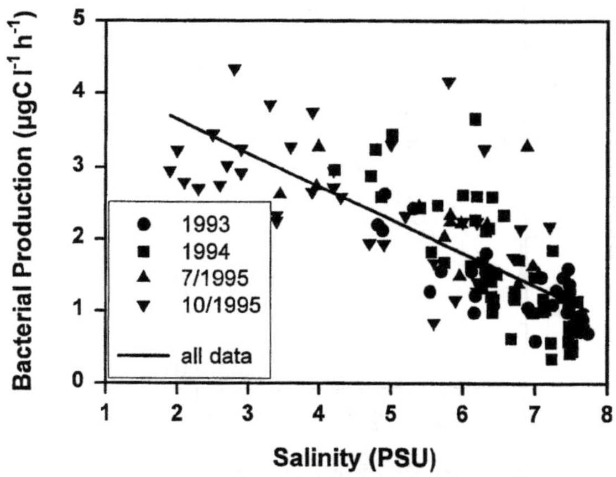

Figure 9. Correlation of bacterial production (normalized to 15 °C; see text) and salinity for all data of the four cruises. The solid line represents linear regression including all data: Bact.Prod. = 4.57–0.46 Sal.; $r = 0.77$; $n = 139$.

similar to that found in community respiration. The decline of bacterial production was accompanied by structural changes. This is indicated by the observation that the relative proportion of the bacterial production in the size fraction > 2 μm decreased from 55% to 7% in the course of mixing (Figure 10). Partial correlation analyses shows that total bacterial production is correlated only with salinity, while bacterial production in the fraction > 2 μm displays a significant partial correlation only with POC.

Discussion

In the light of the observed results we have to reject our initial hypotheses that the processes in the mixing area are dominated by 'new' production properties. The high inorganic nutrient load of the river Oder is already transformed into a highly productive pelagic community during the passage of the inner lagoonal areas (mainly the Szczecin lagoon).

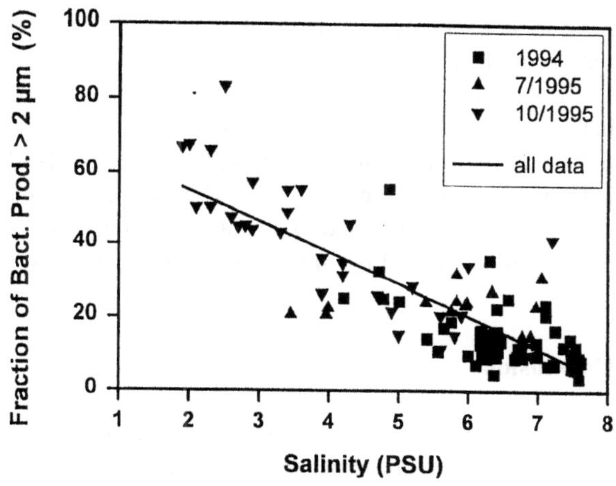

Figure 10. Correlation of the fraction of bacterial production > 2 μm and salinity for all data of three cruises (except 1993, when it was not estimated). The solid line represents regression line including all data: Frac.Bact.Prod = 72.6–8.7 Sal.; $r = 0.82$; $n = 106$.

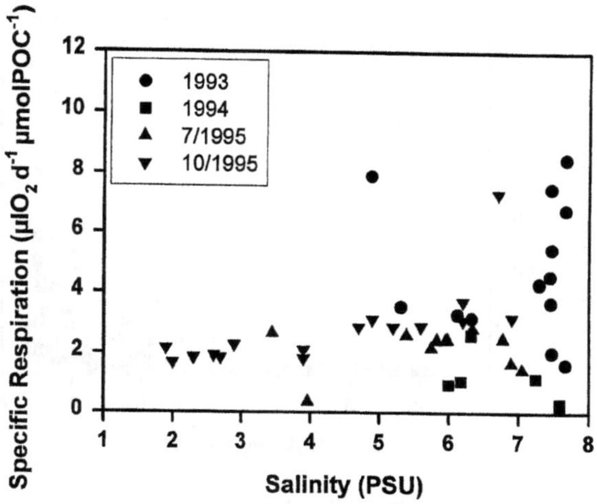

Figure 11. Relationship between specific respiration (quotient from normalized community respiration divided by accombined POC) and salinity at all investigation periods. The large scatter of rates at the highest salinities is rather an effect of the lower accuracy of the method at low rates of oxygen consumption and marks its lower limit.

In the discharge area, the riverine material is mixed into the coastal environment as a living, actively regenerating community of adapted organisms from a preconditioned lagoonal system. This is well illustrated by the strong correlation between POC and chlorophyll *a* (see Figure 3). A median of the C/N ratio between 7 and 8 implies the POC to be a product of a normal pelagic community (Redfield et al., 1963) with only minor fractions of detrital riverine material. So, in addition to the high proportion of phytoplankton carbon, we have to assume another significant fraction of the POC-pool to be heterotrophic biomass.

The negative linear regression between salinity and POC and also with chlorophyll *a* suggests simple conservative mixing (see Figure 2). Kirchman et al. (1984) could explain most of the observed variation in bacterial abundance in a tidally mixed estuary by conservative mixing, according to a significant linear regression between these values and salinity. Similar observations for the distribution of stock parameters were reported by other authors (e.g. Lochet, F. & M. Leveau, 1990; Revelante, N. & M. Gilmartin, 1992).

The observed regular conservative mixing pattern would, however, require a passive behaviour of all components, which is by no means supported by the rate measurements and changes in community structure during mixing. These observations point towards a fast turnover of all components during the mixing process. Turnover of the total particulate carbon pool by primary production (POC/primary production) was calculated for mean carbon values and production data that were normalized to a mean light profile of the mixing gradient between 2 and 7.5 PSU salinity. The turnover time was in the range of 4.5 days and phytoplankton carbon contributed to less than 50% to the C-pool. The algal doubling time was in the range of 1 to 2 days, which is an indication of a very active algal community (Eppley, 1981).

Within the mixing gradient no reduction in metabolic activity normalized to POC could be detected in spite of qualitative changes of the populations, as indicated by the succession of algal groups and size classes. The shift between species is faciliated by the fast turnover, whereby the overall functional properties of the system are not visibly affected. Shifts in size classes of producers were matched by equivalent changes in size classes of bacterial production. Pollehne et al. (1995) detected as well a shift among the protozoans from larger ciliates to heterotrophic nanoflagellates. Thus, the relatively constant specific activity per biomass indicates a dilution of a well adapted pelagic system, as the pooled data on POC-specific respiration rate show in Figure 11. Both community respiration and bacterial production point towards a viable and active heterotrophic community in the mixing zone, which does not display a distinct reduction or even a gap of activity at any point in the gradient.

The river is apparently not a direct source of dissolved substrates for bacterial growth. Figure 12 shows the turnover time of easily degradable model substrates

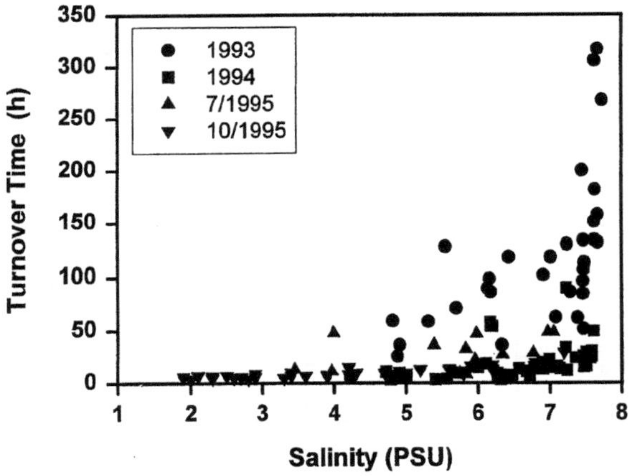

Figure 12. Relationship between turnover time (mean from that of glucose and aspartic acid) and salinity at all investigation periods.

(mean of glucose and aspartic acid turnover time) in the salinity gradient. Low turnover times in the lower salinity range show a rapid turnover. This contradicts a substantial riverine input of such substances. The trend to increased turnover times with higher salinities may on the contrary reflect a better supply of labile dissolved organic carbon in the outer range of the mixing gradient. This indicates that the bacterial substrates are primarily a product of a strong coupling between auto- and heterotrophs on short temporal and spatial scales, which is in good agreement with Coffin et al. (1990) who reported the DOM derived from phytoplankton to be the dominant carbon source for estuarine bacteria.

The slight accumulation of these products also suggests that the net system balance turns more to autotrophy in the later stages of mixing. This fits to the observation of an increase in integrated water column primary productivity due to better light conditions in the less turbid outer area. The relation between increase in underwater light and areal primary production, however, is not linear. So, the general picture of the observed auto-heterotrophic relationship is much more one of a closed, well geared regenerating community with only minor losses. As, however, the biomass level is extremely high, even these losses are able to satisfy the benthic energy demand in the area (Powilleit et al., 1995).

A biological system that functions like this, will have different effects on the two main problems caused by human activities in coastal areas:

I. It can not counteract eutrophication, as material is constantly recycled and not permanently stored. Due to the complete vertical mixing of the water column the benthic biota are closely involved in this cycle and can only provide temporal sinks.

II. As the physiological activity of both auto- and heterotrophic components is high, there is a high potential for the modification of compounds that get involved in the physiology of the organisms. Therefore, this type of system can be expected to play an important role in the transformation of organic or inorganic pollutants.

Conclusions

During the growing season in the Oder estuary, the 'new' production phase occurs already in the lagoonal system. Most of the matter exported further into the Pomeranian Bight is bound in an active and well balanced biological system. As this balance is maintained over the whole salinity gradient, the observed distribution of all bulk parameters follows the pattern of a conservative mixing process. The maintenance of a balanced system in this gradient is due to the preconditioning of the populations in the lagoonal system, and the gradual way of mixing, which leaves a few generation times for adaptational processes. The basic setting for this type of transition is provided by the overall low salinity gradient between the lagoonal area and the open Baltic Sea, as compared to oceanic estuaries.

Acknowledgement

We gratefully acknoweldge the engaged support of capitain and crew of R/V 'Prof. Albrecht Penck'. We thank colleagues at the IOW for supplying data and performing analyses, particularly S. Busch, E. Kerstan, H. Siegmund, D. Setzkorn and I. Topp. The study was funded by German Ministry for Education, Research and Technology (03F0105B). This is publication no. 240 of the Baltic Sea Research Institute.

References

Bjørnsen, P. K., 1986. Automatic determination of bacterioplankton biomass by image analysis. Appl. envir. Microbiol. 51: 1199–1204.

Coffin, R. B., D. J. Velinsky, R. Devereux, W. A. Price & L. A. Cifuentes, 1990. Stable carbon isotope analysis of nucleic acids to trace sources of dissolved substrates used by estuarine bacteria. Appl. envir. Microbiol. 56: 2012–2020.

Cole, J. J., S. Findlay & M. L. Pace, 1988. Bacterial production in fresh and salt water ecosystems: A cross-system overview. Mar. Ecol. Prog. Ser. 43: 1–10.

Ducklow, H. W., 1982. Chesapeake Bay nutrient and plankton dynamics. 1. Bacterial biomass and production during spring tidal destratification in the York River, Virginia, estuary. Limnol. Oceanogr. 27: 651–659.

Ducklow, H. W. & C. A. Carlson, 1992. Oceanic bacterial production. Adv. Microb. Ecol. 12: 113–181.

Ducklow, H. W. & D. L. Kirchman, 1983. Bacterial dynamics and distribution during a spring diatom bloom in the Hudson River plume, USA. J. Plankton Res. 5: 333–355.

Eppley, R. W., 1981. Relations between nutrient assimilation and growth in phytoplankton with a brief review of estimates of growth rates in the ocean. In T. Platt (ed.), Physiological Bases of Phytoplankton Ecology. Can. Bull. Fish. aquat. Sci. 210: 251–263.

Findlay, S., M. L. Pace, D. Lints & J. J. Cole, 1991. Weak coupling of bacterial and algal production in a heterotrophic ecosystem: The Hudson River estuary. Limnol. Oceanogr. 36: 268–278.

Fuhrman, J., J. W. Ammerman & F. Azam, 1980. Bacterioplankton in the coastal euphotic zone: distribution, activity and possible relationships with phytoplankton. Mar. Biol. 60: 201–207.

Jacobsen, T. R., L. R. Pomeroy & J. O. Blanton, 1983. Autotrophic and heterotrophic abundance and activity associated with a nearshore front off the Georgia coast, USA. Estuar. coast. Shelf Sci. 17: 509–520.

Jonas, R. B., J. H. Tuttle, D. L. Stoner & H. W. Ducklow, 1988. Dual-label radioisotope method for simultaneously measuring bacterial production and metabolism in natural waters. Appl. envir. Microbiol. 54: 791–798.

Kirchman, D., B. Petersen & D. Juers, 1984. Bacterial growth and tidal variation in bacterial abundance in the Great Shippewissett Salt Marsh. Mar. Ecol. Prog. Ser. 19: 247–259.

Lochet, K. & M. Leveau, 1990. Transfers between a eutrophic ecosystem, the River Rhone, and an oligotrophic ecosystem, the northwestern Mediterranean Sea. In D. J. Bonin & H. L. Goltermans (eds), Fluxes Between Trophic Levels and Through the Water-Sediment-Interface 207: 95–103.

Pastuszak, M., K. Nagel & G. Nausch, 1996. Variability in nutrient distribution in the Pomeranian Bay in Septermber 1993. Oceanologia 38: 195–225.

Pollehne, F., S. Busch, G. Jost, B. Meyer-Harms, M. Nausch, M. Reckermann, P. Schaening, D. Setzkorn, N. Wasmund & Z. Witek, 1995. Primary production patterns and heterotrophic use of organic material in the Pomeranian Bay (Southern Baltic). Bull. Sea Fish. Inst., Gdynia 3(136): 43–60.

Postel, L., N. Mumm & A. Krajewska-Sołtys, 1995. Metazooplankton distribution in the Pomeranian Bay, (southern Baltic) – species composition, biomass and respiration. Bull. Sea Fish. Inst., Gdynia 3: 61–73.

Powilleit, M., J. Kube, J. Masłowski & J. Warzocha, 1995. Distribution of macrobenthic invertebrates in the Pomeranian Bay (Southern Baltic) in 1993/1994. Bull. Sea Fish. Inst., Gdynia 3: 75–87.

Redfield, A. C., B. H. Ketchum & F. A. Richards, 1963. The influence of organisms on the composition of seawater. In M. N. Hill (ed.), The Sea. John Wiley & Sons, New York, 2: 26–77.

Revelante, N. & M. Gilmartin, 1992. The lateral advection of particulate organic matter ftom the Po delta region during summer stratification, and its implications for the northern Adriatic. Estuar. coast. Shelf Sci. 35: 191–212.

Riemann, B., P. K. Bjørnsen, S. Newell & R. Fallon, 1987. Calculation of cell production of coastal marine bacteria based onmeasured incorporation of (^{3}H)thymidine. Limnol. Oceanogr. 32: 471–476.

Riemann, B. & M. Søndergaard, 1984. Measurements of diel rates of bacterial secondary production in aquatic environments. Appl. envir. Microbiol. 47: 632–638.

Rybinski, J., E. Niemirycz & Z. Makowski, 1992. Pollution load. In A. Trzosinska (ed.), Marine Pollution (2) An Assessment of the Effects of Pollution in the Polish Coastal Area of the Baltic Sea 1984–1989. National Scientific Committee on Oceanic Research PAS, Gdansk: 21–52.

Shia, F.-K. & H. W. Ducklow, 1994. Temperature regulation of heterotrophic bacterioplankton abundance, production, and specific growth rate in Chesapeake Bay. Limnol. Oceanogr. 39: 1243–1258.

Shia, F.-K. & H. W. Ducklow, 1995. Multiscale variability in bacterioplankton abundance, production, and specific growth rate in a temperate salt-marsh tidal creek. Limnol. Oceanogr. 40: 55–66.

Siegel, H., M. Gerth & T. Schmidt, 1996. Water exchange in the Pomeranian Bight investigated by satellite data and shipborne measurements. Continental Shelf Res. 16: 1793–1817.

Simon, M. & F. Azam, 1989. Protein content and protein synthesis rates of planktonic marine bacteria. Mar. Ecol. Prog. Ser. 51: 201–213.

UNESCO, 1994. Protocols for the Joint Global Ocean Flux Study (JGOFS) core measurements. IOC/SCOR Manual and Guides 29: 128–134.

Valiela, I., 1995. Marine Ecological Processes, 2. Springer, New York, 686 pp.

von Bodungen, B. & B. Zeitzschel, 1996. Die Ostsee als Ökosystem. In G. Rheinheimer (ed.), Meereskunde der Ostsee, 2. Springer-Verlag, Heidelberg: 230–244.

White, P. A., J. Kalff, J. B. Rasmussen & J. M. Gasol., 1991. The effects of temperature and algal biomass on bacterial and specific growth rate in freshwater and marine habitats. Microb. Ecol. 21: 99–118.

Hydrobiologia **363:** 117–126, 1998.
T. Tamminen & H. Kuosa (eds), Eutrophication in Planktonic Ecosystems: Food Web Dynamics and Elemental Cycling.

Variability of nutrient limitation in the Archipelago Sea, SW Finland

T. Kirkkala, H. Helminen & A. Erkkilä
Southwest Finland Regional Environment Centre, Inkilänkatu 4, FIN-20300 Turku, Finland

Key words: nutrient limitation, eutrophication, nitrogen, phosphorus, Archipelago Sea

Abstract

Eutrophication is the most acute environmental problem in the Archipelago Sea, SW Finland. When analysing the factors behind this escalating eutrophication the determination of limiting nutrient at a given time is essential. Besides experimentations, nutrient limitation of plankton has been extensively studied by direct chemical analyses. We used the latter approach in this work. Nutrient limitation was studied by calculating different nutrient ratios – total nitrogen:phosphorus, inorganic nitrogen:phosphorus, and nutrient balance ratio. Results showed that phosphorus usually limited primary production only near the coast line. In the middle zone of the Archipelago Sea the limiting factor varied temporally. Outer in the open sea nitrogen limited primary production during most of the year. Phosphorus limited phytoplankton growth especially in spring and in summer and nitrogen in late summer and in autumn. Our results suggested that nitrogen is an important limiting nutrient in the Archipelago Sea. In recent years when the eutrophication has proceeded there has been a shift from production limitation by both nutrients to limitation by nitrogen alone. But if we want to define and characterize the nutrient limitation of the entire ecosystem of the Archipelago Sea, budgets have to be calculated for both N and P and internal recycling must be taken into account as well as external supply of nutrients and loss processes.

Introduction

Nutrient enrichment and eutrophication are regarded as the most acute environmental problems in the Archipelago Sea, SW Finland (e.g. Jumppanen & Mattila, 1994) as in the whole Baltic Sea (e.g. Larsson et al., 1985; Wulff et al., 1990). When analysing the factors behind this escalating eutrophication the determination of limiting nutrient is essential. Several studies have shown that phosphorus (P) and nitrogen (N) are the nutrients potentially most limiting plankton growth in fresh and brackish waters (see e.g. Wetzel, 1983; Hecky & Kilham, 1988). It has been frequently stated that N limits phytoplankton growth in the sea and P in freshwater; consequently the question of which of them is limiting at intermediate salinities is receiving increasing attention (Paasche & Erga, 1988).

In the Archipelago Sea nutrient discharges from fish farming and diffuse loading have increased during the last 20–25 years. Although phosphorus load from industrial and municipal waste waters has decreased effectively (Pitkänen, 1994), total nitrogen load has remained high and environmental authorities must soon make decisions concerning nitrogen reduction in this area. Therefore it is important to know how the role of limiting nutrient varies in different parts of the Archipelago Sea.

The limiting nutrient concept was originally developed by Liebig (1855) as the 'Law of the Minimum'. Simply paraphrased, it states that the yield of any organism will be determined by the abundance of the substance that, in relation to the needs of the organism, is least abundant in the environment (Wetzel, 1983).

Besides experimentation, nutrient limitation of plankton has been extensively studied by direct chemical analyses (Tamminen, 1990). This chemical evaluation is usually based on the comparison of observed nutrient ratios (inorganic, particulate or total) to corresponding Redfield ratios (Redfield et al., 1963) or local derivatives (Tamminen, 1990). The relative shortage – in relation to Redfield or others ratios – of a given nutrient is interpreted as indication of potential limitation by that nutrient (e.g. Forsberg et al., 1978).

In this study we analyse temporal and spatial variability of nutrient concentrations, describe long-term changes of water quality and finally evaluate the question of nutrient limitation in the Archipelago Sea. Thus, this paper summarizes a large number of chemical data from the coastal waters of southwestern Finland, but does not give any results of experiments for physiological nutrient limitation. Therefore, some limitations of our approach should be taken up here. Nutrient ratios cannot be taken as indicative of limitation if ambient nutrient concentrations are high. Yet nutrients control the algal growth rate only if supply falls short of demand over the timescale of cellular growth and reproduction (Harris, 1986), and thus low concentrations do not necessarily imply nutrient limitation. So in this – as in other similar investigations (eg. Paasche & Erga, 1988) – the word limitation has been used rather loosely to indicate a state where the lack of one nutrient elicits potential reduction in the algal growth rates.

Study area

The Archipelago Sea is located between the Baltic proper and the Bothnian Bay (See Figure 1). The Archipelago Sea is characterized by an enormous topographic complexity, including about 25 000 islands. The average water depth is only 23 m and the deepest trench reaches 146 m. The total coastal drainage area is about 8900 km^2 (of which lakes cover under 2% and fields 28%). The total area of the Archipelago Sea is 9436 km^2 and the water volume is 213 km^3. Water flows mainly from the Baltic Sea basin to the Bothnian Bay through the Archipelago Sea and back to the Bothnian Sea mainly along the Swedish coast. This means an eastward net transport in the south, northwards along the eastern coasts, and a southward transport along the western coast of the Baltic Sea (HELCOM, 1993). Eight rivers run to the Archipelago Sea. The mosaic morphology and the environmental gradients (salinity, temperature, exposure etc.) create several biotopes and complicated ecological webs (Blomqvist & Bonsdorff, 1993; von Numers, 1995). As the topography is complex, and the water shallow, the area acts as a buffer or filter between the coastline and the open sea, and also between the Baltic proper and the Bothnian Bay. A great deal of suspended matter and nutrients settles down to the bottom, but most part of the nutrients is used for primary production (Jumppanen & Mattila, 1994).

The archipelago is affected by nutrients coming from several sources; from industrial and municipal waste waters and as a diffuse load from agriculture and forestry. The coastal areas have been affected by a heavy nutrient load since the 1960s. Fish farming was introduced in the 1970s as a new source of nutrient load. Now the Archipelago Sea is the major fish farming area in Finland. In the 1990s the annual total phosphorus load from industrial and municipal waste waters, fish farming and agriculture and forestry has been about 500 t and the total nitrogen load about 7000 t.

Material and methods

The Southwest Finland Regional Environment Centre (SFREC) has monitored the water quality in the Archipelago Sea since the 1960s. In all, water samples have been collected at 24 monitoring stations 2–4 times a year in late winter and late summer. The monitoring station Seili (established in 1983) is the most intensively studied station in the Archipelago Sea having over twenty sampling times a year. Time series for this study were obtained from the Seili station and the stations Airismaa and Nötö (See Figure 1). We concentrate here on the samples taken during May–October since biological production mostly happens during the open water season. Spatial variability of water quality has been studied in May and in August at 50–60 monitoring stations covering the whole study area.

All samples used here were analyzed in the laboratory of the SFREC where all physical and chemical analyses have been performed using standard methods (Koroleff, 1976, 1979; National Board of Waters, 1981). Total phosphorus (TP), phosphate-phosphorus (DIP), total nitrogen (TN), ammonium-nitrogen (NH_4-N), nitrate- and nitrite-nitrogen (NO_2-N + NO_3-N) (NH_4-N and NO_2-N + NO_3-N together DIN) have been analyzed from discrete unfiltered samples (1, 5, 10, 20, 30 ... 1–2 m above the bottom). DIP has been analyzed by an ammonium molybdate method with ascorbic acid as the reducing agent. In TP determination the sample was digested by $K_2S_2O_8$ before it was analyzed with ammonium molybdate. NH_4-N has been analyzed colorimetrically with hypochlorite and phenol. The sum of NO_3-N and NO_2-N has been determined by reduction of NO_3 followed by NO_2-

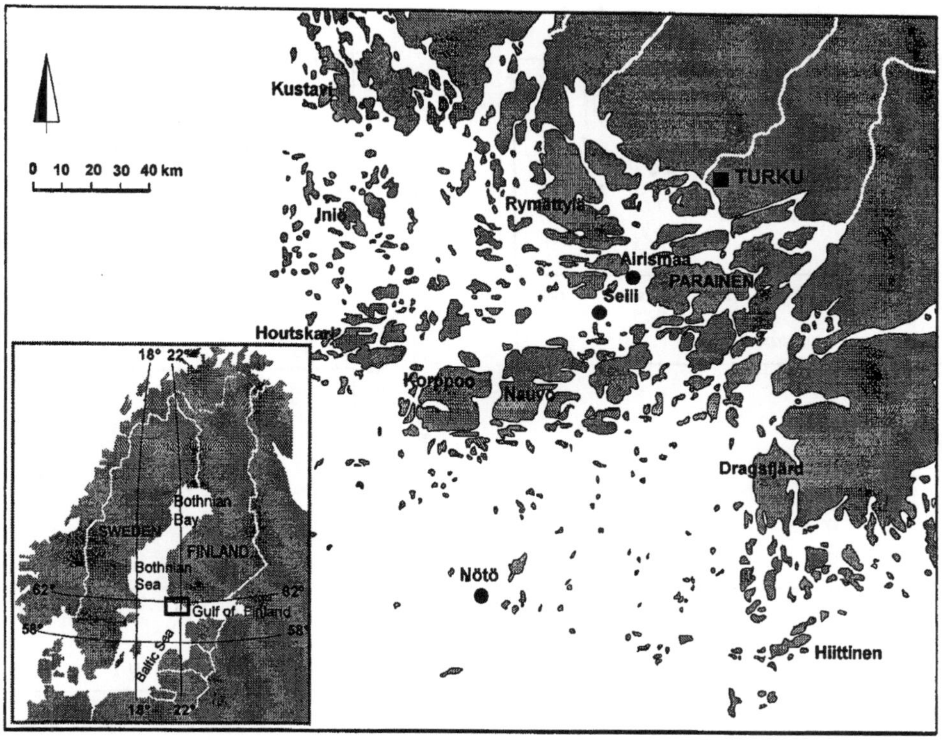

Figure 1. Study area.

N determination. TN has been analyzed as NO_3-N after the digestion of the sample with $K_2S_2O_8$. Filtrations for chlorophyll-*a* analyses were performed from composite samples (2∗Secchi depth collected at 2 m intervals). The algal pigments have been extracted with ethanol and chlorophyll-*a* concentration has been measured spectrophotometrically. Primary productivity has been estimated with ^{14}C technique.

The limits of detection in nutrient analyses have been determined in the SFREC by comparing to known concentrations following the guidelines given by the Environmental Agency of Finland (Mäkinen et al., 1996). These instructions are based on e.g. 'Guide to Analytical Quality Control for Water Analysis' (prENV ISO/CD 13530).

Here, nutrient limitation was studied by calculating different nutrient ratios in the surface layer (0–10 m): ratios between total nitrogen and phosphorus, between inorganic nitrogen and phosphorus, and finally between these two ratios, which is referred to as the 'nutrient balance ratio' according to Tamminen (1982). The following criteria were used to determine the limiting nutrient:

(1) If the total nutrient ratio TN:TP is over 7, phosphorus limits primary production and if TN:TP <7, nitrogen is the limiting factor (Redfield et. al., 1963).

(2) The inorganic nutrient ratio DIN:DIP has been evaluated according to two criteria. According to Forsberg et al. (1978) when DIN:DIP is below 5, nitrogen is the limiting factor and if the ratio is over 12, phosphorus limits primary production. If the ratio is between 5–12, the limiting factor may be either nitrogen or phosphorus or both. According to Ryther & Dunstan (1971) values below 10 represent situations where nitrogen is likely to be the principal limiting nutrient and values over 10 represent phosphorus limitation.

(3) Nutrient limitation has also been characterized by the nutrient balance ratio (Tamminen, 1982). If (TN/TP):[DIN/DIP]>1, nitrogen limits primary production and otherwise phosphorus is the limiting factor. The nutrient balance ratio compares the readily utilizable fractions of nutrients to respective total pool sizes (Tamminen, 1982).

Trend analyses were conducted by simple regression methods and probabilities (p) and coefficients of determinations (R^2) are given in figures, if p-values were statistically significant.

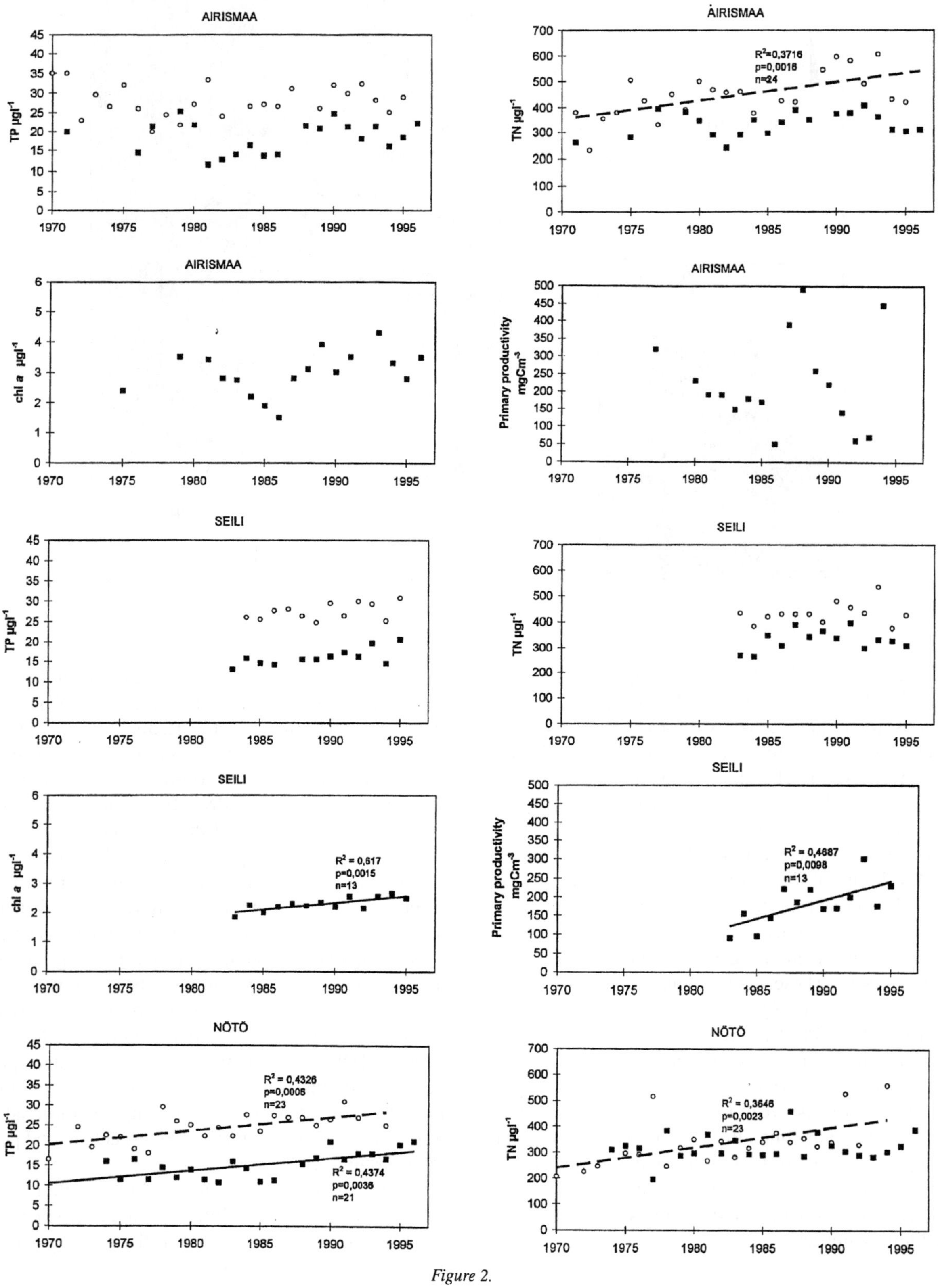
AIRISMAA
TP µgl-1
AIRISMAA
TN µgl-1
R²=0,3716
p=0,0018
n=24
AIRISMAA
chl a µgl-1
AIRISMAA
Primary productivity mgCm-3
SEILI
TP µgl-1
SEILI
TN µgl-1
SEILI
chl a µgl-1
R² = 0,617
p=0,0015
n=13
SEILI
Primary productivity mgCm-3
R² = 0,4687
p=0,0098
n=13
NÖTÖ
TP µgl-1
R² = 0,4326
p=0,0006
n=23
R² = 0,4374
p=0,0036
n=21
NÖTÖ
TN µgl-1
R² = 0,3646
p=0,0023
n=23
1970
1975
1980
1985
1990
1995

Figure 2.

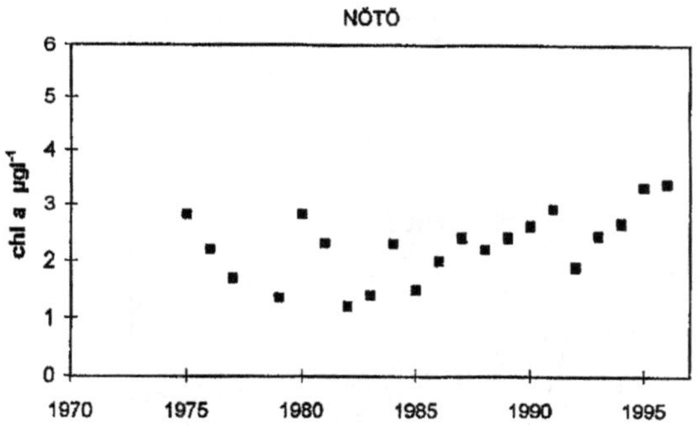

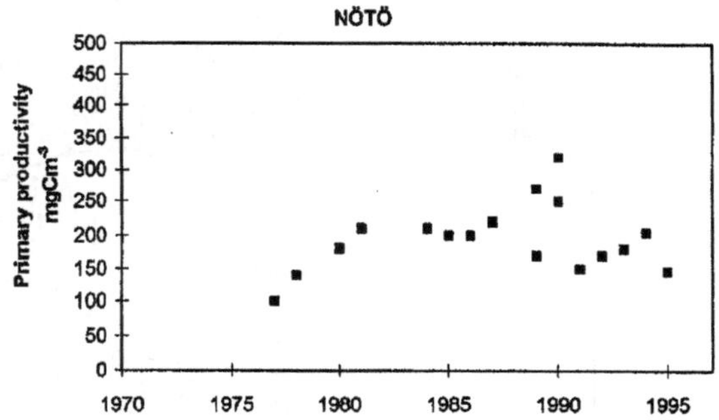

Figure 2. Total phosphorus (TP), total nitrogen (TN), chlorophyll-*a* (chl *a*) concentrations ($\mu g\ l^{-1}$) and primary productivity (mgC m^{-3}) in the Archipelago Sea 1966–1995 (black squares and solid regression line represent late summer values, circles and broken regression line represent winter values). Probabilities (p) and coefficients of determinations (R^2) are given in figures, if p-values are statistically significant.

Results

*Nutrient and chlorophyll-*a *concentrations and primary productivity*

Total phosphorus concentrations of sea water have clearly increased in the outer zone of the Archi- pelago Sea during the last two decades (see Figure 2, the monitoring station Nötö). The increase is especially apparent in the winter season data. The wintertime nutrient concentrations do not, however, directly affect the level of primary production of the following summer, because a large part of nutrients is transferred from the euphotic layer to deeper waters and to the bottom due to the vernal bloom (Pitkänen et al., 1993; Leppänen, 1988). At the monitoring station Seili, which represents the middle zone of the Archipelago Sea, the increase of the nutrient concentrations is not statistically significant. Near the coast, eg. at the monitoring station Airismaa, phosphorus concentrations decreased during the late 1960s and in the early 1970s due to notable improvements in sewage treatment techniques. After that phosphorus concentration has varied considerably.

Nutrient levels showed that the sea-areas near Turku, Rymättylä and Parainen are clearly eutrophied. Also Nauvo, Houtskari and Dragsfjärd-Hiittinen areas could be characterized as eutrophied. Other areas of the Archipelago sea are slightly eutrophied. In the late 1970s in summertime phosphorus concentrations over 20 $\mu g\ l^{-1}$ were observed only in Rymättylä. Concentrations varied between 15-16 $\mu g\ l^{-1}$ near the coast and 10-15 $\mu g\ l^{-1}$ in the outer sea. In the 1990s values over 20 $\mu g\ l^{-1}$ could be observed in more widespread areas nearby Iniö and Rymättylä sea-areas, and in Turku-Parainen-Nauvo-Korppoo sea-areas.

Total nitrogen concentrations have varied considerably, but an increasing trend has been obvious especially in winters. However, late summer concentrations have stayed at the same level (around 300 $\mu g\ l^{-1}$) during the last two decades.

Both the chlorophyll-*a* concentrations and the primary productivity have varied considerably at the Airismaa station and no trend can be observed. At the Seili station chlorophyll-*a* values and the primary productivity are clearly rising and the increase is statistically significant. Trends of chlorophyll-*a* and primary productivity seem to be rising also at the Nötö station, but the regressions are not statistically significant.

Nutrient limitation

When compared to the Redfield ratio, the total nutrient ratios (TN:TP) showed that there would exist phosphorus limitation in every analysed water sample. On the contrary, the ratios of inorganic nutrients, when compared to the criteria by Forsberg et al. (1978), showed that phosphorus usually limited primary production only near the coast line in August 1989 and during the earlier years (See Figure 3). In the middle zone of the Archipelago See the limiting factor varied temporally. Outer in the open sea nitrogen limited primary production during most of the year. If we compare the ratios of inorganic nutrient to the criteria by Ryther & Dunstan (1971), nitrogen seems to be the limiting factor also in the middle zone of the Archipelago Sea. In recent years, when the eutrophication has proceeded, there has been a shift from production limitation by both nutrients to limitation by nitrogen alone. For example, in August 1995, nitrogen limited the primary production in the whole study area.

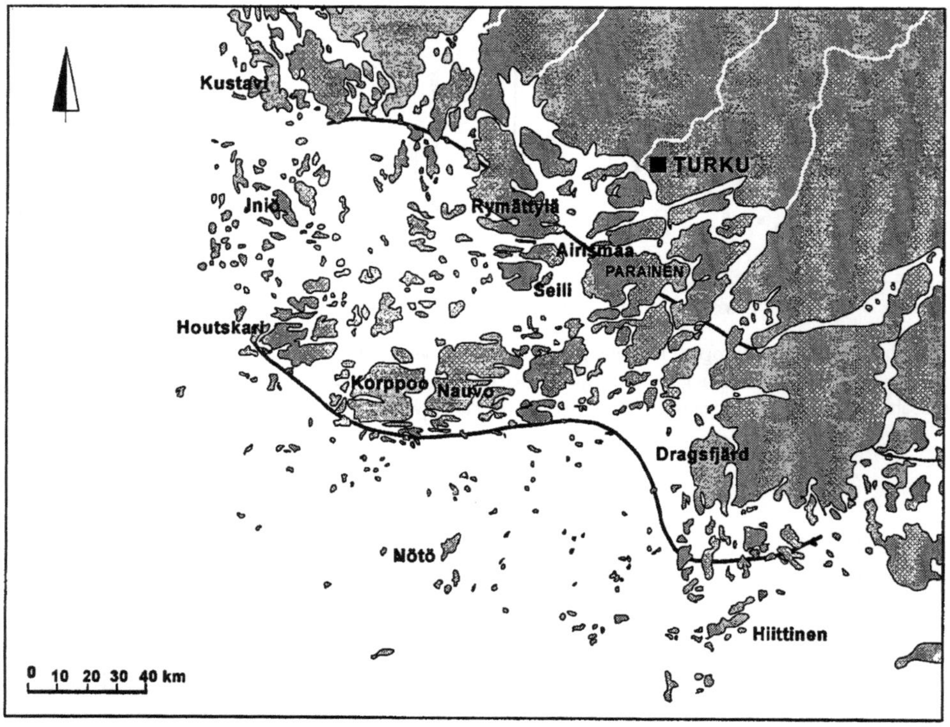

Figure 3. Inorganic nutrient ratios in August 1989. Inside the line near the coast DIN:DIP>12, between the two lines 5<DIN:DIP<12, outside the outer line DIN:DIP<5.

Frequently, a relatively sudden transition from limitation by one nutrient to limitation by the other occurs during open water seasons. According to the results (inorganic nutrients) of the monitoring station Seili, phosphorus limited phytoplankton growth especially in spring and in summer, while nitrogen was the limiting factor in late summer and in autumn when compared to the ratio by Ryther & Dunstan (1971). Results showed that there has been a change in nutrient limitation during the study period. After 1989 no phosphorus limitation could be found during the summer season (from July onwards) and in autumn (See Figure 4). When compared to the ratios presented by Forsberg et al. (1978), the DIN:DIP-ratio at the Seili station most often showed limitation situations in which the limiting factor may have been either nitrogen or phosphorus or both, especially in the 1980s. In the 1990s the DIN:DIP ratio showed mainly nitrogen limitation. The nutrient balance ratio has shown nearly exclusive nitrogen limitation during the whole monitoring period (See Figure 5). Moreover, in recent years all limitation cases by phosphorus alone were found in early May. Thus, during the 13 study years the proportion of the cases in which nitrogen was the limiting factor has increased (See Figures 4, 5a, b and c).

Discussion

Our results suggested that nitrogen is an important limiting nutrient in the Archipelago Sea. During the last decades the eutrophication has proceeded and phosphorus concentrations of the sea water have increased. At the same time a shift from production limitation by both nutrients to limitation by nitrogen alone has occured. Similar results have been found in other studies from the Archipelago Sea (e.g. Tamminen 1990). However, no general agreement on the question of the limiting nutrient at intermediate salinities has been found.

For example, ecosystem-level experiments in temperate coastal marine environments have shown significant responses to enrichment with nitrogen alone but not with phosphorus alone, indicating limitation by nitrogen (Oviatt et al., 1995; Taylor et al., 1995). In a simulated estuarine gradient bioassay, however, both inorganic nutrient concentrations and N:P ratios indicated that phosphorus was the limiting nutrient

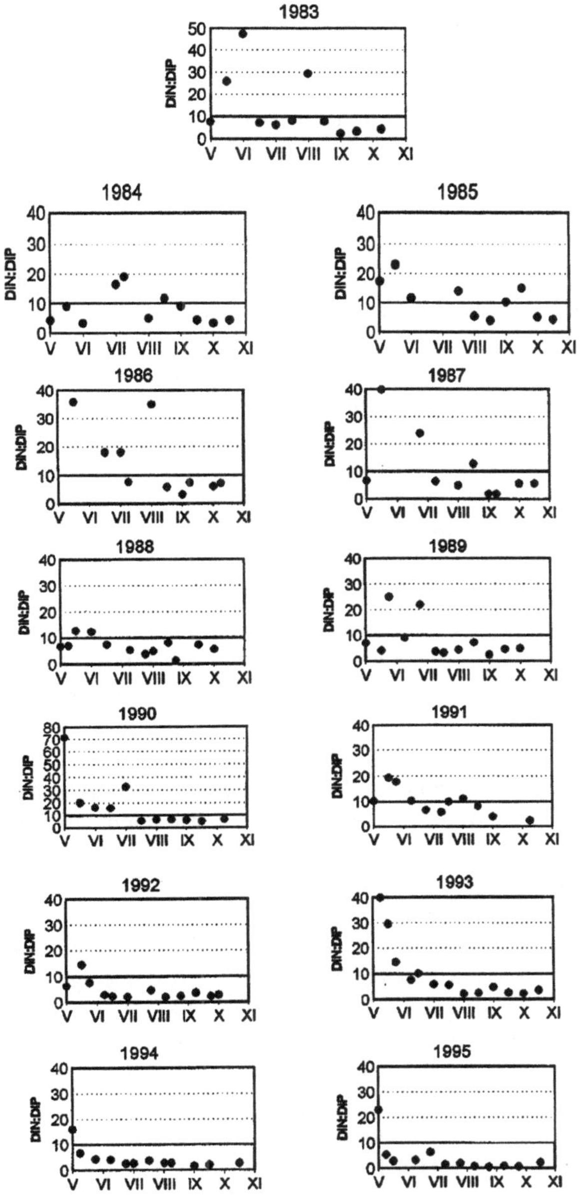

Figure 4. Inorganic nitrogen-phosphorus ratios at the station Seili in 1983–1995.

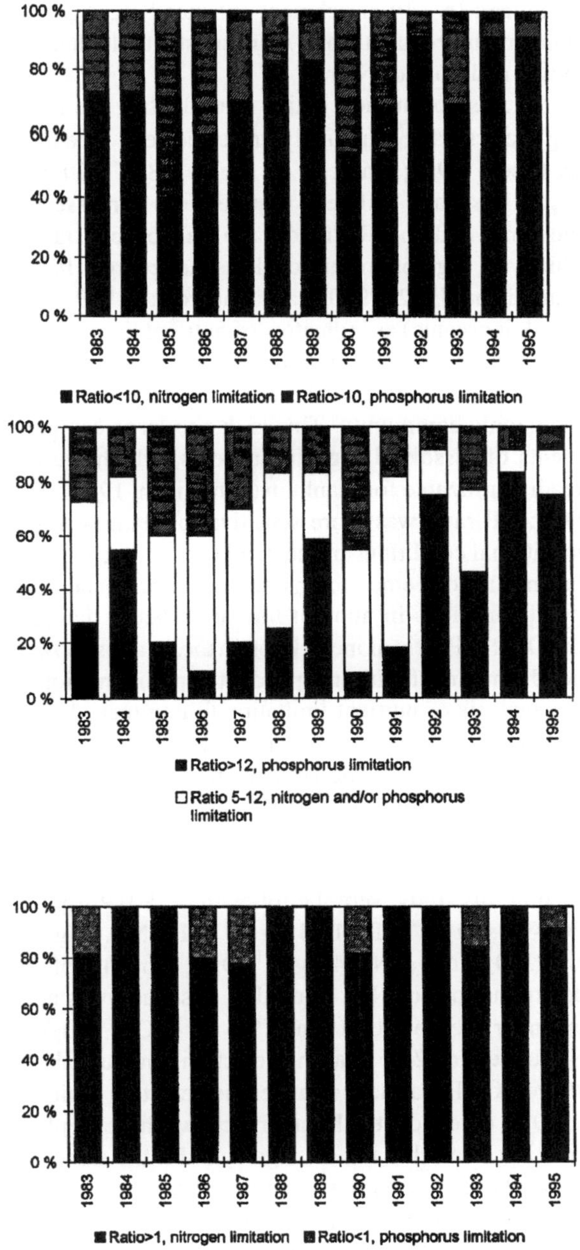

Figure 5. Proportion of cases when nitrogen, phosphorus or both are the limiting factors of primary production at the station Seili in 1983–1995. According to (a) criteria by Ryther & Dunstan (1971)(b) criteria by Forsberg (1978); (c) nutrient balance ratio; see page 119

at salinities lower than 10 ppt, while nitrogen was limiting at salinities of 25 ppt (Doering et al., 1995).

In the oceans, particulate C:N:P adheres to the Redfield ratio at which strong nutrient limitation by one element would not be expected (Hecky et al., 1993). The characterization of marine waters as N limited is based primarily on nutrient enrichment bioassays in bottles which may mispresent in situ conditions, and the N-limitation paradigm might require scrutiny (Hecky & Kilham, 1988). Elser et al. (1990) have also criticized enrichment assays and found that combined (N + P) enrichment enhanced algal growth much more frequently and more substantially than did single additions of N or P.

The use of nutrient ratios in evaluating the limiting factors has been widely criticized (e.g. Sakshaug & Olsen, 1986; Paasche & Erga, 1988). Different organisms have different requirements for essential elements and the Redfield ratio can be considered chiefly as a global mean (cf. Paasche & Erga, 1988). Tamminen (1982, 1990) has compared the N:P ratios (total, inorganic) as indicators of nutrient limitation in the brackish water and found that the TN:TP ratio was the most insensitive to the annual succession of the planktonic community. This is due to large compartments of dissolved organic nutrients, which are not readily utilizable for plankton (Tamminen, 1989). The DIN:DIP ratio was more sensitive to changes in the nutritional conditions of the plankton, but the nutrient balance ratio seemed to characterize spatial and temporal variations in nutrient limitation better than did the DIN:DIP ratio alone. Although these ratios showed good agreement with independent experimentation in the analysis of nutrient limitation (Tamminen, 1982), they are quite sensitive to analytical errors and imprecision especially in the summer periods when DIN- and DIP-concentrations are very low. We had in our study an opportunity to deal with precise analytical determinations. The limits of detection for the variables used in this study were the following: TP 0.9 $\mu g\ l^{-1}$, PO_4-P 0.6 $\mu g\ l^{-1}$, TN 10.3 $\mu g\ l^{-1}$, NH_4-N 1.1 $\mu g\ l^{-1}$ and ($NO_2 + NO_3$)-N 1.4 $\mu g\ l^{-1}$. Only about 4% of the determinations of NH_4-N used in this study was near or under the detection limit, 9% of ($NO_2 + NO_3$)-N was under and 26% near the detection limit and about 7% of PO_4-P was near or under the detection limit.

One objection to the approach we have applied here has been that not all algae exhibit the same elemental ratio when grown in nutrient-replete medium, and this is further compounded by the fact that elemental ratios may vary through the natural light-dark cycle (Paasche et al., 1984; Brzezinski, 1985). Nutrient deficiency tends to be most pronounced when biomass reaches its culmination (Sakshaug & Olsen, 1986). In such situations nutrient demand greatly outstrips nutrient availability. Sakshaug and Olsen (1986) consider that conclusions based on nutrient status measurements are best when algal biomass is nearly monospecific. Balance point for N and P varies species by species (see e.g. Brzezinski, 1985; Paasche et al., 1984; Sakshaug & Olsen, 1986), which indicates that different species have different strategies for nutrient competition. Unfortunately, we have no detailed information on the composition of the algal biomass in the Archipelago Sea.

Natural phytoplankton populations are also continuously exposed to gradients of the limiting nutrients (Sakshaug & Olsen, 1986). This may include nutrient-rich pulses of different sizes, concentrations, and life times. In the Archipelago Sea, background load varies considerably especially during open water season, mainly because of highly variable wind conditions. Because these pulses have more P and N but lower N:P ratios (TN:TP = 16) than the waters in inner coastal zones (TN:TP = 20), rapid changes in phytoplankton species composition might emerge. Laboratory experiments (e.g. Sommer, 1984; Sakshaug & Olsen, 1986) have clearly demonstrated how such shifts occur. The absence of detailed phytoplankton data in our study may create some pitfalls in our interpretation because different phytoplankton species have different strategies for nutrient competition.

Paired N and P budgets including all sink and source elements would be useful when considering relative availability of these nutrient elements for plankton production (Paasche & Erga, 1988). In the Archipelago Sea only tentative estimations can be made because of the paucity of the data concerning for example the water inflow and outflow rates. Mean annual riverine fluxes of N and P to the Archipelago Sea were 5500 t and 491 t in 1986–1990 (TN:TP = 11). The inorganic N:P-ratio was 13 in the rivers discharging to the Archipelago Sea. The direct municipal inputs, including fish farming, of nitrogen and phosphorus to the Archipelago Sea were 1569 t a^{-1} of N and 89 t a^{-1} of P (TN:TP = 18) (Pitkänen 1994). The annual atmospheric input to the Archipelago Sea is about 4800 t a^{-1} of N and 82 t a^{-1} of P, thus the TN:TP ratio is 60 (Kirkkala, 1994).

In late summer the contribution of point-sources increases in relation to diffuse loading, because most of the agricultural areas are located in lake-poor coastal regions. Under normal hydrological conditions nutrient leaching from these areas mainly occurs during spring and late autumn. The average point-source inputs (including fish farming) are estimated as 60% of total N and 50% of total P for the Archipelago Sea. Point-sources contribute 74% of the summertime bioavailable N and 76% of the bioavailable P largely due to intensive fish farming, whereas the contribution of agriculture was estimated only as ca. 20% due to small leaching of nutrients from the fields in summer (Pitkänen, 1994).

Despite the fact that the Archipelago Sea receives a strong surplus of N there seems to be a trend towards low inorganic N:P ratios and even N deficiency, indicating important sinks for bioavailable N. Several coinciding internal processes, first of all denitrification of oxidized inorganic N, decrease the N:P ratios of the euphotic layer in coastal ecosystems (e.g. Seitzinger, 1988). There are also several other processes affecting N:P ratios like sedimentation, turbidity and light conditions (e.g. Pitkänen et al., 1994; Alasaarela, 1980).

So, if we want to define and characterize the nutrient limitation of the entire ecosystem of the Archipelago Sea, budgets have to be calculated for both N and P and internal recycling must be taken into account as well as external supply of nutrients and loss processes. We agree with Tamminen (1990) that definite results on nutrient limitation problems can be obtained by manipulating nutrient concentrations in carefully planned experiments. So far, in the absence of budget calculations and spatially and temporally representative field experiments, we can only utilize long-term monitoring data collected routinely by water authorities.

Acknowledgments

This study was conducted at the Southwest Finland Regional Environment Centre (SFREC). The monitoring data were collected from databases in the SFREC and the Finnish Environmental Institute. The authors wish to thank especially the personnel in the laboratory and field of the SRFEC, who has made this examination possible by working hard and conscientiously and performing chemical analyses of highest quality. Also the designers at the SFREC have helped us considerably in designig the pictures. Marita Soini kindly checked the English of the manuscript.

References

Alasaarela, E. , 1980. Phytoplankton and environmental conditions in the northern part of the Bothnian Bay. Acta Univeritatis Ouluensis, Series A, no. 90, 23 pp.

Bonsdorff, E. & E. M. Blomqvist, 1993. Biotic couplings on shallow water softbottoms – examples from the northern Baltic Sea. Oceanogr. Mar. Biol. Annu. Rev. 31: 153–176.

Bonsdorff, E., E. M. Blomqvist, J. Mattila & A. Norkko, 1997. Coastal eutrophication – causes, consequences and perspectives; in the archipelago areas of the northern Baltic Sea. Est. Coast. Shelf Sci. 44 (Suppl. A): 63–72.

Brzezinski, M. A., 1985. The Si:C:N ratio of marine diatoms: interspecific variability and the effect of some environmental variables. J. Phycol. 21: 347–357.

Doering, P. H., C. A. Oviatt, B. L. Nowicki, E. G. Klos & L. W. Reed, 1995. Phosphorus and limitation of primary production in a simulated estuarine gradient. Mar. Ecol. Prog. Ser. 124: 271–287.

Elser, J. J., E. R. Marzolf & C. R. Goldman, 1990. Phosphorus and nitrogen limitation of phytoplankton growth in the freshwaters of North America: A review and critique of experimental enrichments. Can. J. Fish. aquat. Sci. 47: 1468–77.

Forsberg, C., S.-O. Ryding, A. Claesson & Å. Forsberg, 1978. Water chemical analyses and/or algal assay? Sewage effluent and polluted lake water studies. Mitt. int. Verein. Limnol. 21: 352–363.

Harris, G. P., 1986. Phytoplankton ecology. Chapman & Hall, London, 384 pp.

Hecky, R. E. & P. Kilham, 1988. Nutrient limitation of phytoplankton in freshwater and marine environments: A review of recent evidence on the effects of enrichment. Limnol. Oceanogr. 33: 796–822.

Hecky, R. E., P. Campbell & L. Hendzel, 1993. The stoichiometry of carbon, nitrogen, and phosphorus in particulate matter of lakes and oceans. Limnol. Oceanogr. 38: 709–724.

HELCOM-Baltic Marine Environment Protection Commission – Helsinki Commission, 1993. First assessment of the state of the coastal waters of the Baltic Sea. Baltic Sea Environment Proc. no. 54, 155 pp.

Jumppanen, K. & J. Mattila, 1994. The development of the state of the Archipelago Sea and environmental factors. Lounais-Suomen vesiensuojeluyhdistys r.y. Julkaisu 82, 200 pp. (in Finnish with English summary).

Kangas, P., E. Alasaarela, H. Lax, S. Jokela & C. Storgård-Envall, 1993. Seasonal variation of primary production and nutrient concentrations in the coastal waters of the Bothnian Bay and the Quark. Aqua Fennica. 23: 165–176.

Kirkkala, T., 1994. The nutrient load and the state of the Archipelago Sea. In Blomqvist, E. M. (ed.), Lantbrukets och fiskodlingars belastning i kust- och skärgårdsvatten. Nordiska Ministerrådets Skärgårdssamarbete. Rapport 4: 30–35. (in Swedish)

Koroleff, F., 1976. Determination of nutrients. In Grasshoff, K. (ed.), Methods of Seawater Analysis. Verlag Chemie. Weinheim, New York: 117–133.

Koroleff, F., 1979. The general chemical analysis methods of sea water. Institute of Marine Research, Finland. Meri, no. 7, 60 pp. (in Finnish).

Larsson, U., R. Elmgren & F. Wulff, 1985. Eutrophication and the Baltic Sea: Causes and consequences. Ambio 14: 9–14.

Leppänen, J.-M., 1988. Carbon and nitrogen cycles during the vernal growth period in the open northern Baltic Proper. Finnish Institute of Marine Research. Meri, no. 16, 38 pp.

Liebig, J., 1855. Principles of agricultural chemistry with special reference to the late researches made in England. A facsimile excerpt (17–34) reprinted in Pomeroy, L. R. (ed.) 1974. Cycles of Essential Elements: 11–28. Dowden, Hutchinson & Ross, Stroudsburg, Pennsylvania.

Mäkinen, I., A.-M. Suortti, R. Saares, R. Niemi & J. J. Marjanen, 1996. Ohjeita ympäristönäytteiden kemiallisten analyysimenetelmien validointiin. Suomen ympäristökeskuksen moniste. (in Finnish).

National Board of Waters, 1981. The analytical methods used by National Board of Waters and the Environment. National Board of Waters, Finland, Report 213, 136 pp. Helsinki (in Finnish).

Numers, v. M., 1995. Distribution, numbers and ecological gradients of birds breeding on small islands in the Archipelago Sea, SW Finland. Acta Zool. Fenn. 197: 1–127.

Oviatt, C., P. Doering, B. Nowicki, L. Reed, J. Cole & J. Frithsen, 1995. An ecosystem level experiment on nutrient limitation in temperate coastal marine environments. Mar. Ecol. Prog. Ser. 116: 171–179.

Paasche, E., I. Bryceson & K. Tangen, 1984. Interspecific variation in dark nitrogen uptake by dinoflagellates. J. Phycol. 20: 394–401.

Paasche, E. & S. R. Erga, 1988. Phosphorus and nitrogen limitation of phytoplankton in the inner Oslofjord (Norway). Sarsia 73: 229–243.

Pitkänen, H., 1994. Eutrophication of the Finnish coastal waters: origin, fate and effects of riverine nutrient fluxes. National Board of Waters and the Environment, Finland. 18: 1–45.

Pitkänen, H., P. Kangas, P. Ekholm & M. Perttilä, 1986. Surface distribution of total phosphorus and total nitrogen in the Finnish coastal waters in 1979–1983. National Board of Waters, Finland. Publications of the Water Research Institute, 68: 40–54.

Pitkänen, H., T. Tamminen, P. Kangas, T. Huttula, K. Kivi, H. Kuosa, J. Sarkkula, K. Eloheimo, P. Kauppila & B. Skakalsky, 1993. Late summer trophic conditions in the north-east Gulf of Finland and the river Neva estuary, the Baltic Sea. Estuar. coast. Shelf Sci. 37: 453–474.

PrENV ISO/CD 13530, 1995. Guide to Analytical Quality Control for Water Analysis, 80 pp.

Redfield, A. C., B. H. Ketchum & F. A. Richards, 1963. The influence of organisms on the composition of sea–water. In Hill, M. N. (ed.), The Sea, vol. 2. Interscience, New York: 26–77.

Reynolds, C. S. & A. E. Walsby, 1975. Water blooms. Biol. Rev. 50: 437–481.

Ryther, J. H. & W. M. Dunstan, 1971. Nitrogen, phosphorus and eutrophication in the coastal marine environment. Science 171: 1008–1013.

Sakshaug, E. & Y. Olsen, 1986. Nutrient status of phytoplankton blooms in Norwegian waters and algal strategies for nutrient competition. Can. J. Fish. aquat. Sci. 43: 389–396.

Seitzinger, S. P., 1988. Denitrification in freshwater and coastal marine ecosystems: Ecological and geochemical significance. Limnol. Oceanogr. 33:702–724.

Sommer, U., 1984. The paradox of the plankton: fluctuations of phosphorus availability maintain diversity of phytoplankton in flow-through cultures. Limnol. Oceanogr. 13: 633–636.

Tamminen, T., 1982. Effects of ammonium effluents on planktonic primary production and decomposition in a coastal brackish water environment 1. Nutrient balance of the water body and effluent tests. Neth. J. Sea Res. 16: 455–464.

Tamminen, T., 1989. Dissolved organic phosphorus regeneration by bacterioplankton: 5′-nucletidase activity and subsequent phosphate uptake in a mesocosm enrichment experiment. Mar. Ecol. Prog. Ser. 58: 89–100.

Tamminen, T., 1990. Eutrophication and the Baltic Sea: studies on phytoplankton, bacterioplankton and pelagic nutrient cycles. University of Helsinki, 21 pp.

Taylor, D., S. Nixon, S. Granger & B. Buckley, 1995. Nutrient limitation and the eutrophication of coastal lagoons. Mar. Ecol. Prog. Ser. 127: 235–244.

Wetzel, R. G., 1983. Limnology. 2nd edn. 767 pp. Saunders College Publishing, Philadelphia.

Wulff, F. & L. Rahm, 1987. Long-term seasonal and spatial variations of nitrogen, phosphorus and silicate in the Baltic: An overview. Mar. Envir. Res. 26: 19–37.

Wulff, F., A. Stigebrandt & L. Rahm, 1990. Nutrient dynamics of the Baltic Sea. Ambio 19: 126–133.

Hydrobiologia **363**: 127–139, 1998.
T. Tamminen & H. Kuosa (eds), Eutrophication in Planktonic Ecosystems: Food Web Dynamics and Elemental Cycling.

Effects of environmental factors on the phytoplankton community in the Gulf of Finland – unattended high frequency measurements and multivariate analyses

Eija Rantajärvi[1], Vesa Gran[2], Seija Hällfors[1] & Riitta Olsonen[1]
[1] *Finnish Institute of Marine Research, P.O. Box 33, FIN-00931 Helsinki, Finland*
[2] *Finnish Environment Institute, P.O. Box 140, FIN-00251, Helsinki, Finland*

Key words: phytoplankton, Baltic Sea, Gulf of Finland, high frequency measurements, multivariate analysis

Abstract

Relationships between phytoplankton species composition and environmental factors were studied in open areas of the central and eastern Gulf of Finland in late summer 1993. The data was collected using unattended water sampling, as well as spatially and temporally frequent measurements on *in vivo* fluorescence of chlorophyll *a*, temperature and salinity on board a passenger ferry, which plied between Helsinki and St. Petersburg.

The relative abundances of phytoplankton species, concentrations of nutrients (Tot-N, NO_2-N+NO_3-N, NH_4-N, Tot-P, PO_4-P, Si) and chlorophyll *a* were analysed from the water samples. The collected data set enabled the use of various statistical methods in order to explain the phytoplankton community structure in the study area. The multivariate analyses were carried out using SAS software procedures (GLM, CLUSTER, CANCORR). Variability in the phytoplankton community (biomass, species composition) was high, and this paper clearly demonstrates that valuable information regarding pelagial biological dynamics, connected to extreme values, will be lost if the data is incorrectly simplified.

Introduction

The whole Gulf of Finland has been characterized by increased eutrophication during the recent decades, which is reflected in the intensification and increased frequency of phytoplankton blooms (e.g. Grönlund & Leppänen, 1990; Pitkänen, 1991; Pitkänen et al., 1993; Leppänen et al., 1994a; Pitkänen & Tamminen, 1995; Kauppila et al., 1995). In late autumn 1987, the blooms of the freshwater blue-green alga *Microcystis aeruginosa* had spread into the eastern Finnish coastal areas (Niemi, 1988; Pitkänen et al., 1990) and in 1992, a mass mortality of birds was supposedly caused by a toxic phytoplankton bloom in the eastern Gulf of Finland (Kauppi, 1993). These phenomena have been connected to the huge anthropogenic loading from St. Petersburg and the River Neva, which is eutrophying mainly the eastern parts of the Gulf of Finland (Pitkänen et al., 1993). Especially in spring, but also in summer under special weather conditions, the nutrient load may have far-reaching effects along the northern coast of the Gulf of Finland (Leppänen et al., 1994b; Pitkänen et al., 1993). Furthermore, there is no sill separating the Baltic Proper from the Gulf of Finland and thus the phosphorus-rich deep waters of the Baltic can flow freely into the gulf and consequently increase the deep water nutrient reserves.

Since the 1920s the Finnish Institute of Marine Research has collected mainly hydrographical and chemical data from the fixed stations in the easternmost Gulf of Finland. However, this activity was interrupted by the Second World War, and it was not until 1990 that other than Russian scientists were able to investigate this part of the Gulf of Finland (e.g. Pitkänen, 1991; Pitkänen et al., 1993; Viitasalo, 1993; Perttilä et al., 1995). In 1992 high frequency measurements of phytoplankton and of various hydrographical factors were also started (Leppänen et al., 1994a). Recently several papers related to phytoplankton and hydrodynamics in the eastern Gulf of Finland have been published (e.g.

Pitkänen et al., 1993; Leppänen et al., 1994a; Pitkänen & Tamminen, 1995; Kauppila et al., 1995).

Data for this study was collected by unattended high frequency sampling aboard a ferry between Helsinki and St. Petersburg. We wanted to test the appropriateness of this kind of data set for phytoplankton community structure studies using multivariate statistical analyses. The concordance between predetermined hydrographical areas and species community structure was tested with analysis of variance; the species community structure based on individual samples with cluster analysis; and the relationships between phytoplankton taxa and environmental factors with canonical correlation analysis.

Study area

The study area was divided into six subareas mainly on the basis of geomorphological and hydrographical features (see Figure 1 and Figure 2). The main characteristics of hydrography in the Gulf of Finland have been described elsewhere (Haapala & Alenius, 1994, Alenius et al., submitted manuscript). The hydrography of the eastern areas has been described in more detail by e.g. Pitkänen et al. (1993) and Alenius et al. (submitted manuscript). The Russian names of the islands are used in this text, but also the Finnish names are shown in parenthesis when first mentioned.

1. The inner Neva Estuary (INE), the area west of the flood protection barrier, is characterized by strong outflow from the River Neva and by sharp depth changes (10–36 m) as well as by steep vertical and horizontal salinity gradients (Figure 3). The area is affected by strong vertical mixing and by drastic changes in hydrodynamics which are due to the two-layer dynamics of estuaries (e.g. Pitkänen & Tamminen, 1995).

2. The outer Neva Estuary (ONE) covers the area from the Seskar Island (Seiskari) to the inner Neva Estuary. Water depth ranges from 20 to 40 m and the area is affected by small-scale upwellings and unstable hydrographical conditions. The whole Neva Estuary is characterized by high phytoplankton biomasses throughout the growing season (e.g. Pitkänen et al., 1993; Leppänen et al., 1994a).

3. The first transition zone (T-1) between the Neva Estuary and the open eastern Gulf of Finland extends from the Motshnyi Island (Lavansaari) to the island of Seskar. The water depth varies from 20 to 60 m and shallow sills surround the islands.

4. The second transition zone (T-2) extends from the island of Motshnyi to the island of Gogland (Suursaari), and the maximum water depth in the area is ca 80 m.

5. The central area of the eastern Gulf of Finland (OGF) extends from the Gogland Island to a longitude of 25° 35′ E. The water depth ranges from 40 to 80 m.

6. The central part of the Gulf of Finland, off Helsinki (HKI) covers the area between longitude 25° 00′ E and 25° 35′ E. The water depth varies from 30 to 80 m.

Material and methods

High frequency measurements and water sampling

The high frequency measurements were carried out on board a passenger ferry from Helsinki to St. Petersburg (Figure 1) using an unattended flow-through system. The study period extended from August to October in 1993. The late summer and autumn blooms formed by blue-green algae are common in the whole Gulf of Finland, which made this period ecologically interesting.

The flow-through method has been described in detail by Rantajärvi and Leppänen (1994) and Leppänen et al. (1994c). Two weekly cruises were made with a spatial resolution of 100 to 200 metres. The continuously measured parameters were *in vivo* fluorescence of chlorophyll *a* (Turner Design Model AU-10 fluorometer), temperature and salinity (Aanderaa thermosalinograph). Automated water sampling (ISCO water sampler) was carried out along transects on the 15th, 26th and 29th of August, on the 5th, 12th, 19th and 26th of September and on 10th of October. The spatial resolution of the samples was 10 to 30 km along the ship route.

The water for both high frequency recordings and for water sampling were pumped through an inlet situated in the middle of the ship at a depth of 5 metres. The ship creates turbulence and therefore the water used in these measurements represents the upper surface layer. The depth of the euphotic layer in the Gulf of Finland varies from 20 to 5 m, decreasing eastwards. Consequently, the water sampled from the upper surface layer is considered to be representative for the phytoplankton in the research area.

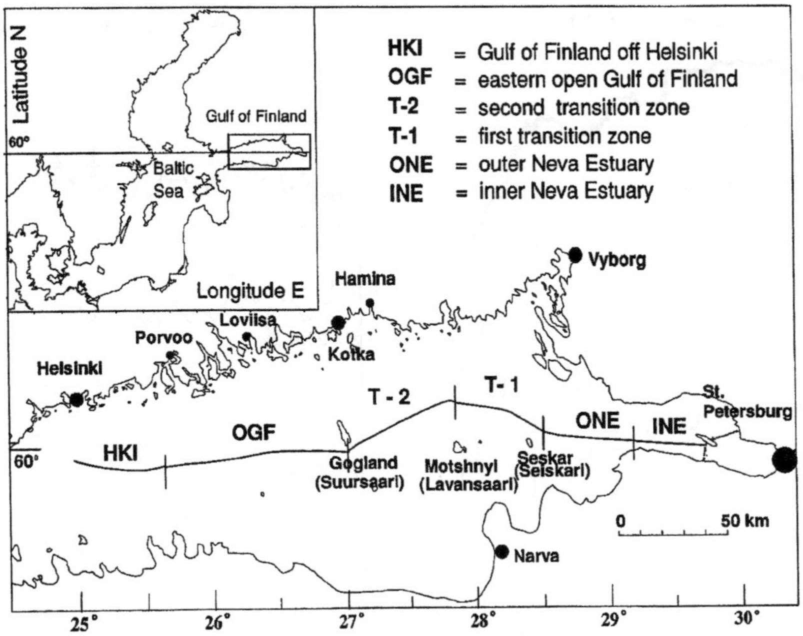

Figure 1. A map of the study area in the Gulf of Finland, Baltic Sea and the route of the ferry.

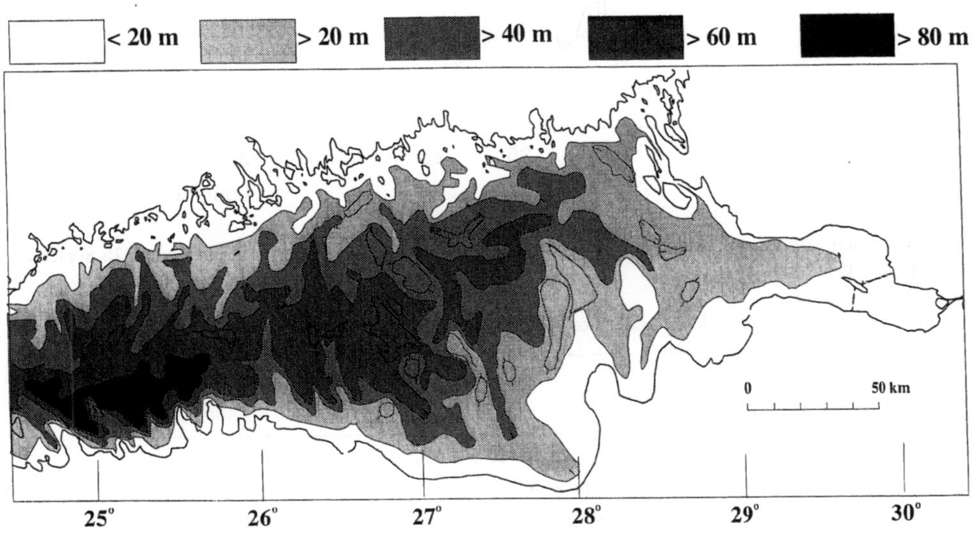

Figure 2. A bathymetric map of the mid and eastern Gulf of Finland.

Analyses of nutrients and chlorophyll a

Nutrient concentrations (Tot-N, NO_2-N+NO_3-N, NH_4-N, Tot-P, PO_4-P, Si) were analysed on the 15th and 29th of August, on the 12th and 26th of September and on the 10th October as described in HELCOM (1988). All the collected water samples were analysed for concentrations of chlorophyll *a* according to HELCOM (1988) and for phytoplankton species composition. The conversion factor for *in vivo* fluorescence of chlorophyll *a* was obtained weekly by comparing these results with chlorophyll *a* concentrations extracted and analysed as described in HELCOM (1988) from simultaneously taken water samples.

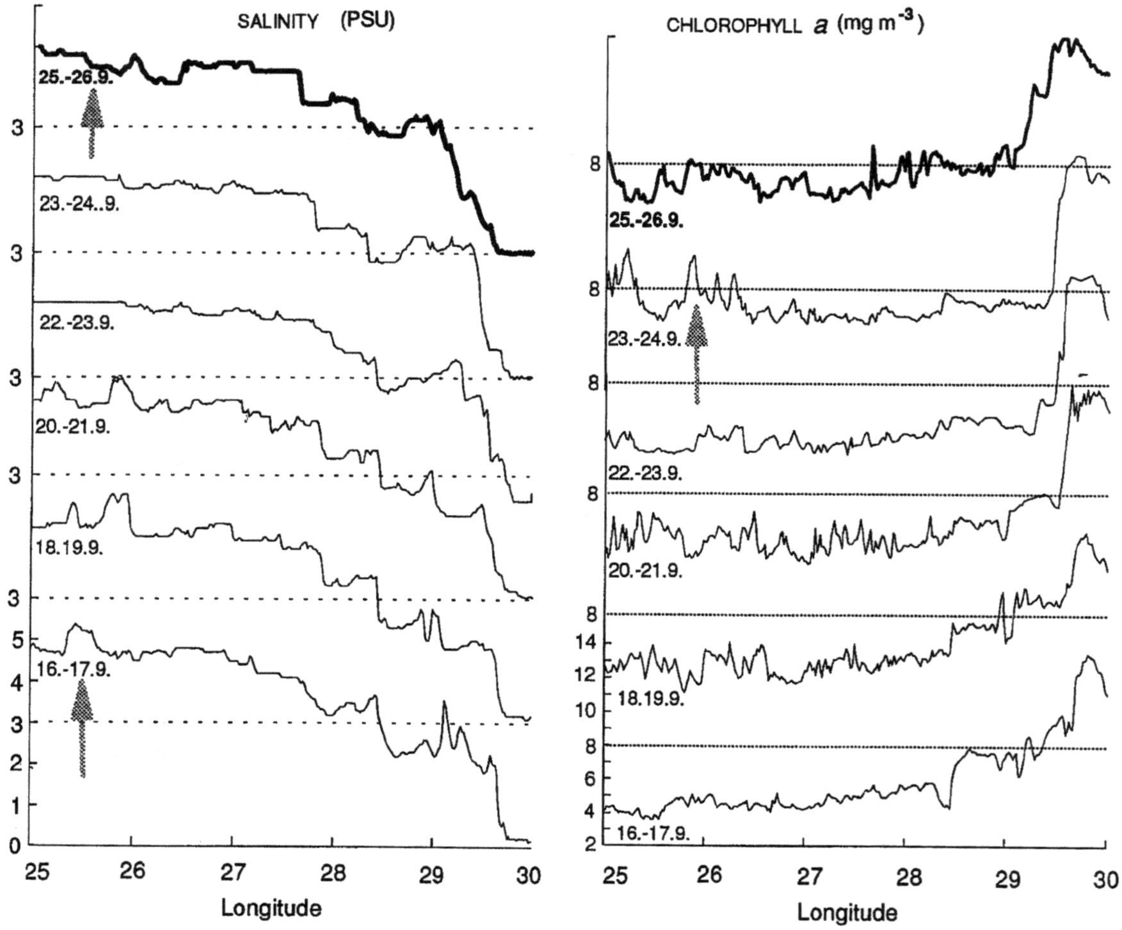

Figure 3. Variability in salinity (PSU) and chlorophyll *a* (mg m^{-3}) along the route of the ferry between 16 and 26 September in 1993. The route is presented in Figure 1.

Table 1. Abundance ranking of phytoplankton species

Relative Abund.	Description
0.1	very sparse, one or a few cells or units in the microscopied areas
1	sparse, slightly more cells or units in every viewed strip or field
10	scattered, irrespective of the magnification several cells or units in many viewed fields
100	abundant, irrespective of the magnification several cells or units in most viewed fields
500	dominant, irrespective of the magnification many cells or units in every viewed field

Phytoplankton analyses

Phytoplankton samples were preserved with the Lugol AA solution (Willén, 1962) and the taxa were identified using the inverted microscope (Wild M40) technique (Utermöhl, 1958). A semi-quantitative counting method was used, where the relative abundance of species were estimated using five ranking classes

(Table 1). The abundance was sample specific i.e. it was not connected to the chlorophyll concentration of the sample but merely to the abundance relations between the species. The sedimented volume was 50 ml, except for the Neva Estuary samples where the volume was reduced to 10 ml because of the high number of cells and detritus.

Statistics

Distributions of all variables were skewed and consequently all the values were logarithmically transformed for all three statistical analyses. First we wanted to test whether statistically significant differences could be found in the phytoplankton community structure between the subareas. This was tested by two-way analysis of variance (Sokal & Rohlf, 1994), where abundance values were weighted according to the ratio of samples with species present to the total number of samples collected in the subarea. In the cluster analysis (Mardia et al., 1979) the abundance of a taxon in the individual samples was used, whereas for the canonical correlation analysis (Kshirsagar, 1972) the values were weighted according to the concentrations of chlorophyll *a*, which was used as a measure of the phytoplankton biomass. In the canonical correlation analysis two sets of variables (species, environmental (Tot-N, NO_2-N+NO_3-N, Tot-P, PO_4-P, salinity, temperature)) were used. The analyses were done using SAS software procedures (SAS/STAT User's Guide Volume 1 & 2, 1989).

The original number of phytoplankton taxa (189) had to be reduced in all analyses because it exceeded the total number of samples (114) and, in the canonical correlation analysis the number of nutrient samples (70). In the analysis of variance and cluster analysis, the number of taxa was 98, and in the canonical correlation analysis 62 (Table 2). The selection of taxa was done according to prevalence and abundance. Some of the species were treated at genus level when it was considered ecologically reasonable.

Results

Hydrodynamics and wind effects

Temperature of the surface waters decreased from 19 to 6 °C during the study period and several rapid fluctuations were detected in the research area.

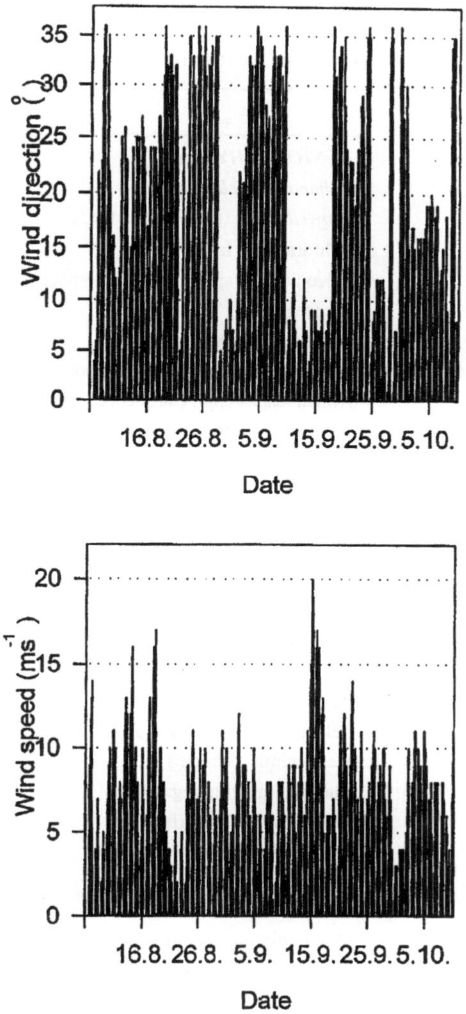

Figure 4. Wind speed and direction at Kalbådagrund Island, situated in the middle of the Gulf of Finland, off Helsinki.

At the end of August, a slight increase in salinity values (0.5–1 PSU) was observed in the westernmost areas (HKI, OGF), simultaneous with fluctuations in the western transition zone (T-2). These small-scale upwellings were preceded by strong north-westerly winds (up to 17 m s^{-1}) during a one week period (Figure 4). In the beginning of September, a clear increase in salinity was redetected in the westernmost areas and in the western transition zone (T-2), after which easterly winds prevailed generating lenses of less saline water in the outer Neva Estuary (ONE).

In the middle of September, strong upwelling along the northern coast of Estonia resulted in decreased temperature (Figure 5) and increased salinity (Figure 3), probably induced by preceded strong (12–20 m s^{-1})

Table 2. List of phytoplankton taxa (62) included in the canonical correlations analysis. Correlations of the phytoplankton taxa with the first three canonical variables (CE1, CE2, CE3) are shown. Values greater than 0.4 are printed in bold

	Abbr.	CE1	CE2	CE3
NOSTOCOPHYCEAE				
Aphanothece clathrata	N1	**0.413**	0.210	**–0.664**
Snowella spp. (incl. S. lacustris, S. septentrionalis, S. litoralis, S. fennica)	N2	–0.263	0.213	0.003
Woronichinia spp. (incl. W. compacta, W. karelica, W. naegeliana, W. fusca)	N3	–0.380	0.014	0.024
Microcystis aeruginosa	N4	0.395	0.307	0.049
Microcystis spp. (incl. M. reinboldii, Lemmermanniella spp., Aphanocapsa spp., Chroococcus spp.)	N5	0.354	0.261	–0.180
Anabaena lemmermannii	N6	0.026	0.134	0.201
Aphanizomenon sp.	N7	–0.314	–0.179	–0.141
Nodularia spumigena	N8	0.052	–0.145	–0.106
Oscillatoriales spp. (incl. Limnothrix planctonica)	N9	**0.614**	0.245	–0.179
Planktothrix agardhii	N10	**0.674**	0.118	0.049
Pseudanabaena limnetica	N11	**0.698**	0.123	–0.374
Picoplanktonic unicells 0.2-2μm (incl. Nannochloropsis sp.)	N12	–0.033	–0.238	–0.084
CRYPTOPHYCEAE				
Cryptomonadales spp.	Cr1	–0.219	–0.060	–0.107
Cryptomonas spp.	Cr2	0.345	0.047	–0.130
Teleaulax spp. (T.acuta, T.amphioxeia)	Cr3	–0.323	–0.048	0.061
Hemiselmis virescens	Cr4	**–0.404**	–0.085	–0.229
Plagioselmis prolonga	Cr5	**–0.558**	0.012	0.040
Rhodomonas lacustris	Cr6	**0.537**	0.163	–0.130
DINOPHYCEAE				
Dinophysis acuminata	D1	**–0.679**	–0.183	0.076
Protoceratium reticulatum	D2	–0.370	0.191	0.039
Gymnodinium spp. (autotr.)	D3	–0.192	–0.049	0.151
Heterocapsa rotundata	D4	–0.342	–0.050	–0.015
PRYMNESIOPHYCEAE				
Chrysochromulina spp.	Chr	0.132	0.334	0.077
CHRYSOPHYCEAE				
Pseudopedinella spp. (incl. P. elastica, P. tricostata)	Ch1	0.053	–0.037	0.156
Ochromonas spp.	Ch2	0.115	0.077	0.131
DIATOMOPHYCEAE				
Actinocyclus octonarius	Di1	–0.166	–0.092	0.211
Chaetoceros ceratosporus	Di2	–0.171	–0.280	–0.002
Chaetoceros wighamii	Di3	**–0.437**	–0.222	0.029
Coscinodiscus granii	Di4	–0.101	–0.143	0.138
Centric diatomae 5–10 μm	Di5	**0.528**	0.307	**–0.429**
Aulacoseira islandica	Di6	**0.501**	**0.414**	–0.328
Rhizosolenia minima	Di7	0.136	0.050	0.207
Skeletonema costatum	Di8	–0.160	–0.120	0.054
Skeletonema subsalsum	Di9	**0.645**	0.273	–0.353
Tabellaria flocculosa	Di10	**0.582**	0.146	–0.302
EUGLENOPHYCEAE				
Eutreptiella gymnastica	E	–0.165	–0.085	–0.107
PEDINOPHYCEAE				
Small pedinophytes (Pedinomonas sp., Resultor mikron)	P1	0.147	0.220	–0.007

Table 2. (continued)

	Abbr.	CE1	CE2	CE3
PRASINOPHYCEAE				
Pyramimonas spp.	P2	–0.016	0.387	0.274
Pyramimonas virginica	P3	–0.316	0.052	0.012
CHLOROPHYCEA				
Botryococcus braunii	Chl1	–0.117	0.036	–0.024
Dictyosphaerium spp. (D.ehrenbergianum, D. elegans, D. pulchellum, D. subsolitarium, D. tetrachotomum)	Chl2	**0.538**	0.224	**–0.436**
Monoraphidium contortum	Chl3	0.392	0.152	–0.197
Monoraphidium spp. (incl. M. griffithii, M. minutum, M. mirabile, M. komarkovae)	Chl4	**0.438**	0.113	**–0.444**
Oocystis borgei	Chl5	–0.199	–0.192	0.050
Oocystis lacustris	Chl6	–0.337	0.116	0.271
Pediastrum spp. (P. boryanum, P. duplex, P. privum, P. tetras)	Chl7	**0.541**	0.274	–0.354
Scenedesmus spp. (S. acuminatus, S. gutwinskii, S. opoliensis, S. obtusus, S. spinosus)	Chl8	**0.573**	0.149	–0.377
Scenedesmus obliquus	Chl9	**0.449**	0.226	–0.252
Scenedesmus armatus	Chl10	**0.526**	0.316	–0.299
Scenedesmus communis	Chl11	0.349	0.245	**–0.455**
Chlorococcales spp. (Golenkinia radiata, Kirchneriella contorta, Micractinium pusillum, Selenastrum capricornutum, Lagerheimia genevensis, Tetraedron caudatum, Tetrastrum staurogeniaeforme, Treubaria triappendiculata)	Chl12	**0.542**	0.075	**–0.424**
Planktonema lauterbornii	Chl13	0.231	0.062	–0.157
HETEROTROPHIC FLAGELLATES				
Quadricilia rotundata	H1	0.127	0.310	0.084
Ebria tripartita	H2	–0.264	–0.109	0.160
Katablepharis ovalis	H3	0.230	0.268	–0.190
Katablepharis sp.	H4	–0.178	0.013	–0.053
Paraphysomonas spp.	H5	0.153	0.125	–0.388
Protoperidinium brevipes	H6	0.043	–0.226	0.011
Gymnodinium spp.	H7	–0.117	0.119	0.204
Oblea rotunda	H8	–0.065	0.045	0.191
Telonema subtile	H9	0.045	–0.004	0.016
PHYTOCILIATA				
Mesodinium rubrum	Mes	**–0.704**	–0.259	0.213

north-easterly winds (Figure 4). During upwelling, the wind direction changed towards north-west. Dissolved inorganic nutrient concentrations in the surface waters increased after upwelling, e.g. nitrate ($N\text{-}NO_2{+}NO_3$) from 0.4 to 1.2 mmol m^{-3}, in the area west of the Gogland Island (OGF).

At the end of September and beginning of October, several lenses formed by less saline water (0.5–1 PSU lower than the surrounding waters) were recorded in the inner and outer Neva Estuary to the eastern transition zone (T-1).

Phytoplankton species succession and community structure

Variability of phytoplankton biomass was highest in the easternmost areas during the whole study period (Figure 6). In August, in the Neva Estuary the biomass

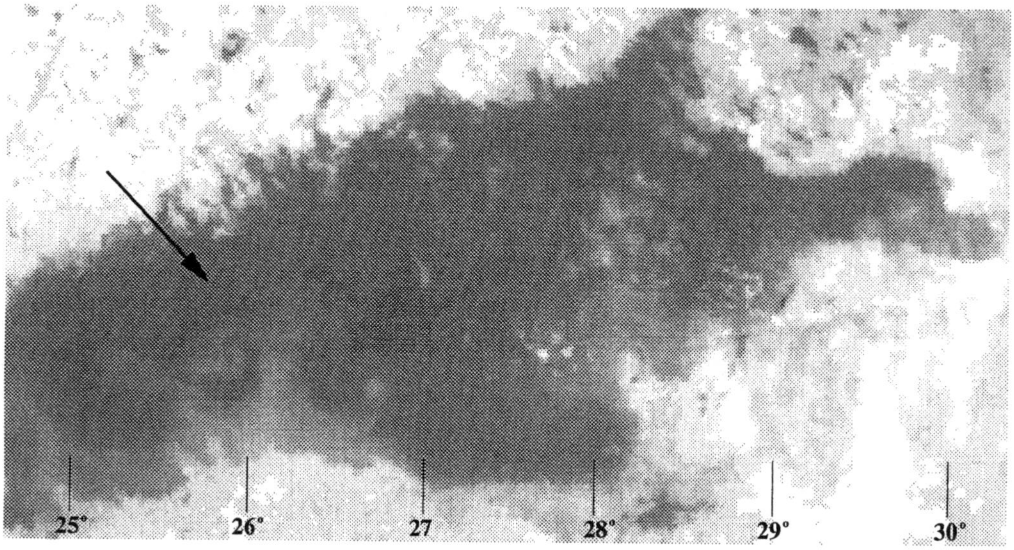

Figure 5. A satellite image (NOAA/AVHRR) of the temperature distribution on the 20th of September 1993 in the Gulf of Finland. The lighter colors represent low temperature values. The image was provided by Ove Rud (University of Stockholm).

was considerably higher compared to that found in the other parts of the study area.

The strong upwelling detected in the middle of September, in the eastern open Gulf of Finland (OGF) was reflected with few days delay as fluctuations in chlorophyll *a* concentrations and with a one week delay as clearly increased concentrations (Figure 3). The last water samples were taken on the 10th of October when the growth of phytoplankton was already limited by temperature and light.

In the Neva Estuary, the diatoms and blue-green algae were the dominant groups during the whole study period. In the inner Estuary, also the chlorophytes formed a significant group. In the western areas, primarily the blue-green algae and the autotrophic ciliate *Mesodinium rubrum* were dominant, although occasionally, the prasinophytes showed high abundances. The proportion of marine dinoflagellates increased towards the west. The phytoplankton biomass was more evenly distributed between the different groups in the western areas.

The freshwater blue-green alga *Planktothrix agardhii* was dominant in the Neva Estuary. The other blue-green algae, such as *Oscillatoriales* spp., *Microcystis* spp., *M. aeruginosa, Pseudanabaena limnetica* and *Aphanothece clathrata* were abundant, but were found in very low abundances in October. All these species showed a steeply decreasing trend towards the west in the transition zones (T-1, T-2). The blue-green algae *Woronichinia* spp., *Snowella* spp., *Aphanizomenon* sp., *Anabaena lemmermannii* and picoplanktonic unicells were common in the whole study area. The blue-green alga *M. aeruginosa* increased in abundance towards the end of September in the area between the inner and outer Neva Estuary, and concurrently the freshwater blue-green algae *Anabaena circinalis* and *Limnothrix planktonica* were spread westwards. The toxic blue-green alga *Nodularia spumigena* was a frequent and common species in all the areas except the Neva Estuary.

In the Neva Estuary, the diatoms consisted mainly of small (5–10 μm) centric species (Table 2). In the western areas, *Actinocyclus octonarius, Chaetoceros wighamii* and *C. ceratosporus* increased in abundance and spread eastwards towards the end of the study period. *Rhizosolenia minima* had the highest abundances in September, mainly in the transition zones (T-1, T-2), but also in the Neva Estuary.

The chlorococcalean *Scenedesmus* spp., *Dictyosphaerium* spp., *Monoraphidium* spp. and the taxon *Chlorococcales* spp. (Table 2) occurred occasionally in high numbers in the Neva Estuary and in the eastern transition zone (T-1). These species showed decreasing trend towards the end of the study period. The chlorophyte *Oocystis lacustris* occurred in all the areas, but had higher abundances in the western areas. The prasinophyte *Pyramimonas virginica* occurred only in the western areas to the western transition zone (T-2).

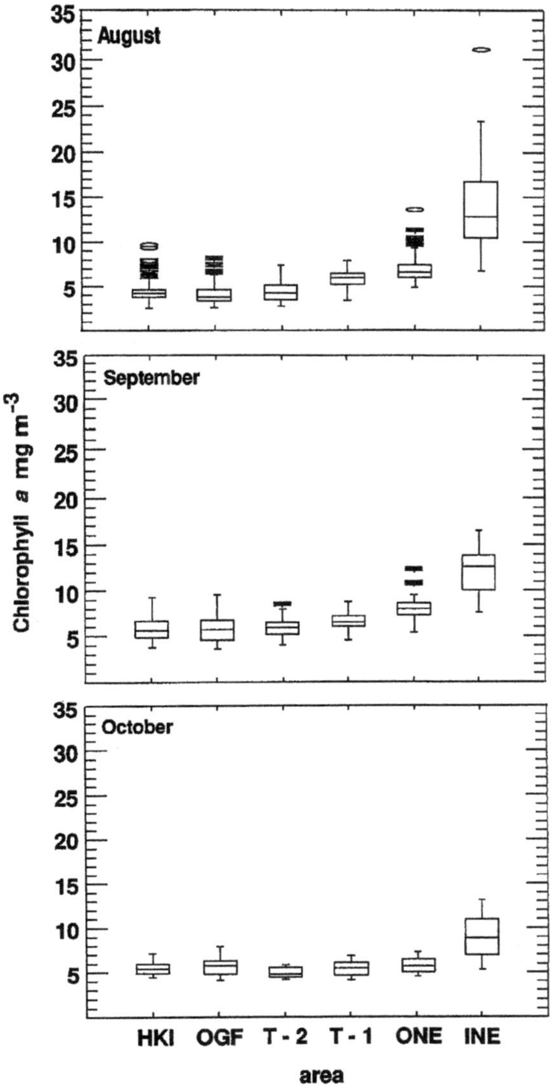

Figure 6. Monthly variability in chlorophyll *a* concentrations (mg m $^{-3}$) in different subareas, based on high-frequency recordings during the study period. The box plot divides the data into four parts of equal frequency. The box itself encloses the middle 50% of the values and the median is drown as a vertical line inside the box. The segments of the lines extents from the first and third quartile to the smallest and largest data point within 1.5 inter quartile ranges. Extreme values are indicated with oval circles. For area division see Figure 1.

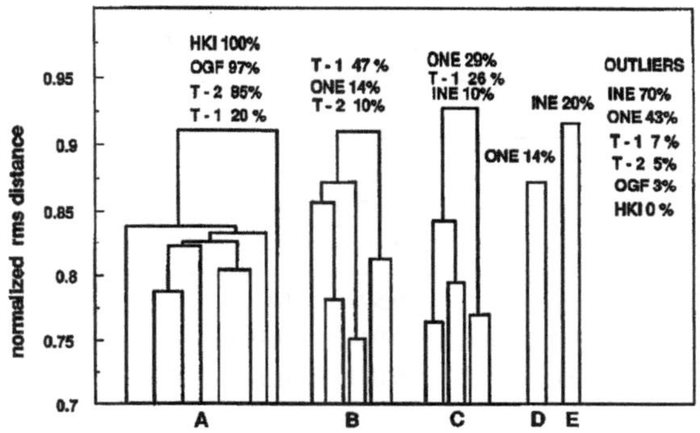

Figure 7. The result of the cluster analysis (CLUSTER) on the species community structure. The vertical axis stands for normalized root-mean-square distance between clusters.The proportion (%) of samples from the specific subareas in each cluster is indicated. The proportions of outliers in various subareas are also shown. For area division see Figure 1.

Cryptomonas spp. and *Rhodomonas lacustris* were the only cryptophytes common in the Neva Estuary. *Teleaulax* spp. and *Hemiselmis virescens* were common in the whole research area, but were more constant in the other areas besides that of the Neva Estuary. *Cryptomonadales* spp. had high abundances mainly in the western areas. *Plagioselmis prolonga* showed high abundances in the westernmost areas and occurred only once in the outer Neva Estuary.

The dinoflagellate *Dinophysis acuminata* had its highest abundances in the western areas in September and spread eastwards towards the end of the study period. *Protoceratium reticulatum* was common only in the open Gulf of Finland and in the western transition zone (T-2), but had disappeared completely by October.

The autotrophic ciliate *Mesodinium rubrum* showed high abundance only in the western areas, but spread eastwards towards the end of the study period.

Hydrographical subareas and species community structure

The analysis of variance detected significant differences ($p < 0.01$) in the phytoplankton species community structure between all the consecutive subarea pairs, except between the two westernmost subareas (HKI, OGF).

The cluster analysis, which was based on the values from the individual samples, resulted in three major groups A, B and C (Figure 7). These groups included 73% of all samples. Group A alone included 50% of the samples and most of the samples collected in the three westernmost areas (HKI, OGF, T-2) were included in this group. Group B was mainly formed from the eastern transition zone samples (T-1). The proportion of samples collected in area T-1 was also high in group C, which besides included a marked part of the samples collected from the outer Neva Estuary. Most of the samples collected from the Neva Estuary (INE,ONE)

Table 3. Correlations of the environmental variables with the first three canonical variables (CE1, CE2, CE3). Values greater than 0.45 are printed in bold

	Canonical variables of Environmental variables		
	CE1	CE2	CE3
Salinity	**−0.524**	0.030	**0.718**
Temperature	0.044	**0.632**	−0.141
PO4	0.071	0.012	**−0.471**
TP	0.431	0.169	−0.231
NO3+NO2	**0.510**	0.441	−0.001
TN	**0.819**	−0.211	−0.293
Proportion of the total variation (%)	**61**	**22**	**10**

were included in the two very small groups D and E, or were outliers.

Response of the species to the environmental factors

The first three canonical variables formed from the environmental variables explained 93% of the total variance in phytoplankton species ordination (Table 3). The response of the species to the environmental variables with the first three canonical variables (CE1, CE2, CE3) are presented in Table 2 and in the ordination diagrams in Figure 8.

The freshwater blue-green algal species *Planktothrix agardhii, Oscillatoriales* spp., *Pseudanabaena limnetica, Microcystis* spp., *M. aeruginosa* and *Aphanothece clathrata* showed a positive correlation with total nutrients. *Aphanizomenon* sp. showed negative correlation to total nitrogen as did also *Woronichinia* spp. *Snowella* spp. correlated positively with temperature.

The small freshwater chlorophytes and diatoms were considered intermediate in their nitrogen and phosphate preferences. The diatoms *Chaetoceros wighamii* and *Skeletonema costatum* preferred saline and cold water.

The cryptophytes *Rhodomonas lacustris* and *Cryptomonas* spp. had a negative correlation with salinity whereas the converse was true for *Plagioselmis prolonga. Cryptomonas* spp. and *R. lacustris* showed slight positive correlation with nitrogen.

The marine dinoflagellates *Dinophysis acuminata* and *Heterocapsa rotundata* preferred slightly cold water while *Protoceratium reticulatum* preferred warm water.

The prymnesiophyte *Chrysochromulina* spp. correlated positively with temperature as did the prasinophyte *Pyramimonas* spp. and small pedinophytes. The prasinophyte *Pyramimonas virginica* favoured saline conditions.

The heterotrophic flagellates *Quadricilia rotundata* and *Katablepharis ovalis* showed a positive correlation with temperature. *K. ovalis* and *Paraphysomonas* spp. showed a negative correlation with salinity. *Paraphysomonas* spp. favoured dissolved inorganic phosphate while *K. ovalis* was situated in the middle concerning its inorganic nitrate and phosphate preferences. The heterotrophic flagellate *Ebria tripartita* preferred saline water.

The ciliate *Mesodinium rubrum* showed a strong positive correlation with salinity and a negative one with temperature.

Discussion

In the eastern Gulf of Finland the phytoplankton biomass is high throughout the growing season and includes the occurrence of unpredictable and toxic blooms (e.g. Kononen et al., 1993). These phenomena have been connected to the huge and varying extent of the discharge of nutrient-rich freshwaters from the River Neva (e.g. Pitkänen et al., 1993; Leppänen et al., 1994a).

The complex hydrodynamics of the eastern Gulf of Finland is closely linked to the outflow from the River Neva, to the geomorphology of the basin and to the prevailing weather conditions. The basic mechanisms of the upwellings have been studied in several papers (e.g. Hela, 1976; Haapala, 1994; Talpsepp et al., 1994). In our study, the strong northwesterly or northeasterly winds generated the clear coastal upwellings detected in the western areas (Figure 3 & Figure 5). Here the bottom topography favours the formation of upwellings, which has also been reported earlier (e.g. Talpsepp et al., 1994). The upwellings were seen as fluctuations in salinity up to the western transition zone (T-2). This chain of events, together with the strong riverine outflow, created marked variability in hydrography and was seen as several fairly stabile less saline water lenses in the eastern areas up to the first transition zone (T-1). These physically created water lenses could promote and maintain the growth of freshwater phytoplankton species (e.g. *Microcystis aeruginosa, Anabaena circi-*

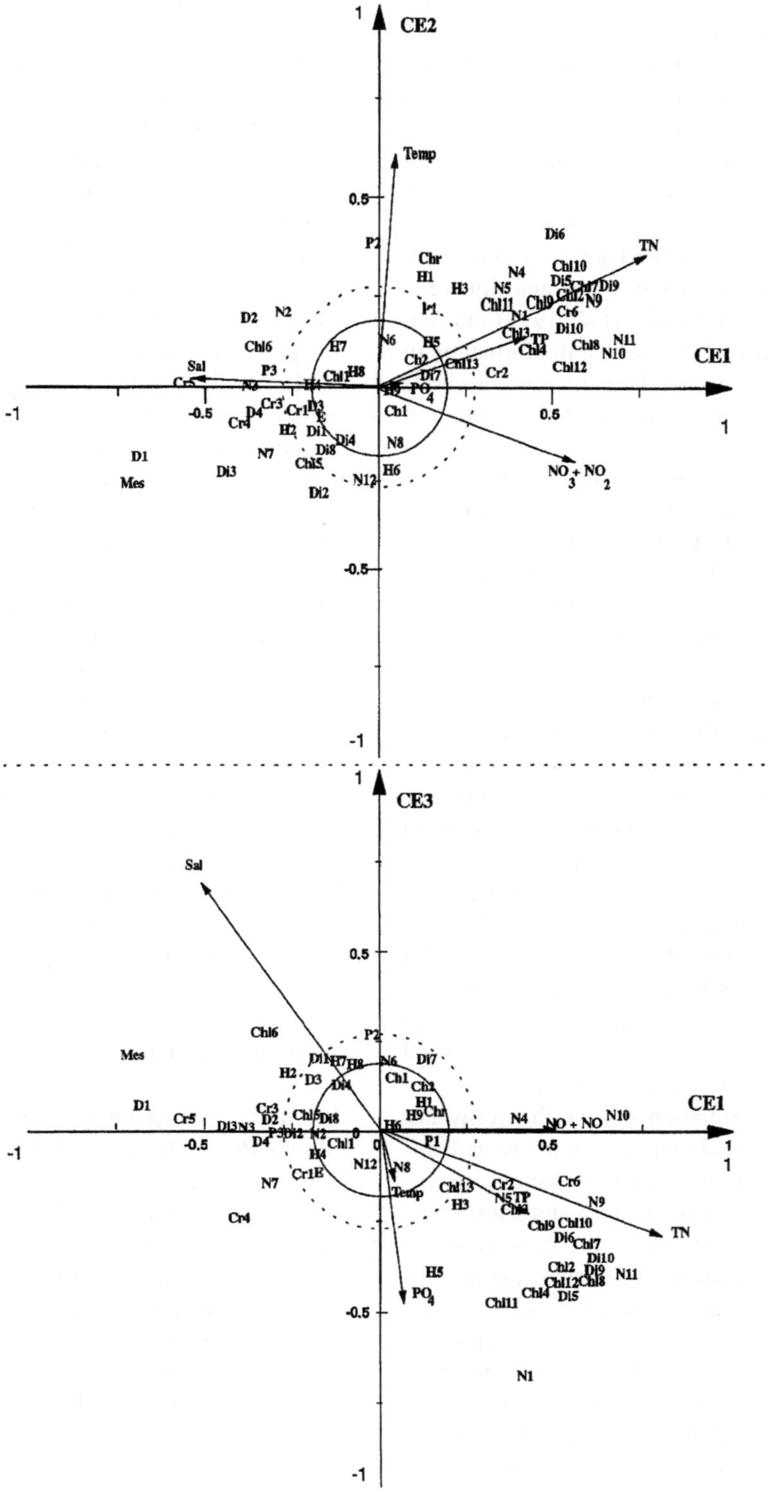

Figure 8. Correlations of the phytoplankton taxa and the environmental variables with the first three canonical variables (CE1, CE2, CE3). The inner origocentered circle denotes the projection of the 90% confidence level and the outer circle the 99% level. For abbreviations see Table 2.

nalis, Limnothrix planktonica), which originated from the inner Neva Estuary. Furthermore, the eastward shift in occurrence of several species which favour high turbulence (e.g. *Actinocyclus octonarius, Chaetoceros wighamii*) was connected to the series of events that originated in the upwellings.

The analysis of variance showed that the variability in phytoplankton community structure was higher between the consecutive subareas than within an area. However, phytoplankton assemblages most often consist of discontinuous patches in space and time (e.g. Kononen et al., 1992; Moisander et al., 1997) and consequently the cluster analysis, where the formation of groups was totally unrestricted gave a more accurate picture of the phytoplankton community structure in the area.

In the cluster analyses the variability in the phytoplankton species assemblages collected from the transition zones clearly reflected the varying extent of the outflow from the River Neva. The large amount of outliers in the Estuary itself is linked to the rapid variability of water masses with direct effects on the phytoplankton community.

Vertical mixing of the water column is reflected in nutrients concentration dynamics. Their release from sediments and the subsequent entrainment from the lower water layers to the euphotic layer have marked effects on phytoplankton dynamics. In the western study areas (HKI, OGF), the number of outliers was low. These outliers were due to the upwellings with subsequent changes in species composition. The outlier detected in the open Gulf of Finland (OGF) on the 26th of September was linked to the upwelling that had occurred ten days earlier (Figure 3). In this case the prevailing dominance of blue-green algae turned into the dominance of flagellates ($< 10\ \mu$m).

In the canonical correlation analysis the freshwater species which had high abundances in the Neva Estuary show mainly opportunistic characteristics concerning their preference for both nitrogen and phosphorus nutrients, or preference for only their nitrogen component. This can be explained by the dynamic interplay of the various inorganic nutrients, which is linked to high and rapid hydrodynamics as well as to the oversupply of nitrogen into the area (e.g. Pitkänen & Tamminen, 1995). According to our results, the species that can adapt quickly to the rapidly changing nutrient ratios have a competitive advantage in the easternmost areas.

The dominance of the non-heterocystic blue-green alga *Planktothrix agardhii* in the Neva Estuary during our study has also found in other studies (e.g. by Pitkänen et al., 1993; Basova & Lange, 1998). According to our results, the dominance of *P. agardhii* is mainly connected to the increased nitrogen load of the area. Basova & Lange (1998) reported that in the Neva Bay and in the inner Estuary this species has become common during the late 1980s, concurrent with the decreased frequency of the nitrogen-fixing blue-green alga *Aphanizomenon flos-aquae*. The oversupply of nitrogen from the St. Petersburg area and the River Neva, as well as the steady increase of inorganic N:P ratio in the deep waters of the Gulf of Finland during the last decades (Kahma & Voipio, 1990) can promote the growth of non-heterocystic phytoplankton species.

In conclusion, the unattended measurements and sampling of the surface water parameters are suitable for phytoplankton studies especially in areas where no distinct and permanent pycnoclines exist in the euphotic layer. We used semi-quantitative counting for phytoplankton species abundances, and chlorophyll *a* concentrations of samples as a measure of total biomass. This enabled the analyses of a large amount of samples. The dynamics of phytoplankton is affected by the fluctuating hydrographical conditions of the water column i.e. the present phytoplankton community is a reflection of the earlier physical and chemical conditions in the area. This study revealed that the time-lag between a nutrient pulse and the response of phytoplankton was few days. Consequently, these temporally and spatially frequent measurements, in combination with multivariate analyses, were a prerequisite to enable the study of preceeding conditions in order to explain the causalities in pelagic phytoplankton dynamics.

Acknowledgements

We thank the personnel of the Finnish Institute of Marine Research, especially Ms. Tuovi Vartio ans Ms. Soili Saesmaa for analyses of chlorophyll *a* and Mr. Tapani Juntunen and Ms. Pirjo Tikkanen for analyses of nutrients. We greatly appreciate the valuable advise of Drs Juha-Markku Leppänen, Heikki Pitkänen and Kaisa Kononen during the processing period of this manuscript. We thank the anonymous referees for their comments in order to improve our paper, as well as MSc. Maria Ekman-Ekebom checking the language.

References

Alenius, P., K. Myrberg & A. Nekrasov. Physical oceanography of the Gulf of Finland: A review. Submitted to Borealis and Environment Research.

Basova, S. & E. Lange, 1998. Trends in late summer phytoplankton in the Neva Bay and eastern Gulf of Finland during 1978 to 1990. Memoranda Soc. Fauna Flora Fennica 74 No 1.

Grönlund, L. & J.-M. Leppänen, 1990. Long-term changes in the nutrient reserves and pelagic production in the western Gulf of Finland. Finnish Mar. Res. 257: 15–27.

Haapala, J. & P. Alenius, 1994. Temperature and salinity statistics for the northern Baltic Sea 1961–1990. Finnish Mar. Res. 262: 51–121.

Hela, I., 1976. Vertical velocity of the upwelling in the sea. Commentationes Physico-Matematicae, Soc. Scient. Fenn. 46: 9–22.

Helcom, 1988. Guidelines for the Baltic Monitoring Programme for the third stage. Baltic Sea Envir. Proc. 25A: 1–51.

Kahma, K. & A. Voipio, 1990. Elimination of seasonal variations from long term changes. Finnish mar. Res. 257: 3–14.

Kauppi, L. (ed.), 1993. Mass mortalities of seabirds in the eastern Gulf of Finland in spring 1992. National Board of Waters and the Environment, Finland. Publications of the Water and Environment Administration – series A., No 142, 46 pp. (in Finnish with an English summary).

Kauppila, P., G. Hällfors, P. Kangas, P. Kokkonen & S. Basova, 1995. Late summer phytoplankton species composition and biomasses in the eastern Gulf of Finland. Ophelia 42: 179–191.

Kshirsagar, A. M., 1972. Multivariate Analysis. Marcel Dekker, Inc. New York: 247–287.

Kononen, K., K. Sivonen & J. Lehtimäki, 1993. Toxicity of phytoplankton blooms in the Gulf of Finland and Gulf of Bothnia, Baltic Sea. In T. J. Smayda & Y. Shimizu (eds), Toxic Phytoplankton Blooms in the Sea. Elsevier, Amsterdam: 269–273.

Kononen, K., S. Nõmmann, G. Hansen, R. Hansen, G. Breuel & E. Gupalo, 1992. Spatial heterogeneity and dynamics of vernal phytoplankton species in the Baltic Sea in April – May 1986. J. Plankton Res. 14: 107–125.

Leppänen, J.-M., V. Gorbatsky, E. Rantajärvi & M. Raateoja, 1994a. Dynamics of plankton blooms in the Gulf of Finland in 1992 measured using an automated flow-through analyzer. In: CBO 18th Conference of Baltic Oceanographers, St. Petersburg, Russia, 23–27 November, 1992. Proceedings 2: 12–27.

Leppänen, J.-M., S. Nömman & M. Kahru, 1994b. Variability of the surface layer in the Gulf of Finland as investigated by repeated continuous transects between Helsinki and Tallinn: a progress report. – In Patchiness in the Baltic Sea. ICES cooperative research reports 201: 69–72.

Leppänen, J.-M., E. Rantajärvi, M. Maunumaa, M. Larinmaa & J. Pajala, 1994c. Unattended algal monitoring system – a high resolution method for detection of phytoplankton blooms in the Baltic Sea. Proceedings of symposium OCEANS-94 held in conjunction with OSATES 94. Brest, France 13–16 September 1994. IEEE, New York, 1: 461–463.

Mardia, K. V., J. T. Kent & J. M. Bibby, 1979. Multivariate analysis. Academic Press, Inc. London: 360–393.

Moisander, P., E. Rantajärvi, M. Huttunen & K. Kononen, 1997. Phytoplankton community in relation to salinity fronts at the entrance to the Gulf of Finland, Baltic Sea. Ophelia 46: 187–203.

Niemi, È, 1988. Exceptional mass occurrence of Microcystis aeruginosa (Kützing) Kützing (Chlorococcales, Cyanophyceae) in the Gulf of Finland in autumn 1987. Memoranda Soc. Fauna Flora Fenn. 64: 165–167.

Perttilä, M., L. Niemistö & K. Mäkelä, 1995. Distribution, development and total amounts of nutrients in the Gulf of Finland. Estuar. Coast. Shelf Sci. 41: 345–360.

Pitkänen, H., P. Kangas, J. Sarkkula, L. Lepistö, G. Hällfors & P. Kauppila, 1990. Water quality and trophic status in the eastern Gulf of Finland. A report on studies in 1987–88. National Board of Waters and the Environment, Finland. Publications of the Water and Environment Administration – series A. No 50. 137 pp. (in Finnish with an English summary and figure legends).

Pitkänen, H., 1991. Nutrient dynamics and trophic conditions in the eastern Gulf of Finland: the regulatory role of the Neva Estuary. Aqua Fenn. 21,2: 105–115.

Pitkänen, H., T. Tamminen, P. Kangas, T. Huttula, K. Kivi, H. Kuosa, J. Sarkkula, K. Eloheimo, P. Kauppila & B. Skakalsky, 1993. Late summer trophic conditions in the north-east Gulf of Finland and the River Neva estuary, Baltic Sea. Estuar. Coast. Shelf Sci. 37: 453–474.

Pitkänen, H. & T. Tamminen, 1995. Nutrient and phosphorus as production limiting factors in the estuarine waters of the eastern Gulf of Finland. Mar. Ecol. Prog. Ser. 129: 283–294.

Rantajärvi, E. and J.-M. Leppänen, 1994. Unattended algal monitoring on merchant ships in the Baltic Sea. TemaNord 546: 1–60.

SAS/STAT User's Guide, Version 6, 4th edn., Volume 1 & 2, Cary, NC. SAS Institute Inc., 1989. 943, 846 pp.

Sokal, R. R. & J. F. Rohlf, 1994. Biometry. 3rd edn. W.H. Freeman and Company, New York, 179–271.

Talpsepp, L., T. Nöges, T. Raid & T. Kõuts, 1994. Hydrophysical and hydrobiological processes in the Gulf of Finland in summer 1987: characterization and relatioships. Cont. Shelf Res. 14: 749–763.

Utermöhl, H., 1958. Zur Vervollkommnung der quantitativen Phytoplanktonmethodik. Mitt. Int. Ver. Limnol. 9: 1–38.

Viitasalo, M., 1993. Mesozooplankton of the eastern Gulf of Finland in the summers of 1990–1992: community analysis and comparison with data from years 1905–1907. Memoranda Soc. Fauna Flora Fennica 69: 97–106.

Willén, T., 1962. Studies on the phytoplankton of some lakes connected with or recently isolated from the Baltic. Oikos 13: 169–199.

Hydrobiologia **363**: 141–156, 1998.
T. Tamminen & H. Kuosa (eds), Eutrophication in Planktonic Ecosystems: Food Web Dynamics and Elemental Cycling.

On the production, elemental composition (C, N, P) and distribution of photosynthetic organic matter in the Southern Black Sea

Ayşen Yılmaz[1], Süleyman Tuğrul[1], Çolpan Polat[2], Dilek Ediger[1], Yeşim Çoban[1] & Enis Morkoç[3]

[1] *Middle East Technical University, Institute of Marine Sciences, P.O.Box 28, 33731, Erdemli-İçel/Turkey*
[2] *İstanbul University, Institute of Marine Sciences and Management, Müşküle Sok., No 10, Vefa-İstanbul/Turkey*
[3] *Turkish Scientific and Technological Research Council (TÜBITAK), Marmara Research Center, P.O.Box 21, 41470, Gebze-Kocaeli/Turkey*

Received 1997; in revised form 1997; accepted 1997

Key words: Black Sea, dissolved nutrients, optical transparency, seston elemental composition, chlorophyll-a, primary production

Abstract

Chemical oceanographic understanding of the southern Black Sea has been improved by recent measurements of the optical transparency, phytoplankton biomass (in terms of chlorophyll-a and particulate organic matter) and primary productivity. During the spring-autmun period of 1995–1996, light generally penetrated only into the upper 15–40 m, with an attenuation coefficient varying between 0.125 and 0.350 m^{-1}. The average chlorophyll-a (Chl-a) concentrations for the euphotic zone ranged from 0.1 to 1.5 $\mu g\ l^{-1}$. Coherent sub-surface Chl-a maxima were formed near the base of the euphotic zone only in summer. Production rate varied between 247 and 1925 in the spring and between 405 and 687 mgC $m^{-2}\ d^{-1}$ in the summer-autumn period. The average POM concentrations in the euphotic zone varied regionally and seasonally between 3.8 and 28.6 μm for POC, 0.5 and 3.1 μm for PON and 0.02 and 0.1 μm for PP. Atomic ratios of C/N, C/P and N/P, derived from the regressions of POM data, ranged between 7.5 and 9.6, 109 and 165, and 11.2 and 16.6, respectively. In the suboxic/anoxic interface, the elemental ratios change substantially due to an accumulation of PP cohering to Fe and Mn oxides. The chemocline boundaries and the distinct chemical features of the oxic/anoxic transition layer (the so-called suboxic zone) are all located at specific density surfaces; however, they exhibit remarkable spatial and temporal variations both in their position and in their magnitude, which permit the definition of long-term changes in the biochemical properties of the Black Sea upper layer.

Introduction

The Black Sea, a relatively large, deep, landlocked basin, is connected to the Sea of Marmara through the narrow and shallow Bosphorus Strait (Figure 1). Low saline Black Sea waters are transported to the Mediterranean while a counterflow in the Bosphorus introduces more saline Mediterranean waters into the Black Sea via the Sea of Marmara. There exists a permanent and strong halocline at depths of 50–150 m, shoaling in the cyclonic gyres and deepening in the coastal regions. Continuous transport of biogenic particles from the productive surface to the lower layers, combined with limited vertical ventilation through the permanent halocline, provides the major reasons for the anoxic and sulphidic condition of the subhalocline waters.

In addition to these natural processes, the increasing input of nutrients and organic matter from the land via the rivers and the discharge of wastes have, during the last two decades, generated dramatic changes in the Black Sea ecosystem, especially in the wide northwestern shelf (Mee, 1992; Bologa, 1985/1986; Bologa et al., 1995; Cociasu et al., 1996, 1997). Long-term modifications and collapses of the biological structure of the ecosystem have been well documented

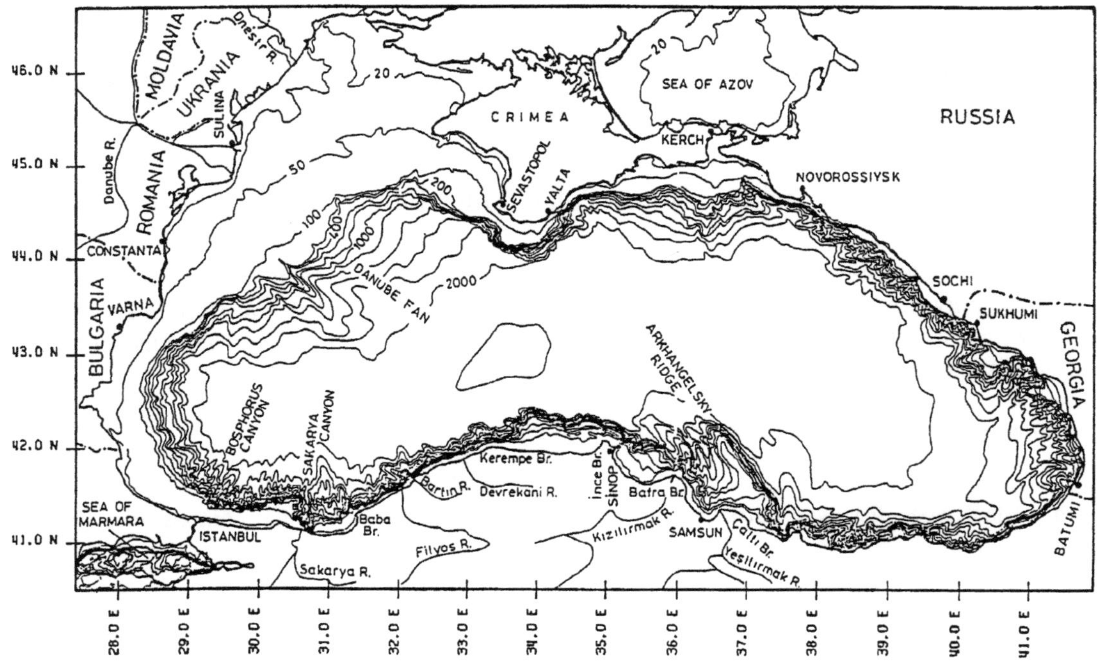

Figure 1. Bathymetry and location map of the Black Sea.

(Bodeanu, 1992; Mee, 1992; Shuskina et al., 1990; Smayda, 1990; Vinogradov et al., 1989; Bologa et al., 1995). However, the lack of good quality historical data of high resolution impairs understanding of how the recent anthropogenic inputs and climatic changes have influenced nutrient and organic carbon pools of the Black Sea. Nevertheless, comparison of the limited earlier measurements with the high-resolution data obtained since 1988 has enabled several workers to address the magnitude of the long-term changes in the nutrient and oxygen profiles from the upper layer down to the sulphide-bearing waters of the deep basin (Murray et al., 1989, 1991, 1994; Kempe, 1991; Cadispoti et al., 1991; Tuğrul et al., 1992; Baştürk et al., 1994). Similar changes have been observed in the nutrient chemistry of the waters of the polluted northwestern shelf (Cociasu et al., 1997).

According to Sorokin (1983) and Vedernikov & Demidov (1993), primary production in the Black Sea displays two phytoplankton maxima throughout the year; the major one occurs in early spring while a secondary peak appears in autumn. Recently, additional summer blooms have frequently been observed in both the coastal and open waters (Hay & Honjo, 1989; Hay et al., 1990, 1991; Sur et al., 1996). Primary production is relatively low in the open sea (50–200 gC m^{-2} y^{-1}) compared to the northwestern shelf area (up to 400 gC m^{-2} y^{-1}) (Vedernikov & Demidov, 1993; Bologa et al., 1985/1986), where there are riverine discharges of nutrients (Cociasu et al., 1997). Since input of nutrients from the anoxic layer through the permanent pycnocline is limited both by denitrification and by oxidation-reduction processes occurring in the oxic/anoxic transition layer, the major nutrient source for the open system is the input from the nutricline (Murray et al., 1995). New production in the open waters of the Black Sea is therefore dominated by the input from the nutricline/coinciding with the oxycline/riverine input via surface circulations and atmospheric transport probably being of secondary importance; consequently the rates of new production in the Black Sea are low (Murray et al., 1995; Oğuz et al., 1996).

In the present paper, we have evaluated high-resolution hydrographic data, light penetration phenomena, dissolved oxygen and hydrogen sulphide, dissolved inorganic nutrients, the elemental composition (C, N, P) of particulate organic matter and the photosynthetic carbon production rates measured seasonally in 1995 and 1996. In this context, we discuss the magnitudes of the spatial and temporal changes in the principal chemical and biological features of the Black Sea upper layer down to the sulphide-bearing waters.

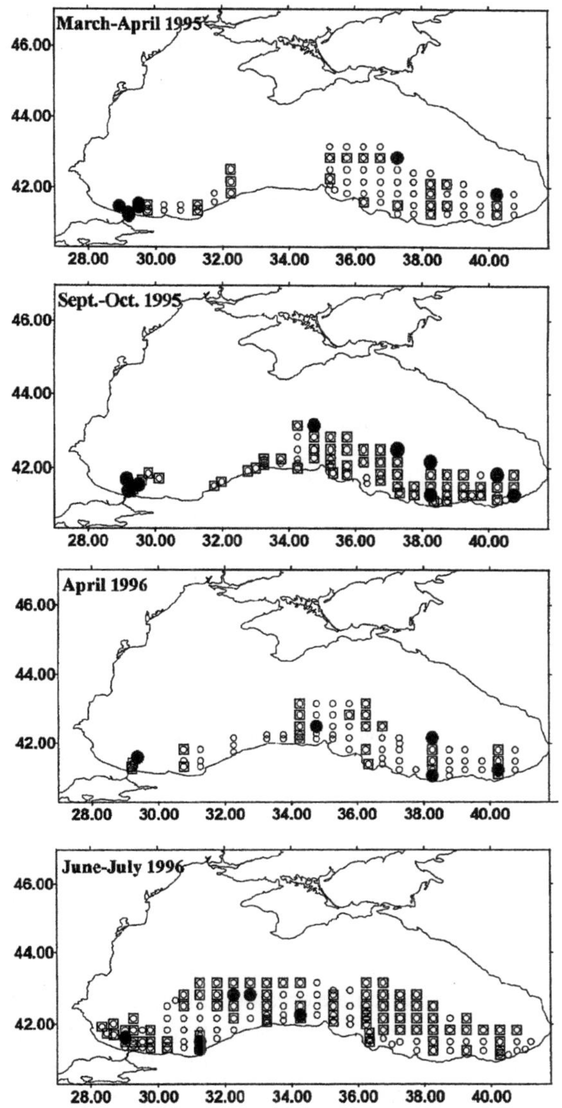

Figure 2. Station networks of the 1995–1996 cruises conducted in the southern Black Sea. Symbols denote different sampling strategies: ○ = Hydrophysical stations, □ = Hydrochemical stations and • = Biological stations.

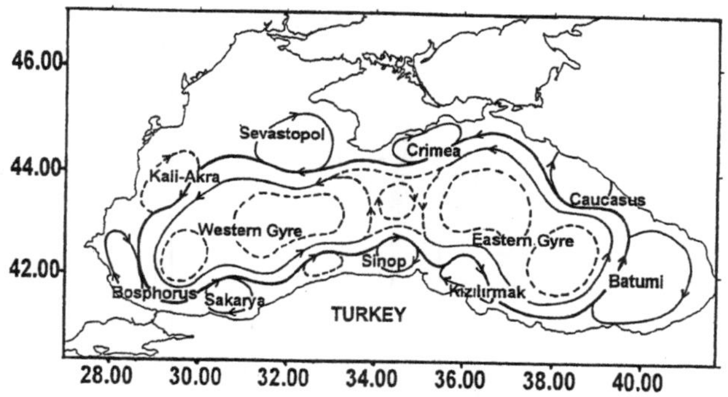

Figure 3. Schematic representation of the main features of the upper layer general circulation of the Black Sea (After Oğuz et al., 1993).

Methodology

Area of Study: Basin-wide cruises in the southern Black Sea surveyed the Turkish Economic Zone in March–April 1995, September–October 1995, April 1996 and June–July 1996. The cooperative marine science programmes (CoMSBlack and NATO TU-Black Sea) between the Black Sea riparian countries were initiated in early 1990s and the basinwide data are published by the joint groups. The station networks of the four cruises are illustrated in Figure 2. Hydrographic measurements were carried out on a regular 0.5° grid system for determining the physical boundaries of the cyclonic and anticyclonic circulations and the sub-basin scale variations of the hydrography. The locations of stations for biochemical studies were selected both by reference to observations from earlier years and by examining real-time CTD measurements. Previous studies clearly demonstrate the physical oceanography of the Black Sea upper layer to be dominated by the quasi-permanent cyclonic gyres in the the eastern and western halves of the basin (Figure 3). The two gyres are separated from a series of anticyclonic eddies in the coastal zone by the cyclonically undulating Rim current (Oğuz et al., 1991, 1993). The influence of the freshwater input, mainly from the Danube, Dnepr and Dnester rivers at the northwestern shelf, can be traced down to the Bosphorus region (Sur et al., 1996).

Sampling and analysis: The hydrographic data were collected using a Sea-Bird Model 9 CTD probe. A Licor 185 Model quantameter was used for solar irradiance measurements in the euphotic zone. This instrument measures Photosynthetically Active Radiation (PAR) within the range of 400–700 nm, in μE m^{-2} s^{-1} unit. Water samples were collected with General Oceanic Go-Flo Rosette bottles attached to the CTD. The analyses of nutrient samples – kept frozen in HDPE for a few weeks – were carried out using a Technicon Model two-channel Autoanalyzer. The colorimetric methods followed were similar to those described by Strickland & Parsons (1972) and Grasshoff et al. (1983). Chlorophyll-a (Chl-a) samples were collected from the euphotic zone down to the depth of 0.1% of surface light. Samples concentrated on GF/F filters were extracted with 90% acetone solu-

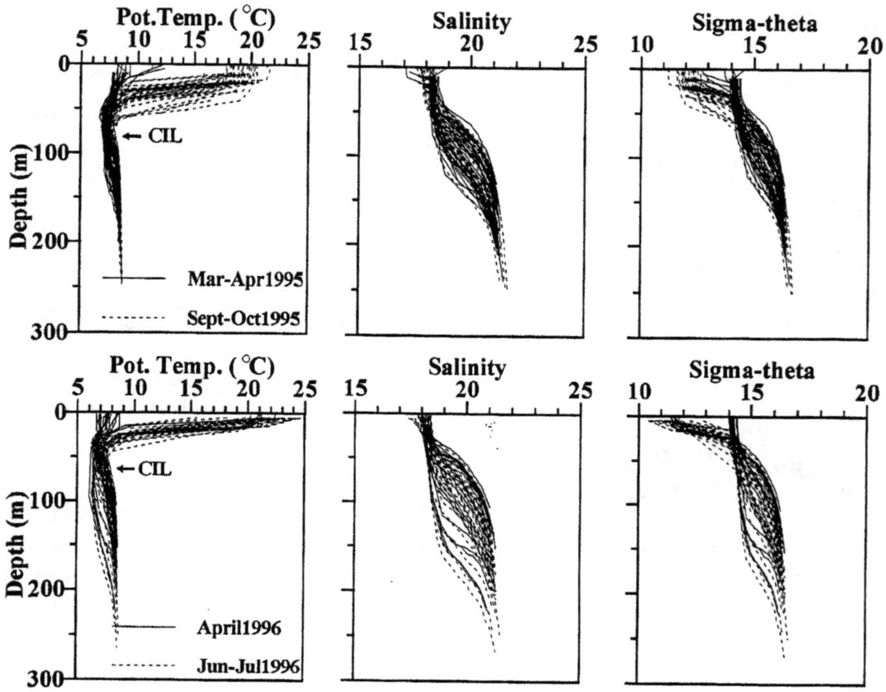

Figure 4. Potential Temperature (°C), Salinity and Sigma-theta profiles in the upper layer of the southern Black Sea for the 1995–1996 period (unpublished data, METU, Institute of Marine Sciences, Physical Oceanography Section).

tion. The fluorescence intensity of clear extracts was then measured (Strickland & Parsons, 1972; Holm-Hansen & Riemann, 1978), using a Hitachi F-3000 Model spectrofluorometer. A commercially available Chl-a standard obtained from Sigma was used to quantify the sample intensities. Particulate organic matter (POM) collected on GF/F filters (pre-combusted at 450–500 °C) for carbon (POC) and nitrogen (PON) and phosphorus (PP) measurements were kept frozen until processing on land. The POC and PON samples were dried at 50–60 °C overnight and then exposed to concentrated HCl fumes to remove inorganic carbonates. Filters were dried again and kept in a vacuum desiccator until analysis by the dry combustion method (Polat & Tuğrul, 1995), using a Carlo Erba Model 1108 CHN analyzer. The PP samples were first exposed to dry combustion at 500 °C for 2 h and then treated with 10 ml of 2N HCl for 10 h and filtered (Polat & Tuğrul, 1995). After the adjustment of pH to 8.0, the oxidized phosphorus in the solution was determined colorimetrically by the routine ortho-phosphate method. Dissolved oxygen (DO) and H_2S concentrations were determined by conventional Winkler and iodometric titration (Baştürk et al., 1994) while low H_2S concentrations were determined by the colorimetric method (Cline, 1969). The rates of carbon fixation by phytoplankton in the samples taken from the surface and from 90%, 75%, 50%, 25%, 10%, 1% and 0.1% surface light depths were determined by tracing the conversion of dissolved inorganic radioactive carbon (^{14}C) into particulate organic carbon. The original methodology (Steemann-Nielsen, 1952) was followed with slight modifications (Richardson, 1991). Incubator experiments were performed under artificial growth conditions.

Results

Hydrographic properties: The composite depth profiles of Potential Temperature, Salinity and Density (Sigma-$\theta = \sigma_t$) from the southern Black Sea demonstrate that throughout the seasons of stratification a nearly isohaline and relatively cool, isothermal water mass exists below the seasonal pycnocline (Figure 4). This prominent and persistent feature of the Black Sea, termed the Cold Intermediate Layer (CIL), possesses a temperature minimum which is characterised by the 8 °C limiting isotherms (Oğuz et al., 1991). The thickness of the CIL is larger (up to 100 m) in the anticy-

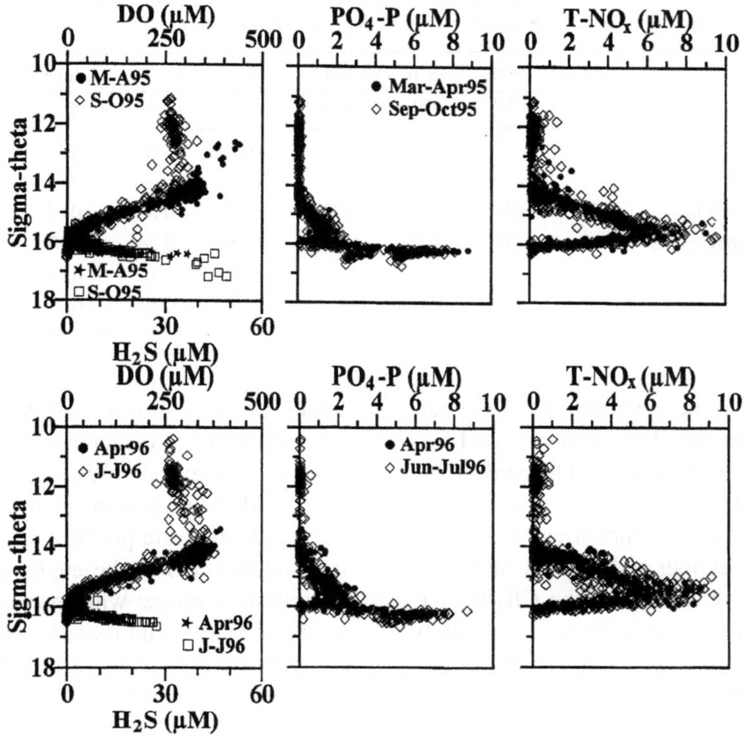

Figure 5. Vertical distribution of Dissolved Oxygen (DO) – H_2S and Dissolved Nutrients (PO_4–P and Total Oxidized Nitrogen = T–NO_x) in the upper layer of the southern Black Sea for the 1995–1996 period. Sigma-theta was used instead of depth on the vertical scale.

clonic regions (ACR) than in the cyclonic regions (CR) (about 50 m). The $\sigma_t = 14.8$ isopycnal surface defines not only the temperature minimum within the CIL but also the upper boundary of the permanent pycnocline in the Black Sea (Buesseler et al., 1994; Murray et al., 1991). In the CIL, the salinity varies slightly from nearly 18.5 to 20.1 ppt. The profiles illustrated in Figure 4 show that, in winter, when the surface waters cool down to 6–7 °C, the upper layer is thoroughly homogenized by convective mixing down to the $\sigma_t = 14.7$–14.8 isopycnal surfaces. With the advent of heating in spring and summer, the surface temperatures rise to 20–25 °C and the CIL becomes topped by a warm surface layer (Figure 4). Below the CIL, the temperature gradually rises from 8 °C to 8.7 °C at the base of the permanent pycnocline; this is observed at different depths (Figure 4). The subhalocline waters possess similar temperatures at similar density surfaces over the entire deep basin though isohalines appear at different depths from the deep to the coastal regions. Composite density profiles demonstrate coherent seasonal changes in the surface layer due to the significant changes in the surface temperature during the year. On the contrary, the density of the water masses below the CIL is mainly determined by their salinity. This salinity-determined density gradient (the permanent pycnocline) is located at shallower depths in cyclonic but at much deeper layers in anticyclonic regions. Below the permanent pycnocline the density increases very slowly with depth, the values being similar at any given depth whatever the region (Figure 4).

Chemical properties: The composite chemical profiles down to the anoxic waters were plotted relative to water density (rather than depth) as a vertical scale (Figure 5). As recently indicated by Tuğrul et al. (1992), Baştürk et al. (1994) and Murray et al. (1995), composite profiles from hydrodynamically different regions exhibit characteristically similar vertical features below the euphotic zone down to the upper anoxic layer.

Dissolved Oxygen (DO): Figure 5 shows the surface layer down to the temperature minimum in the CIL to be nearly saturated with dissolved oxygen (DO = 250–450 μM). The lower DO values in the productive surface waters are due to the decreasing solubility at higher temperatures during stratification. The concentra-

Table 1. Average nutrient [phosphate (PO_4-P) and total oxidized nitrogen (T-NO_x = NO_3 + NO_2-N)] concentrations [(μM), from the surface down to σ_t = 14.5] in the southern Black Sea.

Location	Mar–Apr 1995		Sept–Oct 1995		April 1996		June–July 1996	
	PO_4-P	T-NO_x	PO_4-P	T-NO_x	PO_4-P	T-NO_x	PO_4-P	T-NO_x
Bosphorus	0.04	0.42	0.08	1.07	0.24	0.56	0.12	0.55
Western Cyclone	–	–	–	–	–	–	0.15	0.26
Eastern Cyclone	0.05	0.24	0.06	0.28	0.21	0.63	0.04	0.16
Batumi Anticyclone	0.05	0.55	0.03	0.43	0.35	0.76	0.07	0.98
Off Sakarya	0.09	0.79	–	–	0.08	1.48	0.03	0.53
Off Sinop	–	–	0.05	0.34	0.06	0.52	0.06	0.84

tions decrease steeply in the upper depths of the permanant pycnocline from 200–250 μm at the σ_t = 14.2–14.8 density surfaces to suboxic concentrations of 20–30 μm at the σ_t = 15.4–15.6 surfaces, these surfaces defining the upper and lower boundaries of the main oxycline. Throughout summer and autumn, stratification limits the normal ventilation of the CIL by vertical mixing. Under these circumstances, as recently indicated by Baştürk et al. (1997a), the oxycline commences at greater density surfaces (σ_t = 14.4–14.5) but at shallower depths (35–40 m) in cyclonic regions (CR), than in the frontal zones of the Rim Current (RCFZ) or in anticyclonic regions (ACR) where the onset is located at σ_t = 14.2–14.3 (70–100 m). Below the main oxycline, DO declines slowly to < 5 μm at σ_t = 15.9–16.0, and can no longer be detected at the σ_t = 16.15–16.20 density surfaces where sulphide concentrations are 1–3 μm (Figure 5). This DO-deficient water, formed within the oxic/anoxic transition layer with DO $< 20\,\mu$M and $H_2S < 1$ μM, is called the suboxic zone. Sulphide-bearing waters were consistently observed at density surfaces of $>$ 16.15–16.2 over the entire deep basin. In the upper anoxic layer, the H_2S concentration increased steadily with depth (Figure 5), showing insignificant spatial or temporal variation at any density surface.

Phosphate (PO_4) and T-NO_x (NO_3 + NO_2) Distributions: As previously emphasized by Baştürk et al. (1994), Bingel et al. (1993) and Codispoti et al. (1991), the surface waters of the southern Black Sea are always poor in nutrients during the seasons when these waters are stratified. In the spring-autumn period of 1995–1996, average concentrations in the euphotic zone ranged regionally and seasonally from 0.16 to 1.5 μm for T-NO_x (mainly NO_3) and from 0.03 to 0.35 μm for phosphate (Table 1). The limited nutrient data from previous years (Bingel et al., 1993) together with modelling studies (Oğuz et al., 1996) indicate that intense vertical mixing in winter provides input from the nutricline which may increase surface nitrate concentrations 5–10-fold so that primary productivity becomes light-limited. Composite profiles of T-NO_x and phosphate indicate that, below the euphotic zone, nutrient concentrations increase with increasing density down to the base of the main oxycline (Figure 5). The nitrate concentrations display a well-defined maximum of 5–9 μm between the 15.4–15.6 density surfaces defining the upper boundary of the suboxic zone where DO concentrations decrease to 20–30 μm. In the suboxic zone, due to denitrification, nitrate concentrations decline steadily to 0.1–0.2 μm at the suboxic/anoxic interface. Nitrate then becomes reduced by sulphide in the upper anoxic waters until to undetectable levels. Phosphate concentrations increase within the oxycline to a maximum in the upper suboxic zone or at the σ_t = 15.6–15.7 isopycnal surfaces. Below this broad maximum, phosphate concentrations decline steeply in the cyclonic regions (CR), forming a pronounced minimum (0.05–0.10 μm) at the σ_t = 15.85–15.90 isopycnal surfaces. This feature is less marked in coastal regions and it is nearly imperceptible within the RCFZ. Nevertheless, throughout the deep basin phosphate profiles always increase steeply within the sulphidic water interface and reach peak values of 5–8 μm at σ_t = 16.2 isopycnal surface. Phosphate concentrations decrease slightly in the upper anoxic layer and then increase again slightly with depth. The occurrence throughout the deep basin of the marked maximum at the sulphidic boundary probably results from dissolution of phosphate-associated Fe- and Mn-oxides in the anoxic waters (Shaffer, 1986; Codispoti et al., 1991).

Primary productivity and related parameters: Vertical profiles of POC, PON and PP in the upper water column show remarkable variations with depth, region

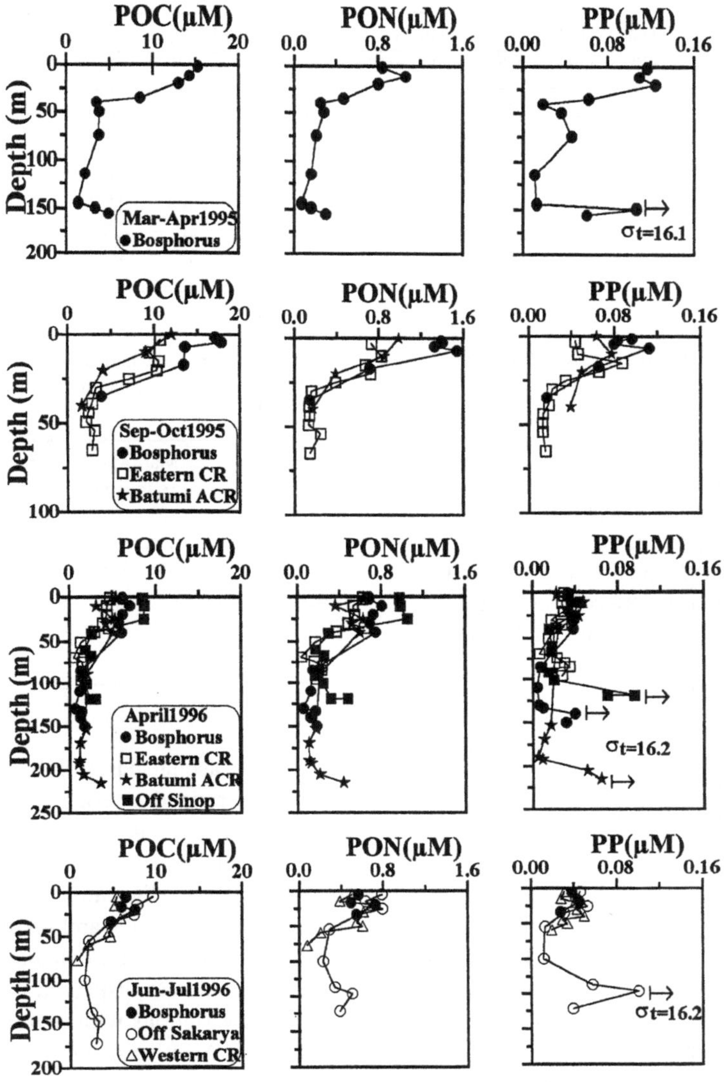

Figure 6. Vertical profiles of Particulate Organic Carbon (POC), Nitrogen (PON) and Total Particulate Phosphorus (PP) obtained in the southern Black Sea for the 1995–1996 period.

Table 2a. Average POM concentrations (µM) in the euphotic zone of the southern Black Sea.

Location	Mar–Apr 1995			Sept–Oct 1995			April 1996			June–July 1996		
	POC	PON	PP	POC	PON	PP	POC	PON	PP	POC	PON	PP
Bosphorus	28.6	3.1	0.115	12.6	1.03	0.075	6.3	0.74	0.036	6.2	0.58	0.038
Western Cyclone	–	–	–	–	–	–	–	–	–	5.5	0.56	0.034
Eastern Cyclone	–	–	–	8.7	0.59	0.046	3.8	0.48	0.024	–	–	–
Batumi Anticyclone	–	–	–	8.2	0.73	0.065	4.6	0.55	0.035	–	–	–
Off Sakarya	–	–	–	–	–	–	–	–	–	7.4	0.65	0.043
Off Sinop	–	–	–	–	–	–	8.6	1.00	0.039	–	–	–

Table 2c. POM elemental composition (molar ratios calculated from individual data) for the euphotic zone of the southern Black Sea.

Location	Mar–Apr 1995			Sept–Oct 1995			April 1996			June–July 1996		
	C/P	C/N	N/P	C/P	C/N	N/P	C/P	C/N	N/P	C/P	C/N	N/P
Bosphorus	231	11.2	22.6	168	12.2	13.7	175	8.5	20.0	168	10.8	15.6
Western Cyclone	–	–	–	–	–	–	–	–	–	131	10.2	16.1
Eastern Cyclone	–	–	–	193	15.6	13.1	158	7.9	20.0	–	–	–
Batumi Anticyclone	–	–	–	126	11.2	11.1	131	8.3	15.7	–	–	–
Off Sakarya	–	–	–	–	–	–	–	–	–	170	10.6	16.2
Off Sinop	–	–	–	–	–	–	220	8.6	25.0	–	–	–

Table 2b. POM elemental composition (derived from regression analysis) for the euphotic zone of the southern Black Sea.

Date	Regression	r^2	n
March–April, 1995	POC = 253.6 PP - 0.6	0.93	12
	POC = 7.8 PON + 4.4	0.98	
	PON = 32.0 PP - 0.6	0.92	
September–October 1995	POC = 109.3 PP + 3.4	0.36	35
	POC = 8.8 PON + 3.3	0.81	
	PON = 11.2 PP + 0.1	0.37	
April, 1996	POC = 111.8 PP + 1.7	0.29	25
	POC = 7.5 PON + 0.3	0.90	
	PON = 14.2 PP +0.2	0.30	
June–July, 1996	POC = 164.5 PP + 0.3	0.82	28
	POC = 9.6 PON + 0.6	0.88	
	PON = 16.6 PP ± 0.0	0.84	

Table 3. Average light indices in the southern Black Sea: Depth [D(m)] of 1% Surface Light and Downward Attenuation Coefficient [K_d (m^{-1})].

Location	Mar–Apr 1995	1995	Sept–Oct 1996	1996	April		June–July	
	D	K_d	D	K_d	D	K_d	D	K_d
Bosphorus	35	0.152	15	0.250	–	–	34	0.139
Western Cyclone	–	–	–	–	–	–	40	0.125
Off Sakarya	–	–	–	–	–	–	35	0.128
Off Sinop	–	–	–	–	21	0.230	–	–

and season (Figure 6 and Table 2a). In 1995–1996, the average particulate concentrations in the euphotic zone estimated for different regions of the southern Black Sea ranged between 3.8 and 28.6 μm for POC, 0.5 and 3.1 μm for PON and 0.02 and 0.1 μm for PP (Table 2a). In April and June, the euphotic zone concentrations were lower in 1996 than in the previous year. On a regional basis, the highest concentrations were observed in the coastal waters near the Bosphorus exit, in April 1995 (Table 2a). Below the euphotic zone, the concentrations declined to their background levels (POC = 1–4 μM, PON = 0.1–0.3 μM, PP = 0.01–0.03 μM) – excluding the coherent maxima which appear in the suboxic/anoxic interface at the density surfaces of σ_t = 16.1–16.2 (Figure 6). It should be noted that the increases in PP concentration within the interface were more pronounced than in POC and PON (Figure 6).

In order to estimate the relative elemental composition of POM collected in the surface water, regression analyses of basin-wide particulate data were carried out on a seasonal basis (Table 2b). The atomic N/P ratio lines ranged seasonally between 11.2 and 16.6 whereas the estimated ratios of C/P and C/N were in the range of 109–164 and 7.5–9.6 for the 1995–1996 period, excluding the anomalously high ratios from the Bosphorus region in March–April 1995. The euphotic zone averages of the particulate ratios, calculated from the individual data displayed in Table 2c, also exhibit marked variations with region and season. In April 1996, the C/N ratio ranged regionally between 7.9 and 8.6, which were less than those for other periods of 1995–1996 (Table 2c). Interestingly, the summer-96 ratios decreased slightly below the euphotic zone to values similar to those in the euphotic zone in April 1996. The estimated C/P ratios (between 126–231) are consistently higher than the Redfield ratio of 106 (Table 2c). In the suboxic/anoxic interface the C/P and N/P ratios of seston were found to be as low as 33 and 4, respectively, due to the less pronounced increases in the POC and PON contents of the interface (Figure 6).

The observed light penetration in the upper water column of the southern Black Sea during 1995–1996 indicated the thickness of the euphotic zone (practically defined as the depth of 1% of the surface light) to range between 15 and 40 m (Figure 7 and Table 3). The less energetic, high wavelength component of the incoming light was absorbed in the upper surface layer (the top 10m), where the highest (downward) attenuation coefficient ($K_d = 0.125$–$0.350\ m^{-1}$) was calculated. Below this layer the solar light penetrated with a constant K_d, which varied seasonally and regionally between 0.125 and 0.250 m^{-1}. The highest estimated K_d values were 0.250 m^{-1} observed in the Bosphorus region during Sept–Oct., 1995 and 0.230 m^{-1} observed in the RCFZ off Sinop in April, 1996, corresponding to a euphotic layer of 15 m and 21 m, respectively. Similar light penetration characteristics were displayed throughout the southern Black Sea in June–July 1996, with K_d estimates in between 0.125 and 0.139 m^{-1}, corresponding to a euphotic zone thickness of 34–40 m (Figure 7, Table 3). The 1% light depth always extended below the seasonal thermocline formed at 30–50 m in September 1995 and 20–30 m in July 1996 (Figures 4 and 7).

Seasonal Chl-a data from different regions of the southern Black Sea are displayed in Figure 7 and Table 4. In June–July 1996, the concentrations in the euphotic zone were generally low ($< 0.5\ \mu g\ l^{-1}$) with the lowest values in the surface mixed layer, and a subsurface Chl-a maximum was formed near the base of the euphotic zone and/or below the seasonal thermocline, corresponding to the depths of 0.5–2% of the surface light (Figure 7). However, the Sept–Oct 1995 profiles indicated an apparent increase in the Chl-a concentration in the surface mixed layer, when the mixed layer depth was greater than that of the euphotic layer (Figures 4 and 7). During the spring–autumn period of 1995–1996 Chl-a concentrations ranged from 0.1 to 0.6 $\mu g\ l^{-1}$ in the Batumi ACR whilst the concentrations in the western CR, the Bosphorus region and off Sinop (RCFZ) ranged from 0.3 to 1.5 $\mu g\ l^{-1}$. Interesingly, relatively low concentrations (0.1 to 0.5 $\mu g\ l^{-1)}$ were determined in the March–April period of 1995–1996, showing almost uniform vertical distributions.

Primary productivity (P) profiles were similar throughout the southern Black Sea; the highest rates, which varied seasonally and regionally between 1 and 10 $mgC\ m^{-3}\ h^{-1}$ (or 10 and 180 $mgC\ m^{-3}\ d^{-1}$), were always determined in the upper euphotic zone down to the 10% light intensity depth or the top 10–20 m of the water column. Below this layer, the rate decreased markedly with depth and dropped to negligible rates at the 1% light intensity depth (Figure 7). In order to determine the maximum rates of production, P(M), under adequate light intensity, samples taken from different depths of the euphotic zone were exposed to the full artificial light conditions in the incubator. The estimated maximum rates, P(M), were comparable or almost constant down to the depths of 1% surface light intensity but varied seasonally and regionally between 1 and 20 $mgC\ m^{-3}\ h^{-1}$. The depth-integrated P rates ranged from 1925 $mgC\ m^{-2}\ d^{-1}$ in the RCFZ off Sinop in April 1996 to 194 $mgC\ m^{-2}\ d^{-1}$ in the Bosphorus region in June–July 1996, yielding an average of 472 $mgC\ m^{-2}\ d^{-1}$ for the stratification seasons of 1995–1996 (Table 5).

Table 4. Euphotic zone average Chlorophyll-a concentrations ($\mu g\ l^{-1}$) in the southern Black Sea.

Location	Mar–Apr 1995	Sept–Oct 1995	April 1996	June–July 1996
Bosphorus	0.66	0.86	0.27	0.82
Western Cyclone	–	–	–	0.54
Eastern Cyclone	0.13	0.99	0.07	0.50
Batumi Anticyclone	0.12	0.30	0.09	0.48
Off Sakarya	–	–	–	0.62
Off Sinop	–	–	0.37	–

Table 5. Daily primary production rates [P(D) ($mgC\ m^{-2}\ day^{-1}$)] integrated for the euphotic zone in the southern Black Sea.

Location	Mar–Apr 1995	Sept–Oct 1995	April 1996	Jun–July 1996
Bosphorus	247	405	–	194
Western Cyclone	–	–	–	687
Off Sakarya	–	–	–	603
Off Sinop	–	–	1925	–

Discussion

The chemocline boundaries and the characteristic features (nitrate, phosphate maxima and phosphate minimum) of the oxic/anoxic transtion layer have consistently appeared at specific density surfaces during the spring-autumn period of 1995–1996 (Figure 5). The oxycline was much thicker in the ACR and RCFZ because of the thicker CIL which forms the upper part of the permanent pycnocline. The suboxic zone formed below the main oxycline (or at $\sigma_t = 15.4$–15.6) extended consistently down to the $\sigma_t = 16.15$–16.2 density surfaces throughout the deep basin. Comparison of present and past data shows the suboxic zone to have displayed no vertical movement since 1988 (Tuğrul et al., 1992; Baştürk et al., 1994 and Murray et al., 1995).

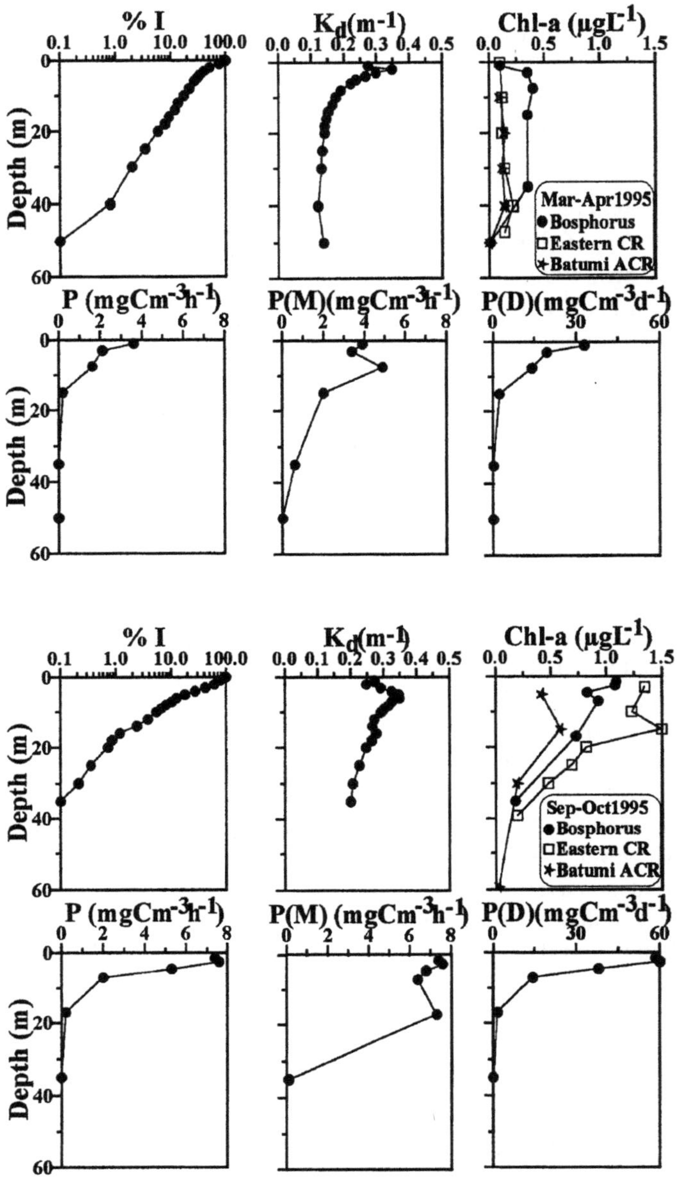

Figure 7a.

Figure 7. Vertical profiles of Light Penetration [as % of surface light (% I) and downward attenuation coefficient (K_d)], Chlorophyll-a (Chl-a) and Primary Production Rates [True noon production = P, Maximum Production = P(M), and Daily Production = P(D)] for the southern Black Sea for the spring–autumn periods of (a) 1995 and (b) 1996.

Nevertheless, its upper boundary deepens slightly during intense winter mixing and then shoals again by 0.1–0.2 density units during the stratification seasons when vertical ventilation of the oxycline is limited (Baştürk et al., 1997a). Unexpectedly, in the summer of 1992, the suboxic boundary was temporarily modified within the western cyclone; it became shifted upwards by nearly 0.3–0.4 density units (about 5–10 m), as compared to its position in the summer of 1991. The characteristic nitrate profile depicted in Figure 5 was also eroded markedly by intense denitrification resulting in an apparent removal of nitrate from the charactersitic

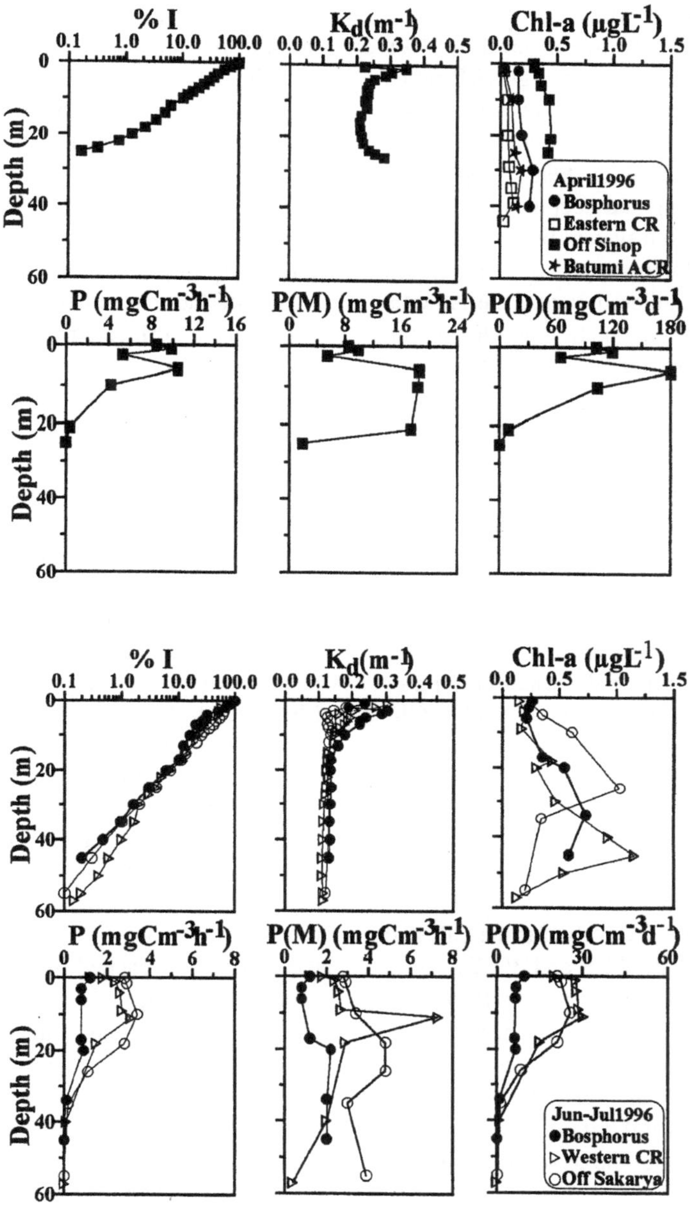

Figure 7b.

chemocline (Baştürk et al., 1997b). In the summer of 1992 the nitrate in the upper layer (down to $\sigma_t = 15.5$) decreased from its characteristic maximum concentration of 5–9 μm, to < 2 μm. In late 1992 the eroded chemical features appeared to recover though with lower nitrate concentrations at the NO_3 maximum depth (S. Konovalov, pers. com.).

High-resolution data obtained in the present study and during the Knorr-88 cruise (Codispoti et al., 1991) indicate significant regional changes in the characteristic features of vertical nutrient profiles. The upper CIL situated in the ACR between the $\sigma_t = 14.1$–14.4 surfaces has a distinctive chemical structure; it is now relatively enriched in nitrate (increasing vertically from 0.5 to 2.5 μm) whereas the phosphate concentrations remain almost constant with depth (but varying regionally between < 0.02 and 0.1 μm). Thus, the atomic N/P ratios (T-NO_x/PO_4) in the P-deficient upper

CIL appear to be anomalously high (between 40–80). This ratio decreases markedly to the levels of 4–8 at the base of the oxycline. Similar high N/P ratios have been reported for the bottom waters of the northwestern shelf (Cociasu et al., 1997), where the CIL, entrapped by the associated eddy fields, appears to be formed by winter cooling and becomes introduced into the southwestern coastal waters by the Rim current (Oğuz et al., 1992). Interestingly, such anomalies in the CIL were not encountered during the Atlantis 1969 cruise (Brewer, 1971); on the contrary, if the debatable ammonia data are ignored, the CIL appears to have been impoverished in nitrate and thus to have possessed anomalously low N/P ratios (between 1–5). It should be noted here that the Levantine Intermediate Water (LIW) which forms during winter cooling of the surface waters and sinks down to intermediate depths in the eastern Mediterranean, is enriched in nitrate but poor in reactive phosphate, resulting in anomalously high N/P ratios (Krom et al., 1992; Yılmaz & Tuğrul, 1997). It is as yet unclear what processes selectively remove reactive phosphate from the water column or cause the accumulation of nitrate in the Black Sea CIL and the Mediterranean LIW.

Because of the formation of a thicker CIL in the ACR, the chemical gradients in the oxycline are nearly a half of those in the CR where the CIL is very thin. Throughout 1995 and 1996 the gradients of DO and nitrate in the oxycline were estimated to be 9.8 μM m^{-1} and 0.26 μM m^{-1}, respectively, for the CR, and 4.2 and 0.1 respectively for the ACR and RCFZ, values which are consistent with those reported by Baştürk et al. (1997a, 1997b) for the early 1990's. Whereas phosphate gradients were comparable, the present nitrate gradient is much higher than that (0.015–0.03 μM m^{-1}) estimated from the 1969 data due to the unexpectedly low nitrate content of the oxycline (Tuğrul et al., 1992).

The phosphate minimum formed in the lower suboxic zone at the $\sigma_t = 15.85–15.90$ is a permanent feature of the CR (Tuğrul et al., 1992; Baştürk et al., 1994; Murray et al., 1995) and it is less distinct towards the coastal margins. Interestingly, this minimum was less pronounced (most probably due to coarse sampling intervals) and located at greater density surfaces – by about 0.1–0.15 density units (or 10-15 m) – in the late 60's (Tuğrul et al., 1992). This apparent shift coincides with the long-term upward expansion of the suboxic zone by about 0.3–0.4 density units since 1969 (Tuğrul et al., 1992; Codispoti et al., 1991; Murray et al., 1995). Long-term changes in the Black Sea ecosystem appear to have resulted from the increasing input of nutrients and labile organic matter by rivers; thus the enhanced primary productivity of the surface waters has led to a decrease in water transparency – the Secchi disk depth – markedly from 9–27 m in the 60's to 2–16 m in the 90's (Vladimirov et al., 1997). The contribution of climatic changes to the deterioration of the Black Sea ecosystem is still poorly understood. The deep phosphate maximum, formed in the sulphidic water interface has remained constant at a precise, specific density surface since the late 60's. This finding suggest that small-scale fluctuations observed in the suphidic water boundary are principally the result of the poor reliability of sulphide data (Tuğrul et al., 1992; Romanov et al., 1997).

The elemental composition of marine POM is relatively uniform (Copin-Montegut & Copin-Montegut, 1983) as compared to those of lakes (Hecky et al., 1993). In fact, the C/N/P ratios of POM in marine waters depend on the hydrography of the different marine regions, nutritional status, growth rates of marine phytoplankton and grazing pressure (Copin-Montegut & Copin-Montegut, 1983). The examination of sestonic ratios is a simple tool for understanding the composition of marine particles as well as the correlation between POM and environmental factors. The C-N and N-P regressions demonstrate that seasonal cycles of the sestonic ratios in the Black Sea euphotic zone take place within a limited range of ratios (C/N = 7.5–9.6; N/P = 11.2–16.6), corroborating the conclusion reached by Copin-Montegut & Copin-Montegut (1983). In other words, the relatively low inorganic nutrient concentrations of the surface waters during stratification seasons (Table 1) modified the sestonic ratios slightly relative to the mean planktonic ratio (N/P = 16). However, the C/P ratios were more variable due to the fast cycling of particulate phosphorus and P-deficient POM synthesis in the P-depleted surface waters. For instance, the C/P and N/P ratios of the Bosphorus region were unexpectedly high in March–April 1995. In June–July 1996, POM was deficient both in P and N elements, whilst the N/P ratio was similar to Redfield ratio, implying the assimilation rate of carbon to exceed those of N and P. Moreover, the small intercepts presented in Table 2b indicate low concentration of carboneceous (N- and P-deficient) organic compounds in the total seston, as well as weak limitation of inorganic nutrients on the chemical composition of POM of mostly regenerative origin as previously suggested by Copin-Montegut & Copin-Montegut (1983). The intercepts of PON/PP regressions are very low, suggesting the increases in particulate nutrients to occur in the same proportions in the living and non-living POM content of the euphotic zone.

In March–April 1995, the Bosphorus region phytoplankton was adapted to low P-supply as confirmed by low production rates and low Chl-a concentrations obtained in this period. These P-limited seston ratios are similar to those of moderately P-limited lakes (Hecky et al., 1993) but higher than the seston ratios of the P-limited eastern Mediterranean surface waters. Nevertheless, the present ratios are comparable to the C/N ratios of the fast-sinking POM collected by sediment traps in summer 1988 by Karl & Knauer (1991).

Altough the POM in the surface waters possesses a mean N/P ratio of about 14, similar to the Redfield ratio of 16, the molar NO_x/PO_4 ratios are always as low as 4–8 in the oxic/anoxic interface. Such anomalous N/P ratios indicate the nitrate loss from the suboxic waters via denitrification which exceeds the phosphate export from this layer to the sulphidic waters via the absorption on to Fe- and Mn oxides and subsequent sedimentation (Shaffer, 1986; Codispoti et al., 1991).

Production (P) rates determined in the present study are similar to those reported for the open part of the northwestern shelf and off the Romanian coast (Bologa, 1985/1986) and exceed those observed in the central Black Sea (Vedernikov & Demidov, 1993). The highest P rates were observed in April 1996 and June–July 1996, although the Chl-a concentrations were less pronounced in April, 1996. The inconsistency between the P rate and the Chl-a concentrations obtained in April 1996 generated relatively high assimilation numbers (30 mgC $mgChl^{-1}$ h^{-1}) suggesting that the light was not a limiting factor since the intensity of solar radiation exceeded the seasonal averages in this period. The estimated assimilation numbers were in the range of 2-15 mgC $mgChl^{-1}$ h^{-1} for the other seasons. Relatively high P rates in June–July 1996 are consistent with the occurrence of short summer blooms in the Black Sea additional to the characteristic spring and autumn blooms (Sorokin, 1983; Bologa, 1985/1986; Vedernikov & Demidov, 1993; Hay & Honjo, 1989; Hay et al., 1990; 1991; Sur et al., 1996). The highest carbon production rates and assimilation numbers were observed in the upper layer of the euphotic zone (above the 10% light depth). In general, the thickness of the euphotic zone was lower (with relatively high K_d) in the CR and the RCFZ than in the ACR due to relative increase of phytoplankton abundance, as also reported by Vidal (1995). The transparency of the water column in the Black Sea is also affected by the presence, besides phytoplankton, of a large variety of particles (inorganic particles, detritus, microzooplankton and even jelly fish) all of which have relatively high stocks in the Black Sea compared to other temperate seas such as the Mediterranean.

The springtime POC concentrations being relatively high or comparable to summer–autumn concentration and low Chl-a concentrations observed in the open waters in the spring, gave rise to relatively high C/Chl ratios (ranging between 400–600, mean = 550 w/w). This indicates the majority of the seston to have been composed of detritus and/or microzooplankton developed after the phytoplankton bloom. Probably the spring cruises took place after the completion of the characteristic bloom in the Black Sea. In summer and autumn both biomass indices (Chl-a and POC) were significantly high and the C/Chl ratios were relatively low (ranging between 100–300 for Sept.–Oct. 1995 and between 100–200 for June–July 1996). The decrease in the C/Chl ratio in the stratification seasons confirms the occurrence of a healthy phytoplankton bloom in this period. Uysal et al. (1997) also observed relatively high phytoplankton (mainly dominated by dinoflagellates) biomass (10^5 cells l^{-1}) in summer 1996 and the populations were comparable to those in spring (10^6 cells l^{-1}).

The Bosphorus exit was always relatively rich in phytoplankton biomass (in terms of Chl-a) because of the nutrient and POM-enriched surface waters from the northwestern shelf entrained by the alongshore currents. Interestingly, during the Sep.–Oct. 1995 and June-July 1996 cruises, relatively high concentrations from the eastern and western cyclones dominated the spatial distribution of Chl-a. This resulted partly from the intermittent influx of nutrients from the nutricline located at the base of the euphotic zone in the CR via diffusive processes and mainly by regenerative processes in the upper layer. Accordingly, in July 1996, the prominent deep Chl-a maximum in the CR was formed at the base of the euphotic zone and nearly coincided with the nutricline onset at the $\sigma_t = 14.5$–14.6 density surface. In the ACR, the nutricline is located much below the euphotic zone and thus the input from the nutricline generally occurs during winter mixing. Therefore, during the stratified seasons, Chl-a increase in the coastal ACRs is dominated by the input from rivers which is transported to the associated eddy fields via the Rim current.

It should be emphasized that the pattern of primary productivity in most of the Black Sea is determined by the material transported via the cyclonic boundary

Rim current and the frontal and jet instabilities between the Rim current and the interior eddy fields. Riverine input contributes to new production mainly in coastal and offshore areas by such a mechanism, though the more common mechanism of nutrient transport from the nutrient rich lower layer to the euphotic zone by winter mixing in the central regions of the Black Sea which is limited by the strong stratification. In comparison, the role of atmospheric sources of nutrients appears to be marginal (Kubilay et al., 1995).

On the other hand, due to the anoxic conditions in deep waters in the Black Sea, source and sink terms for nutrients (mainly for nitrate) immediately below the euphotic zone are unusual compared to conditions in the central cyclones. In the Black Sea anoxic conditions start at shallower depths (50–100 m) and, unlike the open ocean, below the oxic-anoxic interface the waters are enriched in ammonia and devoid of NO_3 and NO_2. Thus the transition suboxic layer laying below the euphotic zone and above the anoxic layer provides an environment condusive to photosynthetic production in the euphotic zone; nutrients are transported from this suboxic layer by diffusional processes as well as by winter mixing. The f-ratio has therefore been estimated to be as low as 0.1 (Murray et al., 1995; Cadispoti et al., 1991) and according to Dortch (1990) in such systems the f-ratio is mainly determined by the availability of ammonia. The one-dimensional physical-biological model of Oğuz et al. (1996) suggest that 40 and 60% of the annual primary production in the Black Sea is supported by NO_3 and NH_4, respectively. In this case the ratio of regenerated nitrogen to total nitrogen utilized is about %75, yielding an effective f-ratio of about 0.25.

Conclusions

High-resolution data reveal that, as a result of as yet undefined processes, the upper CIL down to the temperature minimum depth in the ACR is enriched with nitrate but drastically poor in phosphate. There are thus very high N/P ratios in the upper nutricline and an apparent shift in the nutricline onset in the CIL. Since these P-limited CIL waters are mixed vertically with the surface waters of the ACR in winter and early spring, bloom in such areas is probably limited by phosphorus. Nitrate-limited production occurs in the CR due the low N/P ratios of the chemocline established just below the euphotic zone. The relatively high atomic N/P ratios of POM indicate that the anomalously low ratio of nitrate/phosphate in the oxic/anoxic interface of the entire deep basin is due to nitrate removal via denitrification, greatly exceeding P-export from the suboxic waters.

In addition to characteristic spring and secondary autumn blooms, summer phytoplankton blooms are significantly observed in the Black Sea and such blooms are mainly driven by the transport of material via the cyclonic Rim current and its interaction with the eddy fields in the central parts of the Black Sea. Riverine input contributes to new production mainly in coastal and partly offshore areas. In the central region of the Black Sea new production is also supported by the nutrient transport from the nutricline located below the euphotic zone and from the suboxic zone, such transport being significant during winter mixing.

Acknowledgements

This work was carried out within the scope of the National Black Sea Oceanographic Program (supported by the Turkish Scientific and Technical Research Council, TÜBITAK) and the TU-Black Sea Project (supported by the NATO-Science for Stability Program). The authors are indebted to the participating scientists and technicians of the Institute and the crew of the R/V Bilim. Our thanks also go to Leyla Egesel for the radioactive counting.

References

Baştürk, Ö., C. Saydam, I. Salihoğlu, L. V. Eremeev, S. K. Konovalov, A. Stoyanov, A. Dimitrov, A. Cociasu, L. Dorogan & M. Altabet, 1994. Vertical variations in the principle chemical properties of the Black Sea in the autumn of 1991. Mar. Chem. 45: 149–165.

Baştürk, Ö, S. Tuğrul, S. Konavalov & I. Salihoğlu, 1997a. Variations in the vertical structure of water chemistry within the three hydrodynamically different regions of the Black Sea. In E. Özsoy & A. Mikaelyan (eds), Sensitivity to change: BlackSea, Baltic Sea and North Sea, NATO ASI Series, Kluwer Academic Publishers, in press.

Baştürk, Ö, S. Tuğrul, S. Konovalov, A. Romanov & I. Salihoğlu, 1997b. Effects of circulation on the spatial distribution of principal chemical properties and unexpected short- and long-term changes in the Black Sea. In L. Ivanov & T. Oğuz (eds), NATO TU-Black Sea Project: Ecosystem Modelling as a Management Tool for the Black Sea; Symposium on the Scientific Results, NATO ASI Series, Kluwer Academic Publishers, in press.

Bingel, F., A. E. Kıdeyş, E. Özsoy, S. Tuğrul, Ö. Baştürk & T. Oğuz, 1993. Stock Assessment Studies for the Turkish Black Sea Coast. NATO-TU Fisheries Final Report, Institute of Marine Sciences, Middle East Technical University, Erdemli-Içel/Turkey.

Bodeanu, N., 1992. Algal blooms and development of the main planktonic species at the Romanian Black Sea littoral in conditions of intensification of the eutrophication process. In R. A. Vollenweider, R. Marchetti & R. V. Viviani (eds), Marine Coastal Eutrophication, Elsevier Publ., Amsterdam: 891–906.

Bologa, A. S., 1985/1986. Planktonic primary productivity of the Black Sea: a review. Thalassia Jugoslavica 21/22: 1–22.

Bologa, A. S., N. Bodeanu, A. Petran, V. Tiganus & Y. P. Zaitsev, 1995. Major modifications of the Black Sea benthic and planktonic biota in the last three decades. Bulletin de l'Institut Oceanographique, Monaco Special 15: 85–110.

Brewer, P. G., 1971. Hydrographic and chemical data from the Black Sea. Woods Hole Oceanogr. Inst., Woods Hole, MA, Tech. Rep., Ref. No 71/65.

Buesseler, K. O., H. D. Livingston, L. Ivanov & A. Romanov, 1994. Stability of the oxic-anoxic interface in the Black Sea, Deep-Sea Res. 41: 283–296.

Cline, J. D., 1969. Spectrophotometric determination of hydrogen sulphide in natural waters, Limnol. Oceanogr. 14: 454–458.

Cociasu, A., L. Dorogan, C. Humborg & L. Popa, 1996. Long-term ecological changes in Romanian Coastal Waters of the Black Sea. Mar. Poll. Bull. 32: 32–38.

Cociasu, A., V. Diaconu, L. Popa, I. Nae, L. Buga, L. Dorogan & V. Malciu, 1997. Nutrient stock of the Romanian shelf of the Black Sea in the last three decades. In E. Özsoy & A. Mikaelyan (eds), Sensitivity to change: Black Sea, Baltic and North Sea. NATO ASI Series, Kluwer Academic Publishers 27: 49–63.

Codispoti, L. A., G. E. Friederich, J. W. Murray & C. M. Sakamato, 1991. Chemical variability in the Black Sea: Implications of continuous vertical profiles that penetrated the oxic/anoxic interface. Deep-Sea Res. 38: 691–710.

Copin-Montegut, C. & G. Copin-Montegut, 1983. Stoichiometry of carbon, nitrogen and phosphorus in marine particulate matter. Deep-Sea Res. 30: 31–46.

Dortch, Q., 1990. The interaction between ammonium and nitrate uptake in phytoplankton. Mar. Ecol. Progr. Ser. 61: 183–201.

Grasshoff, K., M. Ehrhardt & K. Kremling, 1983. Determination of nutrients. In Methods of Sea Water Analysis. 2nd edition, Verlag Chemie GmbH, Weinheim: 125–188.

Hay, B. J. & S. Honjo, 1989. Particle deposition in the present and Holocene Black Sea. Oceanography 2: 26–31.

Hay, B. J., S. Honjo, S. Kempe, V. A. Itekkot, E. T. Degens, T. Konuk & E. İzdar, 1990. Interannual variability in particle flux in the southwestern Black Sea. Deep-Sea Res. 37: 911–928.

Hay, B. J., M. A. Arthur, W. E. Dean & E. D. Neff, 1991. Sediment deposition in the Late Halocene abyssal Black Sea: Terrigeneous and biogenic matter. Deep Sea Res. 38(Suppl.): S711–S723.

Hecky, R. E., P. Campbell & L. L. Hendzel, 1993. The stoichiometry of carbon, nitrogen and phosphorus in particulate matter of lakes and oceans. Limnol. Oceanogr. 38: 709–724.

Holm-Hansen, O. & B. Riemann, 1978. Chlorophyll-a determination: improvements in methodolgy. Oikos 30: 438–447.

Karl, D. M. & G. A. Knauer, 1991. Microbial production and particle flux in the upper 350 m of the Black Sea. Deep-Sea Res. 38: 921–942.

Kempe, S., A. R. Diercks, G. Liebezeit & A. Prange, 1991. Geochemical and structural aspects of the pycnocline in the Black Sea (R/V Knorr 134-8 Legl, 1988). In E. İzdar & J. W. Murray (eds), Black Sea Oceanography. NATO-ASI Series C, 351, Kluwer Acad. Publ., Netherlands: 89–110.

Krom, M. D., S. Brenner, N. Kress, A. Neori & L. I. Gordon, 1992. Nutrient dynamics and new production in a warm eddy from the eastern Mediterranean. Deep-Sea Res. 39: 467–480.

Kubilay, N., S. Yemenicioğlu & A. C. Saydam, 1995. Airborne material collections and their chemical composition over the Black Sea. Mar. Poll. Bull. 30: 475–483.

Mee, L. D., 1992. The Black Sea in crisis: The need for concerted international action. Ambio 21: 278–286.

Murray, J. M., H. W. Jannasch, S. Honjo, R. F. Anderson, W. S. Reeburgh, Z. Top, G. E. Friederich, L. A. Codispotu & E. İzdar, 1989. Unexpected changes in the oxic/anoxic interface in the Black Sea. Nature 338: 411–413.

Murray, J. M., Z. Top & E. Özsoy, 1991. Hydrographic properties and ventilation of the Black Sea. Deep-Sea Res. 38: S663–S690.

Murray, J. M., L. A. Codispoti & G. E. Friederich, 1994. Redox environments: The suboxic zone in the Black Sea. In C. P. Huang, C. R. O'Melia & J.J. Morgan (eds), Aquatic Chemistry, Advances in Chemistry Series, American Chemical Society.

Murray, J. M., L. A. Codispoti & G. E. Freiderich, 1995. Oxidation-reduction Environments: The suboxic zone in the Black Sea. In C. P. Huang, C. R. O'Melia & J. J. Morgan (eds), Aquatic Chemistry, ACS Advances in Chemistry Series 244: 157–176.

Oğuz, T., M. A. Latif, H.İ. Sur, E. Özsoy & Ü. Ünlüata, 1991. On the dynamics of the southern Black Sea. In E. İzdar & J. W. Murray (eds), Black Sea Oceanography, NATO-ASI Series C, 351, Kluwer Acad. Publ., Netherlands: 43–63.

Oğuz, T. & P. E. La Violette, 1992. The upper layer circulation of the Black Sea: Its variability as inferred from hydrographic and satellite observations, J. Geophys. Res. 97(C8): 12,569-12,584.

Oğuz, T., V. S. Latun, M. A. Latif, V. V. Vladimir, H.İ. Sur, A. A. Markov, E. Özsoy, B. B. Kotovhchikov, V. V. Eremeev & Ü. Ünlüata, 1993. Circulation in the surface and intermediate layers of the Black Sea, Deep-Sea Res. 40: 1597–1612.

Oğuz, T., H. Ducklow, P. Malanotte-Rizzoli, S. Tuğrul, N. P. Nezlin & Ü. Ünlüata, 1996. Simulation of annual plankton productivity cycle in the Black Sea by a one-dimensional physical-biological model. J. of Geophys. Res. 101(C7): 16,585–16,599.

Polat, Ç.S. & S. Tuğrul, 1995. Nutrient and organic carbon exchanges between the Black and Marmara seas through the Bosphorus strait, Continental Shelf Res. 15: 1115–1132.

Richardson, 1991. Comparison of ^{14}C primary production determinations made by different laboratories. Mar. Ecol. Prog. Ser. 72: 189–201.

Romanov, A., Ö Baştürk, S. Konavalov & S. Gökmen, 1997. A comperative study of spectrophotometric and iodometric back titration methods for hydrogen sulphide determination in anoxic Black Sea waters. In L. Ivanov & T. Oğuz (eds), NATO TU-Black Sea Project: Ecosystem Modelling as a Management Tool for the Black Sea; Symposium on the Scientific Results. NATO ASI Series, Kluwer Academic Publishers, in press.

Shaffer, G., 1986. Phosphate pumps and shuttles in the Black Sea. Nature 321: 515–517.

Shuskina, E. A. & E. I. Musaeva, 1990. Structure of planktonic community from the Black Sea epipelagical and its changes as the result of the introduction of a ctenophore species. Oceanology 30: 306–310.

Smayda, T. J., 1990. Novel and nuisance phytoplankton blooms in the sea: Evidence for a global epidemic. In E. Granelli, B. Sundstroem, L. Elder & D. M. Anderson (eds), Lund, Sweden, 26–30 June, 1989: 9–40.

Sorokin, Yu. I., 1983. The Black Sea. In B. H. Ketchum (ed), Estuaries and Enclosed Seas. Ecosystem of the World. Elsevier, Amsterdam: 253–292.

Steemann Nielsen, E., 1952. The use of radioactive carbon (^{14}C) for measuring organic production in the sea. J. Cons. perm. int. Explor. Mer. 18: 117–140.

Strickland, J.D.H. & T.R. Parsons, 1972. A practical handbook of seawater analysis. 2nd edition. Bull. Fish. Res. Bd Can. 167.

Sur, H.İ., E. Özsoy, Y. P. Ilyin & Ü. Ünlüata, 1996. Coastal/deep ocean interactions in the Black Sea and their ecological/environmental impacts. J. Mar. Systems 7: 293–320.

Tuğrul, S., Ö. Baştürk, C. Saydam & A. Yılmaz, 1992. Changes in the hydrochemistry of the Black Sea inferred from water density profiles. Nature 359: 137–139.

Uysal, Z., A. E. Kideyş, L. Georgieva, D. Altukhov, L. Kuzmenko, L. Manjos, E. Mutlu & E. Eker, 1997. Phytoplankton patches formed along the southern Black Sea coast in spring and summer 1996. In L. Ivanov & T. Oğuz (eds), NATO TU-Black Sea; Symposium on the Scientific Results, NATO, ASI Series, Kluwer Academic Publishers, in press.

Vedernikov, V. I. & A. B. Demidov, 1993. Primary production and chlorophyll in the deep regions of the Black Sea. Oceanology 33: 229–235.

Vidal, C. V., 1995. Bio-optical characteristics of the Mediterranean and the Black Sea. M.S. Thesis, 134 pp., Institute of Marine Sciences, Erdemli, İçel, Turkey.

Vinogradov, M. Ye., E. A. Shushkina, Eh. I. Musaeva & Yu. I. Sorokin, 1989. The comb-jelly *Mnemiopsis leidyi* (A. Agassiz) (Ctenophora:Lobata); a newly introduced species in the Black Sea. Oceanology 29: 293–299 (In Russian).

Vladimirov, V. L., V. I. Mankovsky, M. V. Solov'ev & A. V. Mishonov, 1997. Seasonal and long-term variability of the Black Sea optical parameters, In E. Özsoy & A. Mikaelyan (eds), Sensitivity to change: Black Sea, Baltic and North Sea. NATO ASI Series, Kluwer Academic Publishers, 27: 33–48.

Yılmaz, A. & S.Tuğrul, 1997. The effect of cold- and warm-core eddies on the distribution and stoichiometry of dissolved nutrients in the Northeastern Mediterranean. J. Mar. Systems (in press).

Hydrobiologia **363**: 157–167, 1998.
T. Tamminen & H. Kuosa (eds), Eutrophication in Planktonic Ecosystems: Food Web Dynamics and Elemental Cycling.

Elemental composition of seston and nutrient dynamics in the Sea of Marmara

S. C. Polat[1], S. Tuğrul[2], Y. Çoban[2], O. Basturk[2] & I. Salihoglu[2]
[1] *I.U.-Inst. of Marine Scs.and Management, Muskule Sok., No:1, 34470 Vefa-Istanbul, Turkey*
[2] *M.E.T.U.-Inst. of Marine Sciences, P.O.Box 28, 33731 Erdemli-Icel, Turkey*

Key words: Seston, nutrients, elemental ratios, The Sea of Marmara, Black Sea

Abstract

The Sea of Marmara, an intercontinental basin with shallow and narrow straits, connects the Black and Mediterranean Seas. Data obtained during 1991–1996 have permitted the determination of the elemental composition of seston in the euphotic zone and the N : P ratio of the subhalocline waters of the Marmara Sea. Since primary production is always limited to the less saline upper layer (15–20 m), of the Marmara Sea, the subhalocline waters of Mediteranean origin are always rich in nutrients ($NO_3 + NO_2 = 8–10\ \mu m$, $PO_4 = 0.8–1.2\ \mu m$) but depleted in dissolved oxygen (30–50 μm) throughout the basin, yielding an - O_2 : N : P ratio of 178 : 9 : 1. Pollution of the surface waters since the 60s has modified the subhalocline nutrient chemistry slightly. In the euphotic zone, the N : P ratio of the seston changes from 5.9 to 9.5 between the less and more productive periods. Though the biology of the Marmara has changed significantly during the previous two decades, the close relationship observed between the elemental composition of the surface seston and the NO_3 : PO_4 ratio of the subhalocline waters strongly suggests that during the whole year primary production throughout the basin and POM export to the lower layer remain nitrogen-limited. This suggestion needs to be confirmed by bio-assays, biological studies and sediment trap data from the upper subhalocline depths. Nonetheless, the counterflows in the Marmara basin possess relatively low N : P ratios in both dissolved and particulate nutrients and extend as far as the adjacent seas.

Introduction

Particulate organic matter (POM) produced photosynthetically in surface waters of marine environments possesses carbon, nitrogen and phosphorus in the 'Redfield ratio' of C : N : P = 106 : 16 : 1. This conventional relationship represents the average chemical composition of ocean phytoplankton growing with maximum growth rates (Goldman et al., 1979). However, the C : N : P ratio is well known to vary markedly in marine, brackish and fresh waters, depending on the species composition and the nutrient composition of the productive surface waters (Vinogradov, 1953; from Redfield et al., 1963; Goldman et al., 1979; Sakshaug et al., 1983). POM in the surface waters, moreover, is not solely composed of phytoplankton; throughout most of the year, excluding the bloom period, bacteria, microzooplankton, protozoa and detrital material constitute a major fraction of the total POM inherent in the euphotic zone (Vostokov & Vedernikov, 1988). Furthermore, the total suspended matter in surface waters, the so-called seston, generally – and especially in coastal margins and enclosed seas – contains inorganic material (Vostokov, 1996). Thus, the chemical composition of seston collected from surface waters may differ from the conventional Redfield ratio. As Copin-Montegut & Copin-Montegut, 1983 clearly stated, the stoichiometric ratios of seston from the world ocean vary between 5.5–12.7 for C : N, 42–168 for C : P and 6.5–23.0 for N : P. The POM initially in the euphotic zone is either consumed by herbivores, decomposes in the productive zone, or it sinks to the oxygenated aphotic layer where it decays in the water column and in the sediment with the release of carbon dioxide and nutrients into the ambient water (Redfield et al., 1963; Richards, 1965; Brewer & Murray, 1973).

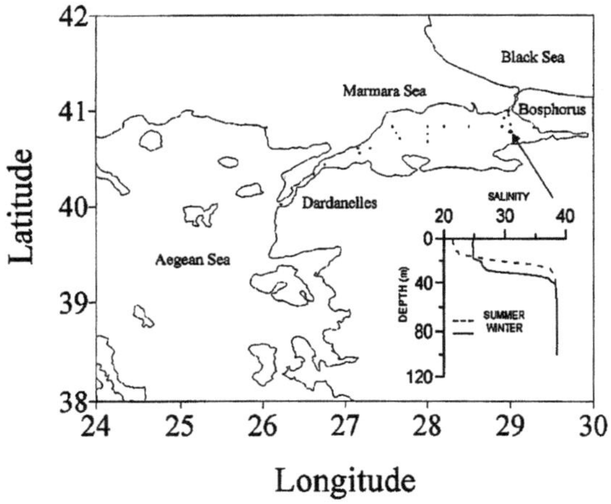

Figure 1. Sampling stations in the Sea of Marmara and typical summer, winter salinity profiles at a reference station with 1230 m depth.

In this paper, we present the elemental stoichiometry of seston collected in the surface waters of the Marmara Sea during 1991–1996. We then discuss the seasonal variation in the elemental ratios of seston and the possible reasons for deviations from the Redfield ratio (C : N : P = 106 : 16 : 1). We also discuss the relationship between the elemental composition of the surface seston and the mean N : P ratio of the subhalocline waters, based on long-term unpublished data obtained in the Sea of Marmara since 1990.

Study area

The Sea of Marmara, a small intercontinental basin with a surface area of about 11500 km^2 and a total volume of 3378 km^3, connects the less saline Black Sea (S = 17–18 ppt) to the more saline Mediterranean (S = 39 ppt) via the Turkish Straits, the Bosphorus and Dardanelles (Figure 1). Because of the large salinity difference between the adjacent seas, two-layer exchange flows are established in the two straits, resulting in the formation of a permanent two-layered stratification of the entire basin of the Marmara Sea. The basin consists of three topographic depressions (the max. depth being 1390 m) in the north and a wide shallow area (shelving to 100 m) in the southern margin. Summer and winter salinity profiles (Figure 1) show the brackish water of Black Sea origin to occupy merely the upper 20–25 m with a renewal time of about 5–6 months (Besiktepe et al., 1994). This upper layer, whose salinity and temperature vary seasonally between 22–26 ppt and 7–24 °C respectively, is separated throughout the year from the underlying saline Mediterranean water by a steep halocline at 25–30 m. The subhalocline waters of the Marmara basin which possess nearly constant salinity and temperature (~ 38.5–38.6 ppt and 14.5–15.0 °C) throughout the basin are renewed on avearge in 6–7 years by the Mediterranean inflow via the Dardanelles undercurrent (Besiktepe et al., 1994). In the Sea of Marmara, primary production occurs only in the upper layer, extending down to the halocline during summer–autumn when the surface nutrient concentrations are low (Ediger & Yilmaz, 1996). The total annual primary production, derived from chlorophyll-a measurements, is of the order of 100 gC m^{-2} for the entire basin (Ergin et al., 1993). Morkoc (1995) estimated an annual production of 170–190 gC m^{-2} y^{-1} for a coastal embayment, based on long-term measurements of carbon assimilation.

During the last three decades dramatic changes have occurred in the Black Sea ecosystem as a result of both human and natural pressures (Mee, 1992; Kideys, 1994; Bologa et al., 1995). Since the Marmara upper layer water is renewed at least twice a year by the Black Sea inflow, similar modifications have been observed in the Marmara ecosystem (Kocatas et al., 1993). Its recent state has been studied by Sorokin et al. (1995) and Shiganova et al. (1995), who have emphasized the contributions of microheterotrophs, carnivorous organisms, and of the chetonopore *Mnemiopsis leidyi* to the lower trophic levels of the marine ecosystem. Unfortunately, as emphasized by Uysal (1996) in his study of some of the planktonic species, there exists a lack of long-term biological data.

Although biological studies are still very limited, a national programme has been investigating the chemical oceanography of the Marmara Sea and Turkish straits since 1986. From the data collected in the two straits, Polat & Tuğrul (1995, 1996) and Tuğrul & Polat (1995) evaluated the exchange of nutrients and organic carbon between the adjacent seas via the Bosphorus and Dardanelles. They showed that the nutrient budget of the Marmara upper layer (and thus the primary production in the sea) is dominated by inputs both from natural and land-based sources; these being the Black Sea inflow, the input from the subhalocline waters by vertical mixing and the domestic/industrial waste discharged directly into the coastal surface waters, especially by the city of Istanbul (Orhon et al., 1994). The inputs from these sources vary seasonally and regionally. In winter, the Black Sea inflow is enriched with dissolved inorganic nutrients whereas its particulate nutrients are depleted. The nitrate + nitrite concentrations increase from 0.1–0.2 μm in spring–autumn to 4.5–7.5 μm in winter but the seasonal change in particulate nutrients is less pronounced due to the background of seston in the inflow. Similar winter increases observed in Romanian coastal waters (Bologa, 1985) suggest that the inorganic nutrient-enriched northwestern shelf waters of the Black Sea reach as far as the Bosphorus region via alongshore currents and may enter the Sea of Marmara through the Bosphorus (Sur et al., 1994). In late spring-summer, when the Danube discharge to the northwestern Black Sea is maximal, the Black Sea inflow occasionally contains high nitrate concentrations (Tolmazin, 1985; Serpoianu et al., 1992; Cociasu et al., 1997). The POM content of the inflow increases markedly during the late winter–spring bloom in the southwestern Black Sea (Oguz et al., 1996). Accordingly, the POC concentration ranges from background levels of 12–17 μm in the less productive period to 25–40 μm during the bloom. Similarly, the PON varies seasonally between 1.0 and 5.5 μm, whereas the seasonality in PP is less pronounced, ranging merely between 0.1–0.2 μm between 1991 and 1994 (Polat & Tuğrul, 1995).

Before they enter the deep Marmara Sea, the saline Mediterranean waters possess relatively low nutrient and organic carbon concentrations but saturated levels of dissolved oxygen (DO) as clearly demonstrated at the Dardanelles-Aegean Sea junction (Polat & Tuğrul, 1996). The phosphate concentration ranged seasonally between <0.02 and 0.1 μm, with an annual mean of about 0.05 μm; nitrate + nitrite concentrations varied between <0.1 and 2.4 μm (annual mean: 1.0 μm), the highest values always occurring in winter. Average annual particulate concentrations (POC = 1.2–7.0 μm, PON = 0.04–0.7 μm, PP = 0.01–0.04 μm) are comparable to dissolved nutrient values Polat & Tuğrul (1996). Both old and new chemical data from the deep Marmara basin (Figure 2) demonstrate that during its stay of 6–7 years (on average) in the basin, the inorganic nutrient content of the Mediterranean inflow increases nearly ten-fold whereas its DO content declines from saturated levels to 30–50 μm since only limited ventilation of the subhalocline waters is possible through the permanent halocline (Besiktepe et al., 1993). The TOC content of the Mediterranean inflow (68 μm), some 3 times lower than that of the Black Sea inflow (210 μm), changes little in the Marmara basin, because POM sinking from the surface waters becomes oxidized in the subhalocline waters and makes only a small contribution to the DOC pool of the system (Tuğrul, 1993; Polat & Tuğrul, 1995).

Table 1. Cruise dates of R/V Bilim in the Sea of Marmara during 1991–1996.

	1991	1992	1993	1994	1995	1996
Jan.	10–13					
Feb.			11–17			
Mar.	1–13	15–21				
Apr.	30.3–1.4			11–14	28–30	6
May						
June	19–23					
July			4–10			
Aug.						
Sep.					23–25	
Oct.	1–7	17–19				
Nov.						
Dec.			5–16			

For a better understanding of the nutrient chemistry of the Marmara Sea, typical depth profiles have been displayed in Figure 2, based on unpublished data collected with R/V Bilim and on measurements reported previously by Miller (1970; R/V Chain), Sen Gupta (1971; R/V Phillsbury), Anderson & Charmack (1974; R/V Thompson), Friederich et al. (1990; R/V Knorr). Figure 2 shows nutrient concentrations to be relatively low in the brackish upper layer throughout the year; the ranges estimated from the unpublished measurements of 1990–1996 are 0.2–1.5 μm for nitrate + nitrite and 0.04–0.25 μm for phosphate, with atomic N : P ratios commonly of <2–4. The indicated seasonalities are much less pronounced than those reported for the Black Sea inflow (Polat & Tuğrul, 1995). Since primary production only occurs in the upper layer, subhalocline nutrient concentrations are as high as 8–10 μm for nitrate and 0.8–1.2 μm for phosphate, changing little with either depth or season. Thus, on an annual time scale, the lower layer system approximates to a steady state. Comparison of the old and new profiles also reveals that long-term changes are more pronounced in the upper subhalocline; the nutricline, which in the past extended below the halocline, is now established within the halocline. Although nutrient concentrations in the deep water have increased slightly since 1965, the atomic N : P ratio has remained almost constant at 8–10. This strongly suggests that, although the biology of the Marmara upper layer has changed since 1960s, the biogenic POM exported from the euphotic zone has retained its chemical composition.

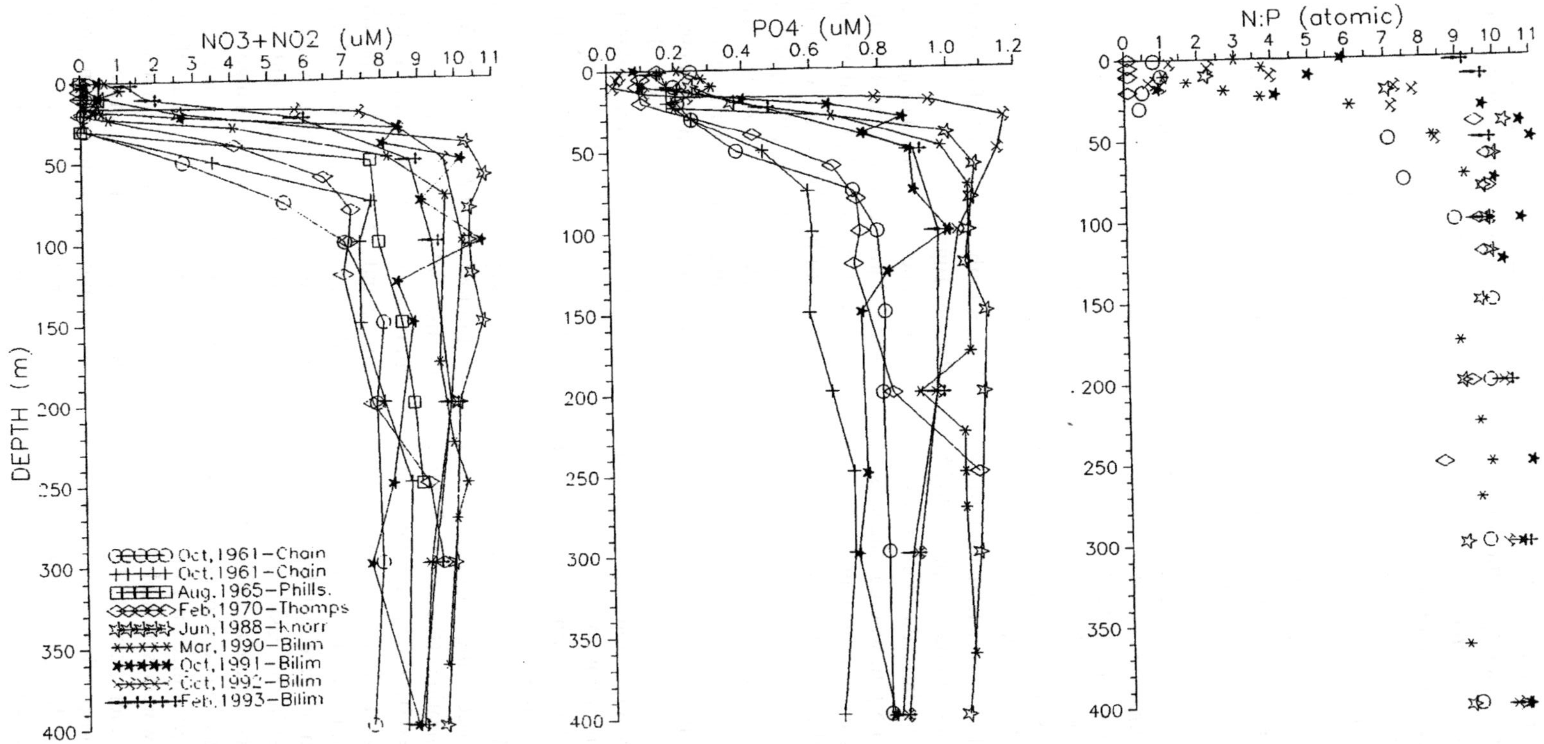

Figure 2. Vertical profiles of $NO_3 + NO_2$, PO_4 and N : P of the upper 400 m at the eastern deep basin of the Sea of Marmara.

Material and method

During 1986–1996, a series of oceanographic cruises covered the Sea of Marmara and the two straits, at the station network being given by Besiktepe et al. (1994). The locations of the chemical stations visited during the 1991–1996 surveys (see Table 1) are illustrated in Figure 1. CTD data were obtained by a Sea-Bird model probe attached to a remote-controlled, 12-bottle (5 l capacity) rosette system. At selected stations at least 6–7 samples for chlorophyll-*a* (Chl-*a*) analysis were collected down to the depth where the light was 0.1% as intense as at the surface; these samples were filtered immediately through Whatman GF/F filters and kept frozen. Nutrient and dissolved oxygen samples (nitrate + nitrite, phosphate) were collected using the 12-bottle rosette. DO samples were taken into a 50 or 150 ml glass bottle; DO fixing chemicals were then added and the bottle was kept in the dark for 1–2 hr. Nutrient subsamples in high density polyethylene bottles were kept frozen; depth intervals of nutrient samples can be realized from the profiles depicted in Figure 2. POM was sampled discretely from at least 5–6 depths from the surface to just below the halocline using 5 l Niskin bottles attached to the rosette. Approximately 2–5 l of subsamples were filtered through 47 (or 25) mm GF/F filters (pre-combusted at 450–500 °C for 3 h). Subsamples were prefiltered through 200 μm nylon mesh placed in a funnel, in order to remove the larger particles (e.g. meso-zooplankton) which would otherwise have caused errors. Occasionally, this step was skipped and forceps were used to remove visible particulates retained on filters. Filters for POM analysis were kept frozen until processed on land. Before analysis, they were dried at low temperature (40–50 °C) and fumed with concentrated hydrochloric acid to remove any carbonate retained on the filters. About 20 mg of each filter was placed in a pre-cleaned tin capsule and the POC and PON content of the samples were determined quantitatively using a Carlo Erba 1108 model CHN analyzer. For particulate phosphorus (PP) analysis, the dry combustion + weak acid dissolution method given by Karl et al. (1991) was followed; the concentrations of the oxidized phosphorus in the samples were then measured colorimetrically as for reactive phosphate. Nutrient (nitrate + nitrite, phosphate) analyses were carried out using a two-channel Technicon autoanalyzer; the methods were similar to those given by Strickland & Parsons (1972). Chl-a samples were first extracted with 90% acetone and then determined fluorometrically (Holm-Hansen et al., 1965). Dissolved oxygen was measured by the automated Winkler titration method.

Results

Surface water

Particulate and chlorophyll-*a* (Chl-*a*) data displayed in Figures 3 and 4 represent euphotic zone concentrations of POC, PON or PP and Chl-a at different stations in the deepest western, central and eastern basins of the Marmara Sea (Figure 1). These stations were visited 14 times at different months between 1991 and 1996 (see Table 1 for cruise dates). The layer-averaged (upper 25 m) data from the three sub-basins for each cruise illustrate the spatial variations in POM. The restricted number of data sets (only 14) prevents reliable deductions of the seasonal trends in POM for 1991–1996. Nevertheless, the basic feature of Figure 3 exhibits coherent increases of POC, PON and PP concentrations of the euphotic zone during late winter-early spring, as a result of the pronounced biomass increases (in terms of chl-a data) during the early spring bloom (Figure 4). The highest particulate concentrations, ranging between 15 and 45 μm for POC, 1.8 and 4.5 μm for PON and between 0.20 and 0.45 μm for PP, were obtained in the February–April period of 1991–1996. During the June–January period variable background concentrations of POM were always observed; for example, in June 1991 and July 1993, unexpectedly high POM values were recorded in the surface waters, accompanied by apparent decreases in the water transparency. In fact, in summer, the surface mixed layer is separated by a well defined thermocline which inhibits the input of nutrients from the nutricline and it was unfortunate that the species responsible for the episodic backgrounds were not identified. The autumn-winter period generally corresponds to the POM-poor seasons in the Marmara upper layer; however, slight increases were recorded in October, when the seasonal thermocline was weak enough to permit wind-induced nutrient input from the nutricline.

The general trend of the euphotic POM data depicted in Figure 3 for 1991–1996 enables one to divide all the data into two subgroups corresponding to high (February to April) and low (June to January) periods of production, irrespective of the sampling locations. The validity of this classification is corroborated by the similar seasonality observed in the chl-a data sets for 1991–1996 (Figure 4). The elemental stochiometry of

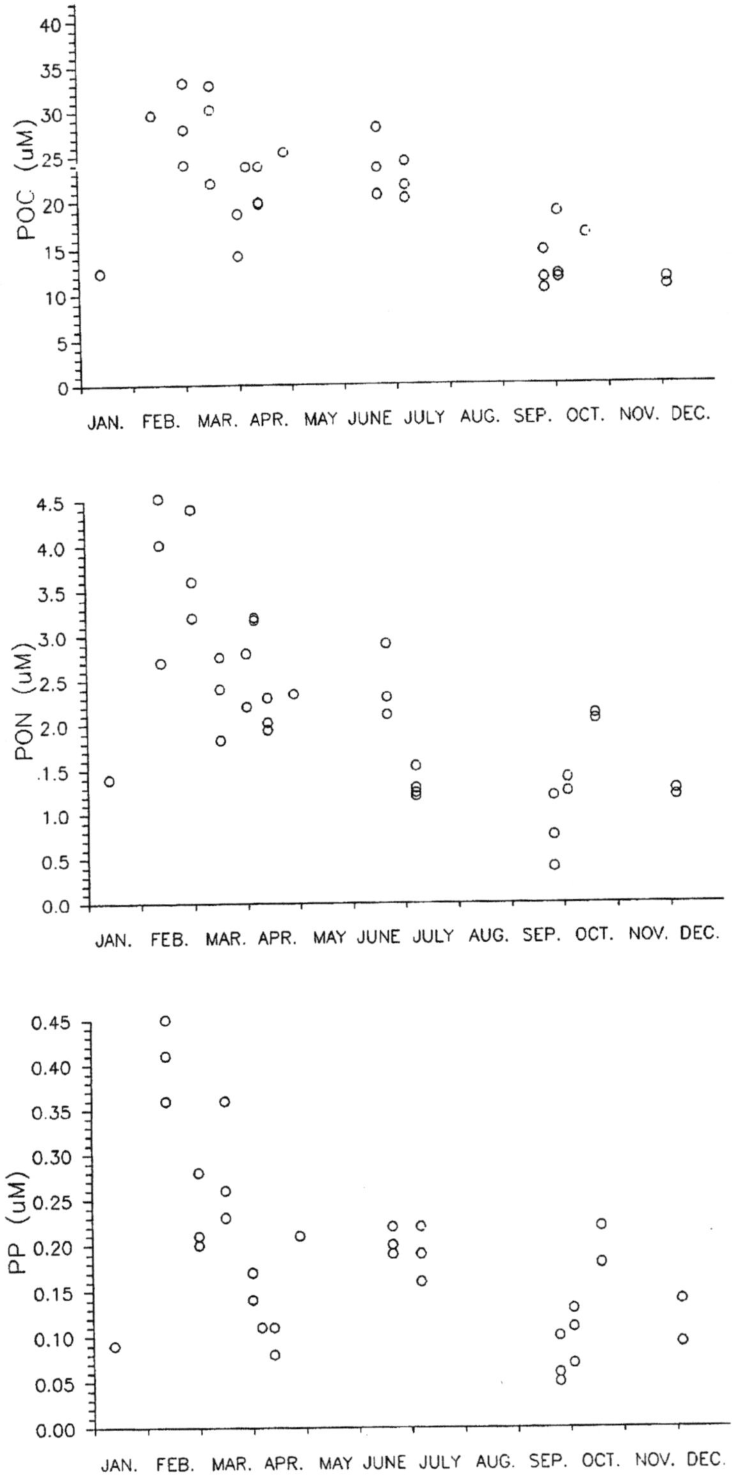

Figure 3. Variation of POC, PON and PP between 1991 and 1996 obtained from depth averaged data of the upper 20–30 m of the Sea of Marmara.

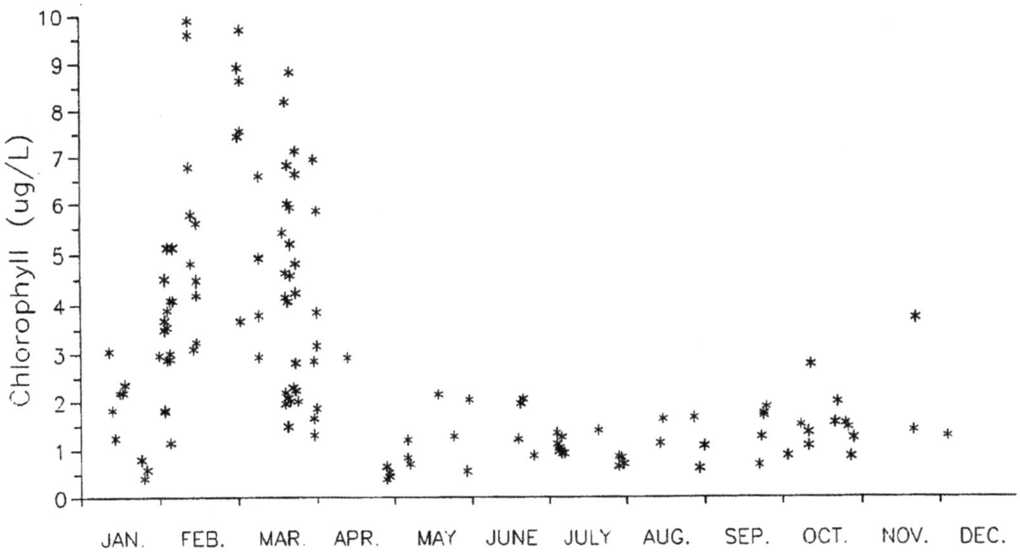

Figure 4. Variation of chlorophyll-a in the surface waters of the eastern Marmara basin from the discerete depths in the euphotic zone (upper 30 m).

Table 2. Elemental (atomic) and POM vs. Chl-a ratios achieved with linear regression equations for the surface waters of the Marmara Sea.

	High				Low			
	b	a	*r*	*n*	b	a	*r*	*n*
POC : PON	7.4 ± 1.16	5 ± 13.0	0.87	54	8.3 ± 1.68	3.5 ± 10.3	0.80	62
POC : PP	98 ± 24.5	6 ± 17.7	0.75	53	78 ± 18.6	5 ± 11.4	0.73	63
PON : PP	9.5 ± 3.43	0.85 ± 2.48	0.62	53	5.9 ± 2.02	0.7 ± 1.27	0.60	62
POC : Chl	38.4 ± 7.30	172 ± 210	0.81	60	37.5*	174*	0.28*	62
PON : Chl	4.7 ± 1.35	21 ± 38.9	0.67	61	8.0 ± 3.18	14 ± 18.4	0.54	63
PP : Chl	0.8 ± 0.24	4.4 ± 5.39	0.71	43	2.8 ± 1.23	2.2 ± 5.35	0.59	41

* Insignificant correlation at $p = 0.05$.

POM and POM to Chl-a ratios, derived from regression analyses of the sub-grouped 1991–1996 data sets are summarized in Table 2. The high correlation coefficients of the reported linear regressions indicate that the bulk of the seston consisted principally of organic material composed of living phytoplankton, heterotrophs and detritus. The estimated POC : PON ratio of the high production period was 7.4, slightly lower than the ratio of 8.3 for the low production period (Table 1). The POC vs PON intercept values are high and variable, suggesting throughout the year the presence of background concentrations of carbonaceous compounds (low in organic nitrogen) in the seston of the Marmara upper layer.

During the period of high production the estimated N : P ratio of the seston, 9.5, was lower than the Redfield ratio of 16 for ocean phytoplankton whereas the C : P and C : N ratios were 98 and 7.4, respectively (Table 2). The estimated POC : Chl-*a* ratio (w/w) for the same period is 38.4. The intercept of the regression line, found to be as high as 172, indicates a relatively high detrital carbon content of seston pool as the Chl-*a* decreases to low levels. The organic carbon pool, except the photosynthetically produced organic material, possibly covers the microzooplankton which may be dominating the heterotrophic community in the period of high production. The estimates from the linear regression equation in Table 2 strongly suggest that the seston was composed of significant amounts (as high as 50%) of living phytoplankton during the period of high production. The living phytoplankton carbon concentration, estimated from the POC : Chl ratio (38.4) and the total POC values, is in the range of 60–370 $\mu g\ l^{-1}$ (mean = 193 $\mu g\ l^{-1}$) in the period of high production. This is 25–100% (mean = 50%) of POC pool in the surface waters. Moreover, the bulk of the seston was enriched in phosphorus relative to nitrogen. The ratios of N : Chl-*a* and P : Chl-*a* derived from the regressions of data sets, are 4.7 and 0.8 for the period of high productivity, yielding a C : N : P ratio of 124 : 13 : 1 for living particulate matter. The relatively low N : P ratios of both seston and living phytoplankton, accompanied with the low molar ratios of nitrate to phosphate observed in the surface waters for most the year, suggest primary production in the whole of the basin to be nitrogen limited.

During the period of low productivity, the PON : PP was as low as 5.9 though the estimated POC : PP ratio was 78 (Table 2). The seasonal difference observed between the PON : PP ratios of the high and low productivity periods is significant (two tailed *t*-test, $p=0.1$). The difference becomes highly significant ($p=0.01$) if the April data of the last two years (see Table 1) are excluded from the regressions. On the other hand, there is no significant correlation between POC and chlorophyll during these years; this finding suggests low carbon synthesis in the surface waters, low algal biomass as seen in Figure 4. The heterotrophic community developed during this period may also provide a low phytoplankton biomass and enhance regenerated production. On the other hand, PON : Chl and PP : Chl correlations are still significant in the period of low production, and the estimated PON : PP from chlorophyll relations is around 6.3 and close to the value (5.9) given in Table 2.

Subhalocline water

The ratio (nitrate + nitrite) : phosphate in the subhalocline waters of the Marmara Sea has been estimated from the linear regression of the long-term, basin-wide data. All the unpublished data obtained in the deep subhalocline waters of the entire basin from the Dardanelles to the Marmara-Bosphorus exit in 1990–1994 have been combined (see Figure 1), excluding nutrient data from heavily polluted coastal areas. This regression analysis is based on the following assumptions:

(a) nutrient increases in the subhalocline waters are principally the result of aerobic oxidation of labile biogenic particles exported from the productive upper layer; nitrate loss via denitrification is insignificant;

(b) the Mediterranean water enters the basin with low and seasonally constant nutrient concentrations (the so-called preformed values).

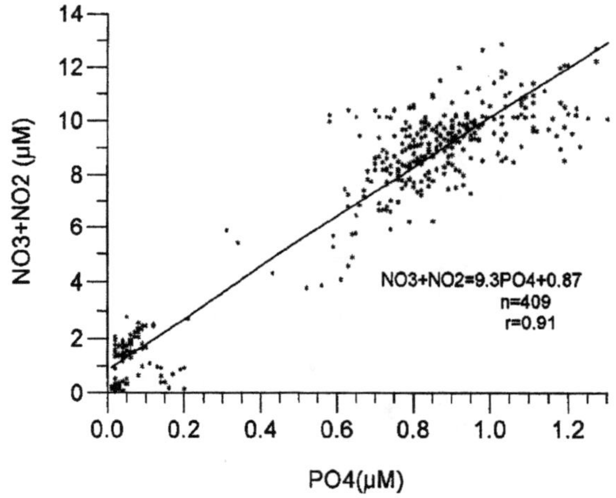

Figure 5. The regressions between NO_3+NO_2 and PO_4 concentrations ($n=409$, $r=0.91$) from the subhalocline waters of the Marmara Sea between 1990 and 1994.

In fact, as described in the 'Study Area' section, the latter assumption is weakened by the data at the Dardanelles-Aegean Sea Junction where the subhalocline water has low nutrient concentrations which increase noticeably in winter. The NO_3+NO_2 vs. PO_4 regression for the subhalocline water mass yields a N:P ratio of 9.3 (Figure 5). Data with low concentrations ($PO_4<0.2$ μm and $NO_3+NO_2<2.5$ μm) represent the slightly diluted Mediteranean inflow in the Dardanelles near its exit to the Marmara Sea; the second group of data having higher concentrations represents the aged Marmara subhalocline waters including the Bosphorus-Marmara Junction where the fresh Mediterranean inflow is diluted rapidly with aged subhalocline waters having modified chemical properties. Data from moderately diluted or less aged inflow is scarce. The DO vs. NO_3+NO_2 and DO vs. PO_4 regressions demonstrate that the measured nutrient concentrations increased linearly with decreasing DO concentrations in the subhalocline waters; the molar ratios derived from the regression lines are 178 for $-O_2:PO_4$ ($n=668$, $r=0.89$) and 19.7 for $-O_2:NO_3+NO_2$ ($n=664$, $r=0.90$), yielding an atomic N:P ratio of nearly 9.0 for the entire subhalocline water. This ratio is very similar to that, 9.3, derived from the $NO_3+NO_2:PO_4$ regression.

Discussion

The atomic ratio of nitrate to phosphate (9.0–9.3) in the subhalocline waters of the Marmara Sea is very close to the mean elemental ratio of 9.5 estimated for the total seston – about 50% of which, especially during the high productivity period (February–April), is composed of phytoplankton living in the euphotic zone. Unfortunately no data from sediment traps exist which would corroborate this close relationship between the mean N : P ratios of the surface seston and the subhalocline nutrient concentrations. The C : N : P ratio obtained from seston during the period of high production is 98 : 9.5 : 1, which is less than that (124 : 13 : 1) estimated from the slopes of particulate data vs. Chl-a concentrations for the same period; the living particulate ratio being similar to those of diatoms (Tett et al., 1975; Epply et al., 1977; Oguz et al., 1996). This finding suggests that the majority of the fast sinking POM (diatom) produced during the late winter-spring period and exported from the euphotic zone possesses an N : P ratio comparable to that of the surface seston but less than the Redfield ratio of 16 for ocean phytoplankton. In the adjacent Black Sea, POM produced during the spring and autumn bloom dominates the particle flux to the lower layer (Izdar et al., 1987). Uysal (1996) has found that during spring in the Marmara Sea the centric diatom species (*Skeletonema costatum, Thalassiosira spp. and Coscinosira spp.*) dominate the phytoplankton population. Diatoms generally have high sedimentation rates; their important role in the vernal export flux has been discussed previously by Smetacek (1985) and several scientists have studied the physical and biological mechanisms controlling the vertical fluxes of POM (Alldredge & Gotschalk, 1989; Rieseball, 1991a and 1991b). On the other hand, grazing may also play an important role during spring blooms of phytoplankton (Banse, 1994) which is usually known to be terminated by nutrient exhaustion.

In the upper layer, during the period of low production, the elemental composition of seston appears to become modified, yielding low ratios of both N : P (5.9) and C : P (78). The former ratio is very similar to the low N : P ratio (6.3) derived from the slope of the particulate data vs. Chl-a concentrations (see Table 2). This strongly suggests that during the period of low productivity, the low concentrations of P-enriched seston found in the nutrient-depleted euphotic zone comprised living and dead phytoplankton of regenerative origin. Although the mechanism by which P-enriched seston was produced in the surface waters is still unclear, the PP observed during the low productive period was readily soluble in weak acid (Polat, 1995). During the period of low productivity when the Marmara surface waters are relatively warm and infertile, primary production is probably sustained by small phytoplankton species – not diatoms – utilizing nutrients regenerated in the surface layer. Thus, the temporary milky green color and the low transparency of the Marmara upper layer in summer is an indicator of *Emiliana huxleyi* production (Holligan et al., 1993). The contribution of the substantially increased ctenophere populations to the nutrient pool of the Marmara Sea and their possible effects on the other grazers is still understood poorly (Sorokin et al., 1995). However, ctenopheres also play a key role in the trophic structure of the Black Sea, especially in summer and autumn (Vinogradov, 1992).

The high intercepts values of the regression lines of POC vs particulate nutrients (Table 2) indicate the existence of measureable quantities of carbonaceous compounds (especially low in organic nitrogen) in the surface waters of the Marmara Sea throughout the year. In the two-layered ecosystem of the Marmara Sea, nitrogen and phosphorus deficiency of the surface seston relative to the conventional Redfield ratio (C : N : P = 106 : 16 : 1) may have originated from the slow decay of carbonaceous compounds in the productive zone. Selective accumulation of carbonaceous seston in surface waters from different seas was also reported by Copin-Montegut & Copin-Montegut (1983).

Nutrient concentrations in the subhalocline waters are determined both by the biogenic particle influx from the surface layer and the age (renewal time) of the Mediterranean waters in the basin. The molar - $O_2 : PO_4$ ratio, calculated from the regression analysis, was nearly 178, very similar to that estimated by Takahashi et al. (1985) for various deep ocean waters. Takahashi et al. (1985) therefore corrected the measured nutrient and DO concentrations using the initial concentrations of chemicals before they had left the ocean surface. This strongly suggests that the phosphate enrichment of the Marmara subhalocline waters is the result of the oxidation of biogenic particles by biomediated processes, rather than the sedimentation of phosphorus-attached lithogenic and/or biogenic particles and the subsequent release of phosphorus in the lower layer. On the other hand, the mean - $O_2 : NO_3 + NO_2$ ratio is 19.7, much larger than the mean ratio of 10.5 estimated by Takahashi et al. (1985) for the deep oceans. The contribution of denitrification to the anomalous - $O_2 : NO_3 : PO_4$ ratio (178 : 9.0 : 1) of the Marmara subhalocline waters is probably insignificant. According to Polat's (1995) estimate, the annual NO_3 loss by denitrification at the sediment/water interface is only a small fraction (< 5%) of the annual PON influx from the surface layer to the subhalocline waters. Nevertheless, this process is evident in the polluted coastal areas where high concentrations of NO_2 and PO_4 were detected, the DO being < 10 μm in the shallow bottom water. The sediment surfaces were not rich in organic carbon, ranging between 1–1.5% (Ergin et al., 1993). These points strongly suggest that during the high production period, the POM exported from the surface layer has a relatively low N : P ratio. This is consistent with the low N : P ratio found in the seston in the euphotic zone. Nonetheless, the subhalocline water leaves the Marmara basin with relatively low $NO_3 + NO_2 : PO_4$ ratios and is entrained in the surface counterflow, especially at the Bosphorus Maramara Junction, and flows out towards the Black Sea through the Bosphorus undercurrent.

Conclusions

The Sea of Marmara is occupied with the two distintly different water masses thoroughout the year. Since primary production is always restricted to the upper layer, the more saline water of Mediterranean origin is always enriched markedly with inorganic nutrients in 6–7 years but depleted in dissolved oxygen. The N : P ratio of seston in the Marmara euphotic zone, derived from the regression analysis, ranges from 5.9 to 9.5 between the periods of low and high production, which are lower than the Redfield ratio. Thus, the euphotic layer seston of biogenic origin must contain phosphorus-enriched particles which need to be identified. This fact, together with the consistently low ratios of $NO_3 : PO_4$ in the subhalocline waters throughout the basin since the 60's, suggests that during the whole year primary production in the Marmara Sea remains nitrogen-limited, even though the biology of the sea has been modified drastically during the last two decades. The suggestions that primary production is nitrogen limited and the POM exported from the surface is deficient in nitrogen still need to be confirmed by detailed bio-assays and biological studies together with sediment trap data from the subhalocline. Such efforts are essential for a better understanding of the nutrient cycling in the landlocked, two-layer Marmara Sea ecosystem which is now receiving substantial wastes from land-based sources, facilitating primary production and modifing community structure.

Acknowledgements

This long-term research programme has been supported by the Turkish Scientific and Technological Research Council (TUBITAK) and the NATO Scientific Affairs Division – within the framework of the Science for Stability Programme. The presentation of this study in PELAG took place within NATO-B1 research programme, which was supported by TUBITAK-BAYG. We gratefully thank Dr Svein Kristiansen and Dr Alec Gaines for their review of the manuscript and valuable comments.

References

Alldradge, A. L., U. Passow & B. E. Logan, 1993. The abundance and significance of large, transparent organic particles in the ocean. Deep-Sea Res. 40: 1131–1140.

Andersen, J. J. & E. C. Carmark, 1974. Observations of chemical and physical fine-structure in a strong pycnocline, Sea of Marmara. Deep-Sea Res. 21: 877–886.

Banse, K., 1994. Grazing and zooplankton production as key controls of phytoplankton production in the open ocean. Oceanography 7: 13–20.

Besiktepe, S. T., E. Özsoy & Ü. Ünlüata, 1993. Filling of the Marmara Sea by the Dardanelles lower layer inflow. Deep-Sea Res. 40: 1815–1838.

Besiktepe, S. T., H. I. Sur, E. Özsoy, M. Abdul Latif, T. Oguz & Ü. Ünlüata, 1994. The circulation and hydrography of the Marmara Sea. Prog. Oceanog. 34: 285–334.

Bologa, A. S., 1985. Planktonic primary productivity of the Black Sea: a review. Thallassia Jugoslavica 21/22: 1–22.

Bologa, A. S., N. Bodeanu, A. Petran, V. Tiganus & Y.P. Zaitsev, 1995. Major modifications of the Black Sea benthic and planktonic biota in the last three decades. Bulletin de l'Institut oceanographique, Monaco special, 15: 180–185.

Brewer, P. G. & J. W. Murray, 1973. Carbon, nitrogen and phosphorus in the Black Sea. Deep Sea Res. 20: 803–818.

Cociasu, A., V. Diaconu, L. Teren, I. Nae, L. Popa, L.Drogan & V. Malciu, 1997. Nutrient stocks on the western shelf of the Black Sea in the last three decades. In E. Ozsoy & A. Mikaelyan (eds), Sensitivity to Changes: Black Sea, Baltic Sea and North Sea, NATO ASI Series, Kluwer Academic Publishers.

Copin-Montegut, C. & G. Copin-Montegut, 1983. Stoichiometry of carbon, nitrogen, and phosphorus in marine particulate matter. Deep Sea Res. 30: 31–46.

Ediger, D. & A. Yilmaz, 1996. Variability of light transparency in physically and biochemically different water masses: Turkish Seas. Fresenis envir. Bull. 5: 133–140.

Ergin, M., M. N. Bodur, D. Ediger, V. Ediger & A. Yýlmaz, 1993. Organic carbon distribution in the surface sediments of the Sea of Marmara and its control by the inflows from the adjacent water masses. Mar. chem. 41: 311–326.

Epply, R. W., G. Harrison, S. W. Chisholm & E. Stewart, 1977. Particulate organic matter in surface waters off southern California and its relationship to phytoplankton. J. mar. Res. 35: 671–696.

Friederich, G. E., L. A. Codispoti & C. M. Sakamoto, 1990. Bottle and pumpcast data from the 1988 Black Sea expedition. Moneterey Bay Aquarium Res. Inst., Tech. rep. No. 90–3.

Goldman, J. C., J. J. McCarthy & D. G. Peavey, 1979. Growth rate influence on the chemical composition of phytoplankton in oceanic waters. Nature. 279: 210–215.

Holligan, P. M., E. Fernandez, J. Aiken, W. M. Balch, P. Boyd, P. H. Burkill, M. Finch, S. B. Groom, G. Malin, K. Muller, D. A. Purdie, C. Robinson, C. C. Trees, S. M. Turner & P. van der Wal, 1993. A biogeochemocal study of the coccolithophere, *Emiliania huxleyi*, in the north Atlantic. Global Biogeochemical Cycles 7: 879–900.

Holm-Hansen, O., C. J. Lorenzen, R. W. Holms & Y. M. H. A. Strickland, 1965. Fluorometric determination of chlorophyll. J. cons. perm. int. explore Mer. 30: 3–15.

Izdar, E., T. Konuk, V. Ittekott, S. Kempe & E.T. Degens, 1987. Particle flux in the Black Sea: Nature of organic matter in the shelf waters of the Black Sea. In E. T. Degens, E. Izdar & S. Honjo (eds), Particle Flux in the Ocean. Mitt. Geol.-Paleont. Inst., Univ. Hamburg, SCOPE/UNEP Sonderband, No: 62, 1–18.

Karl, D. M., J. E. Dore, D. V. Hebel & C. Winn, 1991. Procedures for particulate carbon, nitrogen, phosphorus and total mass analyses used in the US-JGOFS Hawaii Ocean Time-series Program. In D. C. Hurd & D. W. Spencer (eds), Marine Particles: Analysis and Characterization, Geophysical Monograph 63: 71–77.

Kideys, A. E., 1994. Recent dramatic changes in the Black Sea ecosystem: the reason for the sharp decline in Turkish anchovy fisheries. J. mar. Syst. 5: 171–181.

Kocatas, A., T. Koray, M. Kaya & O.F. Kara, 1993. Fisheries and environment studies in the Black Sea system. Part. 3: A review of the fishery resources and tehir environment in the Sea of Marmara. Studies and Reviews. General Fisheries Council for the Mediteranean. No: 64, Rome, FAO: 87–143.

Mee, L. D., 1992. The Black Sea in crisis. Ambio 21: 278–286.

Miller, A. R., P. Tcheria & H. Charnock, 1970. Medireranean Sea Atlas of Temperatute, Salinity, Oxygen Profiles and Data from the R/V Atlantis and R/V Chain with distribution of Nutrient Chemical Properties.

Morkoç, E. & S. Tu´δrul, 1995. Atýksu kirlili´δinin Ýzmit Körfezinin fiziksel ve biyokimyasal özelliklerine etkisi. Tr. J. enginer. & enviz. Scienc. 19: 87–96 (in Turkish).

Oguz, T., H. Duclow, P. Malanotte-Rizolli, S. Tuğrul, N.P. Nezlin & Ü. Ünlüata, 1996. Simulation of annual plankton productivity cycle in the Black Sea by one-dimensional physical-biological model. J. geophy. Res. 101: 16,585–16,599.

Orhon, D., O. Uslu, S. Meriç, I. Salihoglu & A. Filibeli, 1994. Wastewater management for Istanbul: Basis for treatment and disposal. Envir. Pollut. 84: 167–178.

Polat, S. Ç. & S. Tuğrul, 1995. Nutrient and organic carbon exchanges between the Black and Marmara seas through the Bosphorus strait. Cont. shelf Res. 15: 1115–1132.

Polat, S. Ç., 1995. Nutrient and Organic Carbon Budgets in the Sea of Marmara: A Progressive Effort on the Biogeochemical Cycles of Carbon, Nitrogen and Phosphorus. Ph.D. Thesis, Erdemli-Turkey, 215 pp.

Polat, S. Ç. & S. Tuğrul, 1996. Chemical exchange between the Mediterranean and the Black Sea via the Turkish straits. In F. Briand (ed.), Dynamics of Mediterranean Straits and Channels, CIESM Science Series N°2. Bulletin Oceanographique Monaco, N° special 17: 167–186.

Rieseball, U., 1991a. Particle aggregation during a diatom bloom. I. Physical aspects. Mar. ecol. prog. Ser. 69: 273–280.

Rieseball, U., 1991b. Particle aggregation during a diatom bloom. I. Biological aspects. Mar. ecol. prog. Ser. 69: 281–291.

Redfield, A. C., B. H. Ketchum & F. A. Richards, 1963. The influence of organisms on the composition of sea-water. In M. N. Hill (ed.), The Sea, ideas and observations on progress in the study of the seas. Interscience 2: 26–77.

Richards, F.A., 1965: Anoxic basins and fjords. In J. P. Riley & G. Skirrow (eds), Chemical Oceanography, V.1, Academic Press: 611–645.

Sakshaug, E., K. Andresen, S. Myklestad & Y. Olsen, 1983. Nutrient status of phytoplankton communities in Norwegian waters (marine, brackish, fresh) as revealed by their composition. J. Plankton Res. 5: 175–196.

Sen Gupta, R., 1971. Oceanography of the Black Sea: Inorganic nitrogen compounds. Deep-Sea Res. 18: 457–475.

Serpoianu, G. I. Nae & V. Malciu, 1992. Danube water influence on sea water salinity at the Romanian littoral. Rapports et Procés verbaux des Reunions de la CIESMM. 33: 233.

Shiganova, T., A. N. Tarkan, A. Dede & M. Cebeci, 1995. Distribution of the Ichthyo-Jellplankton Mnemiopsis leidyi (Agassiz, 1865) in the Marmara Sea (October 1992). Turk. J. mar. Sci. 1: 3–12.

Smetacek, V. S., 1985. Role of sinking in diatom life-history cycles: ecological, evolutionary and geological significance. Mar. Biol. 84: 239–251.

Sorokin, Yu. I., A. N. Tarkan, B.Ozturk & M.Albay, 1995. Primary production, bacterioplankton and planktonic protozoa in the Marmara Sea. Turk. J. mar. Sci. 1: 37–54.

Strickland, J. D. H. & T. R. Parsons, 1972. A practical handbook of seawater analysis, 2nd edn., Ottowa, Fisheries Research Board.

Sur, H. I., E. Özsoy & Ü. Ünlüata, 1994. Boundary current instabilities, uppwelling, shelf mixing and eutrophication processes in the Black Sea. Progr. in Oceanogr. 33: 227–264.

Takahashi, T., W. S. Broecker & S. Langer, 1985. Redfield ratio based on chemical data from isopycnal surfaces. J. geophy. Res. 90: 6907–6924.

Tett, P., J. C. Cottrell, D. O. Trew & B. J. B. Wood, 1975. Phosphorus quota and the chlorophyll : carbon ratio in marine phytoplankton. Limnol. Oceanogr. 20: 587–603.

Tolmazin, D., 1985. Changing coastal oceanography of the Black Sea. Progr. in Oceanogr. 15: 217–276.

Tuğrul, S., 1993. Comparison of TOC concentrations by persulfate-UV and HTCO techniques in the Marmara and Black Seas. Mar. Chem. 41: 265–270.

Tuğrul, S. & S. Ç. Polat, 1995. Quantitative comparison of the influxes of nutrients and organic carbon into the Sea of Marmara both from antropogenic inputs and from the Black Sea. Water Sci. Techn 32: 115–121.

Uysal, Z., 1996. A net-plankton study in the Bosphorus junction of the Sea of Marmara. Tr. J. Bot. 20: 321–327.

Vinogradov, M. E. & E. A. Shushkina, 1992. Temporal changes in community structure in the open Black Sea. Oceanology 32: 485–491.

Vostokov, S. V. & V. I. Vedernikov, 1988. Living and nonliving particulate organic matter in the euphotic zone of the epipelagic ecosystem. Oceanology 28: 100–105.

Vostokov, S. V., 1996. Suspended matter as a measure of productivity in the western Black Sea. Oceanology 36: 241–247.

Hydrobiologia **363:** 169–177, 1998.
T. Tamminen & H. Kuosa (eds), Eutrophication in Planktonic Ecosystems: Food Web Dynamics and Elemental Cycling.

Impact of increased river discharge on the phytoplankton community in the outer Oslofjord, Norway

Svein Kristiansen
Section for Marine Botany, Department of Biology, University of Oslo, P.O. Box 1069 Blindern, N-0316 Oslo, Norway
(tel: 4722854529; fax: 4722853844. e-mail: svein.kristiansen@bio.uio.no)

Key words: bloom, flood, harmful, toxic, *Skeletonema costatum*

Abstract

An extraordinary flood event occurred in Southern Norway in May–June 1995. The impact of fresh water discharge from the river Glomma on phytoplankton species composition and on primary production during the flood was studied. The phytoplankton community was dominated by species usually found in the fjord, the diatom *Skeletonema costatum* dominated the community together with cyanobacteria. Potentially harmful species (e.g. *Chrysochromulina polylepis*) were found in low cell densities. No large increase in nutrient concentrations were found. Primary production and uptake rates of nitrate and ammonium were very low in low-salinity water near the river outlet. Nutrients from the river water were efficiently utilized in the open part of the fjord. The flood trigged a phytoplankton bloom dominated by harmless species commonly observed in the fjord plankton.

Introduction

Seasonal patterns in fresh water discharge are well known to influence both production and species composition in the marine environment (Mann & Lazier, 1991). Norwegian rivers usually have a peak in fresh water discharge in May–June, after the spring bloom in the coastal water (Tollan, 1976; Rey, 1981). Glomma is the largest river in Norway, and it discharges large quantities fresh water into an archipelago in the outer Oslofjord. In May–June 1995, an exceptionally large flood occurred in southern Norway. Large snow quantities and intense melting after a late spring combined with rain caused the flood (Tollan, 1995). The freshwater discharge from Glomma was noticeable outside the normal influence area in the outer Oslofjord and in the northern part of Skagerrak (Danielssen et al., 1996; pers. observ.). Increased freshwater discharge may enhance or reduce productivity in fjords (Skreslet, 1986). Residence time of the productive layer is obviously important in this respect. When the residence time is less than the phytoplankton generation time, phytoplankton (and nutrients) will be flushed out of the fjord. Productivity will often increase in river plumes in coastal waters during periods of peak discharge (Mann & Lazier, 1991; Cloern, 1996). Similar increase will be expected in coastal water surrounding outlets of fjords with short residence time.

An increased frequency of phytoplankton blooms has accompanied nutrient enrichment of coastal waters (Smayda, 1990) , and several harmful prymnesiophycean blooms in Scandinavian coastal waters during the last 5-10 years have been connected with exceptional hydrographical conditions (Granéli et al., 1993). During the flood, large quantities of fresh water carrying dissolved and particulate matter were discharged into the fjord. We wanted to study the effect of freshwater on the phytoplankton in the Oslofjord and addressed the following questions: Did the phytoplankton species composition change during and after the flood? Did any potentially harmful algae bloom? Were primary production and nitrogen uptake rates light-limited in the influence area?

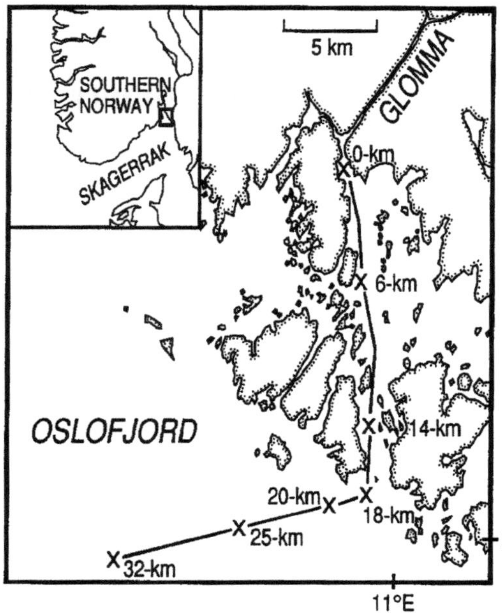

Figure 1. Station map. Samples were collected along a transect from mouth of the river Glomma (Station 0-km) and into the outer Oslofjord (Station 32-km) on 14 June and 5 July 1995. A few additional samples were collected at selected stations on 13 & 26 June, 12 & 27 July and 22 August 1995.

Methods

Samples were collected along a transect from the river mouth and into the outer Oslofjord using 1.7-liter Niskin bottles at 0.5, 2, 4, 6, 10 and 20 m depth in June–August 1995 (Figure 1). Salinity (continuous measurements) and Secchi-depth readings were about similar in a grid 8 × 8 km southwest of Station 32-km during sampling 14 June and the day before (11.9 ± 0.2 psu and 2.2 ± 0.2 m). Station 32-km was therefore a typical open water station in the outer Oslofjord, and was used as the last and outermost station in the transect. Only samples for phytoplankton species composition were collected in August.

Hydrography, irradiance, chlorophyll, fluorescence and nutrients

Salinity, temperature and *in vivo* fluorescence profiles were obtained using a Niel Brown CTD and an *in situ* fluorometer (Q-meter). The fluorometer was calibrated against measured chlorophyll *a* (Chl *a*, see below) in surface samples, and the calibrated values are called Chl *a*-fluorescence (1 unit ≈ 1 μg Chl *a* l^{-1}). Continuous measurements of surface hydrography and *in vivo* fluorescence were obtained from a 15-l carboy with running surface seawater (20 l min^{-1}) using a S-C-T Meter (YSI Model 33, calibrated against a Mini STD 200) and the *in situ* fluorometer. Irradiance (PAR) was measured using a LI-193SA Spherical Quantum Sensor for vertical profiling and a LI-190SA Quantum Sensor for monitoring surface irradiance, both connected to a LI-1000 DataLogger. All nutrient samples were filtered (1.0 μm polycarbonate filters for silicic acid and precombusted GF/C glass fiber filters for all other nutrients). Ammonium was measured manually less than 3 h after sampling (Solórzano, 1969), the other nutrient samples were frozen. Phosphate was measured 2 weeks after sampling (Strickland & Parsons, 1972). Nitrate, nitrite and silicic acid were measured 2 months after sampling with a Chem-Lab automatic analyzer using methods adapted from Strickland & Parsons (1972). Chl *a* samples were filtered onto 1 μm polycarbonate filters and measured in acetone extracts using a Turner Designs Fluorometer calibrated against a Chl *a* standard.

Phytoplankton

Several methods were used to give the best quantitative and qualitative description of the phytoplankton species composition. Samples mainly for diatoms and dinoflagellates were preserved with neutralized formaldehyde, sedimented in 2-ml counting chambers and counted using an inverted microscope (Sournia, 1978). Samples mainly for flagellates and small diatoms were preserved with glutaraldehyde, filtered onto 0.2 μm black polycarbonate filters and counted with an epifluorescence microscope (Porter & Feig, 1980). Non-preservable species were identified and enumerated using dilution cultures and/or transmission electron microscopy (Sournia, 1978; Moestrup & Thomsen, 1980; Moestrup, 1984). Cell volumes were calculated from own cell size estimates and converted to phytoplankton carbon-biomass using the equations given in Strathmann (1967). Species accounting for $< 0.5\%$ of phytoplankton C are not included in the estimates.

Rate measurements

Photosynthesis and nitrogen (nitrate and ammonium) uptake rates were measured using ^{14}C and ^{15}N isotopes (Strickland & Parsons, 1972; Kristiansen & Paasche, 1989). Samples for rate measurements were collected at 60% light-depth (≤ 0.5 m depth at the stations sampled 14 June and 0.5–1.0 m depth at the stations sampled 5 July). Incubation bottles were covered with layers of neutral-density screening in order to simulate irradiances at selected depths, and placed in deck incubators cooled by circulating surface seawater immediately after sampling. Additions of 740 kBq $NaH^{14}CO_3$, 0.25 μmol l^{-1} $^{15}NH_4$, and 0.25 and 2.0 $^{15}NO_3$ μmol l^{-1} (99 atom%) were done to separate bottles. Nitrate uptake rates were calculated from the high-addition samples (2.0 $^{15}NO_3$ μmol l^{-1}) if ambient nitrate concentration was > 10 μmol l^{-1}. The low-addition samples were used elsewhere. All isotopes were added at the same time, and the samples were incubated for 4–5 h (1400–1900 h local time). Incubations were terminated by filtering onto 1 μm polycarbonate filters (^{14}C) or precombusted GF/C filters (^{15}N).

Results

Fresh water discharge was measured at Solbergfoss (a waterfall ca. 60 km up the river from Station 0-km) and includes most of the discharge from Glomma into the outer Oslofjord 1 day later. Maximum fresh water discharge into the fjord was 3600 m^3 s^{-1} during the first 2 weeks of June 1995 (Figure 2). It was 2.6 times higher than median daily discharge from Glomma in June during the last 30 years. The flood set up a pronounced low salinity surface layer in the top 1–5 m of the water column. Salinity was low at the inner stations (< 0.5 psu at Station 0-km in June) and increased with increasing distance from the river mouth (Figure 3). Salinity at Station 32-km was 11.9 psu in June and increased to 24.0–25.0 psu in July. Surface current velocities were not measured, and residence time of the productive layer was difficult to calculate because of a complex topography in the area (Figure 1). It is obvious that residence time increased with increasing distance from the river mouth. Residence time during the flood peak has been estimated to < 1 day at the inner stations and > 2 days at the outer station (B. Bjerkeng, Norwegian Institute of Water Research, pers. com.).

Nutrient concentrations were generally high at the inner stations and decreased with increasing distance from the river mouth (Table 1). In July, however, phosphate concentration decreased to ≤ 0.07 μmol l^{-1} in the top 6 m at Station 20-km and then increased to 1.0–1.4 μmol l^{-1} in the high salinity water at Stations 25-km and 32-km. Dilution plots (nutrient vs salinity, plots not given) show that nitrate and silicic acid concentrations decreased with increasing salinity

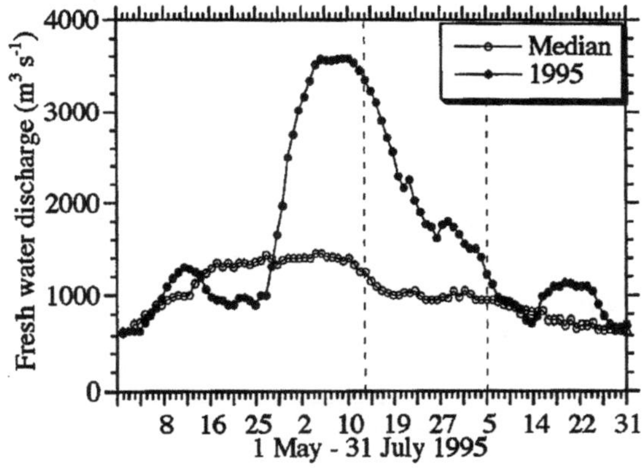

Figure 2. Fresh water discharge (m^{-3} s^{-1}) from Glomma (at Solbergfoss) into the outer Oslofjord 1 May–31 July 1995, and median discharge during the last 30 years. Sampling dates (14 June & 5 July) are indicated by broken lines. Values from Glommens and Laagens Brukeierforening, Oslo.

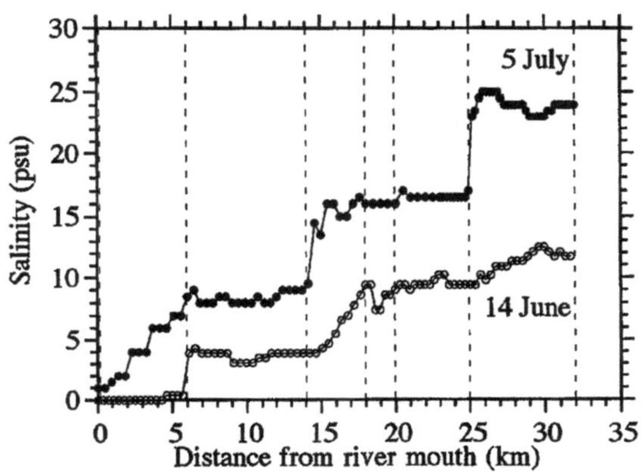

Figure 3. Salinity (psu) in the surface layer (0-3 m depth) on 14 June and 5 July. Stations are indicated by broken lines.

up to 10–15 psu. Phosphate concentration apparently did not change with salinity. The limited number of samples do not allow us to calculate dilution of the river-borne nutrients (only 4 samples with salinities < 5 psu). It seems, however, that the nutrient concentrations at 0.5, 2 and 4 m depths mainly decreased by dilution at stations with salinities < 10–15 psu (i.e. $\leq$ 25 km from the river outlet in June and $\leq$ 15 km from the outlet in July, Figure 3). Inorganic nutrient ratios (N:P & Si:N) were variable, more so in July than in June (Table 2). Totally 54 samples were collected, the N:P ratio was > 16 in 6 samples (1 sample in June and 5 samples in July), and the Si:N ratio was < 0.5 in 2 samples (both in July). The highest N:P ratios (71 and 130) were both found in phosphate-depleted water in July (Stations 6-km and 20-km).

Table 1. Nutrient concentrations (μmol l^{-1}) in the surface layer (0.5–2 m depth) and in 4–20 m depth on 14 June and 5 July. Highest concentrations at Stations 0-km & 6-km[i].

	Depth	Nitrate	Phosphate	Silicic acid
14 June				
	0.5–2 m	0.8–19	0.36–1.03	6.4–25
	4–20 m	0.1–8.0	0.36–1.23	0.9–16
5 July				
	0.5–2 m	0.1–18	0.05–1.44[i]	0.4–31
	4–20 m	0.1–5.6	0.05–1.13[i]	0.2–4.3

[i]Highest phosphate concentrations on 5 July were found at Station 32-km.

Table 2. Ratios between dissolved nitrate, phosphate and silicic acid (mol mol^{-1}) in the surface layer (0.5–2 m depth) and in 4–20 m depth on 14 June and 5 July.

	Depth	N:P	Si:N
14 June			
	0.5–2 m	2.2–19	0.9–8.3
	4–20 m	<6.5	>0.8
5 July			
	0.5–2 m	<130[i]	>1.8
	4–20 m	<24	>0.4

[i]Highest N:P ratio in PO_4-depleted water.

Table 3. Depth profiles at Stations 6-km and 25-km 5 July. Salinity (psu), silicic acid, phosphate, nitrate and ammonium (all in μmol l^{-1}) and chlorophyll *a* (μg l^{-1}).

Depth	Salinity	$Si(OH)_4$	PO_4	NO_3	NH_4	Chl *a*
Station 6-km						
0.5	5.40	27.8	0.35	6.8	1.5	1.9
2	16.60	9.6	0.08	5.4	0.9	13.9
4	24.76	1.5	0.08	1.0	0.9	10.0
6	26.01	0.7	0.08	0.6	0.8	3.4
10	27.71	0.6	0.11	0.7	1.2	1.7
20	31.50	4.3	0.43	5.6	2.6	0.9
Station 20-km						
0.5	16.50	15.6	0.07	8.6	0.8	4.1
Station 25-km						
0.5	23.24	4.96	0.99	2.2	0.3	3.9
2	24.06	2.04	1.44	0.7	0.1	5.2
4	25.06	0.63	1.08	0.2	0.2	5.0
6	25.88	0.34	0.94	0.1	<0.1	5.4
10	28.26	0.38	1.00	1.0	1.0	2.6
20	32.54	0.97	0.21	1.8	1.5	0.5

Continuous Chl *a*-fluorescence measurements indicated a large increase in Chl *a* at the outer stations in June (Figure 4). Measured Chl *a* concentrations at 0.5 m depth increased from < 1.0 μg l^{-1} between Stations 0-km and 20-km to 5.0 μg l^{-1} at Station 32-km. Only small variations were found in Chl *a*-fluorescence in July. The peaks in Chl *a*-fluorescence found 6–7 km and 12 km from the river mouth in July were part of a pronounced subsurface Chl *a* maximum having its core below the low-salinity surface layer at Station 6-km (Table 3). Salinity increased at Station 14-km (Figure 3), and the subsurface maximum approached the surface and gradually diminished between Stations 18-km and 20-km. Maximum Chl *a* concentration was 13.9 μg l^{-1} at 2 m depth or 7 times higher than the surface concentration (Station 6-km). Nutrient concentrations were high in the low-salinity surface layer above the subsurface Chl *a* maximum at Station 6-km (Table 3), and the N:P and Si:N ratios were 19 and 4.2 respectively. Phosphate concentration was very low (0.08 μmol l^{-1}) in the subsurface Chl *a* maximum, and the N:P and Si:N ratios were 71 and 1.8. Station 25-km was in July located in a salinity front where salinity increased from 17–23 psu within 0.3 km (Figure 3). No pronounced subsurface Chl *a* maximum was found at this station (Table 3), and only remnants of the low-salinity surface layer were left on top of a nitrogen-depleted intermediate water layer at 4–6 m depth (concentrations of nitrate and ammonium were < 0.2 μmol l^{-1}). The N:P ratio was low both in the surface layer (2.3) and in the intermediate layer below (0.1–0.2). Concentrations of silicic acid were higher, and the Si:N ratio was 2.2 in surface layer and 3.6–7.1 in the intermediate layer below.

Totally 49 species or genera were identified in the samples. Species diversity was higher in July than in June, especially close to the river mouth (Figure 5). However, the phytoplankton community was dominated by a few species only, the most abundant being the diatom *Skeletonema costatum* (Greville) Cleve and cyanobacteria (*Synechococcus*-like). Maximum cell density of *S. costatum* were 20.3×10^6 l^{-1} in the subsurface maximum (2 m depth) at Station 6-km in July. Almost as high densities (19.8×10^6 l^{-1} and 20.0×10^6 l^{-1}) were found at 0.5 and 2 m depths at Station 32-km in June. Maximum cell density of cyanobacteria (39×10^6 l^{-1}) was found at 4 m depth at Station 32-km in August. Cyanobacteria cell densities were generally 1–2 order of magnitude higher in July and August than in June. Fresh water diatoms were found in low cell densities ($<0.1 \times 10^6$ l^{-1})

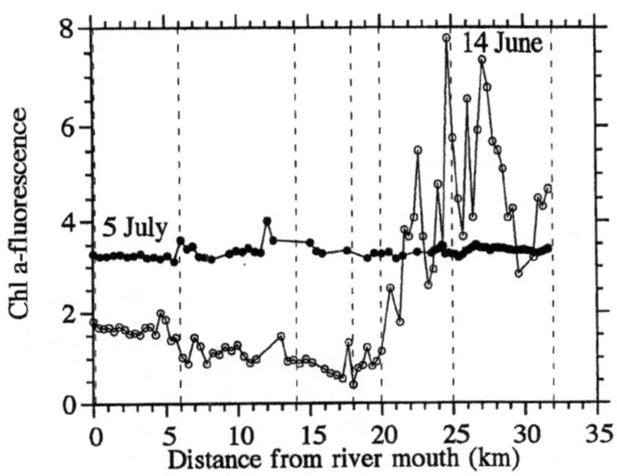

Figure 4. Chl *a*-fluorescence (1 unit ≈ 1 μg Chl *a* l^{-1}) in the surface layer (0-3 m depth) on 14 June and 5 July. Stations are indicated by broken lines.

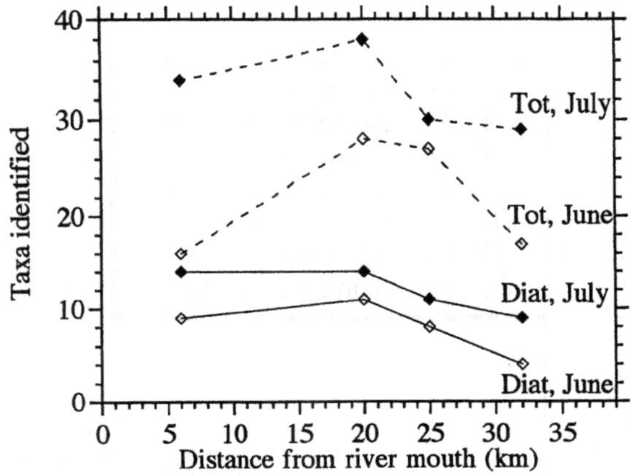

Figure 5. Phytoplankton taxa (total and diatoms) identified in samples collected 14 June and 5 July.

Table 4. Potentially harmful genera or species[i] identified in samples collected in June–August.

Taxon	June	July	August
Pseudo-nitzschia spp.	x	x	x
Dinophysis acuminata	x	x	
D. acuta	x	x	
D. norvegica	x	x	x
D. rotundata (heterotrophic)	x	x	x
Gymnodinium galatheanum	x		
Prorocentrum minimum			x
Chrysochromulina spp.	x	x	x
C. polylepis	x	x	
C. leadbeateri	x		

[i] According to Hallegraeff et al. (1995)

Table 5. Photosynthesis (μmol C l^{-1} h^{-1}), chlorophyll *a*-specific photosynthesis (g C (g Chl *a*)$^{-1}$ h^{-1}) and summed ammonium + nitrate uptake rate (mmol N l^{-1} h^{-1}).

Station	Photosynthesis (μmol C l^{-1} h^{-1})	Chl *a*-specific photosynthesis (g C (g Chl *a*)$^{-1}$ h^{-1})	$NO_3 + NH_4$ uptake rate (nmol N l^{-1} h^{-1})
14 June			
6-km	0.03	0.4	4.0
20-km	0.01	1.0	0.6
25-km	1.08	3.3	46.3
32-km	3.35	8.1	41.0
5 July			
6-km	0.12	1.5	–
20-km	0.90	2.6	–
25-km	0.46	1.8	–
32-km	0.35	1.4	31.8

at the inner stations (e.g. *Asterionella formosa* Hassall, *Aulacoseira* sp., *Diatoma* sp. and *Rhizosolenia longiseta* Zacharias). Ten potentially harmful genera or species (Hallegraeff et al., 1995) were identified in the samples (Table 4). Cell densities of these species were all low (< 10^6 l^{-1}). *Skeletonema costatum* totally dominated phytoplankton C-biomass in June–July (Figure 6). In August, the dinoflagellate *Ceratium lineatum* (Ehrenberg) Cleve and unidentified small *Chaetoceros* species dominated.

Photosynthesis and nitrogen uptake rates were greatly reduced at the inner stations during the flood peak in June. Photosynthesis and nitrogen uptake rates were 1–2 order of magnitude lower at Stations 6-km and 20-km than at Station 32-km in June (Table 5). Three weeks later, in July, photosynthesis and Chl *a*-specific photosynthesis were of the same order of magnitude at all the stations. All the rates were clearly lower at the outer station in July than in June.

Turbidity was high in the flood water, and the euphotic zone depth (1% light depth) in June ranged from 1.2–4.4 m (Stations 6-km & 32-km respectively). The optical properties of the fjord water improved in July, and the euphotic zone depth then ranged from 4.5–13.5 m (same stations as above). Strong currents prevented light measurements at Station 0-km. Secchi disk readings suggest 1% light depth at 0.9 and 4.0 m at Station 0-km in June and July. Surface samples from Stations 6-km and 32-km in June were exposed to a range of irradiances (2–100% of surface irradiance). Photosynthesis was photoinhibited at 100% of surface irradiance at both stations (Table 6). Surface

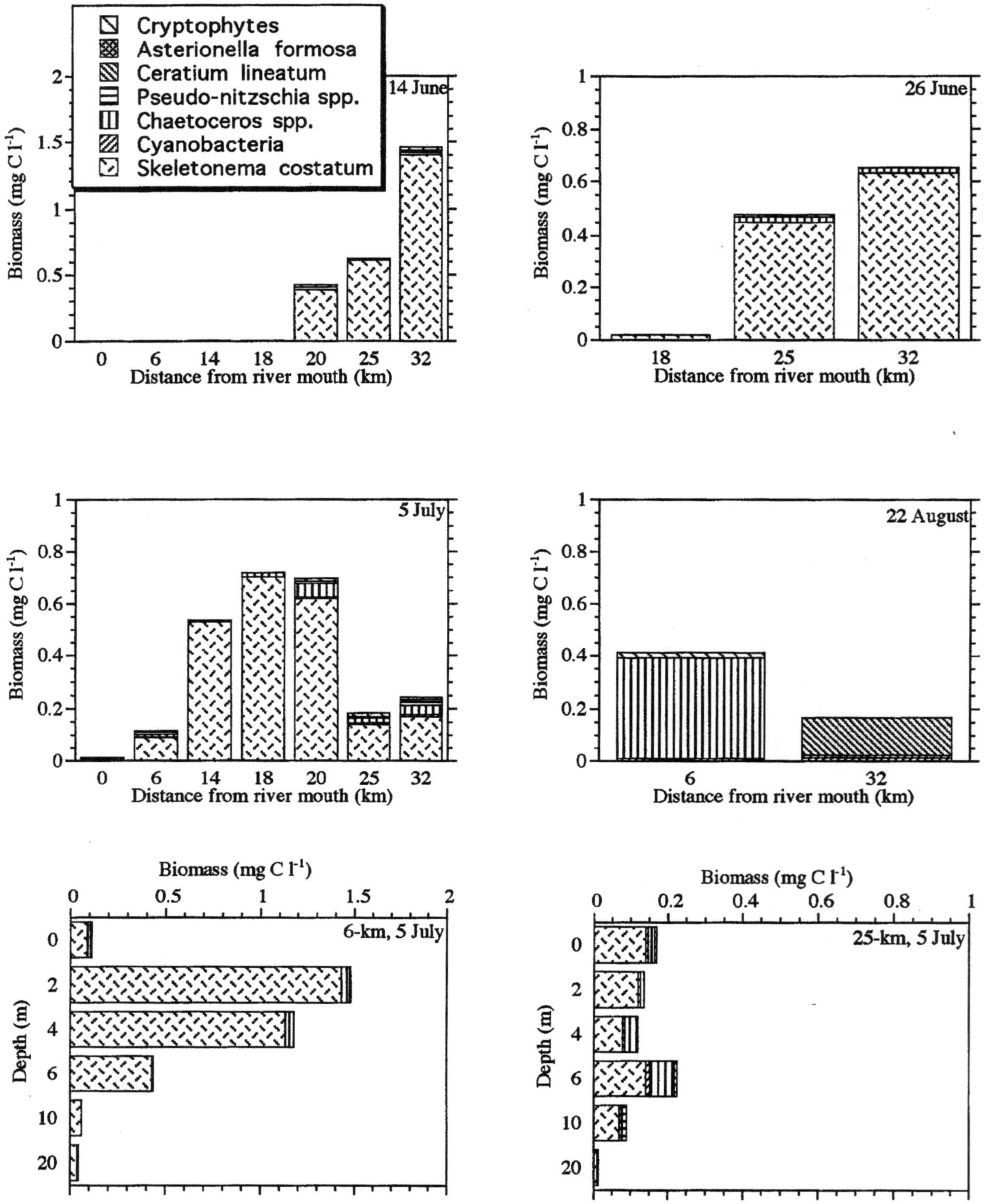

Figure 6. Phytoplankton biomass (mg C l^{-1}) at 0.5 m depth on 14 and 26 June, 5 July and 22 August and depth profiles of the same at Stations 6-km and 25-km on 5 June.

Table 6. Effect of irradiance on primary production and on uptake rates of ammonium and nitrate in samples from 0.5 m depth at Stations 6-km & 32-km in June (relative values). Corresponding irradiance depths are given for both stations.

Irradiance (%)	Depth (m)	Prim. prod. (%)	Ammonium (%)	Nitrate (%)
Station 6-km				
100	0	75	100	90
60	0.2	100	93	100
22	0.4	28	88	83
5	0.8	13	-	76
2	1.0	6	88	72
Station 32-km				
100	0	33	83	100
60	0.3	100	76	82
22	1.0	18	100	-
5	2.0	4	-	14
2	2.6	1	32	5

irradiance during the incubation period was low and variable (0.4–5.9 mol m^{-2} h^{-1}) because of variable clouds. Photosynthesis at both stations was severely light-limited ($\leq$28% of maximum rates) at $\leq$22% of surface irradiance. From measured light-profiles at the 2 stations, 22% of surface irradiance was at 0.4 and 1.0 m depth at Stations 6-km and 32-km. Nitrogen uptake rates were less influenced by irradiance. Unfortunately, some samples from the uptake vs. irradiance experiment were lost. It is, however, reasonable to assume that ammonium and nitrate uptake rates were significantly reduced at $\leq$22% of surface irradiance (below 0.4 m depth) at Station 6-km (72–88% of maximum rates). Similarly, ammonium and nitrate uptake rates would be significantly reduced at $\leq$5% of surface irradiance (below 2.0 m depth) at Station 32-km ($\leq$32% of maximum rates).

Discussion

Unusually large quantities of fresh water containing dissolved and particulate material were discharged into the outer Oslofjord during the flood. The fresh water created a 1–5 m thick low-salinity turbid surface layer at all stations as well as in northern Skagerrak in June (Figure 3; Danielssen et al., 1996). The layer was reduced in July, and normal salinities (24–25 psu) were re-established at the outer stations. Nitrate and silicic acid concentrations were high close to the river mouth (18–31 μmol l^{-1}), and decreased with increasing salinity up to 10–15 psu. Concentrations of the 2 nutrients were 8 times (nitrate) and 16 times (silicic acid) higher at the outer stations in June than in July. Phosphate concentration, however, showed a different pattern. It apparently did not change with salinity, and the highest phosphate concentrations were found at the 5 inner stations in June and at the 2 outer stations in July. Concentrations of all 3 nutrients at the outer stations were, however, similar to or lower than deep water concentrations in the Oslofjord and Skagerrak (Føyn & Rey, 1981; pers. observ.). The high concentrations of nitrate and silicic acid close to the river mouth were 3 - 5 times higher than deep water concentrations.

Novel and nuisance phytoplankton blooms have been correlated with anthropogenic enrichment of N and P and decline in the Si:N and Si:P ratios (Smayda, 1990; Hallegraeff, 1993). A balanced N:P ratio is close to 16 (Redfield et al., 1963). The 2 high N:P ratios (71 and 130) were both from phosphate-depleted surface water. Mean Si:N ratio for culture experiments with 26 different diatoms is 1.1 (range 0.3–4.4; Brzezinski, 1985). All measured Si:N ratios were > 0.4. We conclude that nutrient input from the flood into the outer Oslofjord was significant but low, and neither the concentrations nor the nutrient ratios found during the flood were alarming.

Nutrient input from the flood water to the fjord significantly increased Chl *a* concentrations at the outer stations in June (5–8 μg l^{-1}; Figure 4). Expected concentrations without the flood event would have been < 2–3 μg Chl *a* l^{-1}. Chl *a* concentration may be 4–5 times higher during a spring bloom in the fjord than found during the flood, thus the measured concentrations were not exceptional (Kristiansen, 1987; Baalsrud & Magnusson, 1990; pers. observ.).

The species identified (49 taxa) were all common species in coastal water (Tomas, 1993; Tomas, 1996) and also in the Skagerrak and the Oslofjord (Throndsen, 1980; Lange et al., 1992; Eikrem & Throndsen, 1993; Ypma & Throndsen, 1996). *Skeletonema costatum* which dominated in most of the samples (both in cell abundance and as estimated C-biomass) is one of the most frequently recorded phytoplankton species in the fjord (Lange et al., 1992). In August, *S. costatum* was replaced by other diatoms (*Chaetoceros* spp.) at Station 6-km. Silicic acid concentration was low at Station 32-km already in July (< 0.4 μmol l^{-1} at 0.5–6 m depth), and the *S. costatum* population was decreasing. Finally, a dinoflagellate (*Ceratium lineatum*) replaced *S. costatum* at Station 32-km in August.

Ten of the recorded species were potentially harmful species or genera, all of them found in low cell densities ($< 10^6\ l^{-1}$). These species are all common in the fjord, usually present in low cell densities (see above). To our knowledge, no harmful blooms have been reported after the flood in Glomma, and the presence of the potentially harmful species (Table 4) cannot be attributed to the flood.

Strong vertical stratification and a favourable chemical environment may have been important in development of the *Chrysochromulina polylepis* bloom in Skagerrak and Kattegat during early summer 1988 (Nielsen et al., 1990; Granéli et al., 1993), and in development of the *C. leadbeateri* bloom in northern Norway during early summer 1991 (Rey & Aure, 1991; Granéli et al., 1993). Additional samples were therefore collected at 2 stations in July, at Station 6-km with a pronounced subsurface Chl *a* maximum and at Station 25-km located in a salinity front (Table 3). *Skeletonema costatum* dominated at both stations, and no remarkable features in the species composition were found (Figure 6). *Prymnesium parvum* and *P. patelliferum* are brackish water flagellates. Harmful blooms of these species have occurred several years in a fjord system (Ryfylke) in western Norway (Eikrem & Throndsen, 1993; Granéli et al., 1993). These blooms originated in brackish water (<10 psu) rich in nitrate relative to phosphate. Neither of these two species were found in samples from the Oslofjord (Table 4).

The most noticeable effect of the flood was the high turbidity resulting in a very shallow euphotic zone (0.9–4.4 m) in June. Photosynthesis and N-uptake rates were both light-limited in the uppermost 1–2 m in June, the former probably more severely limited than ammonium and nitrate uptake rates. All rates were low at the inner stations in June, probably because of a combination of light-limitation and a shift in species composition from fresh water to marine species. The low Chl *a*-specific photosynthesis at the inner stations (0.4–1.0 g C (g Chl $a)^{-1}$ h^{-1}) also indicate that phytoplankton growth rates were suppressed. Daily nitrate + ammonium uptake rates at Stations 6-km and 20-km were $< 0.01\ \mu mol\ l^{-1}\ d^{-1}$ while daily rates at the outer stations were one order of magnitude higher. The low rates at the inner stations support the above assumption that decreasing nitrate and silicic acid concentrations with increasing distance from the river mouth were mainly due to simple dilution. Estimated residence time was < 1 day, and the phytoplankton was flushed out of the area inside the archipelago. In July, photosynthesis was less variable and the highest rates (absolute and Chl *a*-specific) were found at Station 20-km. *Skeletonema costatum* dominated also at this station with the highest N:P ratio (130) found in June and July.

Conclusion

Diatoms, mainly *Skeletonema costatum*, dominated the phytoplankton community in the fjord during and immediately after the flood. In August, 2–3 months after the flood, the diatoms were replaced by a dinoflagellate, *Ceratium lineatum*, at the outer station. The flood had apparently no more than a modest and short term effect on the phytoplankton community. The nutrients supplied by the river water were diluted at the innermost stations and later on efficiently utilized in the open part of the fjord. The flood trigged a phytoplankton bloom dominated by harmless species commonly observed in the fjord plankton.

Acknowledgments

This investigation was a joint effort by staff members and students at Section for Marine Botany, Department of Biology. My thanks to P. Backe-Hansen, L. Broch, S. Brubak, W. Eikrem, T. Farbrot, L.-J. Naustvoll, S. Skattebøl and Profs G. R. Hasle, E. Paasche and J. Throndsen. The work was in part funded by the Norwegian Pollution Control Authority.

References

Brzezinski, M. A., 1985. The Si:C:N ratio of marine diatoms: interspecific variability and the effect of some environmental variables. J. Phycol. 21: 347–357.

Baalsrud, K. & J. Magnusson, 1990. Eutrofisituasjonen i Ytre Oslofjord. Hovedrapport. Overvåkingsrapport. No. 427/90. Norsk institutt for vannforskning, Oslo.

Cloern, J. E., 1996. Phytoplankton bloom dynamics in coastal ecosystems: A review with some general lessons from sustained investigations of San Francisco Bay, California. Rev. Geophys. 34: 127–168.

Danielssen, D. S., M. Skogen, J. Aure & E. Svendsen, 1996. Flomvann fra Glomma og miljøforholdene i Skagerrak sommreren 1995. (The Glomma flood and the environmental conditions in the Skagerrak in the summer 1995). Fisken og Havet 4: 1–37.

Eikrem, W. & J. Throndsen, 1993. Toxic prymnesiophytes identified from Norwegian coastal waters. In T. J. Smayda & Y. Shimizu (eds), Toxic Phytoplankton Blooms in the Sea. Elsevier Science Publishers, Amsterdam: 687–692.

Føyn, L. & F. Rey, 1981. Nutrient distribution along the Norwegian Coastal Current. In R. Sætre & M. Mork (eds), The Norwegian Coastal Current. University of Bergen, Bergen: 629–639.

Granéli, E., E. Paasche & S. Y. Maestrini, 1993. Three years after the *Chrysochromulina polylepis* bloom in Scandinavian waters in 1988: Some conclusions of recent research and monitoring. In T. J. Smayda & Y. Shimizu (eds), Toxic Phytoplankton Blooms in the Sea. Elsevier Science Publishers, Amsterdam: 23–32.

Hallegraeff, G. M., 1993. A review of harmful algal blooms and their apparent global increase. Phycologia 32: 79–99.

Hallegraeff, G. M., D. M. Anderson & A. D. Cembella, 1995. Manual on Harmful Marine Microalgae. UNESCO, Paris, 551 pp.

Kristiansen, S., 1987. Nitrate reductase activity in phytoplankton from the Oslofjord, Norway. J. Plankton Res. 9: 739–748.

Kristiansen, S. & E. Paasche, 1989. An improved method for determining relative ^{15}N abundance in ammonium regeneration studies by direct diffusion. Mar. Ecol. Prog. Ser. 54: 203–207.

Lange, C. B., G. R. Hasle & E. E. Syvertsen, 1992. Seasonal cycle of diatoms in the Skagerrak, North Atlantic, with emphasis on the period 1980–1990. Sarsia 77: 173–187.

Mann, K. H. & J. R. N. Lazier, 1991. Dynamics of Marine Ecosystems. Blackwell Scientific Publications, Oxford, 466 pp.

Moestrup, Ø., 1984. Further studies on *Nephroselmis* and its allies (Prasinophyceae). II. *Mamiella* gen. nov., Mamiellaceae fam. nov., Mamiellales ord.nov. Nord. J. Bot. 4: 109–121.

Moestrup, Ø. & H. A. Thomsen, 1980. Preparation of shadow-cast whole mounts. In E. Gantt (ed.), Handbook of Phycological Methods. Developmental and Cytological Methods. Cambridge University Press, Cambridge: 385–390.

Nielsen, T. G., T. Kiørboe & P. K. Bjørnsen, 1990. Effects of a *Chrysochromulina polylepis* subsurface bloom on the planktonic community. Mar. Ecol. Prog. Ser. 62: 21–35.

Porter, K. G. & Y. S. Feig, 1980. The use of DAPI for identifying and counting aquatic microflora. Limnol. Oceanogr. 25: 943–948.

Redfield, A. C., B. H. Ketchum & F. A. Richards, 1963. The influence of organisms on the composition of sea-water. In M. N. Hill (ed.), The Sea. Vol. 2. Interscience Publishers, New York: 26–77.

Rey, F., 1981. The development of the spring phytoplankton outburst at selected sites off the Norwegian Coast. In R. Sætre & M. Mork (eds), The Norwegian Coastal Current. University of Bergen, Bergen: 649–680.

Rey, F. & J. Aure, 1991. Oppblomstringen av *Chrysochromulina leadbeateri* i Vestfjorden, mai–juni 1991. Miljøforhold og mulige årsaker. (The *Chrysochromulina leadbeateri* bloom in the Vestfjord, North Norway, May-June 1991. Environmental conditions and possible causes). Fisken og Havet 3: 13–32.

Skreslet, S. (ed.), 1986. The role of freshwater outflow in coastal marine ecosystems. Springer-Verlag, Berlin, 453 pp.

Smayda, T. J., 1990. Novel and nuisance phytoplankton blooms in the sea: evidence for a global epidemic. In E. Granéli, B. Sundström, L. Edler & D. M. Anderson (eds), Toxic Marine Phytoplankton. Elsevier, New York: 29–40.

Solórzano, L., 1969. Determination of ammonia in natural waters by the phenolhypochlorite method. Limnol. Oceanogr. 14: 799–801.

Sournia, A. (ed.), 1978. Phytoplankton Manual. UNESCO, Paris, 337 pp.

Strathmann, R. R., 1967. Estimating the organic carbon content of phytoplankton from cell volume or plasma volume. Limnol. Oceanogr. 12: 411–418.

Strickland, J. D. H. & T. R. Parsons, 1972. A practical handbook of seawater analysis. 2nd edn. Bull. Fish. Res. Bd. Can. 167: 1–310.

Throndsen, J., 1980. Bestemmelse av marine nakne flagellater. Identification of marine naked flagellates. Blyttia 38: 189–207.

Tollan, A., 1976. River runoff in Norway. In S. Skreslet, R. Leinebø, J. B. L. Matthews & E. Sakshaug (eds), Fresh Water on the Sea. The Association of Norwegian Oceanographers, Oslo: 11–13.

Tollan, A., 1995. Vesleofsen. Vær & Klima Nr. 4: 128–137.

Tomas, C. R. (ed.), 1993. Marine Phytoplankton. Academic Press, San Diego, 263 pp.

Tomas, C. R. (ed.), 1996. Identifying Marine Diatoms and Dinoflagellates. Academic Press, San Diego, 598 pp.

Ypma, J. E. & J. Throndsen, 1996. Seasonal dynamics of bacteria, autotrophic picoplankton and small nanoplankton in the inner Oslofjord and the Skagerrak in 1993. Sarsia 81: 57–66.

Hydrobiologia **363:** 179–189, 1998.

T. Tamminen & H. Kuosa (eds), Eutrophication in Planktonic Ecosystems: Food Web Dynamics and Elemental Cycling.

Vertical migration of autotrophic micro-organisms during a vernal bloom at the coastal Baltic Sea – coexistence through niche separation

Kalle Olli[1], Anna-Stiina Heiskanen[2] & Kaarina Lohikari[3]
[1] *Institute of Botany and Ecology, University of Tartu, Lai st. 40, Tartu EE-2400, Estonia*
[2] *Finnish Environment Institute, P.O. Box 140, FIN-00251, Helsinki, Finland*
[3] *Tvärminne Zoological Station, University of Helsinki, FIN-10900, Hanko, Finland*

Key words: Baltic Sea, *Mesodinium rubrum*, *Peridiniella catenata*, *Scrippsiella hangoei*, spring bloom, vertical migration

Abstract

Vertical migration of two dinoflagellate species (*Peridiniella catenata* and *Scrippsiella hangoei*) and a phototrophic ciliate (*Mesodinium rubrum*) were studied during the peak and decline of a vernal bloom at the SW coast of Finland. During the diel cycle, part of the populations of *P. catenata* and *M. rubrum* were observed in the deeper layers with elevated nutrient concentrations, while *S. hangoei* remained in the upper nutrient depleted mixed layer. Using a correspondence analysis the vertical distribution patterns of the species and chlorophyll *a* were examined over a temporal scale of hours and weeks. The vertical migration was reflected in much higher variability in the depth distribution of *P. catenata* and *M. rubrum* over a diel scale, compared to *S. hangoei*. The analysis revealed also significant differences in species specific depth distribution patterns over both time scales. It is discussed that the co-existence of the two dominant dinoflagellate species during the vernal bloom is due to niche separation through behavioural adaptations.

Introduction

In coastal temperate areas stratification of the water column during spring is commonly accompanied by outburst of phytoplankton development and consequent nutrient depletion from the illuminated layer. This results in an unfavourable situation for the primary producers where light and nutrients are spatially separated, and together with decaying turbulence this is commonly considered to trigger the mass sedimentation of vernal bloom diatoms (Heiskanen & Kononen, 1994; Olesen, 1993; Waite et al., 1992a, 1992b). Several autotrophic micro-organisms are capable of sustained directed swimming, covering a considerable vertical distance during a diel cycle (Kamykowski, 1995; Throndsen, 1973). A common consensus on the competitional advantage of the diel vertical migration (DVM) is that during the day time it enables the micro-organisms to exploit high light levels near the surface where low concentrations of inorganic nutrients would normally limit growth; during night the cells descend to deeper layers where nutrient demands are fulfilled (Cullen, 1985; Eppley et al., 1968; Lieberman & Shilo, 1994). Thus it results in optimised specific growth rate of the cells within the constraints of the ecological limits and can influence the production rates, both seasonally and annually (Heaney & Butterwick, 1985).

The pattern of vertical migration is a result of the physiological and behavioural adaptations of the micro-organisms interacting with a complex of environmental factors. A physiological prerequisite to the nutritional advantage of DVM seems to be the ability of dark nutrient assimilation (Cullen & Horrigan, 1981; Heaney & Eppley, 1981; Olsson & Granéli, 1991), which is an energy demanding process (Cullen, 1985). The triggers for directed swimming could be phototaxis or gravitaxis interacting with the inner rhythm of the cells (Cullen & Horrigan, 1981; Eppley et al., 1968) and modified by the physical environment (Kamykowski & McCollum, 1986). However, when a certain threshold or dose of photons have been absorbed, the upward migration ceases at optimal optical depth (Blasco, 1978; Passow, 1991), which is also dependent on the nutritional status of the cells (Anderson & Stolzenbach, 1985; Cullen & Horrigan, 1981; Heaney & Eppley, 1981). Distinct depth maxima have been noted in many species of marine autotrophic micro-organisms (Lindholm, 1995; Lindholm & Mörk, 1990; Owen et al., 1992; Viner, 1985). In these cases the cells evidently take advantage of the elevated nutrient levels, but to sustain viable populations for longer periods they have to return to the upper layers later. Another type of advantages is to reduce loss rates, either through sinking, predation or physical transport to regions unfavourable for growth (Cullen, 1985). Zooplankton is known to perform diel vertical migration and spend the light time in the deep layers to reduce grazing by visually feeding fish (Iwasa, 1982). Resistance to predation is clearly an adaptive advantage to enhance persistence of population (Fielder 1982). Nutrient starvation has been shown to suppress the vertical migration of dinoflagellates in tank experiments (Eppley et al., 1968). Numerous studies have dealt with the effects of temperature (Kamykowski & McCollum, 1986) the steepness of thermocline (Kamykowski, 1981; Kamykowski & Zentara, 1976), halocline (Olsson & Granéli, 1991; Rasmussen & Richardson, 1989; Tyler & Seliger, 1981) and turbulent mixing of the water column (Blasco, 1978) on the vertical migration of micro-organisms. In estuarine and tidal environments migratory behaviour enables the cells to be carried with subsurface currents to annual bloom areas (Tyler & Seliger, 1978, 1981) or avoid flushing losses from estuary (Crawford & Purdie, 1992). Small scale turbulence is proposed to have greater inhibitory effect on dinoflagellate motility compared to other algal classes (Kamykowski, 1995; Thomas & Gibson, 1990), and also *M. rubrum* has been shown to actively avoid turbulent water layers (Crawford & Purdie, 1992).

It is likely that the environmental triggers determining the vertical distribution patterns in nature are much more complex than the scarce experimental data can explain (Passow, 1991). Majority of the experiments with laboratory cultures have attempted to model the effect of single physical or chemical factors, and the interactions have remained largely uncovered (Heaney & Eppley, 1981), leading to difficulties in interpretation of field data.

In the present study we investigated the vertical profiles and migration of the vernal key species (*Peridiniella catenata* and *Scrippsiella hangoei*) and an autotrophic ciliate (*Mesodinium rubrum*) during two diel cycles of an annual spring bloom event in the SW coast of Finland, the northern Baltic Sea. These species were chosen on the basis of adequate abundance, previous knowledge of migratory behaviour (Heiskanen, 1995; Lindholm & Mörk, 1990; Passow, 1991) and the theoretical size considerations (Sommer, 1988). The focus is on the ability of the species to migrate down to nutrient rich layers during a vernal bloom when light and nutrient availability is vertically separated. The results indicate a considerable difference between the species to vary the vertical distribution as well as species specific vertical distribution patterns over the observed time scales.

Material and methods

The sampling was carried out at the Storgadden station, outer archipelago region, about 6 km off the SW coast of Finland (59°47′ N, 23°20′ E) near the entrance to the Gulf of Finland, the Baltic Sea. The salinity in the study area is generally 5–7 PSU. Salinity and temperature variability can be induced by upwellings, as well as by lateral surface water transport from the inner archipelago sea (Haapala, 1994; Niemi, 1975). The sampling site was approximately 50 m deep, relatively large basin surrounded by shallower areas and small rocky islets.

Two 24-hour sampling periods (first period: May 5–6 starting at 20.00; second period: May 27–28 starting at 8.00) were chosen, to coincide the peak and fall of the vernal phytoplankton bloom in 1994. During both sampling periods the station was visited seven times in 4 hour intervals. Upon arrival vertical profiles of temperature, salinity and density were measured (CTDplus 100). The under water light attenuation was measured with LiCor light meter. Once during both periods a mineral nutrient (NO_3-N, NH_4-N, PO_4-P) profile was taken at vertical intervals of 6 meters. Samples for CHL and mineral nutrients were taken to the shore in plastic bottles placed in a cool box. Nutrients were analysed immediately according to (Grasshoff et al., 1983), CHL samples were filtered on Whatman (GF/F) glass-fibre filters, sonicated and extracted in 94% ethanol for 24 h in total darkness. CHL concentration of the extract was measured with a spectrofluorometer (Shimadzu RF 5001), which was calibrated with pure chlorophyll *a* (Sigma).

Phytoplankton and chlorophyll *a* (CHL) was sampled with a 7 L Limnos water sampler with 3 m vertical intervals, from surface to 36 m depth. Sub-samples (300 ml) for phytoplankton were fixed immediately with Lugol solution. The phytoplankton samples were counted with the inverted microscope technique (Lund et al., 1958); at least 200 cells (or chains of *Peridiniella catenata*) were attempted to count depending on the density of the samples. Biomass was estimated from algorithms between cell volume and wet weight. To evaluate the variability caused by sampling and subsequent sub-sampling four replicate samples were taken from 6 m layer on two occasions during the second period, and four replicate sub-samples from one of the replicate samples.

Wind and radiation data were obtained from Längden stationary weather station, approximately 5 km SW from the sampling station. Current velocity was measured continuously by an Aanderaa RCM 4 current meter, moored at the depth of 20 m at the sampling station.

We analysed the depth profiles of the species and CHL by correspondence analysis (ADE-4 software; http://biomserv.univ-lyon1.fr). The data was arranged in a matrix with 13 columns (corresponding to the 13 discrete depths) and 28 rows consisting of 4 blocks (the three species and CHL), each consisting of 7 rows (corresponding to the sampling time). Correspondence analysis was used to project 7 profiles of each species and CHL with 13 depth variables to a few ordination axis which describe most of the variability in the data set. These projections preserve the original Euclidean distances and enables simultaneous visualisation of the differences of depth profiles on two dimensional space. The profiles were grouped according to species (4 groups) or sampling time (7 groups) and the statistical significance of the between-groups difference was tested with a random permutation test.

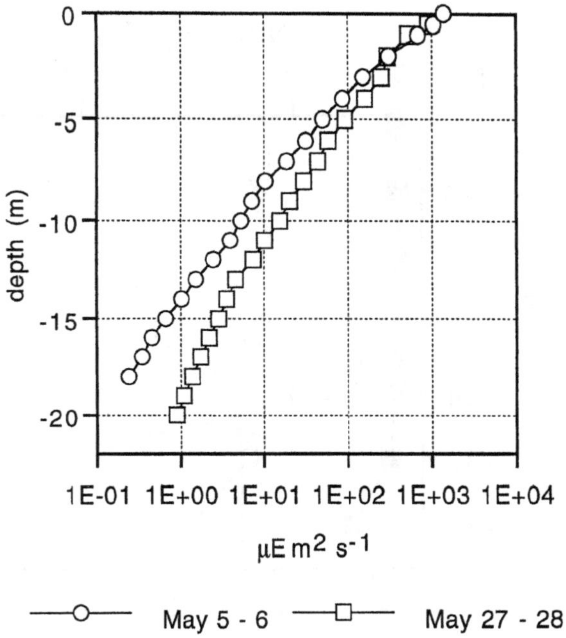

Figure 1. Under water light attenuation during the first (May 5–6) and second (May 27–28) period. Note the log scale of X-axis.

Results

Weather

During both periods a clear and calm weather prevailed. The maximum midday light intensities at the surface were up to 1400 μE m^{-2} s^{-1}. The 1% light level, equal to approximately 10 μE m^{-2} s^{-1}, during midday was measured at the depths of 7 m and 12 m during the first and second period, respectively (Figure 1). According to the Längden weather station the wind magnitude at the area varied between 2–9 m s^{-1}, blowing from N–NE and from N–NW during the first and second period, respectively. Due to the distance between the sampling site and the weather station, these measurements are likely to differ somewhat from the local wind at the sampling station, where on many occasions windless weather with no waves prevailed.

Hydrography

The surface water temperature on May 5–6 increased from 3.4 to 4.8 °C during the 24 hours, creating a shallow thermocline and weak water column stability (Figure 2A). The intermediate layer between 3 and 20 m had a temperature of 3 °C, and layer below that 1 °C. During May 27–28 a surface mixed layer down to 15 m with a temperature of 7.5 °C had developed, below that the temperature decreased gradually to 1 °C at 40–50 m depth (Figure 2B). The halocline had descended from 18 m on May 5–6 to 22 m on May 27–28, which was also reflected in the Brunt-Väisälä buoyancy term (N^2) (Figure 2A, B).

The current velocities at 20 m depth were low during the first period (0–2.5 m s^{-1}) and somewhat higher during the second period (1.5–5 cm s^{-1}). These can be considered as relatively low values, since velocities

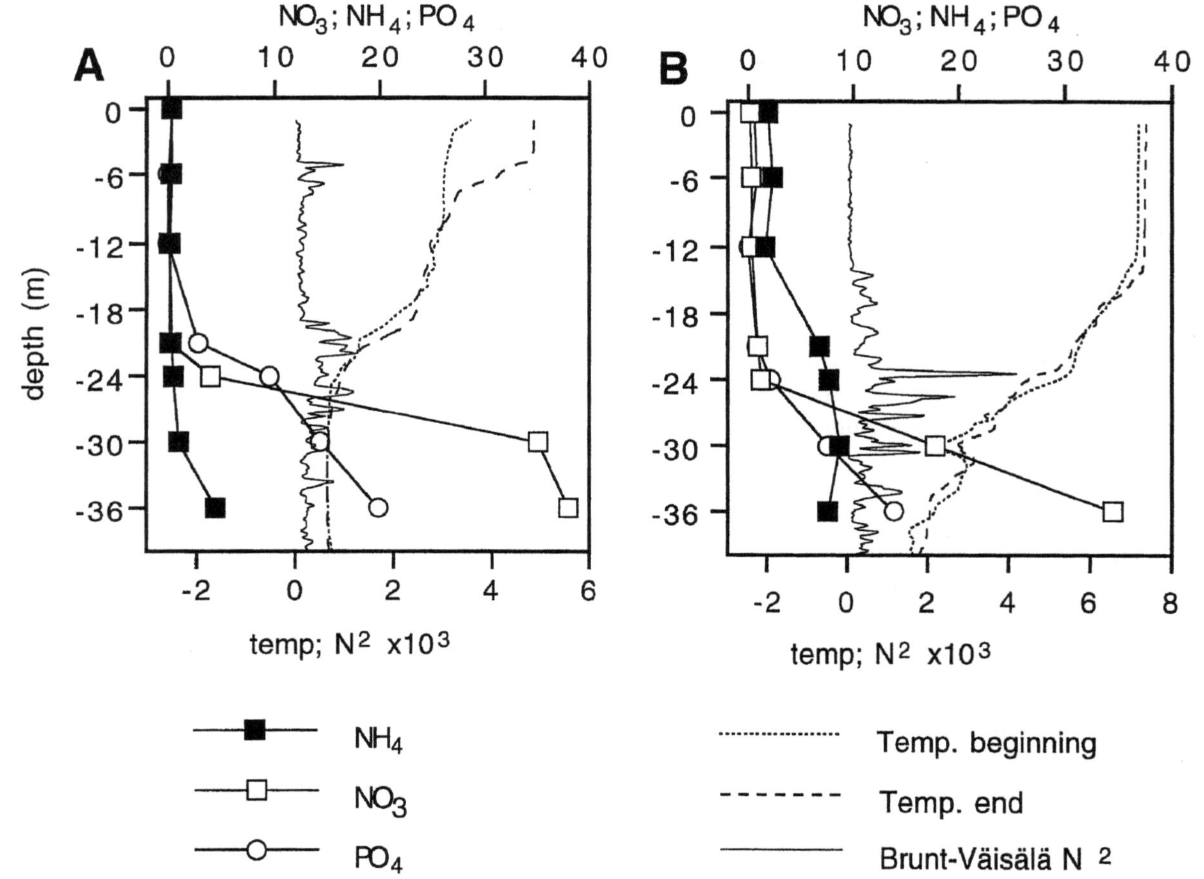

Figure 2. Profiles of mineral nutrient concentrations ($\mu g\ l^{-1}$; scale on the upper X-axis), Brunt-Väisälä buoyancy term (s^{-1}), and water temperature (°C) in the beginning and end of the first (A) and second (B) period (scale on the lower X-axis).

as high as 35 cm s^{-1} were measured during the other periods of the spring bloom.

On May 5–6 the mineral nutrients had been utilized from the upper 12 m. Elevated levels of phosphate were measured at 18 m depth and nitrate at 24 m (Figure 2A). On May 27–28 elevated levels of phosphate and nitrate were found at 30 m depth, however ammonium had increased in the surface layer (0–12 m) (Figure 2B).

Table 1. Variability (% coefficient of variation) of the standing stock of the species and CHL concentration

Species and CHL	I period	II period
Mesodinium rubrum	27	14
Peridiniella catenata	14	19
Scrippsiella hangoei	24	24
CHL	15	20

Phytoplankton abundance

In 1994, the phytoplankton vernal bloom at the western Gulf of Finland was dominated by dinoflagellates and diatoms. According to the data obtained from a weekly monitoring station (59°40.3′ N, 23°14.5′ E), the total phytoplankton biomass in the upper mixed layer (0–10 m) was 5.7 mg l^{-1} on May 3 and 3.2 mg l^{-1} on May 24 (Olli, unpublished data). On May 3, approximately half of the total biomass was formed by the dominant dinoflagellate species *Scrippsiella hangoei* and *Peridiniella catenata* (33% and 12%, respectively), while on May 24 their share was even larger (60% and 15% for *S. hangoei* and *P. catenata*, respectively), mainly due to the decline of the diatom populations.

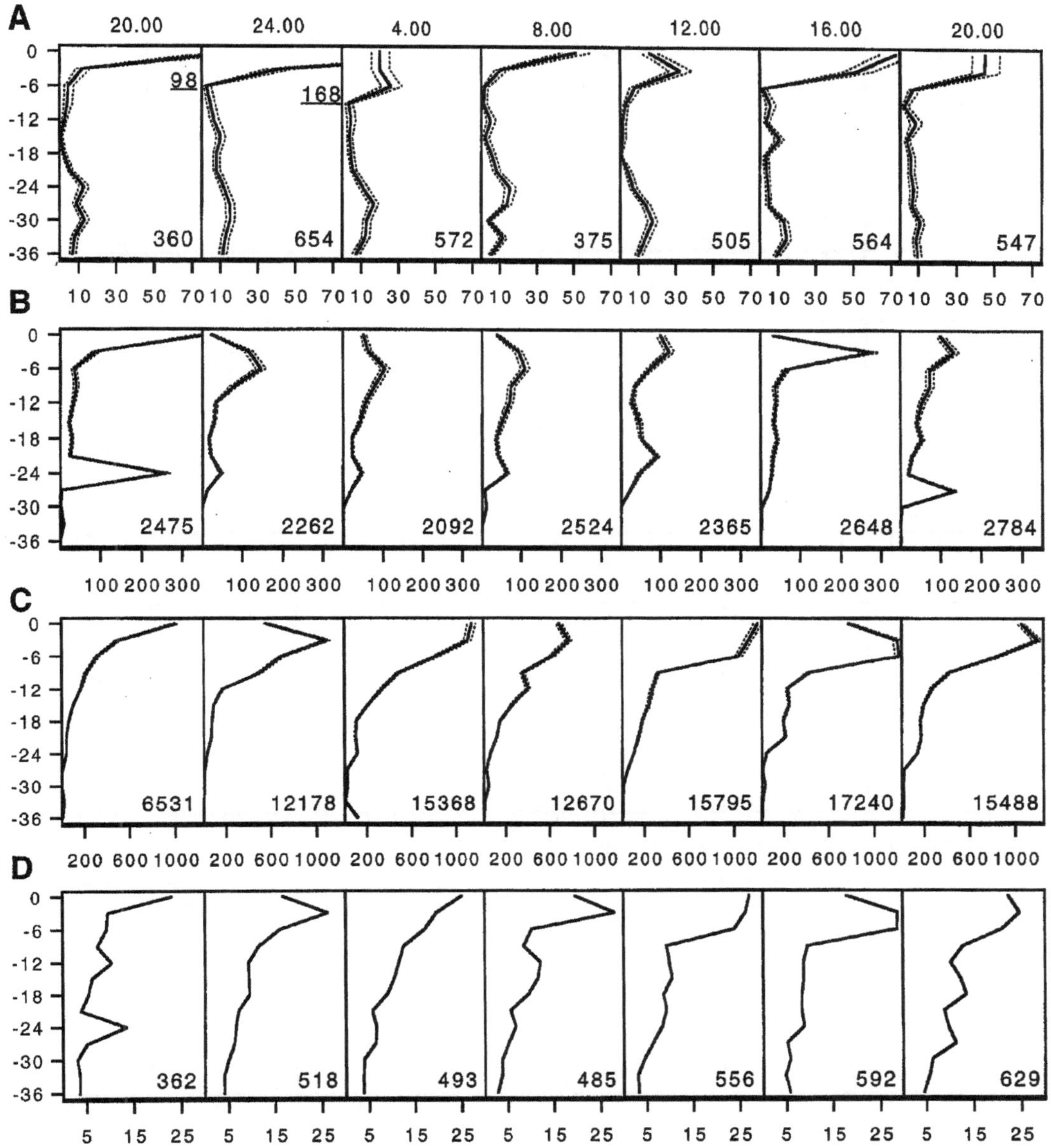

Figure 3. Vertical abundance (cells ml^{-1}) profiles of *Mesodinium rubrum* (A), *Peridiniella catenata* (B), *Scrippsiella hangoei* (C) and CHL concentration ($\mu g\ l^{-1}$) (D) during the first period. Respective sampling times are on A panel. Numbers in the lower right part of graphs depict the standing stock (0–36 m) of phytoplankton (cells $\times 10^6\ m^{-2}$) (A–C) and CHL (mg m^{-2}) (D). Numbers in the upper right part of panel A represent out of scale surface values. Dashed lines show the 95% confidence limits of cell counts.

The abundance of *Mesodinium rubrum* was much lower and the species did not contribute significantly to the total autotrophic biomass. According to subjective estimations these proportions were approximately the same at our study site. The decline of the bloom was reflected in considerable decrease (84%) of CHL concentration and the abundances of *Mesodinium rubrum* (36%), *Peridiniella catenata* (20%) and *Scrippsiella hangoei* (83%), compared to the first period. The variability of species abundances between replicate samples (estimated as coefficient of variation; $n=4$) was between 5–8% and somewhat higher in the case of *P. catenata* (15%). The variability in the abundance of *M. rubrum* between replicate sub-samples was 7%.

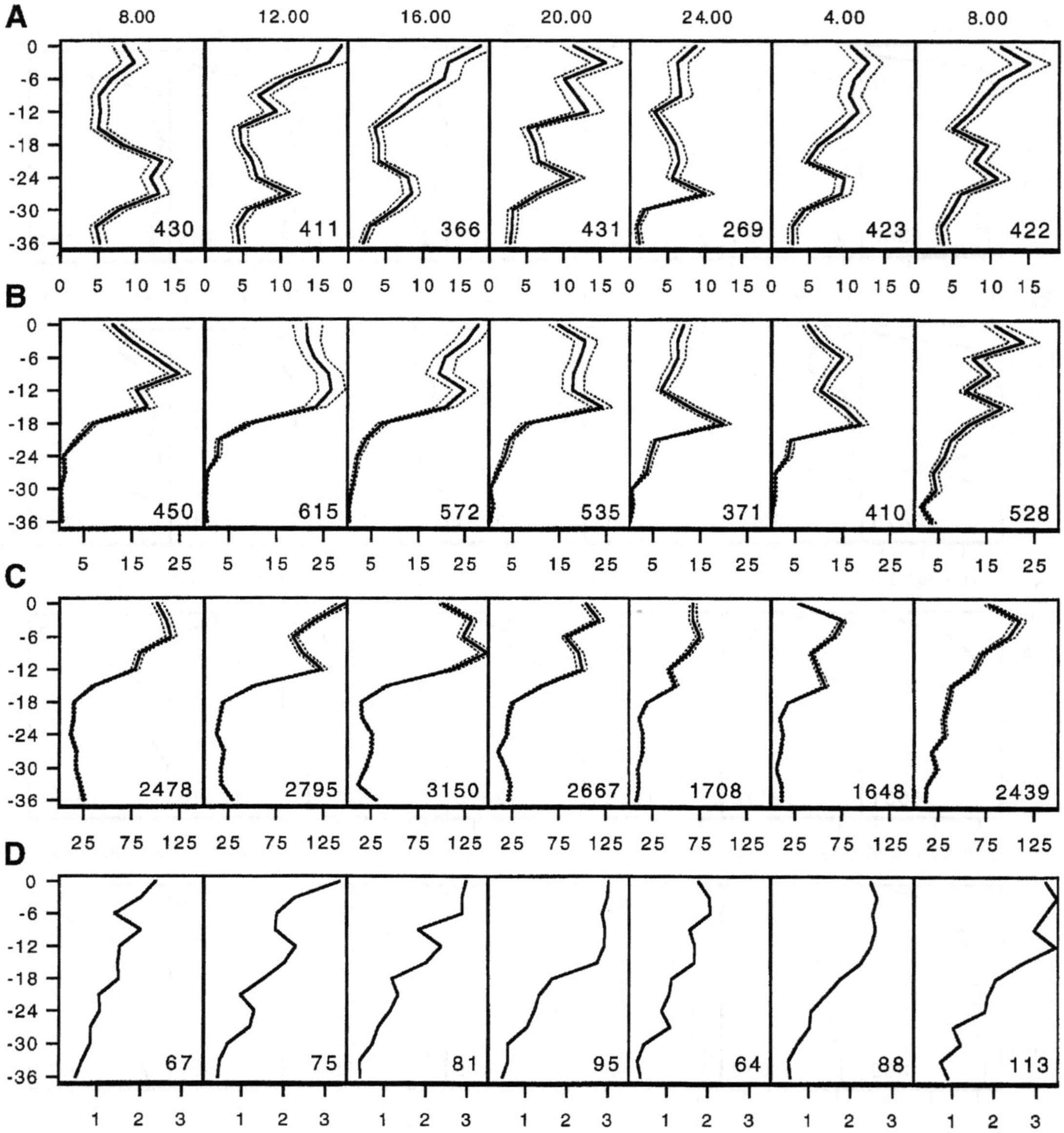

Figure 4. Vertical profiles of the species and CHL during the second period. Notations as in Figure 3.

The change in the standing stock of the species and CHL (0–36 m) is presented in Figures 3 and 4; the variability of the standing stock of species and CHL within both sampling periods is presented in Table 1.

Vertical distribution of CHL and phytoplankton

On May 5 the bulk of *Mesodinium rubrum* cells was concentrated to the upper 0–3 (6) m layer (30–170 cells ml^{-1}), with a less pronounced peak (5–15 cells ml^{-1}) at about 30 m depth (Figure 3A). Also on May 27 a vertical distribution with two depth maxima were observed, upper one (0–12 m) 8–18 cells ml^{-1}, and a lower one (25–33 m) 6–10 (14) cells ml^{-1} (Figure 4A). On May 5 the dinoflagellate *Peridiniella catenata* had a depth distribution characterised by 2 maxima, a pronounced surface peak (100–300 cells ml^{-1}) at 0–6 m and a smaller one at 21–27 m (Figure 3B). On May 27 most of the cells were concentrated above the 21 m depth (10–25 cells ml^{-1}), with peak abundances often

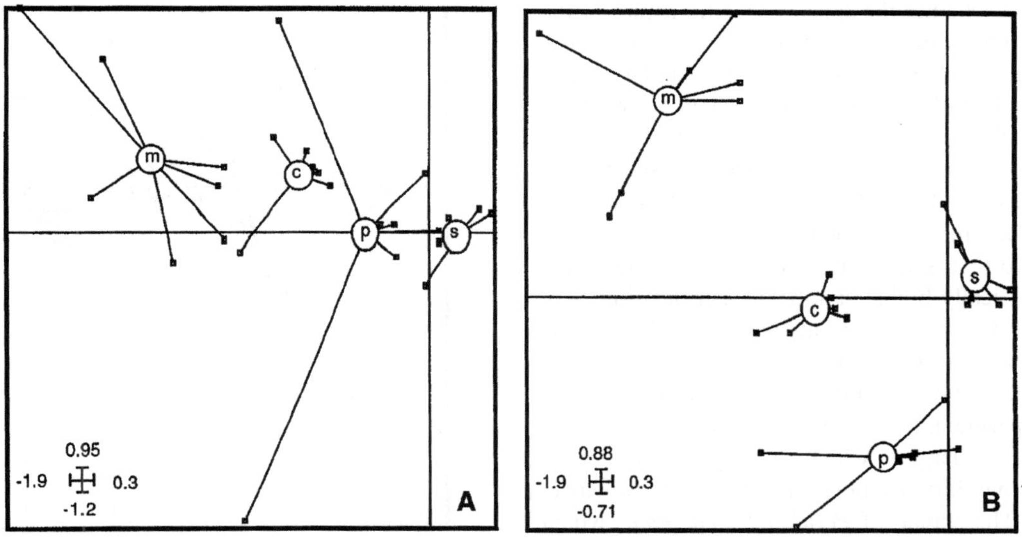

Figure 5. Projection of the vertical distributions of *Mesodinium rubrum* (m), *Peridiniella catenata* (p), *Scrippsiella hangoei* (s) and CHL (c) during the first period onto two-dimensional plane. Groups, defined by the species and CHL are connected to their respective centres of gravity. Axes are defined to maximize the variability between all data points. A – first (X) and second (Y) axis. Note the large spread of *P. catenata*. B – first (X) and third (Y) axis. Numbers in lower left corner delineate the range of the axes.

at 21 m depth (Figure 4B). *Scrippsiella hangoei* was concentrated to the upper 10 m layer (600–1200 cells ml^{-1}) during the first period, while below that depth the cell concentrations decreased to 0–200 cells ml^{-1} (Figure 3C). During the second period the cells were more evenly distributed in the surface water, concentrations of 50–100 cells ml^{-1} reaching down to 18 m depth, and less than 20 cells ml^{-1} below that (Figure 4C). The distribution of the principal phytoplankton species was also reflected in the CHL concentrations. On May 5 CHL concentration in the upper 0–3 m was an average 20 μg l^{-1} (Figure 3D). Below the 10 m depth a sharp decrease in the CHL values down to 5–7 μg l^{-1} was observed. The depth maximum (260 cells ml^{-1}, 24 m layer) of *Peridiniella catenata* is well reflected in the CHL profile on the first sampling occasion (16.00; 13.4 μg l^{-1}). On May 27 the surface values of CHL had dropped to 2–3 μg l^{-1}, decreasing almost constantly to 0.5 μg l^{-1} at 36 m depth (Figure 4D).

Correspondence analysis on the data set from the first period reveals that if the profiles are grouped according to the species (connected to their respective centres of gravity), there is very little overlap between the groups, indicating much more similarity of the depth profiles within the same species than between different species. A reasonably good separation of species groups is achieved with the first two ordination axes (Figure 5A). However, the second axis is highly influenced by the pronounced depth maxima of *Peridiniella catenata* on the first and last profile. The best separation of species groups is achieved on an ordination plane defined by the first and third axes (Figure 5B). This between-species difference was statistically tested by randomly exchanging the data points (depth profiles) between groups (corresponding to species and CHL) and comparing the new between-group variability (i.e. the spread of the centres of gravity) with the original one. Out of the 1000 permutations none resulted in a higher between-group inertia (i.e. with every random exchange the new groups were closer to each other compared to the original one), indicating highly significant differences in the vertical distribution patterns between the species. Symmetrically the groups were defined by the sampling time and the permutation test revealed no significant difference between groups. The same applies also to the data set from the second sampling period (27–28 May). The difference between groups defined by species was statistically highly significant, but not when defined by sampling time. Another effect which is easily seen from Figure 5 is the different degree of spread between the species. The depth profiles of *M. rubrum* and even more those of *P. catenata* are much more different compared to those of *S. hangoei* and CHL which are quite similar during the diel cycle.

To test whether the species specific nature of the depth profiles applies over a temporal scale of 3 weeks (with considerable change in species abundances and hydrography in between), we performed a correspondence analysis on the pooled data of both periods. Between-group (4 groups corresponding to species and CHL, each with 14 depth distribution profiles) analysis resulted in remarkably good separation of the species on the two first ordination axes (Figure 6), which was also statistically highly significant (permutation test). The spread of the data points within species reflects now a change in the depth distribution pattern in the time scale of hours (within period) and weeks (over the two periods). The spread in *Mesodinium rubrum* is clearly the largest, which actually is caused by a considerable shift in depth profile patterns between the periods. The spread of *Peridiniella catenata*, which was clearly the largest when the two periods were considered separately, is now comparable to that of *Scrippsiella hangoei* – characterised by relatively uniform depth distribution within periods. This is due to relatively larger shift of the depth distributions of *S. hangoei* compared to *P. catenata*. The spread of CHL is clearly the smallest, resulting from relatively uniform depth distributions within periods and smallest shift in between periods. Based on row weighted correspondence analysis (the second ordination axis agrees considerably well with the depth gradient), the general direction of the shift in the depth distributions between the two periods appeared to be directed towards greater depths, except in the case of *P. catenata*. The significance of the shift between two periods was tested with a between-group permutation test, the groups defined by the periods. Out of 10000 random permutations only 78 resulted in larger distance between the centres of gravity of the groups.

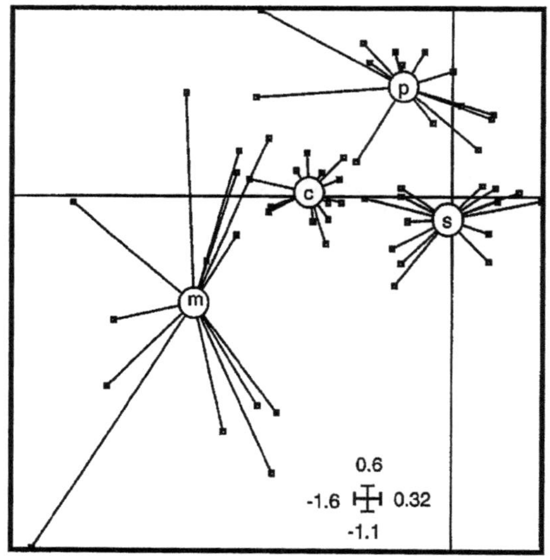

Figure 6. Projection of the pooled data from both periods onto between-group ordination plane. Axes are defined to maximize the variability between centres of gravity. Notations as in Figure 5.

Discussion

The peak and decline phases of the vernal bloom were dominated by dinoflagellates with a potential to perform vertical migration. During both periods the illuminated layer was nutrient depleted. The vertical profiles of the species studied revealed that over the diel cycle part of the populations of *Mesodinium rubrum* and *Peridiniella catenata* were able to reside deep enough to take advantage of the nutrients below the pycnocline, while *Scrippsiella hangoei* seemed to be concentrated to the upper nutrient depleted layer. The depth profiles of *M. rubrum* and especially that of *P. catenata* changed much more during the diel cycle compared to *S. hangoei* and CHL, which was probably caused by the migratory behaviour of the former. The depth profile patterns were significantly different between the species, over a temporal scale of hours as well as weeks. This indicates the importance of species specific behavioural adaptations in the vertical distribution. However, these conclusions could have been affected by artifacts like inaccuracy in sampling, turbulent mixing and horizontal advection of water mass, which will be discussed below.

Sampling and counting errors

The populations of the micro-organisms were assumed to have a random Poisson distribution. The relative precision of the estimated abundance is related to the number of cells counted. In Figures 3 and 4 the estimated mean abundances of the organisms are shown with 95% confidence limits of the direct counts. Additional variation is introduced with the procedure of sampling and sub-sampling (Vernick, 1972). However, the replicability of the counts from parallel samples and sub-samples was relatively good, thus giving a reliable estimate of the population abundance. Yet another aspect of the sampling error is related to the vertical resolution of the discrete samples. The accumulation of autotrophic micro-organisms to depth maxima could occur in relatively thin layers (Lindholm, 1992), making it difficult to uncover the real maxima with routine sampling from too few discrete depths. *Mesodinium rubrum* is known to accumulate on very thin layers (Cloern et al., 1994; Crawford, 1989; Lindholm & Mörk, 1990; Owen et al., 1992). A very pronounced surface accumulation to the thermally stratified surface water was unveiled during the first period in our study (Figure 3A). It can not be ruled out that the high variability of the standing stock of *M. rubrum* during the first period was caused by missing some thin accumulation layers. This problem can be avoided by using *in situ* continuous fluorescence profile measurements (Cloern et al., 1994). The interpretation, however, requires highly monospecific blooms and would have been less valuable in our case, where CHL profiles are clearly a synthesis of the abundance of several species. However, the accumulation layers are probably thinner in strongly stratified environments like wind-sheltered small lakes, saline lakes and sheltered and coastal lagoons (Lindholm, 1992), compared to open coastal areas.

Another likely source of error is turbulent mixing and horizontal patchiness. It is well known that Langmuir rotation can cause horizontal patchiness of micro-swimmers (Reynolds, 1984; Smayda, 1970). Planktonic micro-organisms at the study area are known to have patchy mesoscale horizontal distribution in summer (Kuosa, 1988), yet in the scale of 100–200 m errors caused by sample preparation seem to be more crucial (Kuuppo-Leinikki, 1993). During our study period the horizontal mesoscale variability of CHL surface concentrations (1.5 m depth), measured with *in situ* flow-through fluorometer (Turner design) was approximately 20% (Jukka Seppälä, personal communication). The current velocities at 20 m depth were low and during the course of four hours could cause a water mass transport of less than 300 m and 300–600 m during the first and second period, respectively. Yet, as the physical forcing acts on motile cells and prevents them reaching optimal depth, the actual vertical distribution of the population is a combination of water movements counteracted by the active response of cells to fulfill their physiological requirements (Kuosa, 1988; Sommer, 1985). If the physical forcing overwhelms the ability of the cells to control their vertical position, we would not expect to see significantly different depth distribution patterns between species, which is the case in our study and applies also to time scale of three weeks. The physical mixing in open coastal areas is rather a rule than exception, which might be the reason why classical theoretical day time surface accumulation and night time dispersal or downwards migration is rarely encountered in field studies. Yet it is also likely that the actual triggers and physiological requirements of the cells are more complex than the classical migration theory explains. In our study a depth accumulation of a small proportion of *Mesodinium rubrum* was found during the first period. The ability for diel vertical migration is widely known in *M. rubrum* (Crawford, 1989; Passow, 1991; Williams, 1996), but it is also known that at low temperatures the species can tolerate darkness for several days or weeks (Lindholm, 1985). Vertical migration studies in the Baltic Sea reveal that part of the population remains in deep layers and is not migrating, at least in the diel scale (Lindholm & Mörk, 1990; Passow, 1991). However, the ecology of *M. rubrum* is quite different from the two dinoflagellates studied. In the study area *M. rubrum* does not form annual blooms, but is still quite abundant throughout the summer. In our study the relative decrease of the abundance was lowest in *M. rubrum*. However, it revealed the greatest change of depth distribution between the two periods. The depth distribution was much more uniform on the second period compared to the sharp surface peaks on the first period. Relatively uniform depth distribution (0–30 m) of *M. rubrum* in summer has been also noticed from the Gulf of Riga, the Baltic Sea (Olli, unpublished data).

The vertical migration of the chain forming dinoflagellate *Peridiniella catenata* is also known from previous studies in the Baltic Sea (Heiskanen, 1995; Passow, 1991). In our study *P. catenata* exhibited pronounced depth maxima during the first period and also accumulated in nutrient rich layers below the pycnocline during the second period. It is possible that chain formation has been selected in the evolution as adaptation to migration, since according to hydrodynamical considerations increased cell number in chain increases the swimming speed (Fraga et al., 1989). This would enable long duration of blooms of *P. catenata* based on deep nutrient supply. It is in contrast to the behaviour of *Scrippsiella hangoei*, which in our study did not show migrational behaviour and remained in the upper nutrient depleted layer. In our study area *S. hangoei* is known to form bloom densities in thin layers under the ice in early spring (Larsen et al., 1995). The nutrient supply of *S. hangoei* in the upper water layer during the spring bloom is not clear. High affinity to low nutrient concentrations could explain the competitional success of this species, although no data is available. Dinoflagellate cells are known to accumulate and store excess nutrients which are used when ambient nutrients decrease (Bhovichitra & Swift, 1977; Chapman & Pfiester, 1995; Dortch et al., 1984). Mixotrophy can be an additional way of nutrition, common in many dinoflagellates. Although it has not been proved with *S. hangoei*, it has been noted that pure cultures of the species get axenic over time (Guy Hällfors, personal communication), suggesting that the species can feed on bacteria. If this is true it might be that *S. hangoei* does not need to migrate to fulfill its nutrient requirements.

It is hypothesized here that the co-existence of the two dominant dinoflagellate species during the vernal bloom is due to niche separation, a hypothesis which is supported by the behavioural difference found during this study. Sommer (1982, 1985) has interpreted variations in vertical migration as differential light requirements leading to niche separation and coexistence of closely related cryptomonads in Lake Constance. However, it is necessary to extend this to nutrition strategies, as light preferences are closely related to nutritional status. It has been suggested that vertically migrating dinoflagellates (Fraga et al., 1992) and *Mesodinium rubrum* (Lindholm & Mörk, 1990) can act as nutrient pumps to transport nutrients from deep layers to the euphotic zone, which through regeneration get available to non-migrating species (Prego, 1992). In our study regenerated nitrogen was accumulating during the second period in the surface layer, however, the phosphate depleted layer was even deeper compared to the first period. Generally the vernal bloom phytoplankton in the area does not discriminate between ammonium and nitrate (Tamminen, 1995), although species specific differences can exist. If deep phosphorus taken up by vertically migrating *M. rubrum* and *P. catenata* could be channelled to *S. hangoei* it implicates that the vernal bloom, dominated by dinoflagellates in the area, is not terminated by nutrient depletion after the formation of thermal stratification, as is generally assumed in diatom dominated blooms, but rather due to internal species specific life cycle events. The *S. hangoei* bloom in the area terminates in late May by massive cyst formation followed by rapid sedimentation (Heiskanen, 1993); the fate of *P. catenata* bloom is unknown as cyst formation and large scale sedimentation has not been observed. Further lab experiments on the nutrient uptake kinetics and mixotrophy of *S. hangoei* could clarify the niche separation mechanisms. Unfortunately lab experiments with *M. rubrum* and *P. catenata* are hampered by difficulties in culturing (Crawford, 1989; Lindholm, 1985, Guy Hällfors, personal communication).

Acknowledgements

This study is a contribution to the project PELAG. We are indebted to the staff of the Tvärminne Zoological Station, University of Helsinki. The study was made possible through a personal grant (K. Olli) from Nordic Academy of Advanced Studies (NorFa).

References

Anderson, D. M. & K. D. Stolzenbach, 1985. Selective retention of two dinoflagellates in a well-mixed estuarine embayment: the importance of diel vertical migration and surface avoidance. Mar. Ecol. Prog. Ser. 25: 39–50.

Bhovichitra, M. & E. Swift, 1977. Light and dark uptake of nitrate and ammonium by large oceanic dinoflagellates: *Pyrocystis noctiluca*, *Pyrocystis fusiformis*, and *Dissodinum lunula*. Limnol. Oceanogr. 22: 73–83.

Blasco, D., 1978. Observations on the diel migration of marine dinoflagellates off the Baja California coast. Mar. Biol. 46: 41–47.

Chapman, A. D. & L. A. Pfiester, 1995. The effects of temperature, irradiance, and nitrogen on the encystment and growth of the freshwater dinoflagellates *Peridinium cinctum* and *P. willei* in culture (Dinophyceae). J. Phycol. 31: 355–359.

Cloern, J. E., B. E. Cole & S. W. Hager, 1994. Notes of a *Mesodinium rubrum* red tide in San Fransisco Bay (California, USA). J. Plankton Res. 16: 1269–1276.

Crawford, D. W., 1989. *Mesodinium rubrum*: the phytoplankter that wasn't. Mar. Ecol. Prog. Ser. 58: 161–174.

Crawford, D. W. & D. A. Purdie, 1992. Evidence for avoidance of flushing from an estuary by planktonic, phototrophic ciliate. Mar. Ecol. Prog. Ser. 79: 256–265.

Cullen, J. J., 1985. Diel vertical migration by dinoflagellates: roles of carbohydrate metabolism and behavioural flexibility. In M. A. Rankin (ed.), Migration: Mechanisms and Adaptive Significance. Contributions in Marine Science. Austin: 135–152.

Cullen, J. J. & S. G. Horrigan, 1981. Effects of nitrate on the diurnal vertical migration, carbon to nitrogen ratio, and photosynthetic capacity of the dinoflagellate *Gymnodinium splendens*. Mar. Biol. 62: 81–89.

Dortch, Q., J. R. Clayton, S. S. Thoreson, S. L. Bressler & S. I. Ahmed, 1984. Species differences in accumulation of nitrogen pools in phytoplankton. Mar. Biol. 81: 237–250.

Eppley, R. W., O. Holm-Hansen & J. D. H. Strickland, 1968. Some observations on the vertical migration of dinoflagellates. J. Phycol. 4: 333–340.

Fielder, D. C., 1982. Zooplankton avoidance and reduced grazing responses to *Gymnodinium splendens* (Dinophyceae). Limnol. Oceanogr. 27: 961–965.

Fraga, F., F. F. Pérez, F. G. Figueiras & A. F. Ríos, 1992. Stoichometric variations of N, P, C and O_2 during a *Gymnodinium catenatum* red tide and their interpretation. Mar. Ecol. Prog. Ser. 87: 123–134.

Fraga, S., S. M. Gallager & D. M. Anderson, 1989. Chain-forming dinoflagellates: an adaptation to red tides. In T. Okaichi, D. M. Anderson & T. Nemoto (eds), Red Tides: Biology, Environmental Science, and Toxicology. Elsevier. New York: 281–284.

Grasshoff, K., M. Ehrhardt & K. Kremling, 1983. Methods of Seawater Analysis. Verlag Chemie. Weinheim. 419 pp.

Haapala, J., 1994. Uppwelling and its influence on nutrient concentrations in the coastal area of the Hanko peninsula, entrance of the Gulf of Finland. Estuar. coast. Shelf Sci. 38: 507–521.

Heaney, S. I. & C. Butterwick, 1985. Comparative mechanisms of algal movement in relation to phytoplankton production. In M. A. Rankin (ed.), Migration: Mechanisms and Adaptive Significance. Contributions in Marine Science. Austin: 115–134.

Heaney, S. I. & R. W. Eppley, 1981. Light, temperature and nitrogen as interacting factors affecting diel vertical migrations of dinoflagellates in culture. J. Plankton Res. 3: 331–344.

Heiskanen, A.-S., 1993. Mass encystment and sinking of dinoflagellates during a spring bloom. Mar. Biol. 116: 161–167.

Heiskanen, A.-S., 1995. Contamination of sediment trap fluxes by vertically migrating phototrophic micro-organisms in the coastal Baltic Sea. Mar. Ecol. Prog. Ser. 122: 45–58.

Heiskanen, A.-S. & K. Kononen, 1994. Sedimentation of vernal and late summer phytoplankton communities in the coastal Baltic Sea. Arch. Hydrobiol. 131: 175–198.

Iwasa, Y., 1982. Vertical migration of zooplankton: a game between predatory and prey. Am. Nat. 120: 171–180.

Kamykowski, D., 1981. Laboratory experiments on the diurnal vertical migration of marine dinoflagellates through temperature gradients. Mar. Biol. 62: 57–64.

Kamykowski, D., 1995. Trajectories of autotrophic marine dinoflagellates. J. Phycol. 31: 200–208.

Kamykowski, D. & S. A. McCollum, 1986. The temperature acclimatized swimming speed of selected marine dinoflagellates. J. Plankton Res. 8: 275–287.

Kamykowski, D. & S.-J. Zentara, 1976. The diurnal vertical migration of motile phytoplankton through temperature gradients. Limnol. Oceanogr. 22: 148–151.

Kuosa, H., 1988. Horizontal mesoscale distribution of phytoplankton in the Tvärminne sea area, southern Finland. Hydrobiologia 161: 69–73.

Kuuppo-Leinikki, P., 1993. Horizontal distribution of photo- and heterotrophic micro-organisms on the coastal area of the northern Baltic Sea – a case study. J. Plankton Res. 15: 27–35.

Larsen, J., H. Kuosa, J. Ikävalko, K. Kivi & S. Hällfors, 1995. A redescription of *Scrippsiella hangoei* (Schiller) Comb. nov. – a 'red tide' forming dinoflagellate from the northern Baltic. Phycologia 34: 135–144.

Lieberman, O. S. & M. Shilo, 1994. The physiological ecology of a freshwater dinoflagellate bloom population: vertical migration, nitrogen limitation, and nutrient uptake kinetics. J. Phycol. 30: 964–971.

Lindholm, T., 1985. *Mesodinium rubrum* – a unique photosynthetic ciliate. Adv. aquat. Microbiol. 3: 1–48.

Lindholm, T., 1992. Ecological role of depth maxima of phytoplankton. Arch. Hydrobiol. Beih. Ergebn. Limnol. 35: 33–45.

Lindholm, T., 1995. Green water caused by *Eutreptiella gymnastica* in a stratified Baltic Sea inlet. In P. Lassus, G. Arzul, E. Erard, P. Gentien & C. Marcaillou (eds), Harmful Marine Algal Blooms. Proc., Sixth Int. Conference on Toxic Marine Phytoplankton, 18–22.Oct. 1993. Intercept Ltd: 181–186.

Lindholm, T. & A.-C. Mörk, 1990. Depth maxima of *Mesodinium rubrum* (Lohmann) Hamburger & Buddenbrock – examples from a stratified Baltic sea inlet. Sarsia 75: 53–64.

Lund, J. W. G., C. Kipling & E. D. Le Gren, 1958. The inverted microscope method of estimating algal numbers and the statistical basis of estimations by counting. Hydrobiologia. 11: 143–170.

Niemi, Å., 1975. Ecology of phytoplankton in the Tvärminne area SW coast of Finland. II. Primary production and environmental conditions in the archipelago zone and sea zone. Acta bot. fenn. 105: 1–73.

Olesen, M., 1993. The fate of an early diatom spring bloom in the Kattegat. Ophelia 37: 51–66.

Olsson, P. & E. Granéli, 1991. Observations on diurnal vertical migration and phased cell division for three coexisting marine dinoflagellates. J. Plankton Res. 13: 1313–1324.

Owen, R. W., S. F. Gianesella-Galvão & M. B. B. Kutner, 1992. Discrete, subsurface layers of the autotrophic ciliate *Mesodinium rubrum* off Brazil. J. Plankton Res. 14: 97–105.

Passow, U., 1991. Vertical migration of *Gonyaulax catenata* and *Mesodinium rubrum*. Mar. Biol. 110: 455–463.

Prego, R., 1992. Flows and budgets of nutrient salts and organic carbon in relation to a red tide in the Ria of Vigo (NW Spain). Mar. Ecol. Prog. Ser. 79: 289–302.

Rasmussen, J. & K. Richardson, 1989. Response of *Gonyaulax tamarensis* to the presence of a pycnocline in an artificial water column. J. Plankton Res. 11: 747–762.

Reynolds, C. S., 1984. The Ecology of Freshwater Phytoplankton. Cambridge University Press. Cambridge, 385 pp.

Smayda, T. J., 1970. The suspension and sinking of phytoplankton in the sea. Oceanogr. mar. Biol. ann. Rev. 8: 353–414.

Sommer, U., 1982. Vertical niche separation between two closely related planktonic flagellate species (*Rhodomonas lens* and *Rhodomonas minuta* v. *nannoplanctica*). J. Plankton Res. 4: 137–142.

Sommer, U., 1985. Differential migration of Cryptophyceae in lake Constance. In M. A. Rankin (ed.), Migration: Mechanisms and Adaptive Significance. Contributions in Marine Science. Austin: 166–175.

Sommer, U., 1988. Some size relationships in phytoflagellate motility. Hydrobiologia 161: 125–131.

Tamminen, T., 1995. Nitrate and ammonium depletion rates and preferences during a Baltic spring bloom. Mar. Ecol. Prog. Ser. 120: 123–133.

Thomas, W. H. & C. H. Gibson, 1990. Quantified small-scale turbulence inhibits a red-tide dinoflagellate, *Gonyaulax polyedra* Stein. Deep Sea Res. 37: 1583–1593.

Throndsen, J., 1973. Motility in some marine nanoplankton flagellates. Norw. J. Zool. 21: 193–200.

Tyler, M. A. & H. H. Seliger, 1978. Annual subsurface transport of a red tide dinoflagellate to its bloom area: water circulation pattern and organism distributions in the Chesapeake Bay. Limnol. Oceanogr. 23: 227–246.

Tyler, M. A. & H. H. Seliger, 1981. Selection for a red tide organism: Physiological responses to the physical environment. Limnol. Oceanogr. 26: 310–324.

Vernik, E. L., 1972. The statistics of subsampling. Limnol. Oceanogr. 16: [1971] 811–817.

Viner, A. B., 1985. Thermal stability and phytoplankton distribution. Hydrobiologia 125: 47–69.

Waite, A., P. K. Bienfang & P. J. Harrison, 1992a. Spring bloom sedimentation in a subarctic ecosystem. I. Nutrient sensitivity. Mar. Biol. 114: 119–129.

Waite, A., P. K. Bienfang & P. J. Harrison, 1992b. Spring bloom sedimentation in a subarctic ecosystem. II. Succession and sedimentation. Mar. Biol. 114: 131–138.

Williams, J. A., 1996. Blooms of *Mesodinium rubrum* in Southampton Water – do they shape mesozooplankton distribution? J. Plankton Res. 18: 1685–1697.

Hydrobiologia **363**: 191–205, 1998.
T. Tamminen & H. Kuosa (eds), Eutrophication in Planktonic Ecosystems: Food Web Dynamics and Elemental Cycling.

Group-specific phytoplankton biomass/dissolved carbohydrate relationships in the Gulf of Trieste (Northern Adriatic)

Senka Terzić[1], Marijan Ahel[1], Gustave Cauwet[2] & Alenka Malej[3]
[1] *Center for Marine Research Zagreb, Ruder Bošković Institute, 10000 Zagreb, Croatia*
[2] *Observatoire Océanologique, Laboratoire Arago, Banyuls sur Mer, France*
[3] *National Institute of Biology, Marine Station Piran, Piran, Slovenia*

Key words: carbohydrates, dissolved organic carbon, photosynthetic pigments, phytoplankton, coastal waters, northern Adriatic

Abstract

Distribution of carbohydrates (CHO) and photosynthetic pigments were studied in the Gulf of Trieste, northern Adriatic Sea, during the period of summer stratification with a special emphasis on determining the impact of the taxonomic composition and concentration of phytoplankton biomass on the carbohydrate levels in the water column. Dissolved total carbohydrates (DTCHO), dissolved monosaccharides (DMCHO) as well as particulate carbohydrates (PTCHO) were determined using the colorimetric MBTH-method, while pigment biomarkers of the phytoplankton biomass were determined by reversed-phase HPLC. Concentrations of the total CHO (dissolved + particulate) varied in a wide range from 173 μg C l^{-1} to 1552 μg C l^{-1}. The percentage of PTCHO in the total CHO concentration was relatively low (4–25%), indicating that the main pool of CHO was in the dissolved phase. The contribution of DTCHO to the total dissolved organic carbon (DOC) in late summer was highly variable (10–65%) with an average value of 20 ± 14%, while in early summer this percentage was somewhat lower and less variable (range 11–23%; average 17 ± 3%). Analyses of biomarker pigments revealed a very high diversity and a rather heterogenous vertical and spatial distribution of the phytoplankton biomass during the period of summer stratification. In September 1994, the predominant taxonomic groups of phytoplankton were prymnesiophytes, diatoms, silicoflagellates, cyanobacteria and, especially in the bottom layer, dinoflagellates. A relatively good correlation ($r^2 = 0.51$) found between DTCHO and chl *a* suggested that DTCHO were mainly of phytoplankton origin. Furthermore, a concomitant increase of DTCHO with peridinin and fucoxanthin indicated that dinoflagellates and diatoms had a decisive impact on CHO levels in the water column. By contrast, early summer phytoplankton (June), which was dominated by prymnesiophytes, exhibited a comparatively lower impact on the CHO distribution.

Introduction

Phytoplankton is a major source of organic carbon in the marine environment and there have been numerous laboratory and field studies trying to link concentration and biochemical composition of DOC with composition and concentration of phytobiomass (e.g. Haug and Myklestad, 1976; Ittekkot et al., 1981; Barlow, 1982; Eberlein et al., 1983; Marlowe et al., 1989; Fernandez et al., 1992; Cauwet et al., 1997). It was shown that a substantial fraction of the organic carbon synthesised by phytoplankton is released as high molecular weight dissolved organic carbon (Fogg, 1966). Some microalgae, especially diatoms, produce large amounts of extracellular carbohydrates, mainly in the form of polysaccharides (Myklestad, 1974; Decho, 1990; Myklestad, 1989).

Polysaccharides are involved in the processes of particle interbridging and are considered to play an important role in the formation of organic aggregates in the water column. Very large mucous aggregates (mucilage) which covered several thousand square kilometers of the northern Adriatic in summer 1988, 1989 and 1991 (Stachowitsch et al., 1990; Degobis et al., 1995) were composed mainly of highly interbridged multilayered polymeric carbohydrate or carbohydrate-like material. It is generally accepted that the main reason for the periodical hypertrophic production of mucilage is the accumulation of carbohydrate-like organic material, mainly of the diatom origin (Degobbis et al., 1995). However, the exact conditions under which this process occurs and the mechanisms which are involved are still poorly understood. Faganeli et al. (1995) studied seasonal variation of particulate carbohydrates in the Gulf of Trieste and found a good correlation of PTCHO with phytoplankton biomass. Generally, PTCHO concentration was found to be lower than 100 $\mu g\,l^{-1}$, except for the period characterised by macroaggregate formation. Usual contribution of PTCHO to the total POC in the period of density stratification was estimated at 6% (Posedel & Faganeli, 1991), while macroaggregates contained 39.6% of the carbohydrate-derived organic carbon. Obviously extracellular (dissolved) CHO should play a significant role in the formation of gelatinous aggregates in the northern Adriatic (Deggobis et al., 1995). However, a very few data on their concentration levels and spatial distribution in the water column are available. The concentration of the total CHO in the northern Adriatic throughout 1992 revealed rather low levels (<150–500 $\mu g\,l^{-1}$ expressed as glucose) and showed a weak but statistically significant correlation with the number of diatom cells in the upper 10 m of the water column (Ahel et al., 1995). The observed correlation is in a good agreement with previous studies conducted in other areas which proved that diatom blooms could be an important source of CHO, especially of polysaccharides, in the marine environment (Ittekkot et al., 1981, 1982; Barlow, 1982).

The composition of summer phytoplankton in the northern Adriatic, however, is very complex with nano- and picoplankton contributing an average of 70% to the total biomass (Revelante and Gilmartin, 1992), while only smaller diatom blooms could be triggered by some short-living freshwater pulses or regeneration processes at the bottom (Malej et al., 1995). Consequently, it is very important to examine the correlation of individual groups characteristic for summer phytoplankton and carbohydrate levels in the northern Adriatic. This data gap is one of the important missing links for the proper interpretation of the mucilage phenomenon.

The aim of this paper was to investigate the relationship between photosynthetic pigments as biomarkers of group-specific phytoplankton biomass (see Millie et al., 1993 for review and references) and carbohydrates in the northernmost protrusion of the northern Adriatic (Gulf of Trieste) during the period of density stratification which is characteristic for the occurrence of mucilage. It should be pointed out that there was no mucilage appearance in the Gulf of Trieste during the two subsequent summer periods (1994 and 1995) covered by the present study. Nevertheless, the study describes a small scale spatial and temporal distribution patterns of CHO as a key prerequisite for an early recognition of anomalous accumulation of carbohydrates in the region and for a better understanding of the conditions leading to mucilage formation.

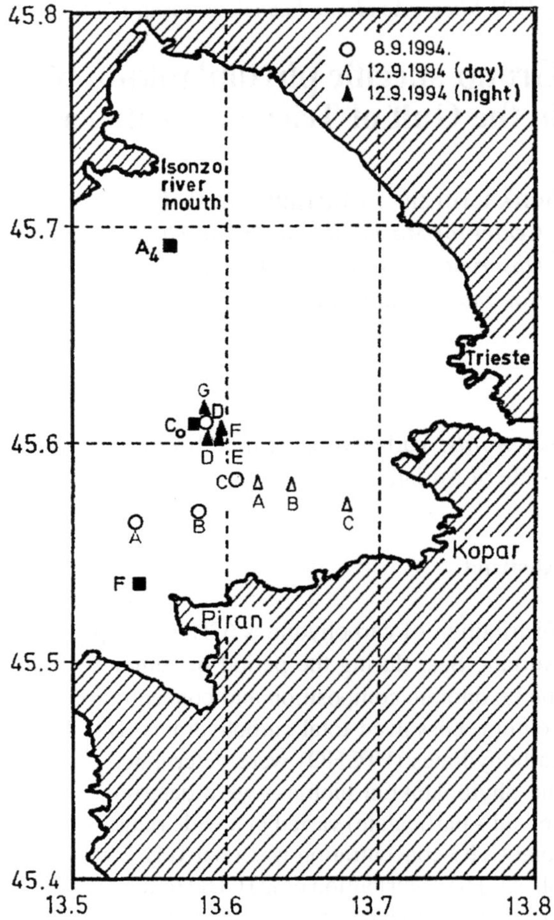

Figure 1. Map of the sampling area with sampling stations in September 1994 and June 1995.

Study area

The Gulf of Trieste is a semi-enclosed shallow basin situated in the northernmost part of the Adriatic Sea (see Figure 1). The surface of the Gulf is approximately 600 km^2, volume 9.5 km^3 and the maximal depth less than 30 m. The structure of the water column has a pronounced seasonal character with temperature ranging from 6 °C in winter to 26 °C in summer periods, and salinity between 31.0 PSU and 38.5 PSU. In spring and autumn, stability of the water column is mainly regulated by increased river water inputs, and in summer by a pronounced thermal stratification (Malej et al., 1995). On the contrary, winter period is generally characterised by a well mixed water column. Concentration of nutrients, especially nitrate, is strongly dependent on the seasonal fluctuations of the Isonzo and Timavo river discharges. The highest nutrient and chlorophyll *a* values in the surface layer are generally found during spring and autumn and are concomitant with the increase of the river discharge (Fonda-Umani et al., 1992). Due to the reduced fresh water inputs and stratification of the water column, summer periods are generally characterised by low nutrient and chlorophyll *a* levels in the surface layer. On the contrary, intensive regeneration processes and very reduced water mixing cause the nutrient enrichment and increase of chlorophyll *a* in the bottom layer during summer (Malej et al., 1995; Terzić, 1996).

Materials and methods

Sampling and hydrographic conditions

Two different sampling strategies were applied to determine short-term temporal and spatial variability of photosynthetic pigments and carbohydrates in the Gulf of Trieste: (a) sampling along isopycnic surfaces, while following a surface drifter with an attempt to sample the same water mass in different time intervals, ie. to measure purely temporal changes (Lagrangian type of experiment) and (b) sampling on fixed stations in order to cover typical horizontal and vertical oceanographic gradients that exist in that ecosystem (Malej et al., 1995; Fonda-Umani et al., 1992). In both cases sampling was performed over 24 hour periods to follow the whole diel cycle.

In September 1994 the sampling was performed by following a drifter and samples were collected over a diel cycle each 4 hours at 4 different depths (0.5 and 5 m as well as above (12–17 m) and below (17–20 m) the thermocline). The field experiment was repeated 3 times during one week (September 8, 10 and 12). The sampling points are presented on Figure 1. Subsurface samples were collected at 0.5 m using a vacuum pump equipped with teflon tubings while sampling at other depths was performed using 5 l Niskin bottles. Water column was well stratified and the surface layer was characterised by a relatively high salinity (36.7 PSU), except during the last day (September 13) when salinity dropped below 35 PSU (Fonda Umani et al., 1997). Nutrient levels were rather low for most of the stations and sampling days. The concentration of phosphate, ammonium and nitrate were found in the ranges of 0.14–0.31 μmol l^{-1}, 0.11–3.73 μmol l^{-1} and 0.21–2.61 μmol l^{-1}, respectively.

During the second cruise (June 1995) the sampling was performed on three fixed stations (see Figure 1). The first station (A_4) was located one mile south of the Isonzo river mouth and was directly influenced by the river plume. Stations C an F were situated in the central part of the Gulf and one mile off Piran (Slovenia), respectively. Pronounced salinity and nutrient gradients were established in the top layer of the water column going from the northern (salinity <30 PSU; NO_3 = 10–12 μmol l^{-1}; SiO_4 = 3.25–4.96 μmol l^{-1}) to the southern part of the transect (salinity >37 PSU; NO_3 = 0.51–4.36 μmol l^{-1}; SiO_4 = 0.96–2.18 μmol l^{-1}). Water samples were collected over a diel cycle and the sampling was repeated twice, on June 13/14 and June 19/20. Three different depths were sampled at each station and the samples were taken using 5 l Niskin samplers.

Determination of photosynthetic pigments

Samples for the photosynthetic pigments analyses (0.5 l) were filtered onto 25 mm GF/F filters (0.7 μm average pore size). Filters were immedately extracted in 3 ml of cold 90% acetone using sonication, centrifuged to clarify the extract, and the chlorophylls and carotenoids separated by reversed-phase HPLC according to Barlow et al. (1993). Briefly, extracts were mixed (1:1; v/v) with 1 M ammonium acetate and injected into a HPLC system incorporating a C_{18} 3 μm Pecosphere column (3.3 × 0.45 cm, Perkin Elmer). A binary linear gradient was used to separate the pigments. Solvent A consisted of 80:20 (v/v) methanol:1 m ammonium acetate, while solvent B contained 60:40 (v/v) methanol:acetone. Chlorophylls and carotenoids were detected by absorbance at 440 nm

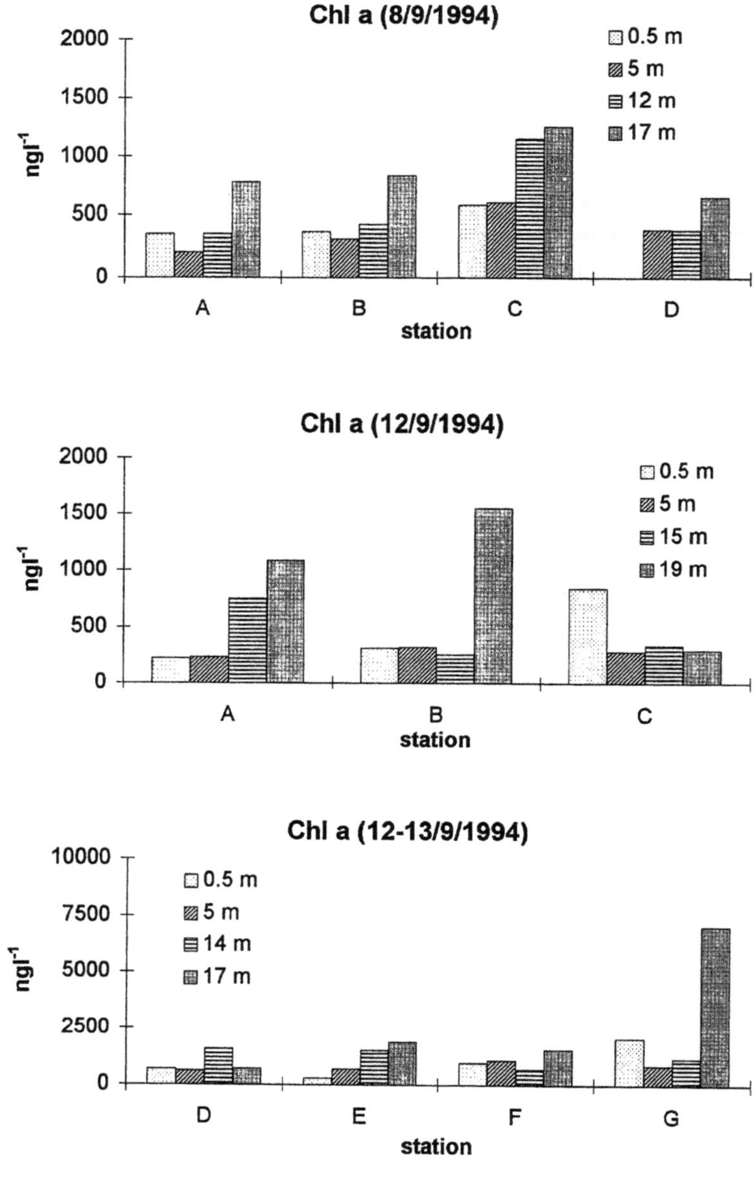

Figure 2. Distribution of chlorophyll *a* (chl *a*) during a field experiments performed in the Gulf of Trieste in September 1994.

(Spectra Physics, Model UV 2000), while phaeopigments were detected by fluorescence (Spectra Physics F 2000) using excitation at 420 nm and emission at 672 nm. Data collection and reprocessing utilised Spectra Physics PC 1000 software. Qualitative identification and quantitative determination of individual pigments was performed according to Barlow et al. (1993) and Terzić (1996).

Determination of dissolved and particulate organic carbon (DOC and POC)

Samples (0.5 to 1 liter) were filtered (vacuum 200 mm Hg; filtration rate approx. 25 ml per min.) onto precombusted (450 °C) and preweighed 25 mm GF/F filters. The first 200 to 500 ml were discarded and 2 × 10 ml from the remaining filtrate were collected in the precombusted glass tubes with teflon lined screw caps. Each tube was poisoned with $HgCl_2$ (final

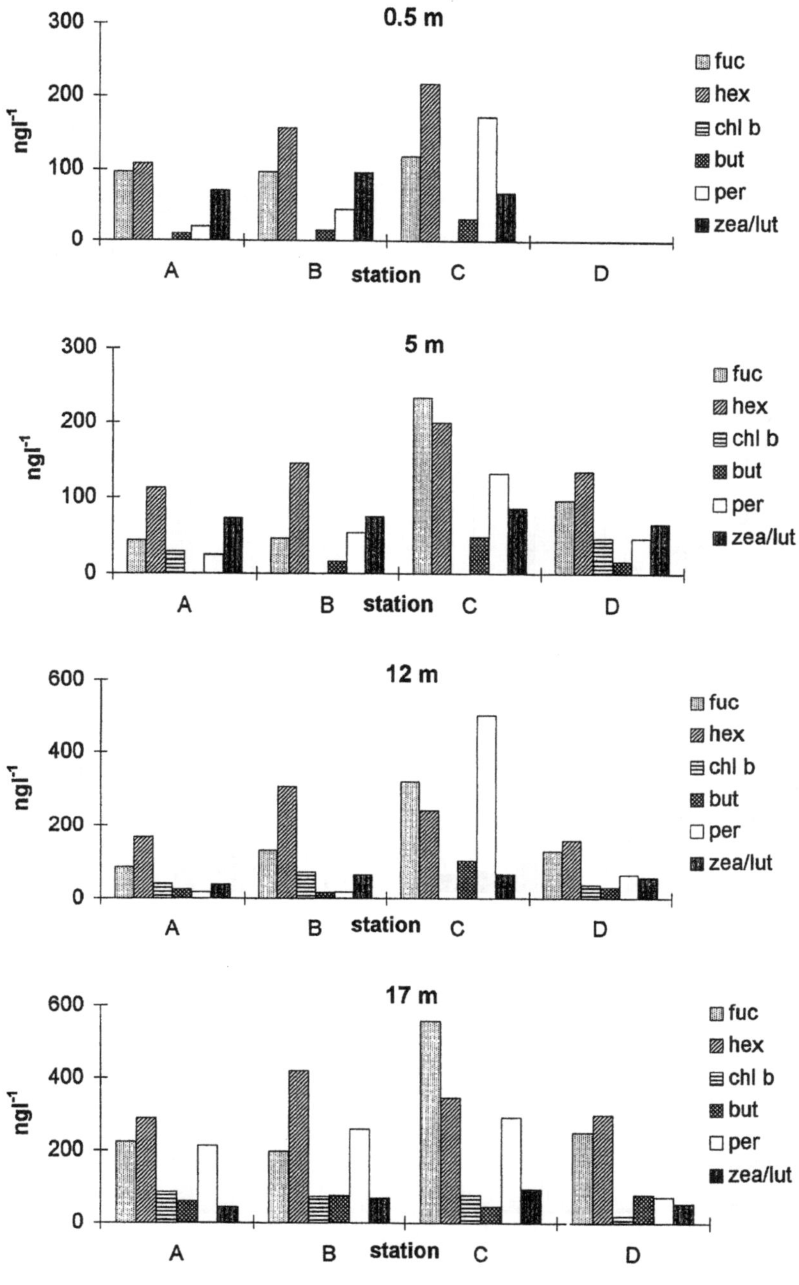

Figure 3. Distribution of biomarker pigments during the field experiment performed in the Gulf of Trieste on 8 September 1994 (fuc = fucoxanthin; hex = 19′-hexanoyloxyfucoxanthin; chl *b* = chlorophyll *b*; but = 19′- butanoyloxyfucoxanthin; per = peridinin; zea/lut = zeaxanthin/lutein).

concentration 5 mg l^{-1}) and stored at room temperature until analysed. After that the filters were rinsed with MilliQ water in order to remove remaining salts, dried overnight at 60 °C and stored in an excicator until analysed for suspended matter, POC and PON content.

DOC was analysed by HTCO method using a Shimadzu TOC 5000 carbon analyser according to Cauwet (1994).

POC was analysed using the SSM module of the Shimadzu TOC 5000 carbon analyser. The filters were fold in a silica boat and put in the inorganic channel of the instrument. Addition of phosphoric acid (10 N) and

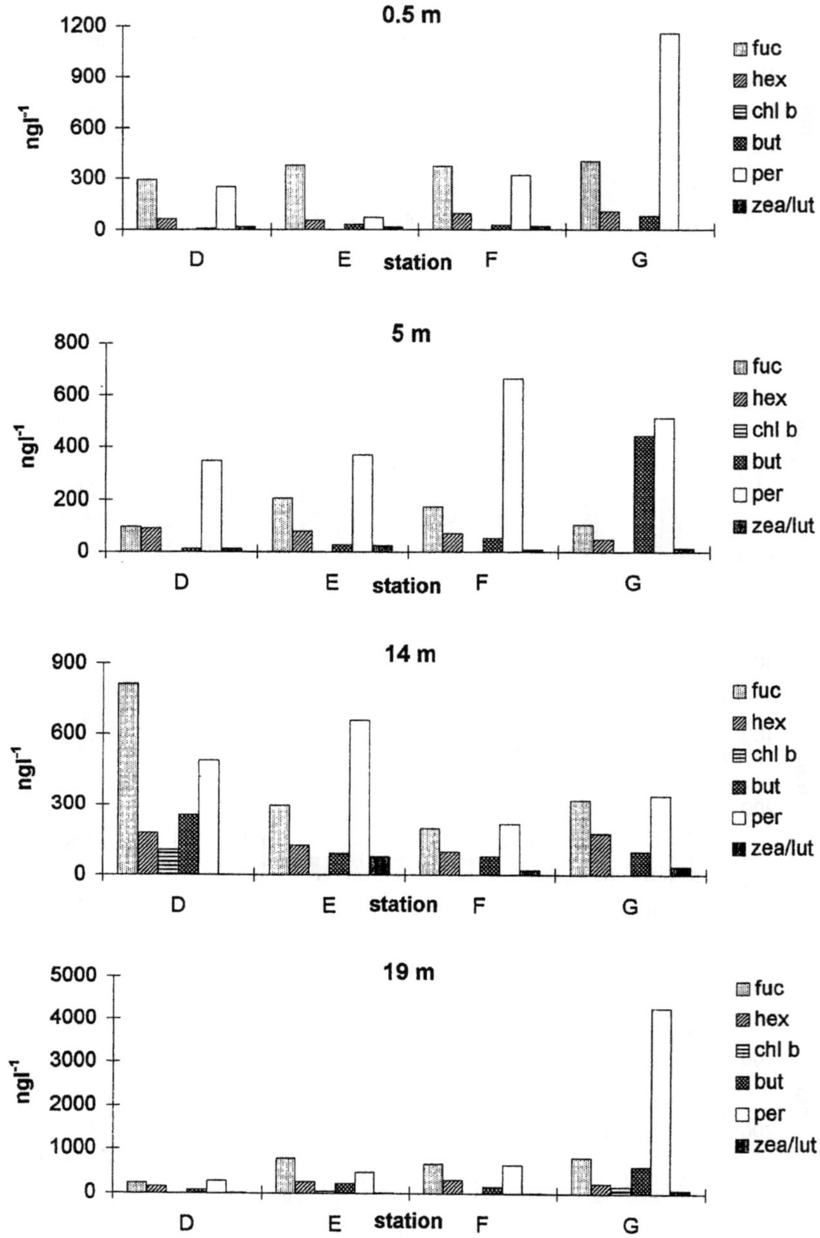

Figure 4. Distribution of biomarker pigments during the field experiment performed in the Gulf of Trieste on 12 September 1994 (fuc = fucoxanthin; hex = 19'-hexanoyloxyfucoxanthin; chl *b* = chlorophyll *b*; but = 19'-butanoyloxyfucoxanthin; per = peridinin; zea/lut = zeaxanthin/lutein).

drying at 200 °C allowed to measure the particulate inorganic carbon (PIC) and eliminate all the carbonates. The filters were afterwards introduced in the total carbon channel (900 °C) to be completely combusted in an oxygen flow (500 ml/mn). CO_2 produced was analysed by the TOC 5000 analyser. The sensitivity of the measurement is 5 μg C and the precision generally better than 2%.

Determination of carbohydrates

Before analysis, samples for the determination of different carbohydrate fractions were treated as described above for DOC and POC (except for omitting the filter rinsing). PTCHO were determined by filtering a sample of 100–500 ml. Dissolved monosaccharides (DMCHO) were determined directly by slightly modified MBTH method (Johnson & Sieburth, 1977), using glucose as a standard. Concentration of DTCHO and PTCHO were determined after hydrolysis with 2 M HCl (100 °C, 3.5 h) (Senior and Chevolot, 1991). The concentration of polysaccharides was calculated subtracting monosaccharide concentration from the total carbohydrate concentration. Standard deviation of repetitive CHO determinations was lower than 5%.

Results

Photosynthetic pigments

Chl *a* concentrations were highly variable in both experimental periods with average values of 832 ± 988 ng l^{-1} and 1390 ± 727 ng l^{-1} in September 1994 and June 1995, respectively.

In September 1994 water column was highly stratified with a pronounced pycnocline below 14 m. Concentration of chl *a* was notably higher below the pycnocline for all of the sampling days and stations except for the station C on September 12, which was situated very close to the coast (see Figure 2). Concentration in the subsurface layer (0.5 m) on September 8 and September 12 varied in the range of 354–372 ng l^{-1} and 224–2053 ng l^{-1}, respectively. The highest chl *a* concentration was determined in the deepest layer (17–19 m) and varied in the ranges of 647–1249 ng l^{-1} and 299–7040 ng l^{-1}, respectively (see Figure 2). Besides a rather heterogenous vertical distribution, a marked spatial variability of chl *a* was observed as well. The heterogeneity of the chl *a* distribution was especially pronounced on September 12/13 (stations D-G; 18–20 m) with chl *a* concentration ranging from 680 to 7040 ng l^{-1} (see Figure 2). It should be noticed that the concentration of chl *a* breakdown products in all of these samples was relatively low (<50 ng l^{-1}) (Terzić, 1996), indicating a relatively good physiological status of the biomass. The determination of different accessory pigments that reflect phytoplankton taxonomic composition (Millie et al., 1993) confirmed a rather heterogenous distribution of phytoplankton in the Gulf of Trieste in September 1994. Composition of biomarker pigments was rather complex and their distribution showed the same pattern as chl *a*, with the highest concentrations found in the deepest layer. The most prominent biomarker pigments on September 8 were 19′-hexanoyloxyfucoxanthin (hex), fucoxanthin (fuc), peridinin (per) and zeaxanthin (zea). According to the hex/chl *a* ratio (0.23–0.50), contribution of hex-containing phytoplankton (prymnesiophytes) to the total biomass was rather uniform. On the contrary, spatial and vertical distribution of fuc-, zea- and per-containing phytoplankton was rather uneven. The zea/chl *a* ratio indicated that cyanobacteria were relatively more abundant in the upper layer (0.5 and 5 m; zea/chl $a = 0.11–0.26$) than in the bottom layer (17 m; zea/chl $a = 0.06–0.08$). By contrast, dinoflagellates (per) were more abundant and less variable in the bottom layer (per/chl $a = 0.23–0.30$). Four days later (September 12) the per/chl *a*

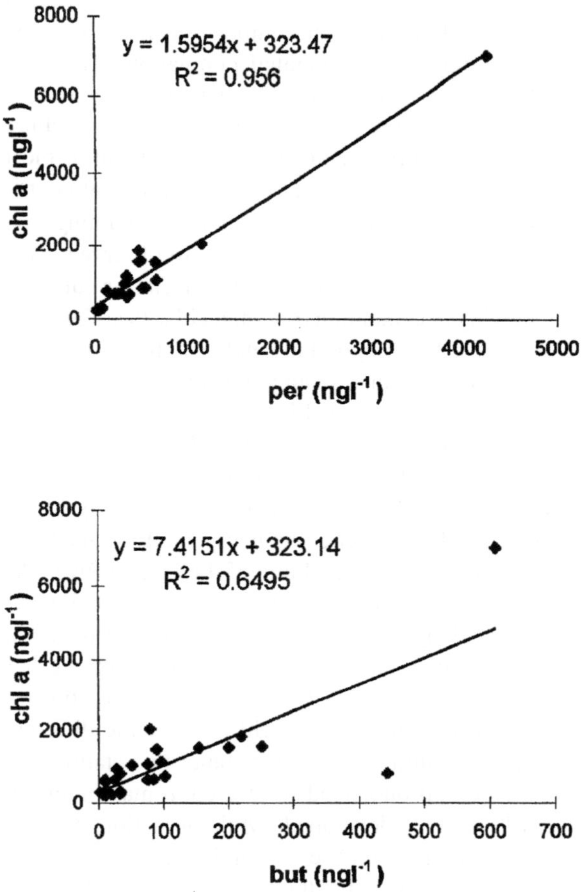

Figure 5. Correlation of chlorophyll *a* (chl *a*) with peridinin (per) and 19'-butanoyloxyfucoxan-thin (but) on 12 September 1994.

ratio significantly increased to 0.32–0.62 indicating a shift in phytoplankton composition from a mixed to a dinoflagellate-dominated population. This was confirmed by a very good correlation between chl *a* and per (see Figure 5; $r^2=0.956$). Although peridinin was the most prominent biomarker pigment at most of the stations (20–4253 ng l^{-1}), an important contribution of 19′-butanoyloxyfucoxanthin-containing phytoplankton (silicoflagellates) was also observed. This was reflected in a significant correlation of 19′-butanoyloxyfucoxanthin (but) and chl *a* (see Figure 5; $r^2=0.650$). An exceptionally high but/chl *a* ratio (0.53) was noticed only in one sample (station G, 5 m), while the average was about 0.15.

The field experiment in June 1995 was performed after a period of heavy rain followed by intensive freshwater run-off which increased nutrient supply to the upper part of the water column, especially on the locations influenced by the plume of the Isonzo River. As a consequence chl *a* concentration at the station A_4 (1283–3729 ng l^{-1}) was significantly higher than that at more distant stations C_0 and F (768–1687 ng l^{-1} and 530–1363 ng l^{-1}, respectively) (see Figure 6). The vertical distribution of chl *a* was quite different in the river plume (A_4) than at other two stations. In the plume, a pronounced surface chl *a* maximum (up to 3729 ng l^{-1}) developed, while at stations C_0 and F the maxima were observed in the bottom layer. No changes were noticed in the chl *a* distribution between day and night sampling periods, except for the station A_4 which could be attributed to the highly changeable hydrographic conditions in the plume. It should be told that, like in September 1994, concentration of all chl *a* breakdown products was rather low (<20 ng l^{-1} of the total phaeopigments), with somewhat higher chlorophyllide *a* concentration at the station A_4 (1 m; 157–277 ng l^{-1}). The composition of phytoplankton, as reflected by characteristic biomarker pigments (see Figure 7), was rather complex which is a typical situation occurring at the beginning of summer stratification period (Malej et al., 1995). The most prominent biomarker pigments were fuc (up to 1600 ng l^{-1}) and hex (up to 1200 ng l^{-1}) (see Figure 7). Their distribution pattern followed closely that of chl *a*, with the highest concentration at station A_4. Some biomarker pigments (per, chl *b*, but, zea) were also determined at a lower concentration (<0.3 μg l^{-1}). The predominance of fuc- and hex-containing phytoplankton was markedly stronger in the plume (A_4) than at more distant stations (C_0, F).

Dissolved and particulate carbohydrates

Total dissolved (DTCHO) and particulate carbohydrates (PTCHO) as well as dissolved monosaccharides (DMCHO) were determined in both field experiments (September 1994 and June 1995) and results are presented in Tables 1 and 2 along with the corresponding DOC and POC concentrations.

Although concentration of DTCHO in September 1994 varied in a rather wide range (137–1512 μg C l^{-1}), most of the values (>80%) were below 350 μg C l^{-1} (as shown in Table 1). The DMCHO/DPCHO ratio was very variable and the contribution of DMCHO and DPCHO to the DTCHO varied in the ranges of 4–71% and 29–96%, respectively. The percentage of DMCHO was generally higher in the samples characterised by low DTCHO concentrations, while in the samples containing high concentration of DTCHO, carbohydrates were found almost exclusively in the form of polysaccharides (up to 96%). In most of the samples DTCHO represented less than 25% of the total DOC (average value 20 ± 14%), however, extremely high contribution (up to 65%) was determined in the samples characterised by enhanced concentrations of DTCHO. Distribution of carbohydrates in June 1995 was more uniform as compared to September 1994 with DTCHO concentrations varying from 170 to 396 μg C l^{-1} (as shown in Table 2). The contribution of DTCHO to the total DOC was 11–23% (average value 17 ± 3%). No significant differences were noticed between day and night distributions of either DOC and DTCHO. DMCHO were determined at concentrations from 38 to 160 μg C l^{-1} and contributed to DTCHO by 12–62%. Like in September 1994, the major part of carbohydrates was present in the form of DPCHO (concentration range = 92–282 μg C l^{-1}; average percentage of DTCHO 66 ± 15%). The concentration of PTCHO contributed a relatively small percentage (4–25%) of the total CHO concentration in both sampling periods (as shown in Tables 1 and 2). Furthermore, PTCHO represented 8–31% and 11–32% of the total POC in September 1994 and June 1995, respectively.

Discussion

Considering the link between phytoplankton production and release of carbohydrates is one of the key prerequisites for a better understanding of carbon cycling in the marine environment because carbohydrates represent a significant fraction of DOC both in

Table 1. Concentrations of carbohydrates, DOC and POC in the Gulf of Trieste in September 1994

Station	Depth (m)	DMCHO μg C l^{-1}	DPCHO μg C l^{-1}	DTCHO μg C l^{-1}	PTCHO μg C l^{-1}	DOC μg l^{-1}	POC μg l^{-1}	DTCHO %DOC	PTCHO %POC
8/9/94									
A	0.5	102	193	295	43	1730	140	17	31
A	5	130	93	223	36	1820	246	12	15
A	12	79	252	331	36	–	–	–	–
A	17	86	–	–	36	–	312	–	12
B	0.5	108	194	302	22	1660	108	18	20
B	5	122	152	274	29	2020	186	14	16
B	12	108	86	194	29	2010	236	10	12
B	17	86	72	158	43	1300	229	12	19
C	0.5	94	96	190	22	1610	163	12	13
C	5	102	301	403	36	1790	224	23	16
C	12	122	51	173	58	1430	224	12	26
C	17	79	115	223	43	1710	214	13	20
D	0.5	–	–	–	–	–	–	–	–
D	5	108	94	202	29	–	210	–	14
D	12	108	50	158	36	1470	194	11	19
D	17	72	65	137	36	–	180	–	20
12/9/94									
A	0.5	58	115	173	14	1750	73	10	19
A	3	72	259	331	14	1700	121	19	12
A	15	72	223	295	14	–	183	–	8
A	18	160	150	310	19	1500	180	21	11
B	0.5	86	94	180	18	–	74	–	24
B	5	65	197	202	15	1840	97	11	15
B	16.5	65	180	245	16	–	120	–	13
B	19	58	122	180	26	1380	179	13	15
C	0.5	79	137	216	39	–	171	–	23
C	5	58	–	–	29	1440	188	–	15
C	17	72	188	260	20	1640	130	16	15
C	19	65	93	158	46	–	209	–	22
D	0.5	94	1418	1512	40	2400	141	63	28
D	5	160	546	706	41	1800	175	39	23
D	14	102	1021	1123	32	1730	–	65	–
D	20	50	152	202	31	1980	–	10	–
E	0.5	79	375	454	23	1860	101	24	23
E	5	65	309	374	34	1700	202	22	17
E	12	72	166	238	30	1840	201	13	15
E	18	65	187	252	32	1560	225	16	14
F	0.5	79	195	274	18	1570	134	17	13
F	6	65	201	266	31	1490	205	19	15
F	14	94	144	238	19	–	139	–	14
F	18	22	432	454	22	1710	194	27	11
G	0.5	65	201	266	45	1780	259	15	17
G	6	115	130	245	30	–	178	–	17
G	14	151	101	252	23	–	162	–	14
G	19	58	1252	1310	69	–	532	–	13

DMCHO: dissolved monosaccharides; DPCHO: dissolved polysaccharides; DTCHO: total dissolved carbohydrates; PTCHO: total particulate carbohydrates; DOC: dissolved organic carbon; POC: particulate organic carbon; –: not determined

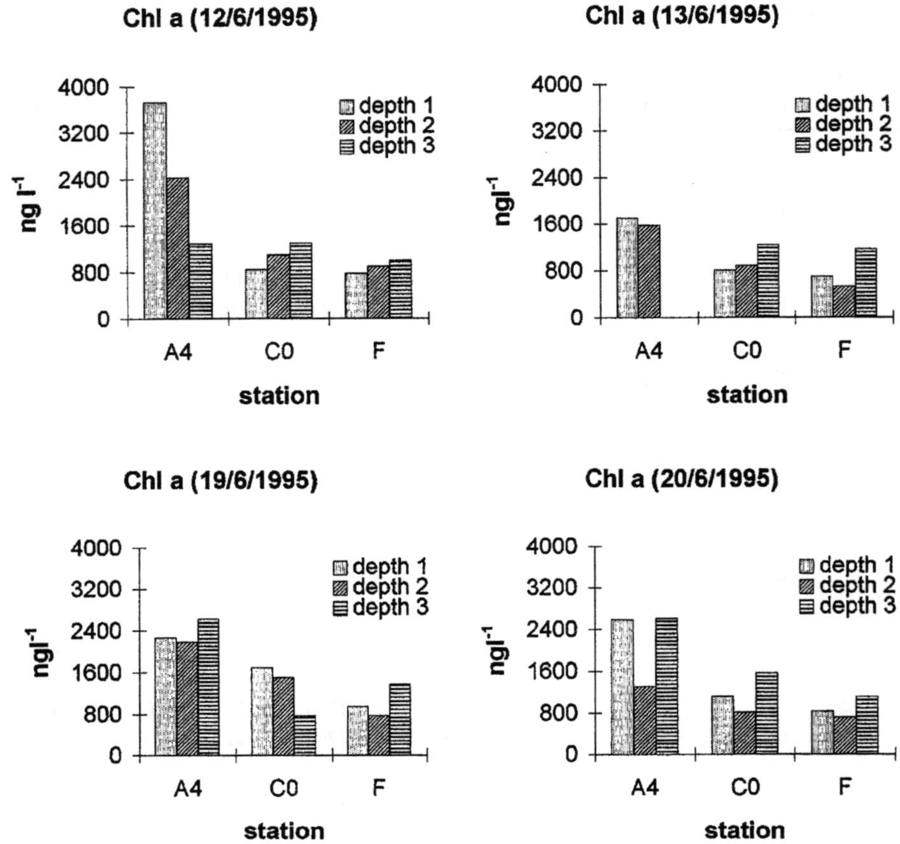

Figure 6. Distribution of chlorophyll *a* (chl *a*) during field experiments performed in the Gulf of Trieste in June 1994. Exact sampling depths for each sampling day and station can be found in Table 2.

the oceans (Pakulski & Benner, 1994) and in coastal waters (Senior & Chevolot, 1991). As opposed to rather numerous laboratory studies on CHO-release by phytoplankton monocultures (see Myklestad, 1995 for review) only few studies, mostly related to phytoplankton blooms, showed a clear relationship between the composition of phytoplankton and CHO levels in the real environment (Ittekkot et al., 1981; Burney et al., 1981). Other reports (e.g. Dhople & Bhosle, 1987; Senior & Chevolot, 1991) usually provided only a correlation of CHO with the total biomass (chl *a*). The specific aim of this paper was an attempt to understand the link between the group-specific phytobiomass, as reflected by biomarker pigments, and carbohydrate levels in temperate shallow coastal waters during the period of density stratification which is characterised by a very complex phytoplankton assemblage (Malej et al., 1995). Moreover, the period of summer stratification is typical season for macroaggregate formation in the northern Adriatic (Stachowitsch et al., 1990; Deggobis et al., 1995). Although CHO were suggested to play a major role in the development of macroaggregates (Kiørboe & Hansen, 1993; Passow et al., 1994), the data on the concentration and distribution of the transparent exopolymeric polysaccharides in the northern Adriatic were completely lacking.

The data obtained in this study showed that the percentage of particulate CHO in the total CHO was rather low and indicated that the main pool of CHO was in the dissolved phase. Most of the DTCHO concentrations determined in this study were in the range below 400 μg C l^{-1}, however, they were significantly higher than those from our earlier report (Ahel et al., 1995) which were obtained by phenol-sulphuric acid method (Dubois et al., 1956). Very probably, concentrations of CHO determined by the MBTH-method are higher due to the more complete hydrolysis of polysaccharides (PCHO). Comparable concentrations of CHO determined by MBTH-method were reported for other seas. Senior & Chevolot (1991) determined concentra-

Table 2. Concentrations of carbohydrates, DOC and POC in the Gulf of Trieste in June 1995

Station	Depth (m)	DMCHO μg C l^{-1}	DPCHO μg C l^{-1}	DTCHO μg C l^{-1}	PTCHO μg C l^{-1}	DOC μg l^{-1}	POC μg l^{-1}	DTCHO %DOC	PTCHO %POC
A_4	1	78	187	265	85	1934	540	14	16
A_4	3	101	163	264	64	1891	324	14	20
A_4	10	68	119	187	45	1846	246	10	18
C_0	1	100	229	329	53	1846	248	18	21
C_0	6	98	175	273	90	1730	280	16	32
C_0	14	75	161	236	36	1916	254	12	14
F	1	103	211	314	57	1727	314	18	18
F	6	110	176	286	52	1720	267	17	19
F	17	76	94	170	45	1573	213	11	21
A_4	1	53	281	334	–	1681	357	20	–
A_4	6	38	278	316	46	1699	300	19	15
A_4	9	44	240	284	39	1652	256	17	15
C_0	1	47	237	284	51	1686	244	17	21
C_0	7	48	282	330	55	1762	268	19	20
C_0	15	41	232	273	44	1502	238	18	18
F	1	85	196	281	59	1805	229	16	26
F	7	85	207	292	57	1819	286	16	20
F	17	100	159	259	39	1532	175	17	22
19/6/95									
A_4	1	114	181	295	112	2029	476	15	24
A_4	3.5	100	195	295	68	1720	233	17	29
A_4	8.5	104	182	286	46	1667	288	17	16
C_0	1	117	257	374	87	1733	408	22	21
C_0	7	95	241	336	38	1711	208	20	18
C_0	20	86	249	335	19	1465	146	23	13
F	1	119	207	326	74	2096	339	16	22
F	5.5	97	193	290	45	1686	247	17	18
F	14.5	92	164	256	45	1366	254	19	18
A_4	1	130	256	386	90	2047	466	19	19
A_4	3	114	199	313	55	2105	260	15	21
A_4	10	134	262	396	50	1942	304	20	16
C_0	1	151	92	243	81	2004	311	12	26
C_0	7	160	196	356	46	1936	–	18	–
C_0	20	104	244	348	47	1502	272	23	17
F	1	103	281	384	43	2320	384	17	11
F	6	88	242	330	46	2119	219	16	21
F	13	63	180	243	51	1800	267	14	19

DMCHO: dissolved monosaccharides; DPCHO: dissolved polysaccharides; DTCHO: total dissolved carbohydrates; PTCHO: total particulate carbohydrates; DOC: dissolved organic carbon; POC: particulate organic carbon; –: not determined

tions of DTCHO in the Elorn estuary (France) in the range of 20–570 μg C l^{-1}, and similar concentrations (84–396 μg C l^{-1}) were reported for the surface layer of the open ocean (Pakulski & Benner, 1994). In addition, in majority of the samples analysed in our study DTCHO accounted for 10–25% of DOC which is very similar to the average percentage of 21 ± 7% found in open ocean by Pakulski & Benner (1994). However, in the samples containing high concentrations of DTCHO carbohydrates comprised up to 65% of DOC. This indicated that occasionally a very strong accumulation of carbohydrates can occur, which dramatically changes the character of DOC. Similar observation was obtained during a diatom bloom in the northern Adri-

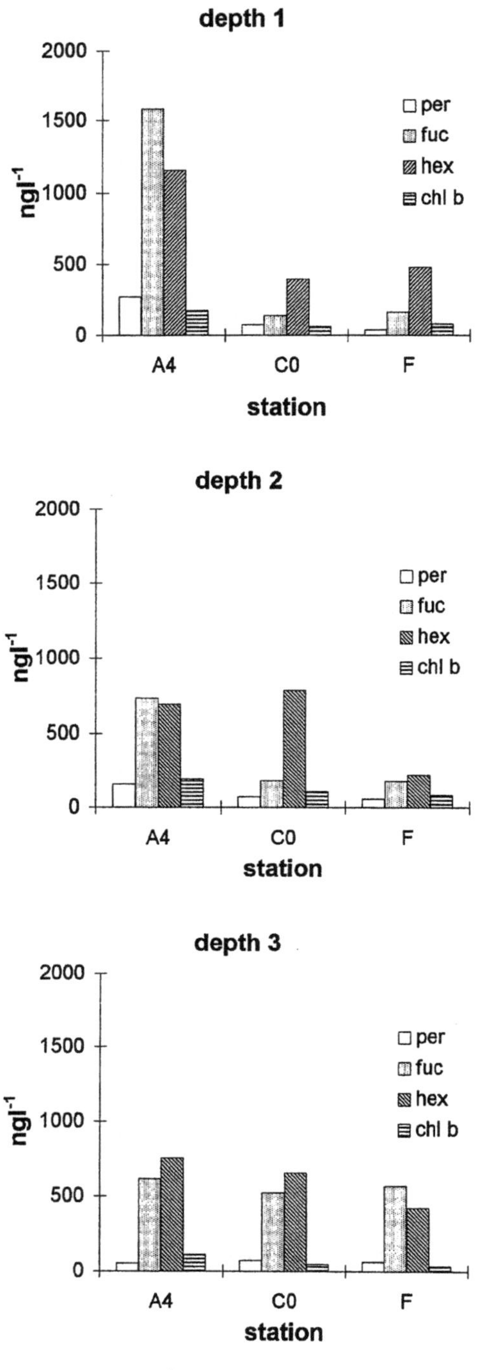

Figure 7. Distribution of biomarker pigments during the field experiments performed in the Gulf of Trieste on 12 June 1994. Exact sampling depths for each sampling station can be found in Table 2. (fuc = fucoxanthin; hex = 19′-hexanoyloxyfucoxanthin; chl *b* = chlorophyll *b*; per = peridinin).

atic (Terzić, 1996) as well as in the enrichment experiment with autochthonous phytoplankton from the Gulf of Trieste (Cauwet et al., 1997).

DMCHO/DPCHO ratios were rather variable and suggested a very dynamic behaviour of CHO in the Gulf of Trieste. Considering that CHO released by phytoplankton are mainly in combined form, it is interesting to note that dissolved monosaccharides often reached high percentages of the total CHO (30–50%). This probably indicated a considerable glycolitic activity of bacteria and/or *in situ* hydrolysis of labile polysaccharides (Ittekkot et al., 1981).

According to the composition of phytobiomass, the biological situation in the Gulf of Trieste during the summer stratification is regularly very complex (Malej et al., 1995). Group-specific distribution of phytoplankton biomass in September 1994 as reflected by phytoplankton pigments revealed a very high diversity and rather heterogenous vertical and spatial distributions. Due to the reduced water mass mixing and intensive regeneration processes, the situation in late summer was characterised by a pronounced increase of chl *a* below the pycnocline, while the concentration in the upper layer was low (see Figure 2). Such a vertical distribution was present during the whole investigated period (September 8–12). However, according to the biomarker pigment analyses (see Figures 3 and 4), a major change in the composition of the phytoplankton population ocurred in that period, resulting in a shift from a mixed population containing prymnesiophytes (hex), diatoms (fuc), cyanobacteria (zea) and dinoflagellates (per) to a dinoflagellate-dominated population. This was confirmed by microscopic observations which indicated an enhancement of the number of dinoflagellate *Lingulodinium polyedra*, especially in the bottom layer (Fonda Umani et al., 1997). Since such a fast succession of phytoplankton was not very likely, the presented data cannot be considered true time series in the same water mass but rather a snapshot of small scale spatial distributions. The composition of phytoplankton in June was also very complex like in September, but it was slightly dominated by prymnesiophytes in the largest part of the Gulf of Trieste, while diatoms predominated in the Isonzo River plume. An increasing hex-to-fuc ratio from the station A_4 towards the station F in the surface layer reflected in fact the existing gradient of salinity and riverborne nutrients in the Gulf with diatoms dominating in the more eutrophic part, near the river plume.

The correlation between the phytoplankton biomass (chl *a*) and DTCHO in early summer (June) was

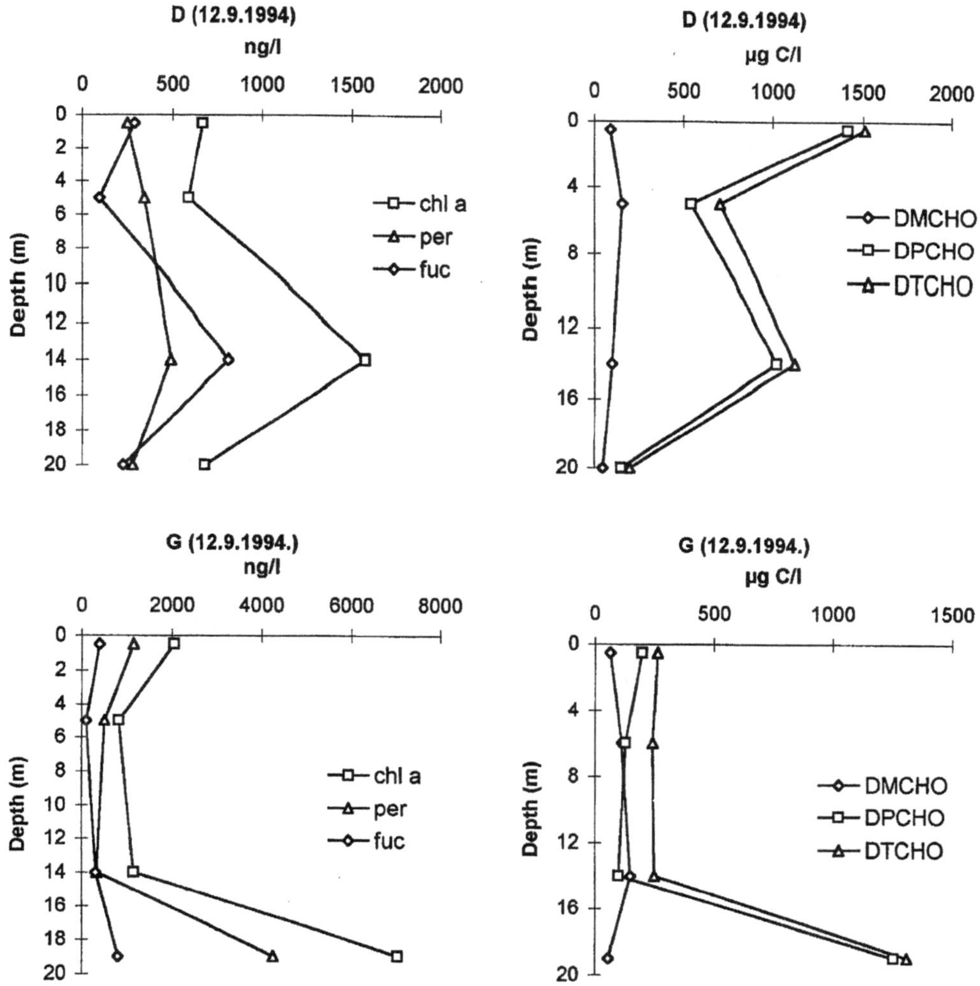

Figure 8. Vertical distribution of biomarker pigments and carbohydrates at two sampling stations (D and G) in September 1994 (chl *a* = chlorophyll *a*; per = peridinin; fuc = fucoxanthin; DMCHO = dissolved monosaccharides; DPCHO = dissolved polysaccharides; DTCHO = dissolved total carbohydrates).

negligible ($r^2 = 0.03$), while in September a moderate positive correlation ($r^2 = 0.51$) was obtained. This difference was interpreted mainly as a consequence of different phytoplankton compositions in June and September, the phytoplankton assemblage in September showing a comparatively higher contribution of microplankton (fuc and per). The fucoxanthin-containing microplankton (diatoms) was shown to be a significant source of dissolved polysaccharides, particularly in the later phase of a bloom (Ittekkot et al., 1981) which is usually characterised by nutrient depleted conditions. In contrast, a significant negative correlation was reported between the phototrophic nanoplankton and dissolved polysaccharides in oligotrophic waters, presumably due to the higher release of CHO by slow growing cells than by fast growing biomass (Burney et al., 1981). Although nanoplankton populations represent a significant percentage in summer phytoplankton in the Gulf of Trieste (Malej et al., 1995), their impact on the CHO levels seems to be negligible or completely masked by the much stronger effects of microplankton. Accordingly, no positive correlation was observed between hex and DTCHO, neither during the June nor September cruises, which suggested that prymnesiophytes, as the main component of summer nanoplankton, had a very small, if any, influence on the concentration levels and distribution of DTCHO. This is in agreement with the observation obtained in the enrichment experiment performed in the same season (June 1994) with autochtonous phytoplankton from the Gulf of Trieste, which was also dominated by prymnesiophytes (Cauwet et al., 1997). The relative abundance of other nanoplanktonic groups such as green algae (chl *b*) as well as of picoplanktonic cyanobacteria were too low to expect any significant contribution of these phytoplankton groups to DTCHO. The temporal distribution of phytoplankton in September was very heterogenous and it is instructive to compare two different situations observed on September 8 and September 12. As mentioned above, during that period a very complex phytoplankton assemblage dominated by prymnesiophytes was replaced by a much simpler population dominated by dinoflagellates and diatoms (per and fuc). Such a change in phytoplankton composition resulted in a relatively good correlation between chl *a* and TDCHO ($r^2 = 0.51$). Moreover, DTCHO distribution on two vertical profiles in the late summer was very well correlated with the distribution of fuc and per (see Figure 8), suggesting that dissolved CHO originated from microphytoplankton, notably from diatoms and dinoflagellates. It is important to notice that in that situation over 90% of DTCHO were in polysaccharide form. Moreover, the percentage of CHO in the total DOC was extremely high, reaching up to 65%. It is known that diatoms may excrete large amounts of DOC, mainly as polysaccharides (Myklestad, 1995), while our results indicate that dinoflagellates could also be an important source of CHO. However, since it is possible that DTCHO were partly formed by bursting of dinoflagellate cells during filtration, we cannot presently be certain that this was primarily a result of active exudation of DPCHO by healthy dinoflagellate cells.

It is interesting to note that increased chl *a* and fuc levels in the river plume (A_4) were not associated with enhanced concentrations of CHO. The reason is, most probably, the fact that enhanced chl *a* in the plume corresponds to the exponential phase of phytoplankton growth, whereas it was shown that release of CHO was usually associated with the senescent phase of a bloom (Ittekkot et al., 1981; Terzić, 1996).

Conclusions

Carbohydrate levels and their relative percentage in the total DOC during a typical summer stratification period in the Gulf of Trieste are very similar to those reported for other coastal and open ocean waters and show no anomaly which would explain the reason why this area is occasionally affected by a large scale formation of mucous aggregates in that season. However, some of the results presented in the paper suggest that the present phytoplankton populations have a large potential to produce dissolved polysaccharide material which could account for up to 65% of DOC. The two main phytoplankton groups which revealed a significant potential to be a source of dissolved carbohydrates in the Gulf of Trieste were diatoms and dinoflagellates, while contribution of prymnesiophytes seemed to be rather insignificant.

Acknowledgements

This work was supported by the Ministry of Science and Technology of the Republic of Croatia. The study was carried out as a part of the European project on 'Production and Accumulation of Labile Organic Carbon in the Adriatic' (PALOMA). Financial support through an UNESCO/IOC Project (contract No. 214.240.5) on 'Eutrophication-Related Processes in Neritic Areas of the Mediterranenan Sea: Phytoplankton/Dissolved Organic Matter Relationships' and by the French Ministry of Foreign Affairs in the framework of the French-Croatian cooperation programme are also greatfully acknowledged. We thank the crew and all the people who did sampling on board and Ivana Jeličić for assisting with carbohydrate analyses.

References

Ahel, M., S. Terzić, A. Malej & R. Precali, 1995. Phytoplankton pigment/carbohydrate relationships in the northern Adriatic. Rapp. Comm. int. Mer Medit. 34: 54 pp.

Barlow, R. G., 1982. Phytoplankton ecology in the southern Benguela current. III. Dynamics of a bloom. J. exp. mar. Biol. Ecol. 63: 239–248.

Barlow, R. G. , R. F. C. Mantoura, M. A. Gough & T. W. Fileman, 1993. Pigment signatures of the phytoplankton composition in the northeastern Atlantic during the 1990 spring bloom. Deep-Sea Res. II 40: 459–477.

Burney, C. M., P. G. Davis, K. M. Johnson & J. McN. Sieburth, 1981. Dependence of dissolved carbohydrate concentrations upon small scalle nanoplankton and bacterioplankton distributions in the western Saragasso Sea. Mar. Biol. 65: 289–296.

Cauwet, G., 1994. HTCO method for dissolved organic carbon analysis in seawater: influence of catalyst on blank estimation. Mar. Chem. 47: 55–64.

Cauwet, G., S. Terzić, M. Ahel, P. Mozetić, V. Turk & A. Malej, 1997. Effect of nutrients addition on microbial plankton and dissolved organic matter variability. Part 2. Biochemical aspect. Proceedings of the International Conference on Physical and Biogeochemical Processes of the Adriatic Sea, Ancona (Italy), 23–27 April 1996. Commission of the European Communities, Ecosystems Research Reports Series, The Adriatic Sea. In press.

Decho, A. W., 1990. Microbial exopolymer secretion in ocean environments: their role(s) in food webs and marine processes. Oceanogr. Mar. Biol. annu. Rev. 28: 73–153.

Deggobis, D., S. Fonda-Umani, P. Franco, A. Malej, R. Precali & N. Smodlaka, 1995. Changes in the Northern Adriatic ecosystem and the hypertrophic appearance of gelatinous aggregates. Sci. tot. Envir. 165: 43–58.

Dhople, V. M. & N. B. Bhosle, 1987. Dissolved Carbohydrate in the Central Arabian Sea. Indian J. mar. Sci. 16: 43–45.

Dubois, M., K. A. Gilles, J. K. Hamilton, P. A. Rebers & F. Smith, 1956. Colorimetric method for determination of sugars and related substances. Analyt. Chem. 28: 350–356.

Eberlein, K., U. H. Brockmann, K. D. Hammer, G. Kattner & M. Laake, 1983. Total dissolved carbohydrates in an enclosure experiment with unialgal Skeletonema costatum culture. Mar. Ecol. Prog. Ser. 14: 45–58.

Faganeli J., N. Kovač, H. Leskovšek & J. Pezdič, 1995. Sources and fluxes of particulate organic matter in shallow coastal waters characterised by summer macroaggregate formation. Biogeochemistry 29: 71–88.

Fernandez, E., P. Serret, I. de Madariaga, D. S. Harbour & A. G. Davies, 1992. Photosynthetic carbon metabolism and biochemical composition of spring phytoplankton assemblages enclosed in microcosm: the diatom-Phaeocystis sp. succession. Mar. Ecol. Prog. Ser. 90: 89–102.

Fogg, G. E., 1966. The extracellular products of algae. Oceanogr. Mar. Biol. annu. Rev. 4: 195–212.

Fonda-Umani, S., P. Franco, E. Ghirardelli & A. Malej, 1992. Outline of oceanography and the plankton of the Adriatic Sea. In Colombo, G. et al. (eds), Marine Eutrophication and Population Dynamics. Olsen & Olsen: 347–365.

Fonda-Umani, S., G. Cauwet, S. Cok, E. Martecchini & S. Predonzani, 1997. Chemical and biological patterns of the Gulf of Trieste: the example of early and late summer. Proceedings of the International Conference on Physical and Biogeochemical Processes of the Adriatic Sea, Ancona (Italy), 23–27 April 1996. Commission of the European Communities, Ecosystems Research Reports Series, The Adriatic Sea. In press.

Haug, A. & S. Myklestad, 1976. Polysaccharides of marine diatoms with special reference to *Chaetoceros* species. Mar. Biol. 34: 217–222.

Ittekkot, V., U. Brockman, W. Michaelis & E. T. Degens, 1981. Dissolved free and combined carbohydrates during a phytoplankton bloom in the Northern North Sea. Mar. Ecol. Prog. Ser. 4: 299–305.

Ittekkot, V., E. T. Degens & U. Brockmann, 1982. Monosaccharide composition of acid-hydrolyzable carbohydrates in particulate matter during a plankton bloom. Limnol. Oceanogr. 27: 711–716.

Johnson, K. M. & J. M. Sieburth, 1977. Dissolved carbohydrates in seawater I. A precise spectrophotometric method for monosaccharides. Mar. Chem. 5: 1–13.

Kiørboe, T. & J. L. S. Hansen,1993. Phytoplankton aggregate formation: observations of patterns and mechanisms of cell sticking and the significance of exopolymeric material. J. Plankton Res. 15: 993–1018.

Malej, A., P. Mozetič, V. Malačić, S. Terzić & M. Ahel, 1995. Phytoplankton responses to freshwater inputs in a small semi-enclosed gulf (Gulf of Trieste, Adriatic Sea). Mar. Ecol. Prog. Ser. 120: 111–121.

Marlow, I. T., L. J. Rogers & A. J. Smith, 1989. Extent and nature of extracellular organic production by marine coccolitophorid *Hymemonas carterae*. Mar. Biol. 100: 381–391.

Millie, D. F., H. W. Paerl & J. P. Hurley, 1993. Microalgal pigment assessments using high-performance liquid chromatography: A synopsis of organismal and ecological applications. Can. J. Fish. aquat. Sci. 50: 2513–2527.

Myklestad, S., 1974. Production of carbohydrates by marine planktonic diatoms. I. Comparison of nine different species in culture. J. exp. mar. Biol. Ecol. 15: 261–274.

Myklestad, S. M., 1995. Release of extracellular products by phytoplankton with special emphasis on polysaccharides. Sci. tot. Envir. 165: 155–164.

Myklestad, S., O. Holm-Hansen, K. M. Varum & B. E. Volcani, 1989. Rate of release of extracellular amino acids and carbohydrates from marine diatom *Chaetoceros affinis*. J. Plankton Res. 11: 763–773.

Pakulski, J. D. & R. Benner, 1994. Abundance and distribution of carbohydrates in the ocean. Limnol. Oceanogr. 39: 930–940.

Passow, U., A. L. Alldredge & B. E. Logan, 1994. The role of particulate carbohydrate exudates in the flocculation of diatom blooms. Deep-Sea Res. 41: 335–357.

Posedel, N. & J. Faganeli, 1991. Nature and sedimentation of suspended particulate matter during density stratification in shallow coastal waters (Gulf of Trieste, northern Adriatic). Mar. Ecol. Prog. Ser. 77: 135–145.

Revelante, N. & M. Gilmartin, 1992. The lateral advection of particulate organic matter from the Po delta region during summer stratification, and its implications for the northern Adriatic. Estuar. coast. Shelf Sci. 35: 191–212.

Senior, W. & L. Chevolot, 1991. Studies of dissolved carbohydrates (or carbohydrate-like substances) in an estuarine environment. Mar. Chem. 32: 19–35.

Stachowitsch, M., N. Fanuko & M. Richter, 1990. Mucus aggregates in the Adriatic Sea: An overview of stages and occurences. P.Z.N.I. Mar. Ecol. 11: 327–350.

Terzić, S., 1996. Biogeokemija autohtone organske tvari u neritičkim područjima Mediterana: fotosintetski pigmenti i ugljikohidrati (in Croatian). [Biogeochemistry of autochtoneous organic matter in the neritic areas of the Mediterranean sea: photosynthetic pigments and carbohydrates]. Ph. D. Thesis. University of Zagreb, 177 pp.

Hydrobiologia **363**: 207–217, 1998.
T. Tamminen & H. Kuosa (eds), Eutrophication in Planktonic Ecosystems: Food Web Dynamics and Elemental Cycling.

The use of spectral fluorescence methods to detect changes in the phytoplankton community

Jukka Seppälä[1,3] & Maija Balode[2]
[1] *Tvärminne Zoological Station, University of Helsinki, FIN-10900 Hanko, Finland*
[2] *Institute of Aquatic Ecology, University of Latvia, Miera 3, LV-2169 Salaspils, Latvia*
[3] *Present address: Finnish Environment Institute, P.O.Box 140, FIN-00251 Helsinki, Finland*

Key words: in vivo fluorescence, spectrofluorometry, phytoplankton pigments, phytoplankton community structure, Gulf of Riga

Abstract

In vivo fluorescence methods are efficient tools for studying the seasonal and spatial dynamics of phytoplankton. Traditionally the measurements are made using single excitation-emission wavelength combination. During a cruise in the Gulf of Riga (Baltic Sea) we supplemented this technique by measuring the spectral fluorescence signal (SFS) and fixed wavelength fluorescence intensities at the excitation maxima of main accessory pigments. These methods allowed the rapid collection of quantitative fluorescence data and chemotaxonomic diagnostics of the phytoplankton community. The chlorophyll *a*-specific fluorescence intensities (R) and the spectral fluorescence fingerprints were analysed together with concentrations of chlorophyll *a* in different algal size-groups, phytoplankton biomass and taxonomic position. The lower level of R in the southern gulf was related to the higher proportion of cyanobacteria relative to total biomass and the lower abundance of small algae. The phycoerythrin fluorescence signal was obviously due to the large cyanobacteria. The basin-wide shift in the shape of chlorophyll *a* excitation spectra was caused by the variable proportions of differently pigmented cyanobacteria, diatoms and cryptomonads.

Introduction

A major goal in biological oceanography is to describe and model the variability of phytoplankton biomass, production and community structure in relation to the physical and chemical environment. Sampling and measuring techniques allowing continuous recording of biological data are needed to understand the couplings of biological, chemical and physical processes at different temporal and spatial scales (e.g. Legendre & Demers, 1984; Harris, 1986). Measuring bio-optical properties of living algae is an efficient tool in high-frequency sensing of the algal community. The *in vivo* fluorescence method presented by Lorenzen (1966) has been widely used to study the distribution of chlorophyll *a* (chl *a*). It is easy to perform and provides high-sensitive on-line information on the distribution of algae. Frequent calibrations are, however, needed against accurate pigment analyses since the fluorescence characteristics of living algae are highly variable. In principle, *in vivo* fluorescence intensity per chl *a* (R) is affected by (*i*) the spectral intensity of the excitation light, (*ii*) the rate of light harvesting by photosystem II (PSII), (*iii*) the energy transfer efficiencies of the photosystems and (*iv*) the quantum yield of fluorescence (Falkowski & Kiefer, 1985; Owens, 1991). In studies of algal cultures or natural phytoplankton communities the variability of R has been interpreted as a function of algal cell size, nutrient status, state of the photosystems, photoacclimation and taxonomic position (Kiefer, 1973; Loftus & Seliger, 1975; Harris, 1980; Alpine & Cloern, 1985; Mitchell & Kiefer, 1988b; Strass, 1990; Althuis et al., 1994; Guo & Dunstan, 1995).

The *in vivo* fluorescence emission of chl *a* (maximum at 682 nm) arises from PSII. The light harvesting pigments in PSII are to some extent class-specific; thus there are differences in the shapes of excitation

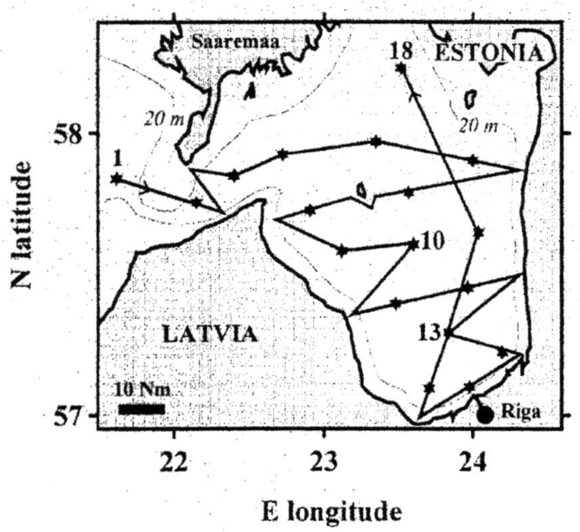

Figure 1. The route (continuous line) taken by R/V Marina during 10.–13.7. 1994 and sampling stations (asterisks) in the Gulf of Riga. The index number is shown for some stations.

spectra of the major phytoplankton classes (Yentsch & Yentsch, 1979). For all algae, light can be absorbed by chl *a* itself (maximum at 438 nm); for chlorophytes also by chl *b* (480 nm) and for chromophytes by chl *c* (460–470 nm) and carotenoids (490–560 nm), affecting the fluorescence excitation specimen. The pigment-system of cyanobacteria differs from that of eukaryotes (Bryant, 1986) in exhibiting only weak fluorescence from chl *a*; cyanobacteria, however, contain phycobilins that fluoresce at specific wavelengths. Emission maxima for phycoerythrin and phycocyanin are at about 570 and 650 nm, respectively. The cryptophytes also have phycobilins and are characterised by both phycobilin and chl *a* fluorescence. (Prézelin, 1981; Anderson & Barrett, 1986)

Use of the spectral fluorescence signal (SFS) to detect the algal community structure was first proposed by Yentsch & Yentsch (1979). Since then, SFS methods have been used in several studies of algal monocultures seeking to establish the effects of light and nutrient stress on algal pigmentation (e.g. Cleveland & Perry, 1987; SooHoo et al., 1986; Mitchell & Kiefer, 1988a; Sakshaug et al., 1991) or in the characterisation of the natural algal community structure (e.g. Yentsch & Phinney, 1985; Oldham et al., 1985; Cowles et al., 1993; Poryvkina et al., 1994). For a given species, the SFS is not stable – although the variable fluorescence can be partly suppressed by treating phytoplankton cells with DCMU (Johnsen & Sakshaug, 1993) – and automated analyses of algal community structure are thus difficult to obtain. SFS measurements are, however, very sensitive for tracking the distribution of fluorescence attributable to class-specific accessory pigments, thus yielding chemotaxonomic information on the phytoplankton community.

To assess the capability of SFS methods to detect changes in the phytoplankton community structure, we made several different types of fluorescence measurements in the Gulf of Riga (Baltic Sea). The gulf is semi-enclosed and exhibits a strong horizontal gradient of salinity caused by the large inputs of fresh water and low level of water exchange with the Baltic Proper. Riverine nutrient inputs have caused heavy eutrophication of the southern and central parts of the gulf in recent decades (Yurkovskis et al., 1993; Balode, 1994; Kostrichkina et al., 1994; Andrushaitis et al., 1995). During our cruise the weather was sunny and the sea calm, leading to strong thermal stratification, cyanobacterial blooms and a subsurface chl *a* maximum. Different phytoplankton communities were detected along the north-south gradient.

Material and methods

In vivo fluorescence measurements were performed at 18 fixed stations in the Gulf of Riga during the cruise of R/V Marina on 10–13 July 1994 (Figure 1). Water samples were collected at 0, 2.5, 5, 7.5, 10, 15, 20 and 30 m using 5–l Niskin bottles. Integrated surface samples (SURF) were obtained by mixing equal sample volumes from 0–10 m. Size-fractionation of the SURF samples was carried out by reverse filtration, using either a 10–μm nylon screen or a 2–μm Nuclepore filter. Samples were processed immediately on board. At the time of sampling, vertical profiling was conducted using a CTD equipped with a chlorophyll fluorometer (Aquapack, Chelsea Instruments). Vertical profiles of the light (PAR) attenuation were measured with a submersible quantum meter (Li-Cor, cosine collector).

The chl *a* content of vertical and fractionated samples was measured by filtering 100–ml aliquots onto Whatman GF/F filters, which were subsequently extracted with 96% ethanol for 24 h at room temperature. Chl *a* was measured with a spectrofluorometer (Shimadzu RF-5001) calibrated using chl *a* standard (Sigma) and an extinction coefficient of 83.4 l g^{-1} cm^{-1} (Wintermans & De Mots, 1965). The phytoplankton community structure was analysed in Lugol-fixed SURF samples using an inverted microscope (Diavert) according to Utermöhl (1958). Biomass (wet weight) transformations to cell volume and carbon content were made following the recommendations of the Baltic Marine Biologists (Edler, 1979). Total nitrogen and phosphorus concentrations were measured from the SURF samples.

Fluorescence measurements

The water samples were maintained in darkness for 1–4 h before fluorescence analysis. *In vivo* spectral fluorescence measurements were made in the onboard laboratory using a PC-controlled spectrofluorometer (Shimadzu RF-5001). Two-dimensional SFSs (2-D SFS) were measured as described by Poryvkina et al. (1994). This technique yields excitation-emission matrices that include all the wavelength combinations at which phytoplankton pigments are capable of fluorescing (see Table 1). Such a matrix also yields signals from DOM and Raman scattering. To obtain more detailed information on the spectral fluorescence of chl *a*, the excitation spectra (emission of 682 nm) was recorded (1-D SFS). Three replicate excitation spectra were measured, of which the average spectrum was used in later analyses. To remove the fluorescence signal of DOM and Raman scattering, aliquots of the samples were filtered through the Whatman GF/F filters and the fluorescence spectra of the filtrate were subtracted from the sample spectra. Quantum correction of SFSs was made using the dye Basic Blue 3 (Kopf & Heinze, 1984).

The measurement of 2-D SFS and 1-D SFS took 30 and 3 min, respectively, and were carried out on SURF samples only. For rapid assessment of the vertical variation in fluorescence characteristics, the *in vivo* fluorescence of all samples was measured at five selected wavelength combinations using an inbuilt program and an automatic sample changer (Table 1).

Results

Basin-wide patterns of hydrochemistry and algal taxonomy

The north-south gradients of surface salinity and nutrient concentrations (Figure 2a, c, d) resemble well those presented earlier for the Gulf of Riga (Yurkovskis et al., 1993). The concentrations of total nutrients increased by about 75% from the Baltic Proper to the most eutrophicated areas in the southern gulf. The elemental N/P ratio did not vary significantly in the study area (data not shown). During our cruise, the horizontal variation in chl *a* (Figure 3a) was similar to the variation in the concentration of total phosphorous ($r = 0.91$).

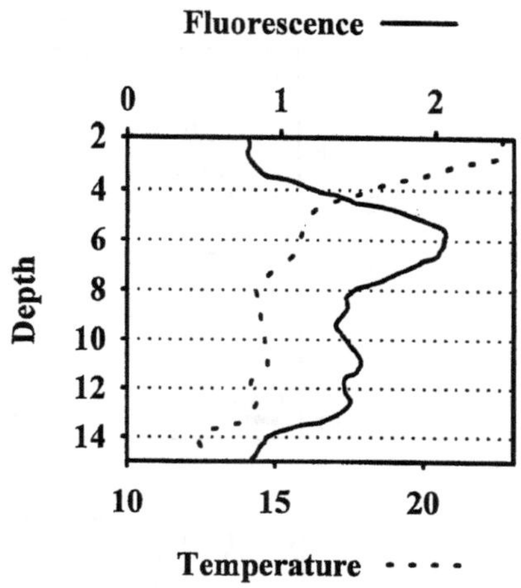

Figure 4. Temperature (°C) and chlorophyll fluorescence (relative units) profiles from station 13 as measured by CTD.

Strong thermal stratification was established shortly before and during our cruise (Figure 4) (Seppälä, unpubl.). The calm conditions allowed of a subsurface chl *a* maximum to form at 5–10 m depth at all stations; chl *a* concentrations were two to three times higher than at the surface. This was also evident in the fluorescence profiles (Figure 4).

In the Baltic Proper (station 1, Figure 1), the phytoplankton community was dominated by picoplankton (< 2 μm) as revealed by chl *a* results (Figure 3b) and characterised by the low biomass of nano- and microphytoplankton (32 mg C m^{-3}). Cryptomonads and diatoms contributed 59% and 36%, respectively, to the total nano- and microphytoplankton biomass (Figure 5). The proportion of picoplankton decreased with increasing algal biomass towards the inner and, especially, the southern gulf (Figures 3b and 5). In the southern part, picoplankton contributed only 17–25% of chl *a*, whereas filamentous cyanobacteria – mainly *Aphanizomenon flos-aquae* and *Nodularia spumigena* – contributed up to 90% of the total nano- and microphytoplankton biomass. The northern part of the gulf exhibited a more diverse phytoplankton community, with pronounced abundances of cryptophytes and diatoms (Figure 5). The Shannon index for biomass in the northern gulf was 2.5 on average, but in central and southern parts, where intensive blooms of cyanobacteria were observed, only 1.4.

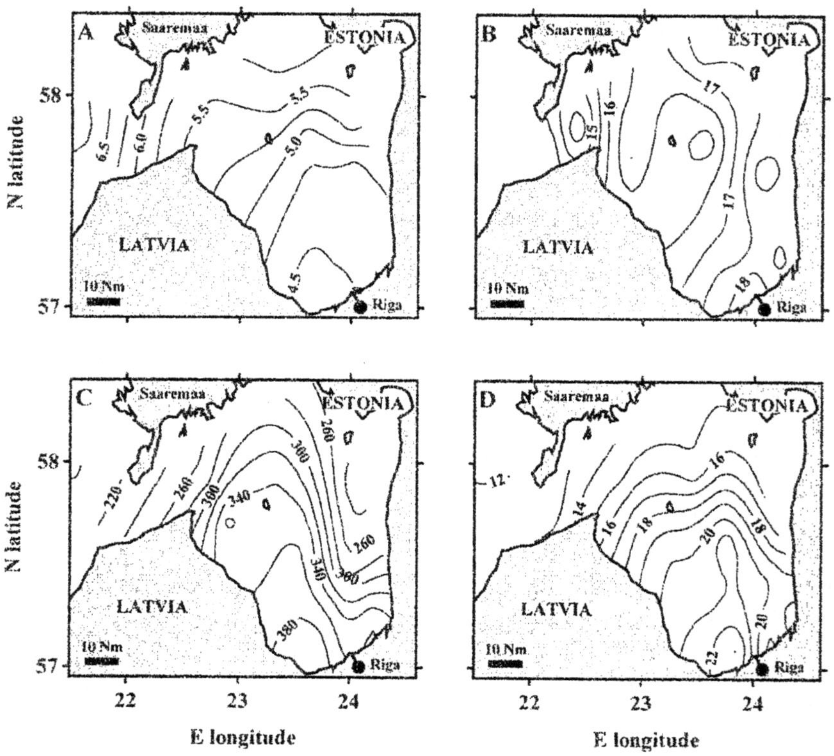

Figure 2. Surface (0–10 m) distribution patterns of a) salinity (PSU) and b) temperature (°C) averaged from the CTD data, and of c) total nitrogen ($\mu g\ l^{-1}$) and d) total phosphorous ($\mu g\ l^{-1}$) from integrated SURF samples.

Table 1. Settings of spectrofluorometer (Shimadzu RF-5001) for detection of *in vivo* fluorescence of phytoplankton pigments with the two-dimensional spectral fluorescence (2-D SFS), one-dimensional spectral fluorescence (1-D SFS) and at several fixed wavelengths

Method	Excitation wavelength nm (slit width)	Emission wavelength nm (slit width)	Light harvesting pigments monitored
2-D SFS			chl's *a, b, c,*
1st spectra	400 (3)	420–700 (5)	carotenoids and
2nd spectra	410 (3)	430–710 (5)	phycobilins
3rd spectra	420 (3)	440–720 (5)	
–	step of 10 nm	step of 10 nm	
25th spectra	640 (3)	660-940 (5)	
1-D SFS	400–660 (3)	682 (10)	chl's *a, b, c* and carotenoids
Fixed wavelengths			
1.	438 (3)	682 (5)	chl *a*
2.	480 (3)	682 (5)	chl's *b* and *c*
3.	520 (3)	682 (5)	carotenoids
4.	560 (3)	580 (5)	phycoerythrin
5.	630 (3)	650 (5)	phycocyanin

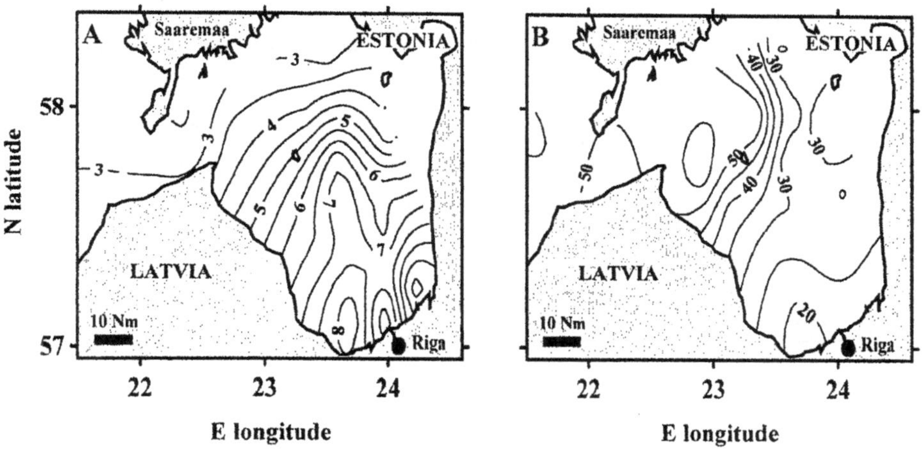

Figure 3. Surface (0–10m) distribution patterns of a) chl *a* (μg l^{-1}) and b) proportion (%) of chl *a* in size-fraction < 2 μm.

Fluorescence measurements

The 2-D SFS were characterised by a strong phycoerythrin signal at an excitation wavelength of 560 nm and an emission wavelength of 570 nm (F570) (Figure 6). The excitation at 560 nm also produced a wide secondary emission peak centered at 650 nm, possibly as a result of energy transfer from phycoerythrin to phycocyanin. No peak for direct excitation of phycocyanin at longer wavelengths (around 630 nm) was, however, observed. The emission line of chl *a* at 682 nm is also detectable in 2-D SFSs, but is even more distinct in excitation spectra, i.e. 1-D SFSs (Figure 7). After correcting for DOM and dividing by the corresponding chl *a* concentration, all 1-D SFSs were quite similar in shape and magnitude. The distinct peaks at 438 (F438) and 570 nm can be attributed to excitation by chl *a* and phycoerythrin, respectively. The 2-D SFSs obviously indicate that the phycoerythrin peak does not represent emission from chl *a*. The excitation reflecting the accessory chlorophylls and carotenoids is responsible for the shoulders at around 460–520 nm. The DOM signal (lower left corner at 2-D SFSs in Figure 6) at shorter emission wavelengths (< 600 nm) was distinct for southern stations influenced by the gelbstoff load from the river Daugava.

The fluorescence intensity had a tendency to rise towards the southern gulf, irrespective of either method or wavelength, thus corresponding to the distribution of chl *a* or carbon biomass (Figure 8, Figures 3a and 5a). The CTD fluorometer profiles or the direct excitation of chl *a* (F438) by SFS methods correlated well with extracted chl *a* (r = 0.87 and 0.79, respectively). For both types of measurements, however, the variability of R showed clear spatial patterns (Figure 8b). The distribution was similar to that of the ratio of excitation of chl *a* directly to the excitation reflecting accessory chlorophylls and carotenoids (F438/F470) (Figure 8d).

The vertical samples revealed strong variability of R; on the surface it was about half that at 5–10 m depth. Similar patterns were recorded at all times of the day, from both *in situ* of CTD-fluorometer profiles and dark-adapted samples (Figure 9). Some indication of the diel vertical changes in SFSs was obtained. In the area where cyanobacterial blooms occurred, the ratio of phycoerythrin fluorescence (F570) to chl *a* fluorescence (F438) exhibited a maximum at midnight on the surface. This maximum moved gradually deeper during the morning (Figure 10). The measurements were not made at the same station (stations 13–16); however, in areas of similar phytoplankton species composition. The ratios of other wavelength combinations failed to show any general patterns.

Discussion

Chl a specific fluorescence, R

In vivo fluorescence is a semiquantitative indicator of the chl *a* content of water, and usually the fluorescence profiles presented have been calibrated against pigment analyses. Linear regression – setting R as a constant – is the easiest way to calibrate fluorescence profiles. Such calibrations yield good results if samples are collected in a wide range of concentrations.

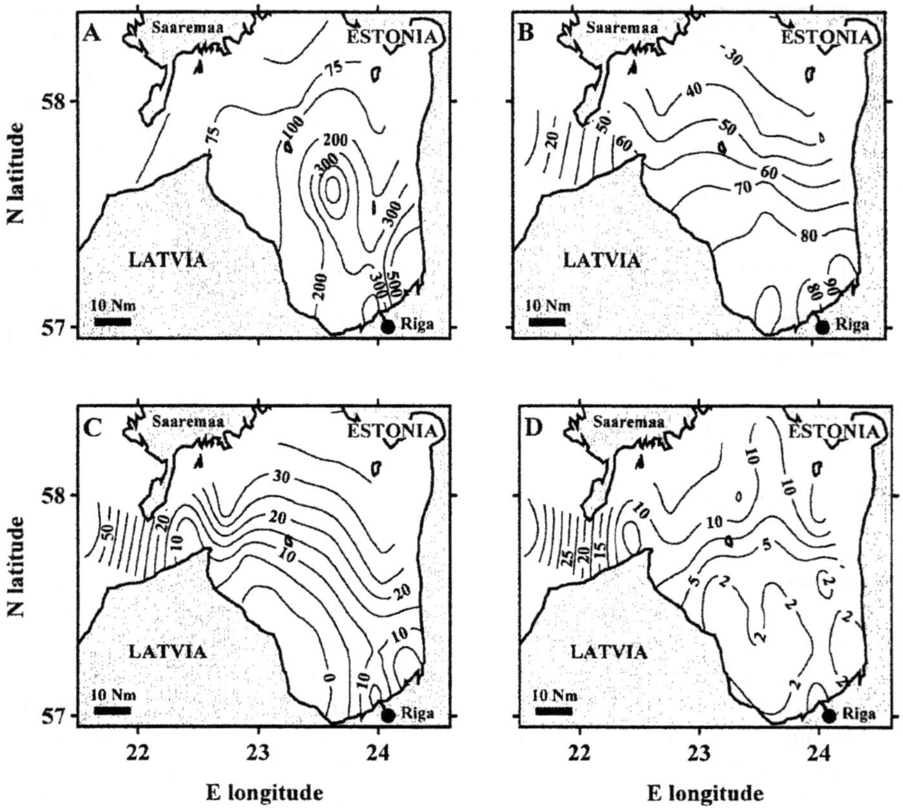

Figure 5. Surface (0–10m) distribution patterns of a) biomass of micro- and nanophytoplankton (mg C m^{-3}) and the proportion (%) of b) cyanobacteria, c) cryptomonads and d) diatoms relative to total biomass. Samples counted using inverted microscope.

R is in reality, however, a variable and it responds, for instance, to phytoplankton species composition, sometimes with detectable spatial and temporal patterns (Loftus & Seliger, 1975; Owens, 1991). The cause of the variability of R is complex – even when single species are concerned – due to the physiology of and physical processes in algal cells (Mitchell & Kiefer, 1988a; Owens, 1991). Hence, the only practical approach is to study statistically the relationship of R to environmental variables such as ambient irradiance, mixing depth, nitrogen availability, species composition and phytoplankton size spectra.

Correlation analysis (not shown) of these variables indicates that the distribution pattern obtained for R during our cruise was mainly due to changes in the phytoplankton community structure. Most significantly, the southward decrease in R was related to the increase in the proportion of filamentous cyanobacteria relative to total phytoplankton biomass (% of total nano- and microphytoplankton), the increase in phycoerythrin fluorescence and the decrease in small algae (as % of chl *a* in < 10 μm fraction) with correlation coefficients of –0.52, –0.77 and 0.84 (Figure 11) respectively ($n = 18$). It is well established that R is much lower (about 10 times using our optical setup, unpubl.) for cyanobacteria than for eukaryotes because most of their chl *a* is connected to photosystem I, which does not fluoresce at normal temperature (Vincent, 1983; Bryant, 1986; Johnsen & Sakshaug, 1996). Furthermore, for small algae – or generally algae with a small package effect – the chl *a*-specific absorption is higher, and thus fluorescence, too, is likely to be higher (assuming a fairly constant ratio of fluorescence to absorption). Theoretical studies of Kirk (1983) indicate that the package effect is a relatively linear function of size for similarly pigmented spherical small cells (0–20 μm). Though the dynamic nature of algal pigmentation and fluorescence yield will cause large deviations from linearity of R and cell size, it was evident that such a relation existed during

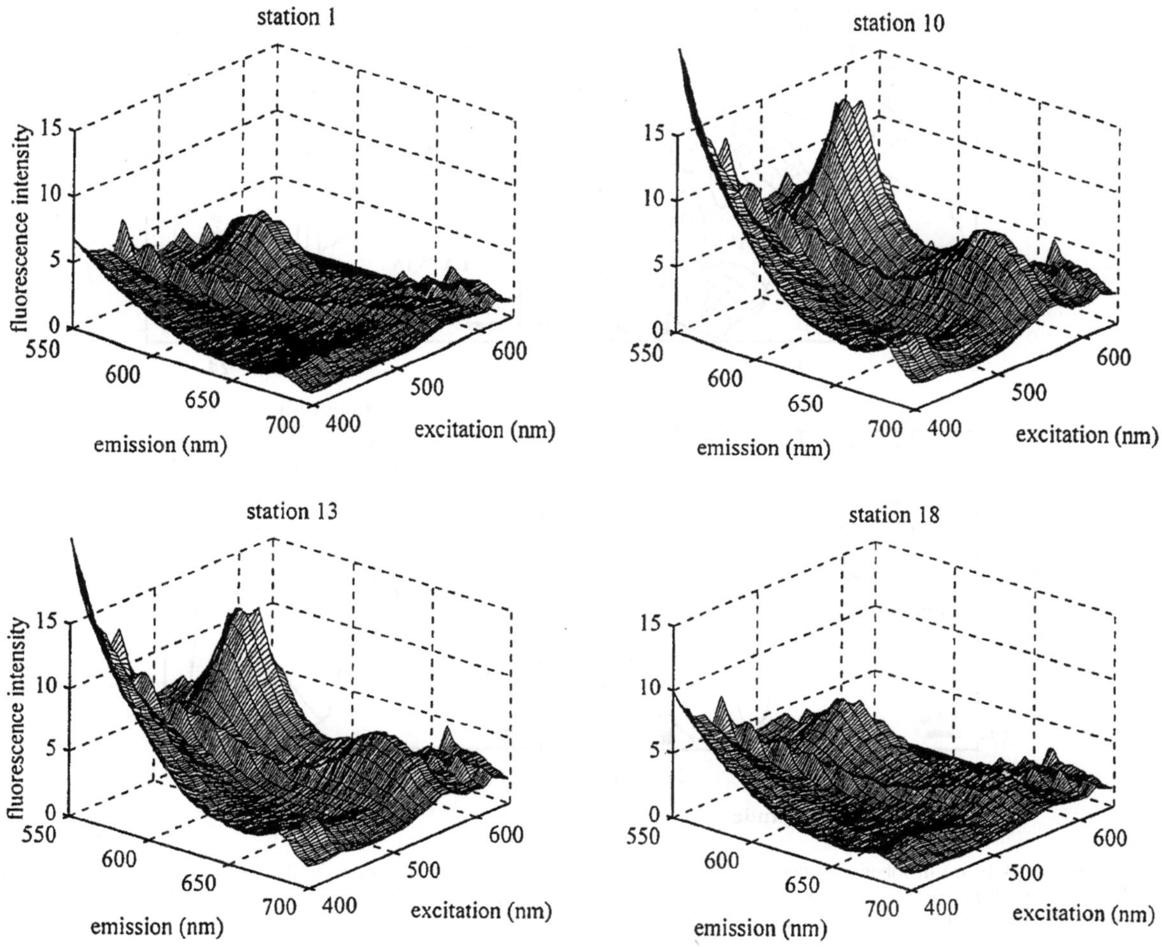

Figure 6. Fluorescence excitation-emission matrices (2-D SFS) of four different phytoplankton communities. Phycoerythrin peak: excitation at 560 nm, emission at 570 nm. Chl *a* excitation: emission at 682 nm. DOM is evident as a rise in fluorescence at low emission wavelengths.

our study (see Figure 11). Similarly Alpine & Cloern (1985) found that the R was highest in the smallest phytoplankton size-class they examined. Yet opposite relation was found for the open ocean phytoplankton (Guo & Dunstan, 1995). They suggested that in estuarine and coastal areas the packaging effect has a more important role in determining R than photosynthetic quantum efficiency, the latter being more important in the oligotrophic oceanic waters.

Photoinhibition is a possible cause for the decreasing *in vivo* chl *a* fluorescence (e.g. Falkowski & Kiefer, 1985). Despite the sunny weather during our cruise (up to 1800 μE m^{-2}s^{-1}), no statistical differences in R were recorded for daytime and night-time samples. If photoinhibition occurs in surface samples it is probably recovered during the dark-adaptation period (see Harris, 1980). Depth-dependent differences in R were observed at all stations, but taking into account the high light attenuation coefficient of the water ($k_d = 0.45$–0.75 m^{-1}) and the weak mixing of the surface layer, these could hardly have been caused by photoinhibition. During the night, the fluorescence properties of the mixed layer should be similar at all depths. The vertical structure of R may have been due to the different proportions of cyanobacteria at different depths (see below), or the fluorescence yield may have increased because of the lower utilization efficiency of light for photosynthesis.

Spectral fluorescence

Phycoerythrin fluorescence (F570) originates from cyanobacteria or cryptophytes. We assume that the observed phycoerythrin signal did not derive

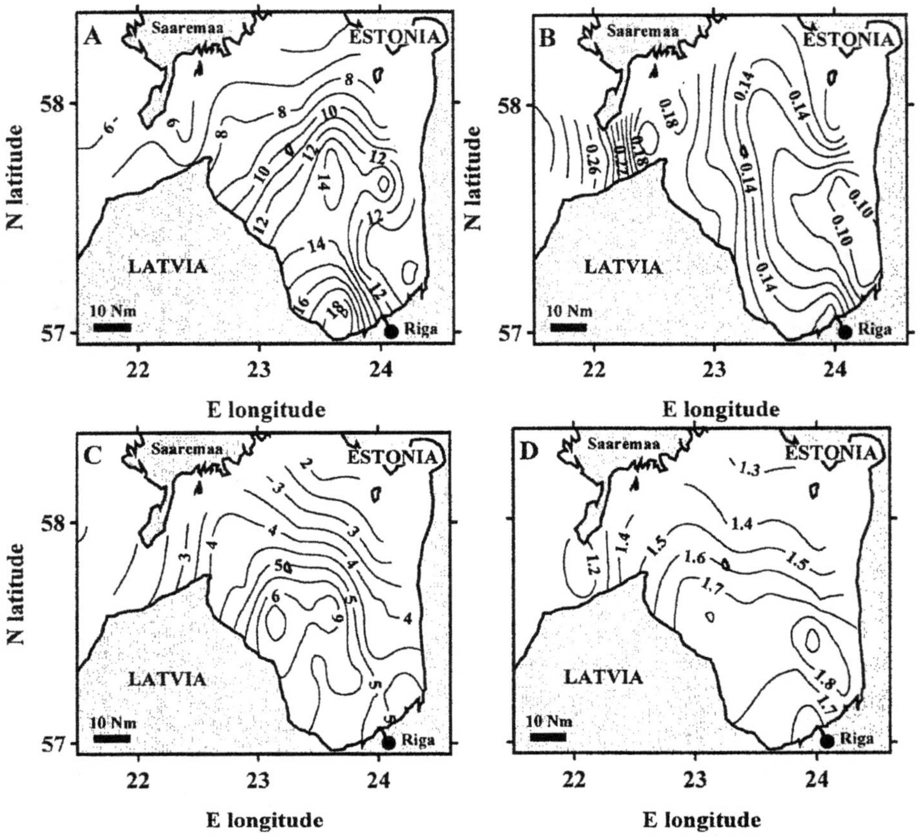

Figure 8. Surface (0–10 m) distribution patterns of (a) chl *a* fluorescence (F438), (b) chl *a*-specific fluorescence (R), (c) phycoerythrin fluorescence (F570), (d) the ratio of direct excitation of chl *a* to excitation through accessory chl's and carotenoids (F438/F470). Values are taken from 1-D SFS, except F570, which is from fixed wavelength measurements.

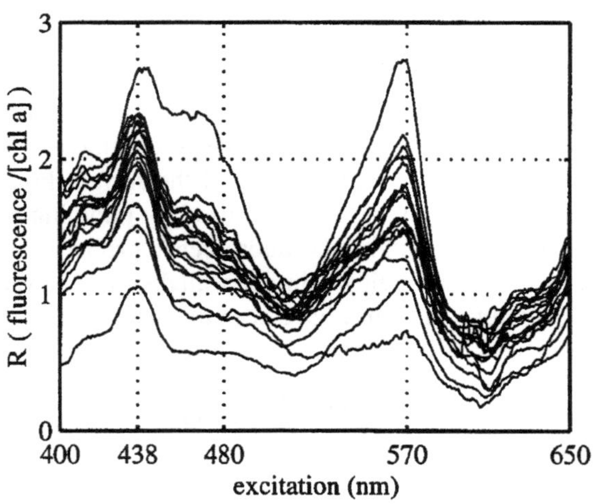

Figure 7. Excitation spectra of chl *a* (1-D SFS) for all stations. The signals from DOM have been subtracted, and the spectra quantum corrected and divided by chl *a* concentration. Excitation peaks: 438 nm (chl *a*) and 570 nm (phycoerythrin).

mainly from the picoplanktonic cyanobacteria (*Synechococcus*-type) cells, because there was no gradient in picoplankton biomass such as in F570 (Figure 8c). The F570 signal was rather related to algae > 10 μm ($r = 0.70$). The pigmentation of picocyanobacteria in the Gulf of Riga has not, however, been studied. There may be a southward shift from phycocyanin to phycoerythrin-dominated strains in the gulf, although phycoerythrin is usually the major phycobilin in oceanic, and phycocyanin in coastal *Synechococcus*-strains (Waterbury et al., 1986). The F570 signal was strongest at the southern stations, where filamentous cyanobacteria dominated and cryptomonads were almost entirely lacking. In the north, where the cryptomonads were more abundant, F570 was low. We therefore suggest that the phycoerythrin signal was mainly due to the large cyanobacteria (Figure 12). The multiple emission peak of phycoerythrin has previously been related to the cryptomonad phycoerythrin Cr-PE_{545} (Rowan,

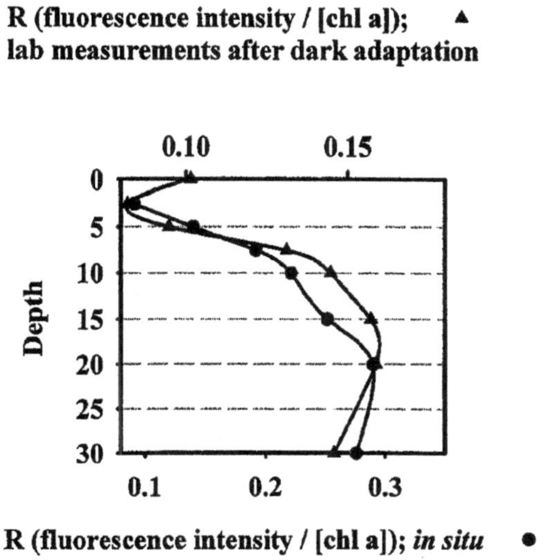

Figure 9. Vertical variability of chl *a*-specific fluorescence (R) at station 13. Fluorescence profiles were measured at midnight by CTD (*in situ*) or, in the discrete samples, in the laboratory after the dark-adaptation period.

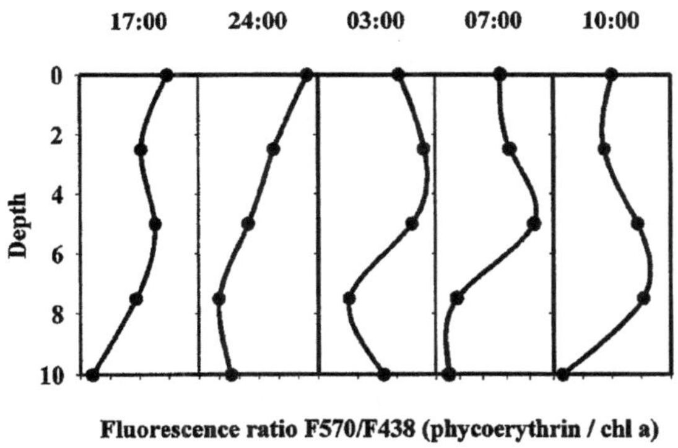

Figure 10. Diel variation in the phycoerythrin to chl *a* fluorescence ratio (F570/F438).

1989; Poryvkina et al., 1994), although such a spectral shape also seems to be normal for cyanobacteria (e.g. Vincent, 1983). We did not detect a fluorescence peak of phycocyanin using 2-D SFSs. As also noted by Poryvkina et al. (1994), this could be due to the lowered production of this pigment in cells grown in strong light.

As for fluorescence and chl *a*, one cannot expect a constant ratio of phycoerythrin concentration to its fluorescence. The pigment-specific variability of phycoerythrin fluorescence has been poorly studied because quantitative analyses are difficult to perform (Wyman, 1992). It may be possible to monitor cyanobacteria by phycobilin fluorescence (e.g. Watras & Baker, 1988).

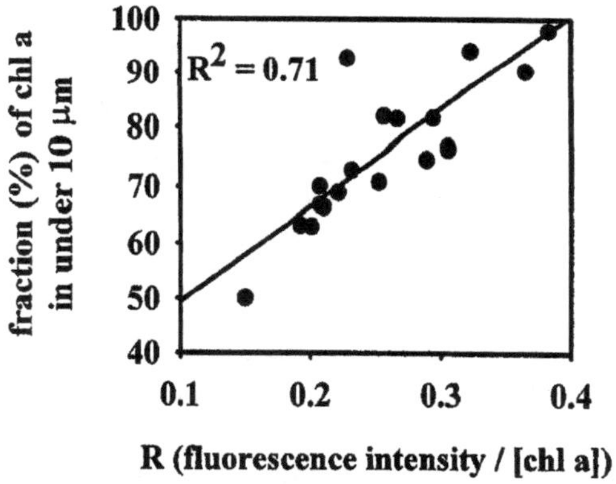

Figure 11. Relationship between chl *a* specific fluorescence and phytoplankton cell size. Line is a fit by linear regression.

Much more information is needed, however, on the variability of this pigment, and its causes in natural waters, especially during periods of cyanobacterial blooms.

Where diatoms and cryptomonads were relatively more abundant, the excitation of chl *a* through transfer of excitons from accessory chlorophylls and carotenoids was higher than in areas of cyanobacterial domination (Figure 8d). Thus the shape of spectra indicates the relative importance of accessory pigments for light harvesting. Yentsh & Phinney (1985) suggested using the CAP ratio as an index for certain phytoplankton groups. This ratio, which is the ratio of chl *a* fluorescence excited at 438 nm to that of carotenoids at 530 nm, could be misleading in cases such as those discussed here, since it does not take phycoerythrin-induced fluorescence in green light into account. Such a ratio should therefore be used with great caution unless the overall fluorescence properties of the samples are known. It might be advantageous to combine the measurements of some fixed wavelengths with more detailed spectral analyses, as with the 2-D SFS method used here.

The only evidence of a systematic vertical change in algal pigmentation was in the ratio of phycoerythrin to chl *a* fluorescence (F570/F438) obtained in the southern part of the gulf. This pattern may have been caused by the active vertical migration of cyanobacteria. No other differences were found in the fluorescence properties of surface and deep populations.

In conclusion, the *in vivo* fluorescence patterns observed in the Gulf of Riga were closely related to general features of the phytoplankton community. The higher proportion of filamentous cyanobacteria and lower abundance of eukaryotes and picoplankton in

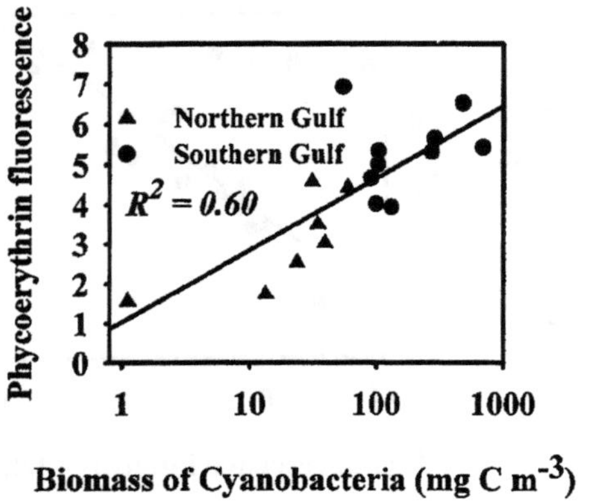

Figure 12. Relationship between logarithm of biomass of cyanobacteria (mg C m^{-3}; picocyanobacteria not included) and phycoerythrin fluorescence. Line is a fit by linear regression.

the southern gulf were accompanied by (1) lower chl *a*-specific fluorescence (R), (2), higher phycoerythrin fluorescence and (3) a shift in the shape of chl *a* fluorescence excitation spectra. Spectral fluorescence signals of other algal groups were masked by signals from the cyanobacterial biomass.

Determining *in vivo* spectral fluorescence is a recent and sophisticated way of making non-manipulating measurements of the algal community. The approach has considerable potential in the field of unattended monitoring of phytoplankton, especially in combination with conventional methods of pigment analysis and phytoplankton microscopy. The responses of bio-optical parameters, e.g. spectral fluorescence, of the algae to fluctuating natural light and nutrient regimes are, however, not yet fully understood. To interpret field data, laboratory studies need to be performed.

Acknowledgements

First and foremost we thank Dr T. Tamminen, E. Salminen, M. Sjöblom and T. Saloranta for participating in the cruise, providing the nutrient results, performing the size-fractionation and helping with sampling and CTD casting. The study was inspired by discussions with Drs S. Kaitala, S. Babichenko, L. Poryvkina and A. Leeben. The great spirit of the crew of R/V Marina is warmly acknowledged, as well as the comments of two anonymous reviewer and the language check by Mrs G. Häkli. This study is a contribution of the projects 'Pelagic Eutrophication and Sedimentation in the Gulf of Riga' financed by Nordic Council of Ministers and 'PELAG III' (Nitrogen Discharge, Pelagic Nutrient Cycles, and Eutrophication of the Coastal Environment of the Northern Baltic Sea).

References

Alpine, A. E. & J. E. Cloern, 1985. Differences in *in vivo* fluorescence yield between three phytoplankton size classes. J. Plankton Res. 7: 381–390.

Althuis, IJ. A., W. W. C. Gieskes, L. Villerius & F. Colijn, 1994. Interpretation of fluorometric chlorophyll registrations with algal pigment analysis along a ferry transect in the southern North Sea. Neth. J. Sea Res. 33: 37–46.

Anderson, J. M. & J. Barrett, 1986. Light-harvesting pigment-protein complexes of algae. In L. A. Staehelin & C. J. Arntzen (eds), Photosynthesis III. Encl. Plant Phys.. Springer-Verlag, Berlin, 19: 269–285.

Andrushaitis, A., Z. Seisuma, M. Legzdina & E. Lensh, 1995. River load of the eutrophying substances and heavy metals into the Gulf of Riga. In E. Ojaver (ed.), Ecosystem of the Gulf of Riga between 1920 and 1990. Estonian Academy Publishers, Tallin, 32–40.

Balode, M., 1994. Long-term changes of summer-autumn phytoplankton communities in the Gulf of Riga. In O. Guelorget & A. Lefebvre (eds), Baltic Sea and Mediterranean Sea. A comparative ecological approach of coastal environments and paralic ecosystems. Montpellier: 96–99.

Bryant, D. A., 1986. The cyanobacterial photosynthetic apparatus: Comparison to those of higher plants and photosynthetic bacteria. In T. Platt & W. K. W. Li (eds), Photosynthetic picoplankton. Can. Bull. Fish. aquat. Sci. 214: 423–500.

Cleveland, J. S. & M. J. Perry, 1987. Quantum yield, relative specific absorption and fluorescence in nitrogen-limited *Chaetoceros grasilis*. Mar. Biol. 94: 489–497.

Cowles, T. J., R. A. Desiderio & S. Neuer, 1993. *In situ* characterization of phytoplankton from vertical profiles of fluorescence emission spectra. Mar. Biol. 115: 217–222.

Edler, L. (ed.), 1979. Recommendations on methods for marine biological studies in the Baltic Sea. BMB Publ. 5: 38 pp.

Falkowski, P. & D. A. Kiefer, 1985. Chlorophyll *a* fluorescence in phytoplankton: relationship to photosynthesis and biomass. J. Plankton Res. 7: 715–731.

Guo, C. & W. M. Dunstan, 1995. Depth-dependent changes in chlorophyll fluorescence number at a Sargasso Sea station. Mar. Biol. 122: 333–339.

Harris, G. P., 1980. The relationship between chlorophyll *a* fluorescence, diffuse attenuation changes and photosynthesis in natural phytoplankton populations. J. Plankton Res. 2: 109–127.

Harris, G. P., 1986. Phytoplankton ecology. Structure, function and fluctuation. Chapman and Hall, London, 384 pp.

Johnsen, G. & E. Sakshaug, 1993. Bio-optical characteristics and photoadaptive responses in the toxic and bloom-forming dinoflagellates *Gyrodinium aureolum*, *Gymnodinium galatheanum*, and two strains of *Prorocentrum minimum*. J. Phycol. 29: 627–642.

Johnsen, G. & E. Sakshaug, 1996. Light harvesting in bloom-forming marine phytoplankton: species-specificity and photoacclimation. In F. L. Figueroa, C. Jimenez, J. L. Pèrez-Llorèns & F. X. Niell (eds), Underwater light and algal photobiology. Sci. Mar. 60: 47–56.

Kiefer, D. A., 1973. Chlorophyll *a* fluorescence in marine centric diatoms: responses of chloroplasts to light and nutrient stress. Mar. Biol. 23: 39–46.

Kirk, J. O. T., 1983. Light and photosynthesis in aquatic ecosystems. Cambridge University Press, Cambridge, 401 pp.

Kopf, U. & J. Heinze, 1984. 2,7-Bis(diethylamino)phenazoxonium chloride as a quantum counter for emission measurements between 240–700 nm. Analyt. Chem. 56: 1931–1935.

Kostrichkina, E., B. Kalveka, B. Jansone & A. Ikauniece-Grunde, 1994. Planktonic communities in the conditions of eutrophycation of the Gulf of Riga. In O. Guelorget & A. Lefebvre (eds), Baltic Sea and Mediterranean Sea. A comparative ecological approach of coastal environments and paralic ecosystems. Montpellier: 110–115.

Legendre, L. & S. Demers, 1984. Towards dynamic biological oceanography and limnology. Can. J. Fish. aquat. Sci. 41: 2–19.

Loftus, M. E. & H. H. Seliger, 1975. Some limitations of the *in vivo* fluorescence technique. Chesapeake Sci. 16: 79–92.

Lorenzen, C. J., 1966. A method for continuous measurement of *in vivo* chlorophyll concentration. Deep Sea Res. 13: 223–227.

Mitchell, B. G. & D. A. Kiefer, 1988a. Chlorophyll *a* specific absorption and fluorescence excitation spectra for light limited phytoplankton. Deep Sea Res. 35: 639–663.

Mitchell, B. G. & D. A. Kiefer, 1988b. Variability in pigment specific particulate fluorescence and absorption spectra in the northeastern Pacific Ocean. Deep Sea Res. 35: 665–689.

Oldham, P. B., E. J. Zillioux & I. M. Warner, 1985. Spectral 'fingerprinting' of phytoplankton populations by two-dimensional fluorescence and fourier-transform-based pattern recognition. J. mar. Res. 43: 893–906.

Owens, T. G., 1991. Energy transformation and fluorescence in photosynthesis. NATO ASI Series G27: 101–137.

Poryvkina, L., S. Babichenko, S. Kaitala, H. Kuosa & A. Shalapjonok, 1994. Spectral fluorescence signature in the characterization of phytoplankton community composition. J. Plankton Res. 16: 1315–1327.

Prézelin, B. B., 1981. Light reactions in photosynthesis. In T. Platt (ed.), Physiological bases of phytoplankton ecology. Can. Bull. Fish. aquat. Sci. 210: 1–43.

Rowan, K. S., 1989. Photosynthetic pigments of algae. Cambridge University Press, Cambridge, 334 pp.

Sakshaug, E., G. Johnsen, K. Andresen & M. Vernet, 1991. Modeling of light-dependent algal photosynthesis and growth: experiments with Barents Sea diatoms *Thalassiosira nordenskioeldii* and *Chaetoceros furcellatus*. Deep Sea Res. 38: 415–430.

SooHoo, J. B., D. A. Kiefer, D. J. Collins & I. S. McDermid, 1986. *In vivo* fluorescence excitation and absorption spectra of marine phytoplankton: I. Taxonomic characteristics and responses to photoadaptation. J. Plankton Res. 8: 197–214.

Strass, V., 1990. On the calibration of large-scale fluorometric chlorophyll measurements from towed undulating vehicles. Deep Sea Res. 37: 525–540.

Utermöhl, H., 1958. Zur vervollkommung der quantitativen Phytoplankton-Methodik. Mitt.int. Verein. theor. angew. Limnol., 9: 1–38.

Vincent, W. F., 1983. Fluorescence properties of the freshwater phytoplankton: three algal classes compared. Br. phycol. J. 18: 5–21.

Waterbury, J. B., S. W. Watson, F. W. Valois & D. G. Franks, 1986. Biological and ecological characterization of the marine unicellular cyanobacterium *Synechococcus*. In T. Platt & W. K. W. Li (eds), Photosynthetic picoplankton. Can. Bull. Fish. aquat. Sci. 214: 71–120.

Watras, C. J. & A. L. Baker, 1988. Detection of planktonic cyanobacteria by tandem *in vivo* fluorometry. Hydrobiologia 169: 77–84.

Wintermans, J. F. G. H. & A. De Mots, 1965. Spectrophotometric characteristics of chlorophylls *a* and *b* and their phaeophytins in ethanol. Biochem. Biophys. Acta 109: 448–453.

Wyman, M., 1992. An *in vivo* method for the estimation of phycoerythrin concentrations in marine cyanobacteria (*Synechococcus* spp.). Limnol. Oceanogr. 37: 1300–1306.

Yentsch, C. S. & D. A. Phinney, 1985. Spectral fluorescence: an ataxonomic tool for studying the structure of phytoplankton populations. J. Plankton Res. 7: 617–632.

Yentsch, C. S. & C. M. Yentsch, 1979. Fluorescence spectral signatures: The characterization of phytoplankton populations by the use of excitation and emission spectra. J. Mar. Res. 37: 471–483.

Yurkovskis, A., F. Wulff, L. Rahm, A. Andrushaitis & M. Rodiquez-Medina, 1993. A nutrient budget of the Gulf of Riga, Baltic Sea. Estuar. coast. Shelf Sci. 37: 113–127.

Hydrobiologia **363:** 219–227, 1998.
T. Tamminen & H. Kuosa (eds), Eutrophication in Planktonic Ecosystems: Food Web Dynamics and Elemental Cycling.

The budgets of nitrogen and phosphorus in shallow eutrophic Lake Võrtsjärv (Estonia)

Peeter Nõges[1,2], Arvo Järvet[3], Lea Tuvikene[1,2] & Tiina Nõges[1,2]
[1] *Institute of Zoology and Botany, Võrtsjärv Limnological Station, EE2454 Rannu, Tartu County, Estonia*
[2] *Tartu University, Institute of Zoology and Hydrobiology, 46 Vanemuise St., EE2400 Tartu, Estonia*
[3] *Tartu University, Institute of Geography, 46 Vanemuise St., EE2400 Tartu, Estonia*

Key words: Nutrient budget, nutrient uptake, shallow lake, sedimentation, resuspension, denitrification

Abstract

The nutrient budget, phytoplankton primary production and sedimentation rate were studied weekly in the large (270 km^2) and shallow (mean depth 2.8 m) eutrophic Lake Võrtsjärv in 1995. The annual external loading was 7.7 g m^{-2} y^{-1} of total nitrogen (TN) and 0.2 g m^{-2} y^{-1} of total phosphorus (TP), including 7% of both inputs as direct atmospheric precipitation. The external budget revealed a retention of 53% of TN and 28% of TP annual input. About 80% of the total loss of nitrogen was accounted for by denitrification (3.3 g m^{-2} y^{-1}) and only 20% was buried into the sediment. Wind-induced sediment resuspension played the major role in the upward nutrient flux formation and, thus, in the formation of the temporal pattern of nutrient concentration during the ice-free season. Other fluxes as the external loading or new sedimentation of autochthonous production were overcome and masked by the powerful resuspension – sedimentation cycle, exceeding the former by one or two orders of magnitude. The intensity of upward flux of nutrients (mainly caused by resuspension) increased in accordance to decreasing water level in autumn and correlated with the weekly average wind speed. The summer population of filamentous blue-greens dominated by *Limnothrix redekei* was light-limited until the minima of TN:TP ratio (< 10) and DIN:TN ratio (~ 0) in July initiated a clear peak of N_2-fixing algae (*Aphanizomenon gracile, Anabaena* spp.) which lasted until September.

Introduction

Nutrient budgets have provided data on a large variety of lakes, from those acting as efficient sinks of phosphorus up to the lakes in which almost continuous phosphorus release occurs from anaerobic (Laugaste, 1994) or aerobic sediments (Löfgren, 1987). The relevance of internal processes, especially of the matter exchange at the sediment-water interface, increases with the area and shallowness of the lake. In large shallow lakes and reservoirs the resuspension-sedimentation fluxes caused by wind-induced mixing can accelerate phosphate release (Søndergaard et al., 1992) and provide a repetitive use of nutrients by phytoplankton (Galicka, 1992) or, on the contrary, enable phosphate to adsorb on fine-grained sediment particles to be carried away from the ecosystem (Wisniewski, 1995). The internal loading of nutrients can be determined as the upward flux (UWF) from the sediment (Boers & Uunk, 1991). In the case of total nutrients this flux consists of the release of soluble forms and resuspension of particulate, mainly organically bound, nutrients. The downward flux of nutrients (DWF) can be measured by sediment traps as gross sedimentation, i.e. as a deposition of primarily produced + resuspended matter. Common mass balance calculation allows to estimate the resultant of UWF and DWF as the net budget. Depending on the polarity of subtracting, the resultant flux is discussed as the net sedimentation (Dudel & Kohl, 1991; Jensen et al., 1992) or net internal loading (Boers & Uunk, 1991). Combining the budget calculations with sedimentation measurements enables to quantify both, DWF and UWF.

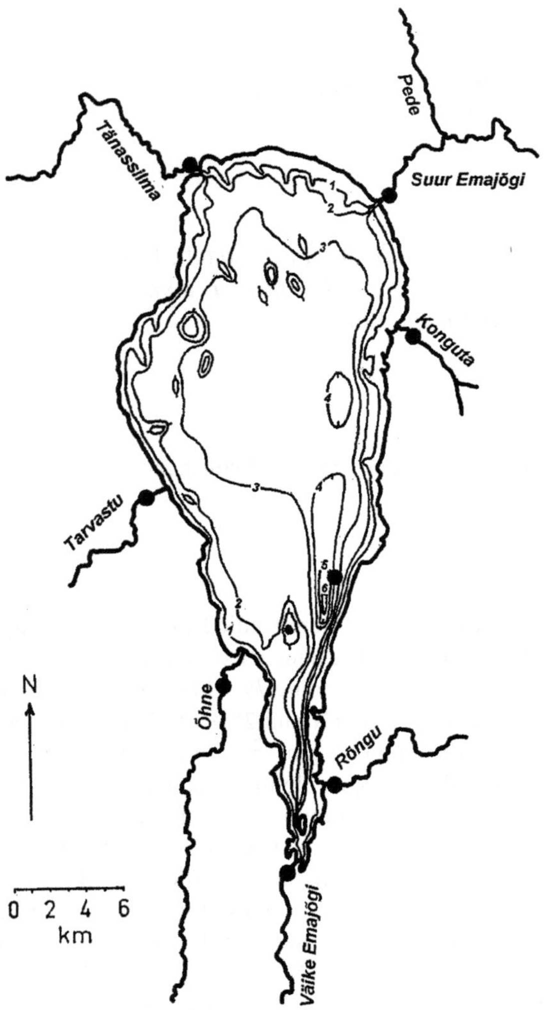

Figure 1. Lake Võrtsjärv and the sampling sites.

Largely variable role of denitrification ranging from 0–10% (Andersen, 1974) up to 86–90% of the nitrogen loss rate (Jensen et al., 1992) has been obtained for lakes. Although the budget assessment of denitrification is not considered a perfect method (Vollenweider, 1989) it is still widely used (Dudel & Kohl, 1991; Jensen et al., 1990, 1992; Ahlgren et al., 1994) as it gives at least a clue of the magnitude of the process.

The aim of the present paper is to draw out the scale, occurrence and magnitude of the main nutrient fluxes determining the trophic features of the shallow eutrophic lake Võrtsjärv.

Description of the study site

Lake Võrtsjärv (Figure 1) is a large and shallow waterbody located in the central part of Estonia. Its area is 270 km^2 while the mean depth is only 2.8 m. The water level is very variable: the annual mean amplitude (1.4 m) is equal to one half of the mean depth. Water renewal proceeds, on the average, once a year, but may differ a lot in dry and rainy years ($\tau_{\omega} = 0.71$–$1.33\ y^{-1}$) (Jaani, 1990).The mainly cultivated drainage area exceeds the area of the lake about 12 times. The lake has 6 main inflows and one outflow which carries the water to Lake Peipsi. Due to the shallowness and the large wind exposed area L. Võrtsjärv is unstratified and turbid (Secchi depth from 0.5 to 1.0 m during ice-free period). According to the mean concentrations of TN (2 mg l^{-1}) and TP (about 50 $\mu g\ l^{-1}$) the lake is regarded as eutrophic. Phytoplankton is represented predominantly by filamentous forms: diatoms of genus *Aulacoseira* in spring and blue-greens *Planktolyngbya limnetica* (Lemm.) Kom. et Cronb., *Limnothrix redekei* (Van Goor) Meffert, *Oscillatoria amphibia* f. *tenuis* (Anissim.) Elenk. and *Aphanizomenon gracile* Lemmermann in summer and autumn. Phytoplankton biomass and chlorophyll *a* often reach their maxima in September–October. In turbid water phytoplankton growth is limited mainly by light (Nõges & Järvet, 1995). Annual primary production was 224 g C m^{-2} in 1995 (Nõges, 1997).

Material and methods

In 1995 water samples for nutrient analyses were taken every week from all 6 main inflows, from one point in the southern part of the lake and from the outflow (see Figure 1). A 1.5 l Ruttner sampler was used and the samples from the lake were taken as integrated over depth. In internal and net budgets the outflow was taken into account as a point integrating the northern part of the lake.

Daily flow values of the rivers Väike Emajõgi, Õhne and Tarvastu, representing 70–75% of the total discharge, and of the outflowing R. Suur Emajõgi, formed the basis for water budget calculations. Flow in smaller inflows was calculated by using earlier regressions. The mean specific runoff of the studied area was applied to calculate the runoff from the shore areas (8.5% of the watershed). The lake was regarded as a turbulent mixed reactor using the mean values of two sampling points (the lake point and the outflow). All calculations were made by one day step. Lake volume was calculated from daily water level data. Daily values for concentrations were obtained by linear interpolation of weekly data.

We chose the three-step budget calculation method. The external budget (EB) showing the retention (sedimentation, transformation, denitrification) was calculated as the input–output difference:

$$EB = L_{in} - L_{out}, \quad (1)$$

where L_{in} = external (riverine + atmospheric) loading; L_{out} – outflow.

The atmospheric loading was taken into account only for NH_4-N, NO_3-N and PO_4-P by applying the 1995 results of measurements in Lahemaa National Park (N-Estonia) (Kört & Roots, 1996). The internal budget (IB) reflecting the changes in water column nutrient storage during the time interval $t(n)-t(n-1)$ was calculated as:

$$IB = C_{t(n)}V_{t(n)} - C_{t(n-1)}V_{t(n-1)}, \quad (2)$$

where C = concentration; V = lake volume.

The internal budget was used as an intermediate step (results not shown) in order to get the net budget, which shows the changes in nutrient amounts in the lake water not explainable by the external budget:

$$NB = IB - EB. \quad (3)$$

Net budget is formed as a resultant of UWF and DWF, hence, the upward flux was calculated as the sum of NB and gross sedimentation rate (= DWF):

$$UWF = NB_{TN;TP} + DWF. \quad (4)$$

The gross sedimentation rate was measured at four stations in the lake from June 19 to October 23. Twin traps of 50 cm length and of 5 cm inner diameter were exposed at 1.5 m from the bottom and emptied every week. The dry weight of the trap catch was measured and the trap material analyzed for TN and TP according to standard methods used by the Finnish National Board of Waters and the Environment.

Primary production of phytoplankton (PP) was measured by ^{14}C method (Steeman-Nielsen, 1952) during the whole year at one station in the lake with one week intervals. PP measurements in 1995 are described in more detail by Nõges (1998). The particulate carbon

Table 1. Hydrological regime of L. Võtsjärv in 1995.

Month	Inflow (10^6 m^3 d^{-1})	Outflow	Lake volume (10^6 m^3)	Mean depth (m)
I	1.9	1.7	680	2.5
II	6.0	1.3	718	2.6
III	6.5	1.7	929	3.2
IV	3.7	3.0	989	3.3
V	4.2	3.6	997	3.3
VI	1.4	3.7	980	3.3
VII	0.6	3.7	869	3.0
VIII	0.5	2.9	746	2.7
IX	0.5	2.4	652	2.4
X	0.8	2.0	601	2.3
XI	0.8	1.6	555	2.2
XII	0.6	1.4	534	2.1
Year	2.3	2.4	771	2.7

fixation was used in the calculations of potential new sedimentation. Nutrient uptake rates by phytoplankton were estimated on the basis of PP and Redfield's ratio (Redfield, 1958) which relates nutrient uptake to carbon assimilation.

The TN:TP ratio in the trap material was assumed to represent the nutrient ratio of the sediment surface. Using this ratio and the net sedimentation of TP, the net sedimentation of TN was calculated as

$$TN_{sed} = TN : TP_{trap} * TP_{sed}. \quad (5)$$

The difference between the annual N removal and the amount of nitrogen buried in the sediment (TN_{sed}) was used as an estimate of denitrification (DN) – N_2-fixation (Nfix) balance as suggested by Ahlgren (1967) and Vollenweider (1969), and practiced still by several authors (Dudel & Kohl, 1991; Jensen et al., 1990; 1992):

$$DN - Nfix = EB_{TN} - TN_{sed}. \quad (6)$$

Data on wind speed and direction were obtained from the hydrometeorological station located in Tõravere (in 20 km distance from the lake).

Results

Water budget

The daily hydrological load from the watershed varied from 0.4 × 10^6 m^3 (Aug. 25) to 16.2 × 10^6 m^3 (May

15) and the outflow from 0 to 3.9 × 10^6 m^3 (May 08). The lake had no outflow between February 25–27 when a weak back flow (0.2 × 10^6 m^3 d^{-1}) occurred in the R. Suur Emajõgi. Water inflow predominated over outflow from January till May (see Table 1). The maximum volume (1031 × 10^6 m^3) and maximum mean depth (3.42 m) were reached on May 24. Due to bad outflow conditions the long-lasting high water level in summer is a characteristic feature of the lake. Although the decrease started from the end of May, the lake depth reached its annual mean value only in the first decade of August. As a result of dry summer and autumn the lake volume decreased continuously and reached the minimum of the year (519 × 10^6 m^3) on December 31, being 22% smaller than in the beginning of the year.

External budget of nutrients

Annually 2086 tons (7.7 g m^{-2}) of total nitrogen entered the lake, both from the watershed and by direct atmospheric precipitation (see Figure 2; Table 2). Two thirds of this loading was in dissolved inorganic form (DIN). Atmospheric input formed 11% of DIN and 7% of the total nitrogen loading. The annual outflow of TN was 986 t and only 1/3 of that was contributed by DIN. As nitrogen of the organic compounds forms the major part of the difference TN–DIN, it was conventionally named as organic nitrogen (N_{org}). The annual loss of DIN was 76% and that of organic nitrogen 6% of the loading, resulting in 53% of TN retention.

In the annual phosphorus loading (61 *t* or 0.23 g m^{-2}) 64% was accounted for by soluble reactive phosphorus (SRP) and 36% by organic phosphorus (P_{org}, calculated as TP–SRP). The atmospheric loading of P (only SRP data vere available) formed 10% of SRP and 7% of TP input. The annual decrease of SRP, forming 74% of the loading, was mostly accounted for by transformation of SRP into organic P, the outflow of which largely exceeded the inflow (negative retention – 55%). This reduced the retention of TP to 28% of the annual loading. Partially the retention ability of the lake was underestimated due to the unbalanced water budget in 1995, as the volume of the lake in the beginning of the year was larger than in the end.

The seasonal dynamics of nutrient load from the watershed was dependent on the dynamics of the hydrological load to a greater extent than on the changes in nutrient concentrations in the inflowing water. Nutrients were accumulated during the increasing phase of the lake volume from January to May and

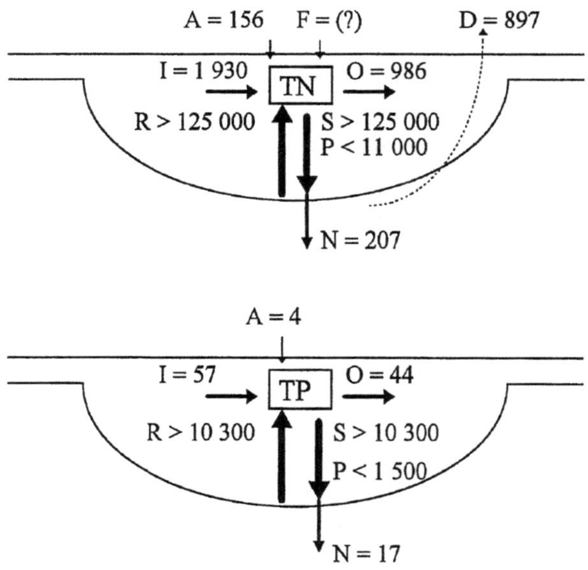

Figure 2. Annual budget of total nitrogen and total phosphorus in L. Võrtsjärv in 1995 (tons per lake). A – atmospheric precipitation, D – denitrification, F – dinitrogen fixation, I – inflow, N – net sedimentation, O – outflow, P – primary sedimentation, R – resuspension + diffusion, S – gross sedimentation.

predominantly flushed out beginning from June (see Table 2). The outflow exceeded the load during the shortest period in the case of DIN (June, July). The external budgets of TN, TP and SRP were negative from June till October, while N_{org} and P_{org} were predominantly leaving the lake till the end of the year. Besides the direct relation between the hydraulic and nutrient loads, water regime affected also nutrient concentrations. During low flow period in dry summer 1995, most of the rivers switched over to groundwater feeding. This reduced drastically the nutrient inflow in August and September. In November and December, the late autumn rainfalls and the increased surface runoff turned the external budgets of inorganic and total nutrients positive again, despite of the continuing decrease in lake volume. In seasonal aspect the atmospheric loading gained a significant role during the low flow period and reached more than a half of DIN input in August and 31% of SRP input in September.

Internal processes in the lake

Nutrient uptake by plankton

Nutrient uptake was calculated in order to estimate the upper limit of potential new sedimentation flux. The

Table 2. External budget of nutrients in 1995.

Flux		Month I	II	III	IV	V	VI	VII	VIII	IX	X	XI	XII	Average
DIN	inflow (mg m^{-2} d^{-1})	15	39	43	19	18	4	1	1	1	3	4	4	13
	atm. input (mg m^{-2} d^{-1})	1.8	2.2	4.5	0.8	1.9	1.7	0.6	1.2	0.7	1.0	2.3	0.4	1.6
	total input (mg m^{-2} d^{-1})	17	41	47	20	20	6	2	2	2	4	6	4	14
	atm. input %	11	5	9	4	10	29	30	52	33	26	35	10	11
	outflow (mg m^{-2} d^{-1})	3	2	6	8	6	7	4	1	1	1	1	1	3
	retention %	82	94	88	58	72	− 26	− 91	41	62	74	91	70	76
N_{org}	inflow (mg m^{-2} d^{-1})	4	10	36	14	9	5	1	0	1	1	2	1	7
	outflow (mg m^{-2} d^{-1})	3	1	9	12	8	11	6	9	7	6	5	3	7
	retention %	39	85	76	17	8	− 141	− 376	− 1760	− 714	− 391	− 186	− 164	6
TN	input (mg m^{-2} d^{-1})	21	51	83	34	28	10	3	3	3	5	8	5	21
	% DIN	80	81	57	58	70	55	60	83	72	77	79	80	67
	% N_{org}	20	19	43	42	30	45	40	17	28	23	21	20	33
	outflow (mg m^{-2} d^{-1})	6	4	14	20	13	18	10	10	8	7	5	4	10
	% DIN	54	62	39	42	41	39	39	13	11	15	10	30	34
	% N_{org}	46	38	61	58	59	61	61	87	89	85	90	70	66
	retention (mg m^{-2} d^{-1})	15	47	69	14	15	− 8	− 6	− 7	− 5	− 2	3	1	11
	retention %	73	92	83	41	52	− 77	− 200	− 265	− 152	− 34	33	22	53
SRP	inflow (mg m^{-2} d^{-1})	0.31	1.56	0.71	0.26	0.65	0.21	0.08	0.05	0.06	0.15	0.16	0.14	0.35
	atm. input (mg m^{-2} d^{-1})	0.02	0.03	0.08	0.01	0.22	0.03	0.02	0.00	0.03	0.01	0.03	0.00	0.04
	total input (mg m^{-2} d^{-1})	0.33	1.59	0.79	0.27	0.88	0.23	0.10	0.06	0.09	0.16	0.19	0.14	0.39
	atm. input %	7	2	11	4	25	11	21	9	31	5	13	1	10
	outflow (mg m^{-2} d^{-1})	0.03	0.03	0.03	0.08	0.12	0.30	0.10	0.11	0.16	0.18	0.04	0.04	0.10
	retention %	91	98	96	70	87	− 29	2	− 87	− 88	− 18	80	69	74
P_{org}	input (mg m^{-2} d^{-1})	0.07	0.40	0.38	0.25	0.83	0.27	0.08	0.06	0.06	0.10	0.12	0.09	0.22
	outflow (mg m^{-2} d^{-1})	0.06	0.09	0.24	0.30	0.60	0.51	0.66	0.59	0.33	0.45	0.20	0.11	0.35
	retention %	2	78	38	− 20	28	− 90	− 741	− 926	− 487	− 336	− 66	− 23	− 55
TP	input (mg m^{-2} d^{-1})	0.40	1.98	1.18	0.52	1.71	0.50	0.18	0.11	0.14	0.26	0.31	0.23	0.62
	% SRP	84	80	68	52	51	46	56	50	60	60	61	61	64
	% P_{org}	16	20	32	48	49	54	44	50	40	40	39	39	36
	outflow (mg m^{-2} d^{-1})	0.09	0.12	0.27	0.38	0.72	0.81	0.75	0.69	0.49	0.63	0.24	0.16	0.45
	% SRP	33	27	12	21	16	37	13	15	33	29	15	28	23
	% P_{org}	67	73	88	79	84	63	87	85	67	71	85	72	77
	retention (mg m^{-2} d^{-1})	0.30	1.86	0.91	0.14	0.99	− 0.31	− 0.58	− 0.58	− 0.35	− 0.37	0.07	0.08	0.17
	retention %	76	94	77	27	58	− 61	− 323	− 508	− 245	− 144	22	33	28

particulate primary production varied between 66 and 2945 mg C m^{-2} d^{-1}. The calculated nutrient uptake by plankton (see Table 3) reached the maximum in May (293 mg N m^{-2} d^{-1} and 41 mg P m^{-2} d^{-1}) and decreased gradually until December (18 mg N m^{-2} d^{-1} and 3 mg P m^{-2} d^{-1}). The influence of different timing of uptake and sedimentation maxima was diminished by calculating the annual means (113 mg N m^{-2} d^{-1} and 16 mg P m^{-2} d^{-1}) used in the general model (Figure 2).

Trap catch and upward flux of nutrients

During the 18 week measurement period (June 19–October 23), the trap catches of nutrients varied within a wide range, from 362 to 4279 mg N m^{-2} d^{-1} and from 30 to 329 mg P m^{-2} d^{-1} (Table 3). Despite of the changing amount of trapped material, its composition was rather stable. Organic matter (ash-free dry weight) formed 25.6 ± 1.3% of the total dry weight, and the TN:TP mass ratio was 12.3 ± 1.2 (± STD). The amount of trapped material was the smallest in July and increased gradually towards autumn. Trap catches exceeded the maximum potential new sedimentation

Table 3. Internal nutrient fluxes in L. Võrtsjärv in 1995.

Flux	Month												
	I	II	III	IV	V	VI	VII	VIII	IX	X	XI	XII	Average
TN phytoplankton uptake (mg m^{-2} d^{-1})	23	26	36	103	293	216	170	152	104	56	23	18	113
trap catch (mg m^{-2} d^{-1})						935*	866	1048	1476	2164**			1287***
upward flux (mg m^{-2} d^{-1})						872*	860	1010	1531	2180**			1295***
net sedimentation (mg m^{-2} d^{-1})	12	−52	42	−15	−5	7	−2	−8	−1	8	32	7	2.5
denitrification – Nfix (mg m^{-2} d^{-1})	4	−53	35	121	24	27	8	45	−54	27	−29	−35	10
denitrification – (% of N removal)													80
TP phytoplankton uptake (mg m^{-2} d^{-1})	3	4	5	14	41	30	23	21	14	8	3	3	16
trap catch (mg m^{-2} d^{-1})						77*	69	89	121	168*			105***
upward flux (mg m^{-2} d^{-1})						77*	69	89	121	168*			105***
net sedimentation (mg m^{-2} d^{-1})	0.95	−4.22	3.42	−1.18	−0.43	0.55	−0.16	−0.64	−0.07	0.69	2.58	0.60	0.17

* Mean for June 19–30.
** Mean for October 01–23.
*** Mean for June 19–October 23

4.3–38.6 times in the case of N and 2.5–21 times in the case of P, i.e. the trapped material consisted for 60–97% of resuspended sediments. The dynamics of gross sedimentation showed no relation to the primary production pattern.

The upward fluxes of nutrients, calculated as the sum of net budget, and the trap catch (Eq. 4) were nearly equal to the latter. The net budget showing the resultant of downward and upward fluxes became extremely small while averaged over month. The upward flux varied from 342 to 4187 mg N m^{-2} d^{-1} and from 28 to 155 mg P m^{-2} d^{-1}. The monthly mean increased gradually together with the decrease in the mean depth of the lake.

A statistical model was developed using multiple linear regression analysis with stepwise variable selection, which explained nearly one half of the upward flux variability by changes in the average wind speed during trap exposition and the mean depth (Table 4). Other examined variables as wind direction, wind fetch and the cumulative sum of PP were rejected by the model. The variables selected by this regression procedure demonstrate that the wind-induced sediment resuspension plays the major role in the formation of nutrient upward flux and, thus, in the formation of the temporal pattern of nutrients during the ice-free season. Other fluxes as the external loading or new sedimentation of autochthonous production are overcome and masked by the powerful resuspension – sedimentation cycle exceeding the formers by one or two orders of magnitude.

Net sedimentation of nutrients and denitrification

The monthly means of TP net sedimentation, calculated as negative net budget, varied from – 4.2 mg m^{-2} d^{-1} in February (release of SRP from the sediment) up to 3.4 mg m^{-2} d^{-1} in March (sedimentation of diatoms) (Table 3). Another peak in TP net sedimentation (2.6 mg m^{-2} d^{-1}) occurred in November after ice formation. The annual net sedimentation of TP was equal to annual retention of phosphorus (0.06 g m^{-2}).

The monthly mean net sedimentation of TN calculated on the basis of TP net sedimentation and N:P ratio in recently sedimented material (trap catch) (Eq. 5) varied between – 52 and 42 mg m^{-2} d^{-1}. The difference between the change in the TN storage (net budget) and net sedimentation should show the N_2-fixation – denitrification balance. Several errors can accumulate in this kind of calculations and the seasonal pattern obtained in this way is probably misleading (unexplainable negative denitrification values in February, September and November referring to N-fixation, see Table 2). A more reliable estimate of the DN – Nfix balance could be achieved for the annual cycle as a difference between N retention and burial (Eq. 6). The discrepancy showed that denitrification was responsible for at least 81% of the annual nitrogen loss (if N_2-fixation was ignored) and only 19% of removed nitrogen was buried in the sediment. The development of the algal community in 1995 showed that N_2-fixation took probably place at least in late July – August when the minima of TN:TP ratio (< 10) and DIN:TN ratio (~ 0) in the water column initiated a clear peak of N_2-fixing algae (*Aphanizomenon gracile, Anabaena* spp.) which lasted until September (Figure 3).

Table 4. The linear multiple regression model for the upward flux of nutrients.

Dependent variable	Independent variable	Coefficient $\pm STE$	Sign. level	R^2 (adjusted)
UWF of TN	Constant	1991 ± 703	0.0054	0.445
(mg m^{-2} d^{-1})	Mean wind speed (m s^{-1})	529 ± 95	0.0000	
	Mean depth (m)	19067 ± 4186	0.0000	
	Log (Mean depth)	− 5.51 ± 1.13	0.0000	
UWF of TP	Constant	117 ± 57	0.0430	0.425
(mg m^{-2} d^{-1})	Mean wind speed (m s^{-1})	47.1 ± 7.7	0.0000	
	Mean depth (m)	1388 ± 340	0.0001	
	Log (Mean depth)	−4012 ± 920	0.0000	

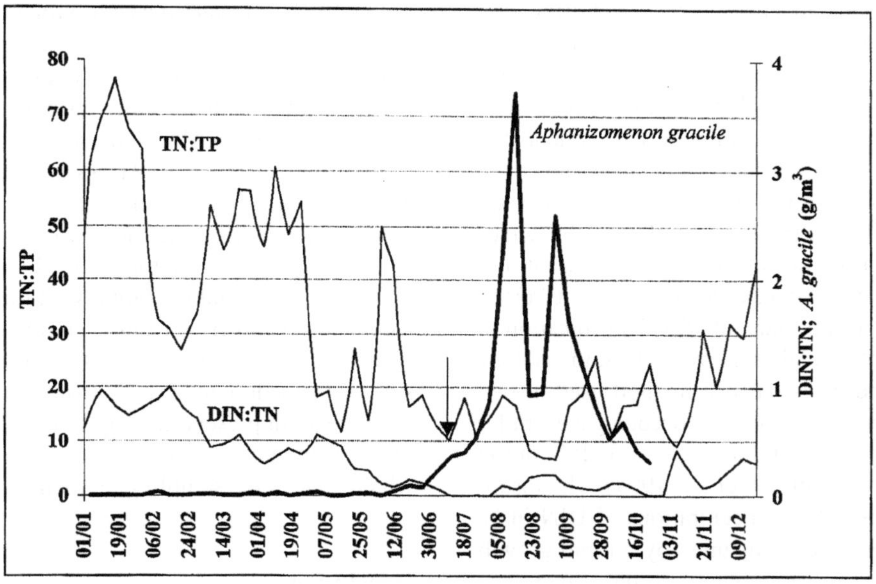

Figure 3. Nutrient ratios and the biomass of N_2-fixing *Aphanizomenon gracile* in L. Võrtsjärv in 1995. Arrow indicates the possible starting point of dinitrogen fixation (TN:TP = 10; DIN:TN = 0).

Discussion

The nutrient input to L. Võrtsjärv in 1995 (7.7 g m^{-2} y^{-1} TN and 0.23 g m^{-2} y^{-1} TP) can be considered rather low. The mean nitrogen load for 69 shallow Danish lakes was 142 g m^{-2} y^{-1} (median 52) (Jensen et al., 1990). Annual TP loadings reaching several grams per square meter have been reported for several German, Polish and Russian lakes (Klein, 1991; Dudel & Kohl, 1991; Galicka, 1992; Martinova, 1993; Kozerski et al., 1993). The loading to L. Võrtsjärv is small due to the large area and long residence time of water but still exceeds the transition range between oligo- and eutrophic conditions (0.056–0.112 g m^{-2} y^{-1} TP) calculated according to Vollenweider (1976). By most of the trophic characteristics (PP, Chl *a*, nutrient concentrations) L. Võrtsjärv can be regarded as eutrophic, but according to Secchi depth and plankton composition it falls into the hypertrophic range. High turbidity is caused by sediment resuspension, and permanent mixing increases nutrient turnover by keeping high diffusive fluxes in the steep concentration gradient at sediment-water interface (Hesslein, 1980, Boström et al., 1982, Löfgren, 1987).

The TN:TP input ratio 34 corresponds to the runoff from fertile soils (Downing & McCauley, 1992). Prevailing nitrogen loss reduced the mean ratio to 22 in the outflowing water. Most of the removed nitrogen was clearly not kept in the sediment as the TN:TP ratio of trapped sediments was only 12. In L. Müggelsee where the geometric mean of TN:TP mass ratio in seston was 8.9 and not higher than 5 on the sediment surface, up to 65% N was lost in the passage from the pelagic waters to the sediment. Denitrification (2–34 mg N m^{-2} d^{-1}) accounted for up to 65% of TN loss and up to 20% of TN load (Dudel & Kohl, 1991). The proportion of denitrification in nitrogen removal in L. Võrtsjärv (81%) exceeded all data cited in the review by Seitzinger (1988), but was very close to the results reported by Jensen et al. (1990) for 58 shallow Danish lakes where denitrification accounted for 77% of the total nitrogen removal. The latter authors have shown, basing on literature data, the discrepancy between denitrification estimates derived from *in vitro* experiments and those reported from mass balance studies of eutrophic lakes. They explain the higher values obtained from mass balance calculations by the impact of sediment disturbance, ignored in experiments with sediment cores. In shallow lakes the high sediment surface to water volume ratio and high frequency and extent of wind-induced resuspension allows contact between nitrate-rich lake water and exposed anoxic sediment layers, which is the most important site for denitrification in shallow, non-stratified lakes. Nevertheless, the isotope (^{15}N) method and mass balance method used in two rather shallow eutrophic lakes in Sweden gave similar low denitrification rates in the range of 5–25% of the external N loading (Ahlgren et al., 1994). The disappearance of measurable concentrations of DIN in L. Võrtsjärv in August, accompanied by a rapid growth of N_2-fixing species in phytoplankton, makes it reasonable to assume the occurrence of dinitrogen fixation. In shallow eutrophic L. Müggelsee having several similarities with L. Võrtsjärv in plankton composition, N_2-fixation is relatively small, but in July-August fits in the range of input by precipitation and dry deposition (4–6 mg N m^{-2} d^{-1}). It is suggested that the maximum dinitrogen input is related to the seasonal nitrate minimum with low (nitrate limited) denitrification rates (Dudel & Kohl, 1991).

As the phytoplankton of L. Võrtsjärv consists mainly of filamental algae not prefered as food for small-sized zooplankton (ciliates, small rotifers and *Chydorus sphaericus* (Müller) being most numerous), the detrital food chain is predominating (T. Nõges, unpublished). Thus, the major part of primary production is sedimented. The trap data could not be used for net sedimentation estimates as the trapped material consisted mainly of resuspended sediments.

The net sedimentation rate of TP could be calculated rather correctly as the negative net budget. As the influence of in- and outflow is eliminated from the net budget, changes in TP reflect only the exchange between water and sediment. Negative values of net sedimentation show the predominance of the upward flux. The large temporal variability of the net budgets, reflected in the dynamics of net sedimentation, is one of the specific features of shallow lakes. Frequent disturbance of the nutrient-rich sediment surface by turbulent mixing and the rapid restauration of the steep oxygen gradient in the sediment surface layer during calm days, even at high bottom water oxygen concentrations (Löfgren, 1987), creates a chemically and physically highly variable environment. The measured gross sedimentation of nutrients exceeded the potential upper level of primary sedimentation (uptake by phytoplankton) about 5 times during summer. Similar results were obtained by Andersen & Lastein (1981) in shallow and eutrophic L. Arreskov where trap catches exceeded the primary sedimentation 6.3 times, as an annual average. In some cases, high resuspension rate can cause erroneous phytoplankton biomass estimates. The high phytoplankton biomass in Lake Võrtsjärv in October (not shown in results) could not be explained by algal growth and was clearly caused by resuspension of the sediment surface layer containing a lot of dead algae. In preserved samples it is almost impossible to distinguish between algae which were dead and which alive at the sampling moment.

Conclusions

1. Lake Võrtsjärv acts as an efficient trap for nutrients which accumulate in quantities more than a half of the external N load and about one third of the external P load during the annual cycle. In 1995, 80% of the annual nitrogen loss was accounted for by denitrification and about 20% buried in the sediment.

2. Sediment resuspension plays the major role in the formation of the temporal pattern of nutrients during the ice-free season. Other fluxes, as the external loading or new sedimentation of autochthonous production, are overcome and masked by the powerful resuspension – sedimentation cycle exceeding these by one or two orders of magnitude.

Acknowledgements

The investigation has been supported by Estonian Science Foundation (grants 1641 and 2017), and by a grant of the Finnish Ministry of Environment through the Environmental Agency of Häme. Partially, data of the Estonian Environmental Monitoring Program were used in the analysis. The authors are thankful to the chemistry laboratory of Võrtsjärv Limnological Station and to South-Estonian Environmental Laboratory for performing hydrochemical analyses.

References

Ahlgren, I., 1967. Limnological studies of Lake Norvikken a eutrophicated Swedish lake. Schweiz. Z. Hydrol. 29: 53–90.

Ahlgren, I., F. Sörensson, T. Waara & K. Vrede, 1994. Nitrogen budgets in relation to microbial transformations in lakes. Ambio 23: 367–377.

Andersen, J. M., 1974. Nitrogen and phosphorus budgets and the role of sediments in six shallow Danish lakes. Arch. Hydrobiol. 74: 528–550.

Andersen, F. Ø. & E. Lastein, 1981. Sedimentation and resuspension in shallow, eutrophic Lake Arreskov, Denmark. Verh. int. Ver Limnol. 21: 425–430.

Boers, P. & J. Uunk, 1991. Lake restoration: estimation of internal phosphorus loading after reduction of external loading from sediment data. In P. Boers, The release of dissolved phosphorus from lake sediments. Diss. Ph.d. Netherlands: 79–90.

Boström, B., M. Jansson & C. Forsberg, 1982. Phosphorus release from lake sediments. Arch. Hydrobiol. Beih. Ergeb. Limnol. 18: 5–59.

Dudel, G. & J.-G. Kohl, 1991. Contribution of dinitrogen fixation and denitrification to the N-budget of a shallow lake. Verh. int. Ver Limnol. 24: 884–888.

Downing, J. A. & E. McCauley, 1992. The nitrogen: phosphorus relationship in lakes. Limnol. Oceanogr. 37: 936–945.

Galicka, W., 1992. Total nitrogen and phosphorus budgets in the Lowland Sulejów Reservoir for the hydrological years 1985–1988. Arch. Hydrobiol./Suppl. 90: 159–169.

Hesslein, R. H., 1980. *In situ* measurements of pore water diffusion coefficients using tritiated water. Can. J. Fish. aquat. Sci. 37: 545–551.

Jaani, A., 1990. Võrtsjärve veerežiim ja -bilanss. Eesti Loodus 11: 743–747.

Jensen, J. P., P. Kristensen & E. Jeppesen, 1990. Relationships between nitrogen loading and in-lake nitrogen concentrations in shallow Danish lakes. Verh. int. Ver. Limnol. 24: 201–204.

Jensen, J. P., E. Jeppesen, P. Kristensen, P. B. Christensen & M. Søndergaard, 1992. Nitrogen loss and denitrification as studied in relation to reductions in nitrogen loading in a shallow, hypertrophic lake (Lake Søbygård, Denmark). Int. Revue ges. Hydrobiol. 77: 29–42.

Klein, G. & I. Chorus, 1991. Nutrient balances and phytoplankton dynamics in Schlachtensee during oligotrophication. Verh. Int. Ver. Limnol. 24: 873–878.

Kört, M. & O. Roots, 1996. Õhu saasteainete kauglevi seire. In O. Roots & R. Talkop (eds), Keskkonnaseire 1995. Keskkonnaministeeriumi Info- ja Tehnokeskus, Tallinn: 12–15.

Kozerski, H.-P., J. Gelbrecht & R. Stellmacher, 1993. Seasonal and long-term variability of nutrients in Lake Müggelsee. Int. Rev. ges. Hydrobiol. 78: 423–438.

Laugaste, R., 1994. The state, the origin of nutrients and the measures necessary for recovering Lake Verevi. In A. Järvekülg (ed), Eesti jõgede ja järvede seisund ning kaitse. Teaduste Akadeemia Kirjastus, Tallinn: 47–63.

Löfgren, S., 1987. Phosphorus retention in sediments – implications for aerobic phosphorus release in shallow lakes. Acta Universitatis Upsaliensis. Comprehensive Summaries of Uppsala Dissertations from the Faculty of Science 100, 24 pp.

Martinova, M. V., 1993. Nitrogen and phosphor compounds in bottom sediments: mechanisms of accumulation, transformation and release. Hydrobiologia 252: 1–22.

Nõges, P. & A. Järvet, 1995. Water level control over light conditions in shallow lakes. Report Series in Geophysics. University of Helsinki 32, 81–92.

Nõges, T., 1997. Different fractions of primary production in the large, shallow eutrophic Lake Võrtsjärv. Ann. Limnol., in press.

Nõges, T., 1998. Zooplankton-phytoplankton interactions in lakes Võrtsjärv, Peipsi (Estonia) and Yaskhan (Turkmenia). Hydrobiologia, 342/343: 175–183.

Redfield, A. S., 1958. The biological control of chemical factors in the environment. Am. Sci. 46: 205–211.

Seitziger, S. P., 1988. Denitrification in freshwater and coastal marine ecosystems: Ecological and geochemical significance. Limnol. Oceanogr. 33: 702–724.

Søndergaard, M., P. Kristensen & E. Jeppesen, 1992. Phosphorus release from resuspended sediment in the shallow and wind-exposed Lake Arreso, Denmark. Hydrobiologia 228: 91–99.

Steeman-Nielsen, E., 1952. The use of radioactive carbon (^{14}C) for measuring primary production in the sea. Journal du Conseil permanent international pour l'exploration del la mer. 18: 117–140.

Wisniewski, R., 1995. The regulatory role of sediment resuspension in seston and phosphorus dynamics in shallow Lake Druzno. Proc. 6th Inrernat. Conf. on the Conservation and Management of Lakes – Kasumigaura'95: 917–920.

Vollenweider, R. A., 1969. Möglichkeiten und Grenzen elementarer Modelle der Stoffbilanz von Seen. Arch. Hydrobiol. 66: 1–36.

Vollenweider, R. A., 1976. Advances in defining critical loading levels for phosphorus in lake eutrophication. Mem. Ist. Ital. Idrobiol. 33: 53–83.

Vollenveider, R. A., 1989. Assessment of mass balance. In S. E. Jørgensen & R. A. Vollenweider (eds), Guidelines of Lake Management I. Principles of Lake Management. ILEC, UNEP, Japan: 53–69.

Hydrobiologia **363:** 229–240, 1998.
T. Tamminen & H. Kuosa (eds), Eutrophication in Planktonic Ecosystems: Food Web Dynamics and Elemental Cycling.

Vertical and seasonal distributions of micro-organisms, zooplankton and phytoplankton in a eutrophic lake

Anne-Mari Ventelä, Vesa Saarikari & Kristiina Vuorio
Laboratory of Ecology and Animal Systematics, FIN-20014 University of Turku, Finland

Key words: ice cover, winter, micro-organisms, zooplankton, phytoplankton

Abstract

The vertical distributions of bacteria, picoalgae, protozoan and metazoan zooplankton, and phytoplankton in the highly eutrophic Lake Köyliönjärvi (SW Finland) were studied monthly during the period of ice-cover in January-April 1996. For comparison, we also provide some data on the distributions of the plankton during the summer. The whole water column remained oxic during the ice-covered period, although the near-bottom oxygen concentrations were always very low. The heterotrophic nanoflagellates were more abundant in winter than in summer, but ciliates, picoalgae and bacteria were more numerous in summer. In general both zooplankton and phytoplankton had low biomass during the ice-covered period. However, the biomass of the diatom *Aulacoseira islandica* ssp. *islandica* was high under the ice in April. The calanoid copepod *Eudiaptomus graciloides* was the dominant zooplankton species from January to March, but had almost disappeared by the beginning of April and did not increase again until in June. The dominant rotifer species in winter were *Keratella cochlearis, Filinia terminalis*, and *Filinia longiseta* in the surface water and *Rotaria neptunia* near the bottom.

Introduction

The chemical and physical conditions in the water column in winter differ from the summer conditions. On ice-covered lakes, snow causes the attenuation of photosynthetically available radiation (PAR, 400–700 nm), with even thin snow cover (2–8 cm) reducing transmittance to 10% or less (Bolsenga & Vanderploeg, 1992). Under ice, low temperatures and low light levels are limiting factors for production (e.g. Tulonen, 1993). In some cases, the reduced photosynthesis in combination with respiratory demand from decomposition processes may lead to anoxia and fish kills (e.g. Pennak, 1968).

The under-ice zooplankton community is commonly characterized by low biomass with dominance of adult and juvenile copepods (Pennak, 1968; Hakkari, 1969; Granberg, 1970; Vanderploeg & Bolsenga, 1992; Agbeti & Smol, 1995) and rotifers (Pennak, 1968; Hakkari, 1969; Granberg, 1970).

Furthermore, the biomass of phytoplankton is also generally low under the ice cover (Pennak, 1968; Granberg, 1970; Arvola & Kankaala, 1989). According to the so-called PEG-model (Sommer et al., 1986), the increasing light and available nutrients in spring stimulate the growth of phytoplankton. If the ice is not covered by snow, the clear ice can transmit even 70–95% of the incoming PAR (Bolsenga & Vanderploeg, 1992), and the phytoplankton biomass may even show its seasonal maximum under the ice (Cleve-Euler, 1951; Willén, 1962; Maede & Ichimura, 1973; Sommaruga & Psenner, 1995).

According to Arvola and Kankaala (1989), who reported the under-ice bacterial abundances in five Finnish lakes, the number of bacteria was low in winter, but increased towards the spring. The highest bacterial densities were observed during the phytoplankton maximum in April and May. The number of ciliates usually remains low under the ice cover (Granberg, 1970; Kivi, 1986; Sommaruga & Psenner, 1995). However, the heterotrophic nanoflagellates have been observed to be abundant in the under-ice plankton community

(Wright, 1964; Arvola & Kankaala, 1989). Also, in some lakes the number of picoalgae has been high under the ice (Pennak, 1968; Arvola & Kankaala, 1989; Sommaruga & Psenner, 1995).

High-latitude lakes are ice-covered for half of the year, yet, under-ice plankton communities have received much less attention than the open water communities, especially in eutrophic lakes. Most of the previous under-ice studies have been done in oligotrophic or mesotrophic environments and they have only investigated specific aspect of the plankton community, most often phytoplankton. The aim of this study was to describe the distributions of all the main components of the under-ice plankton community in the highly eutrophicated and biomanipulated Lake Köyliönjärvi, including the microbial component of the food web. In addition, we provide some data on the distribution of the plankton during the summer for a seasonal comparison.

Material and methods

Study site

Lake Köyliönjärvi (SW Finland) has a total area of 12.5 km^2 and a mean depth of 3.1 m. It has two separate basins. The lake is eutrophic with the total amount of phosphorus concentration of > 100 mg/m^3 in summer. The main part of nutrient loading comes from agriculture. Since 1992 the lake has been biomanipulated by removing 335 500 kg (by April 1996) cyprinid fish by winter seine net fishing from the southern basin (Hirvonen et al., 1993; Hirvonen & Salonen, 1995; Sarvala et al., unpublished). Decreasing fish biomass in Köyliönjärvi has led to increasing crustacean zooplankton biomass and the last two years have showed an increasing share of larger cladoceran species in summer (Sarvala et al., 1998). The lake is normally ice-covered from November to May.

Methods

Vertical plankton distributions were studied monthly during the ice-covered period in January–April 1996. The winter samples were taken through bore-holes in the ice at 1 m depth intervals using a Limnos water sampler (volume 2.6 litres). The sampling site was located in the middle of the southern basin. Vertically (1–6 m) measured variables were temperature, pH, concentrations of oxygen, total nitrogen, nitrate and nitrite nitrogen, ammonium nitrogen, total phosphorus, dissolved total phosphorus, phosphate phosphorus and dissolved phosphate phosphorus. The chemical analyses were made with standard methods in the laboratory of the Southwest Finland Regional Environment Center. Organisms examined were crustacean zooplankton, rotiferan zooplankton, phytoplankton, ciliates, heterotrophic nanoflagellates (HNF), bacteria and picoalgae. The winter time samples taken at 1 m intervals were analyzed separately. For the seasonal comparison, we calculated the mean value for the whole water column. The summer samples were taken from three sampling points in the southern basin in 1 m intervals and pooled both vertically and horizontally, because there is no permanent stratification in this lake during the summer. Samples from each site were taken from surface to bottom at 1 meter intervals.

Zooplankton samples were preserved with ice-cold ethanol (final concentration ca. 70%) and subsampled when required. About 200 copepodids or adults, 50 nauplius stages and 200 rotifers were identified, counted and measured in each subsample using an inverted light microscope. The carbon content for each species was estimated using values from Latja & Salonen (1978) and Salonen (1981). Phytoplankton samples were preserved with acid Lugol's solution (final concentration 0.25%) and counted with inverted light microscope using the Utermöhl technique. Ciliate samples were counted by concentrating one litre of Lugol-preserved sample (final concentration 1%) to a small volume by sedimentation for 3 d and enumerating under inverted light microscope using the Utermöhl technique. Ciliate volumes were measured using geometric formulas and multiplied with the factor 1.4 to correct the shrinkage effect caused by Lugol preservation (Müller & Geller, 1993). A conversion factor of 140 fg μm^{-3} was used for estimating biomass (Putt & Stoecker, 1989; cited in Sherr & Sherr, 1993).

Microbial samples were preserved with glutaraldehyde (final concentration 2%). Separate preparations were made for HNF, bacteria and picoalgae by filtering 1 ml sample to black stained 0.2–μm pore-sized Millipore-filter with gentle vacuum. For picoalgae preparation no stain was used due to their autofluorescence (MacIsaac & Stockner, 1993). For bacteria and HNF preparations DAPI staining was used (Porter & Feig, 1980). All the microbial preparations were stored at –24 °C before analysis. The picoalgae were counted with Leica Dialux epifluorescence microscope using Leica M2 filter set with green excitation rate (BP 546/14) . HNF and bacteria were counted using Leica

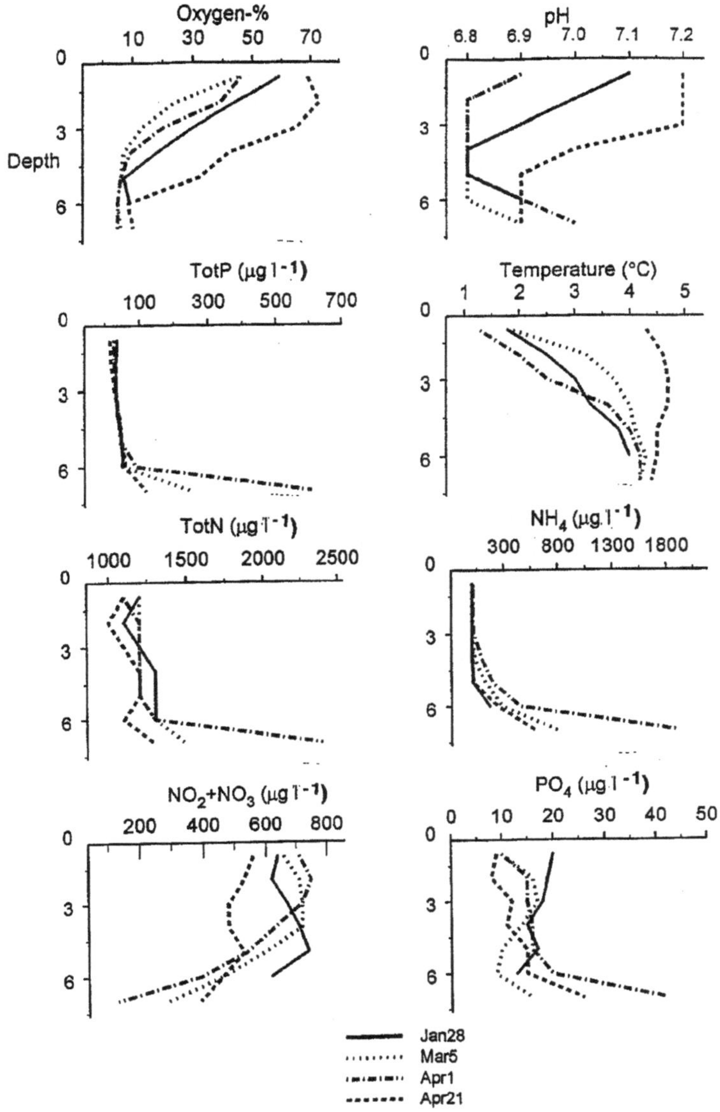

Figure 1. Vertical profiles of oxygen, pH, total phosphorus, temperature, total nitrogen, ammonium, nitrite + nitrate and phosphate in Lake Köyliönjärvi during the ice-covered period January–April 1996.

A filter set with ultraviolet excitation rate (BP 340–380).

Results

Chemical and physical conditions – In 1996 the lake was ice-covered from the beginning of December to the beginning of May. The mean ice thickness was 0.54 m. The snow cover varied between 0–35 cm. During the ice-covered period water temperature varied between 1.3–4.7 °C, but in deepest water temperature remained above 4 °C throughout the winter (Figure 1). Due to cold spring the water temperature was still low (15 °C) in the beginning of June.

The winter time secchi depth was highest (3.5 m) in the beginning of April and lowest (1.5 m) just before the ice break at the end of April. During the cyanobacterial bloom in August, the secchi depth was as low as 0.25 m.

The whole water column remained oxic during the ice-covered period, although the near-bottom oxygen concentrations were always very low (Figure 1). The oxycline declined during the diatom bloom in April. The diatom bloom also caused a change in the

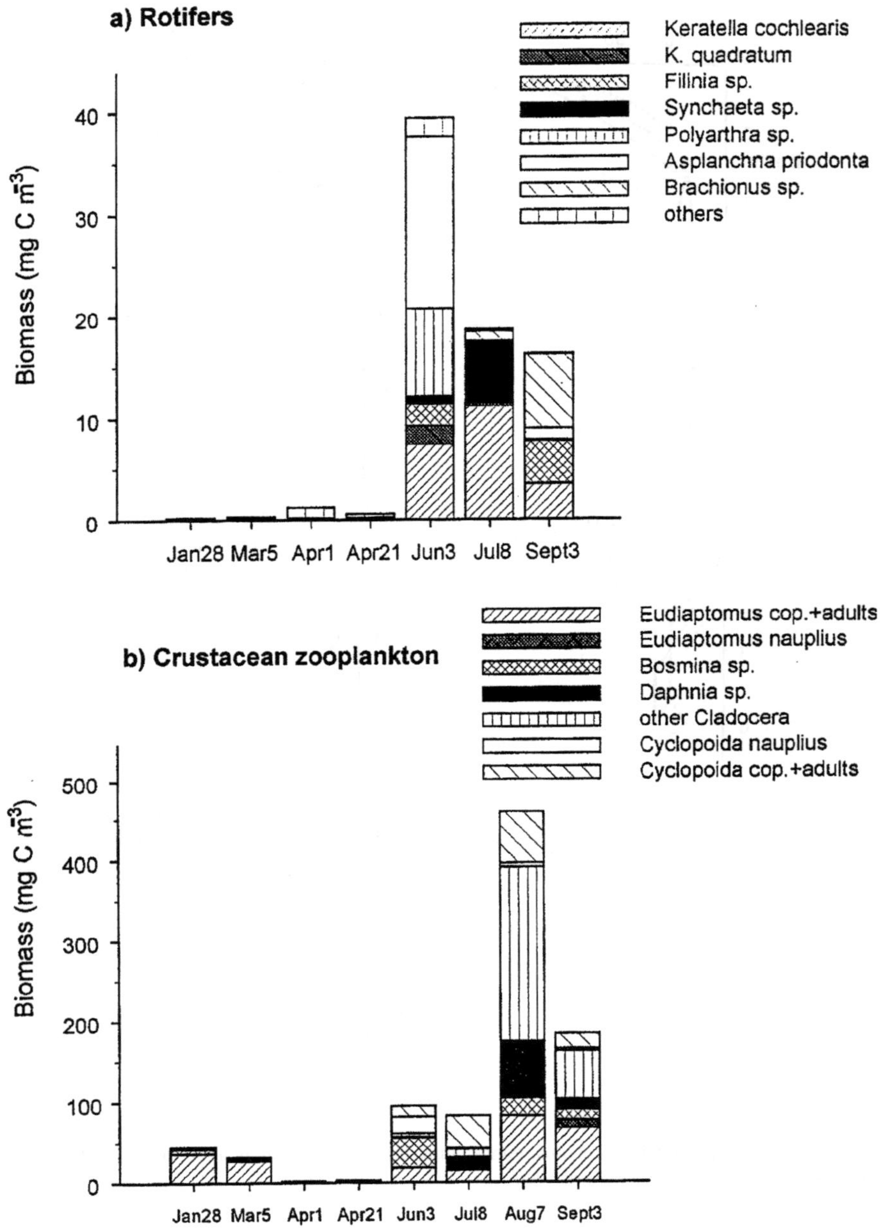

Figure 2. Seasonal variation in the abundance of a) rotifers and b) crustacean zooplankton in Lake Köyliönjärvi in January-September 1996. The ice-covered period lasted from January 28 to April 21. The winter time values are mean values including the samples taken in 1 m intervals from the depths of 1–6 m from one sampling site. The summer time values are pooled samples including three sampling sites.

pH-profile (Figure 1). Even though the metalimnion remained oxic, the increased values of ammonium nitrogen with decreased values of nitrate and nitrite near the bottom indicated that the anoxic bacterial processes were in function (Figure 1). The total nitrogen levels did not differ from the summer values (average total N in summer 1100 μg l^{-1}), but the amount of total phosphorus was lower in winter (average total P in summer 80 μg l^{-1}).

Crustacean zooplankton – The crustacean zooplankton biomass was relatively low during the winter (Figure 2b). In January, crustacean zooplankton biomass was small at the depths of 1–4 m (Figure 3). How-

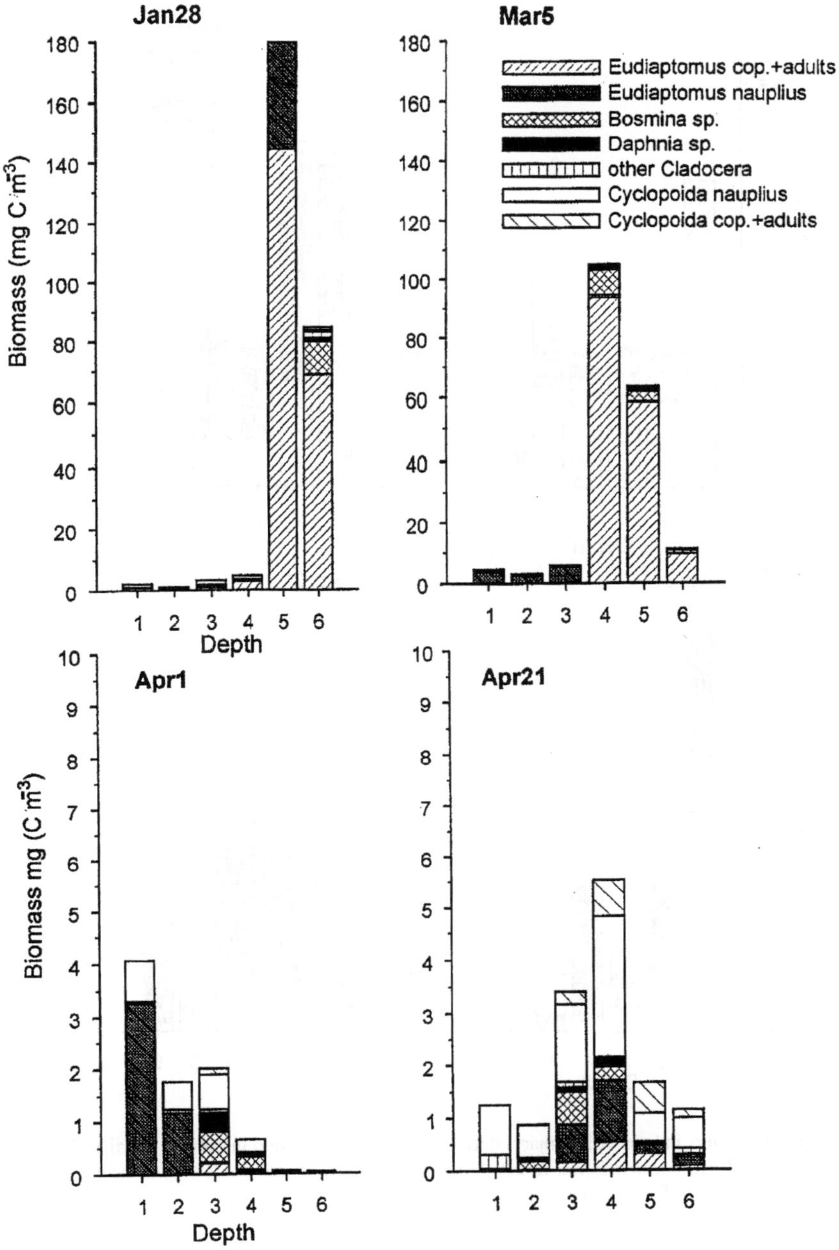

Figure 3. Vertical distribution of crustacean zooplankton in Lake Köyliönjärvi during the ice-covered period, in January-April 1996. Note the changing scale in X-axis.

ever, the egg-carrying females of the calanoid copepod *Eudiaptomus graciloides* (Lilljeborg, 1888) were abundant at 5–6 m. In the beginning of March, *E. graciloides* was still the dominant species with increasing share of nauplii, which were mainly observed at 1–3 m while the adults stayed in deeper water. In the beginning of April, the crustacean zooplankton community had dramatically changed. The crustacean zooplankton biomass maximum was only 4 mg C m^{-3}. The total *E. graciloides* biomass was 0.88 mg C m^{-3}, consisting only of the nauplius stages. At the end of April, the crustacean zooplankton consisted of the cyclopoid and calanoid nauplius stages. In June and July, the crustacean zooplankton community was dominated by *Bosmina coregoni* Baird 1857, *Bosmina longirostris* (Müller, 1785) and *Mesocyclops leuckarti*

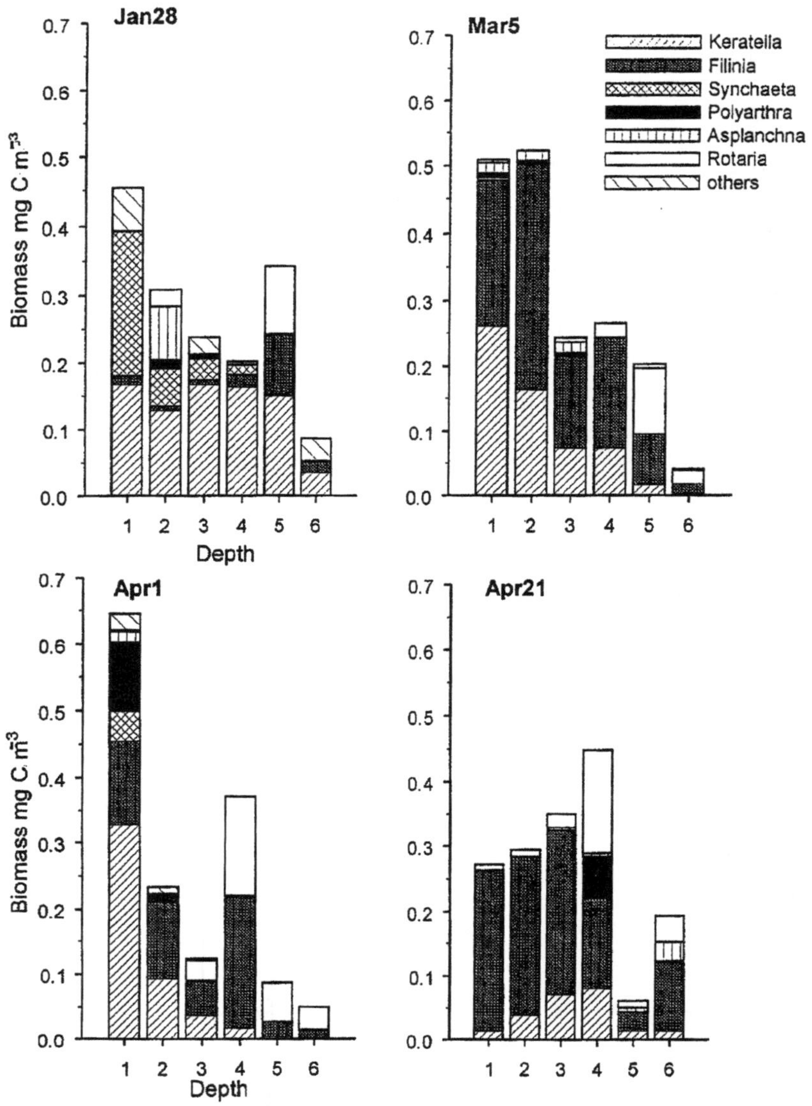

Figure 4. Vertical distribution of rotifers in Lake Köyliönjärvi during the ice-covered period in January-April 1996. The observed groups were *Keratella (K. cochlearis var. hispida, K. cochlearis var. tecta), Filinia (F. longiseta, F. terminalis), Synchaeta* sp., *Polyarthra (P. vulgaris, P. dolichoptera), Asplanchna (A. priodonta), Rotaria (R. neptunia)* and others.

(Claus, 1857) and its nauplius stages (Figure 2b). Also the number of *E. graciloides* started to increase again. The seasonal zooplankton maximum was in August when *Chydorus sphaericus* (Müller, 1785) dominated the system (Figure 2). *E. graciloides, Daphnia cucullata* Sars 1862, *Daphnia cristata* Sars 1862 and *Mesocyclops leuckartii* were also relatively high in number. *E. graciloides* and *Chydorus sphaericus* were abundant also in September. The seasonal maximum of *Eudiaptomus graciloides* was in August, which was early compared to previous years (Saarikari, unpublished data) and it consisted of both adults and copepodids.

Rotifers – Rotifers were very scarce during the ice-covered period compared to summer time abundance. (Figure 3a). The dominant rotifer species in winter were *Keratella cochlearis* (Gosse, 1851) *var. hispida, Filinia terminalis* (Plate, 1886), and *Filinia longiseta* (Ehrb., 1834) in the surface water and *Rotaria neptunia* near the bottom. Seasonally, the largest rotifer biomass was detected in June (Figure 2a). The dominant species in biomass in June were *Asplanchna priodonta* Gosse

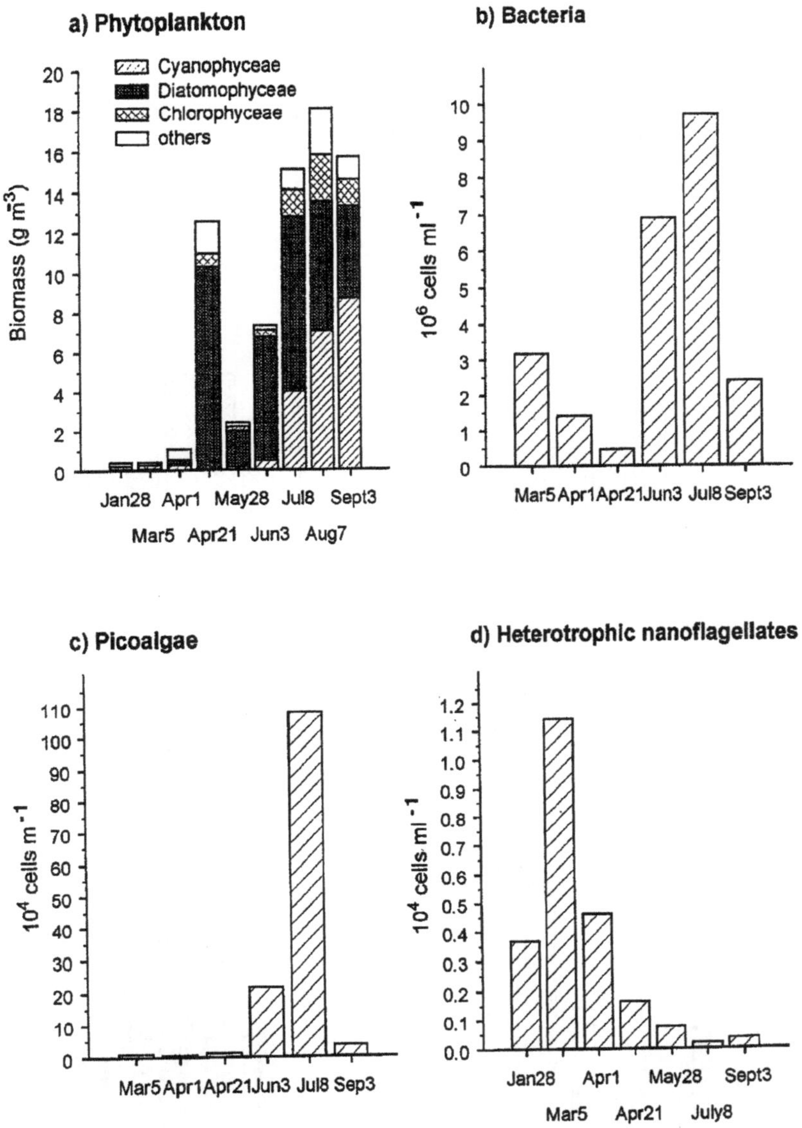

Figure 5. Seasonal variation in the abundances of a) phytoplankton, b) bacteria, c) picoalgae and d) heterotrophic nanoflagellates in Lake Köyliönjärvi in January-September 1996. The ice-covered period lasted from January 28 to April 21. The winter time values are mean values including the samples taken at 1 m intervals from the depths 1–6 m from one sampling site. The summer time values are pooled samples including three sampling sites.

1850 and *Polyarthra* spp. *Keratella cochlearis* (Gosse, 1851) *var. tecta* was common throughout the summer. *Synchaeta* sp. was abundant in July, *Brachionus* and *Filinia* species in September.

Phytoplankton – The under-ice phytoplankton biomass was low from January to early April (Figure 5a). The community consisted of small species like chlorophyceans and cyanophyceans (Figure 6). A strong diatom bloom was observed at the end of April (Figure 5a), when the diatom biomass almost approached the late summer cyanobacterial bloom. The dominant diatom species was *Aulacoseira islandica* ssp. *islandica* (O. Müller) Simons. It was most abundant at 3 m (Figure 6). In June and early July, diatoms were still the dominant group (Figure 5a). Cyanobacteria were most abundant from late July to September. The dominant species were *Microcystis aeruginosa* (Kützing) Kützing, *Microcystis wesenbergii* (Komárek) Starmach and *Anabaena flos-aquae* Brébisson, which all form

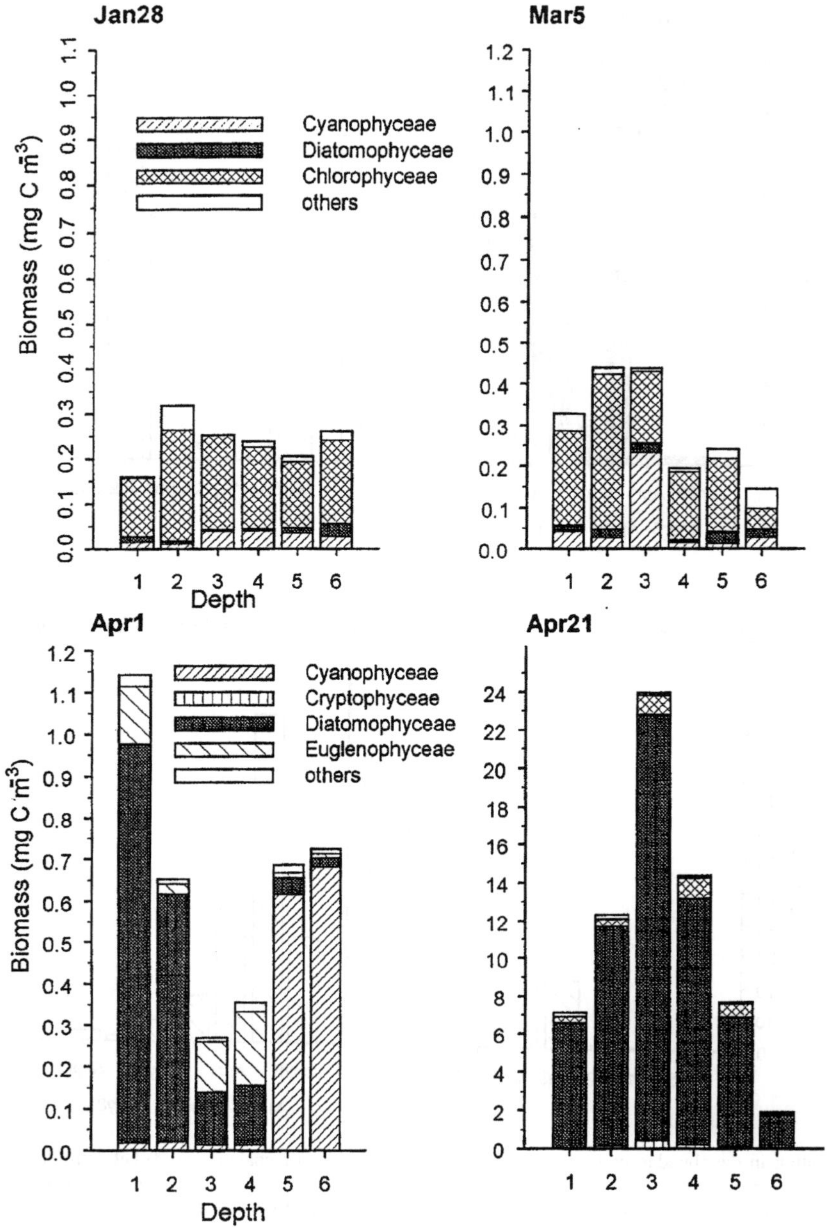

Figure 6. Vertical distribution of phytoplankton in Lake Köyliönjärvi during the ice-covered period in January-April 1996. Note the changing scale in X-axis.

surface blooms. The diatoms were, however, quite common also during the period of cyanobacterial abundance.

Heterotrophic bacteria and picoalgae – The bacterial abundance was lowest during the diatom bloom in April (Figure 5b). The highest numbers of bacteria were observed during the summer months, but their number seemed to decrease during the cyanobacterial bloom. The number of autotrophic picoalgae was low under the ice and highest in July. During the cyanobacterial bloom their number was strongly decreased. The vertical distribution of bacteria and picoalgae was variable (Figure 7).

Heterotrophic nanoflagellates – The HNF abundance

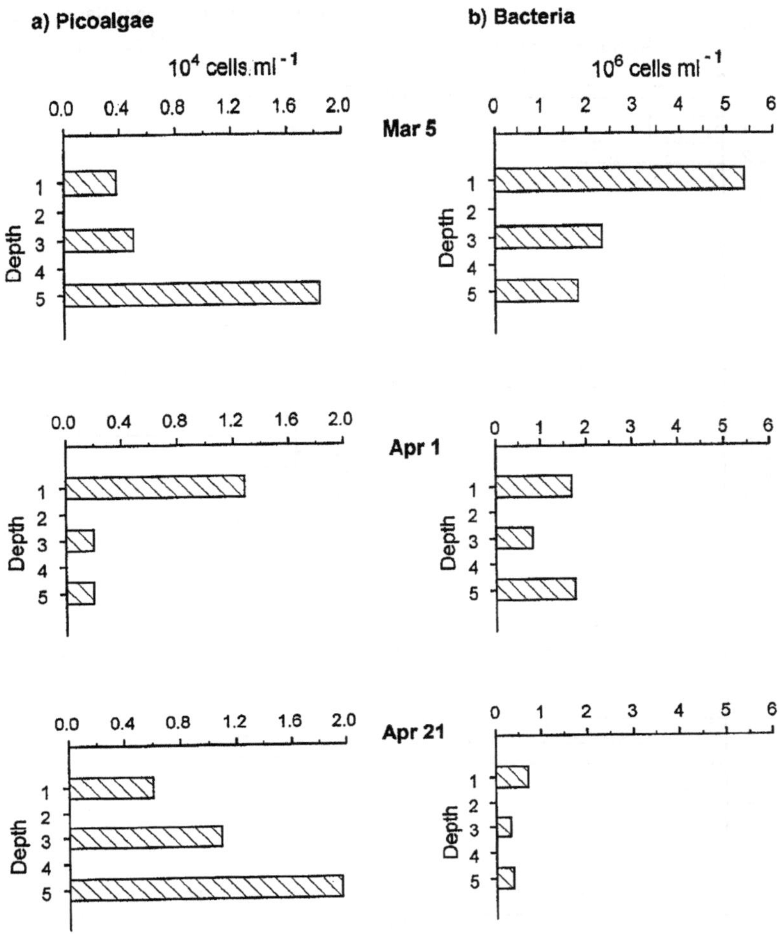

Figure 7. Vertical distribution of a) picoalgae and b) bacteria in Lake Köyliönjärvi during the ice-covered period in January-April 1996.

was highest in March and lowest during the summer months (Figure 5d). The vertical pattern changed during the ice-covered period. In January, the HNFs were most abundant in deeper water, but in March, they were most numerous near the surface (Figure 8b).

Ciliates – The winter biomasses were low compared to summer values, but relatively high compared to other zooplankton groups in winter. The ciliate abundance showed a shifting vertical pattern (Figure 8a). In January, ciliates were abundant in the surface water but much less numerous in deeper water, while in the beginning of March there were few ciliates in the whole water column. One month later, in the beginning of April, ciliates were very abundant especially at 3–4 m, but during the diatom bloom their number decreased again. The size group 20–35 μm was dominant in winter (Figure 9a). The highest ciliate biomass was observed in the beginning of June (Figure 9b) and the number of ciliates was greatest in July. The size group > 35 μm was dominant in June, while the size groups < 20 μm and 20–35 μm were dominant in July (Figure 9a).

Discussion

The under-ice metazooplankton community in eutrophic Lake Köyliönjärvi in winter 1996 was dominated by the calanoid copepod *Eudiaptomus graciloides* in January-March but had almost disappeared by April. The heterotrophic nanoflagellates were abundant under the ice, while ciliate biomass remained low during the ice cover, but peaked in spring. The diatom *Aulacoseira islandica* ssp. *islandica* was very abundant in April. The picoalgal and bacterial

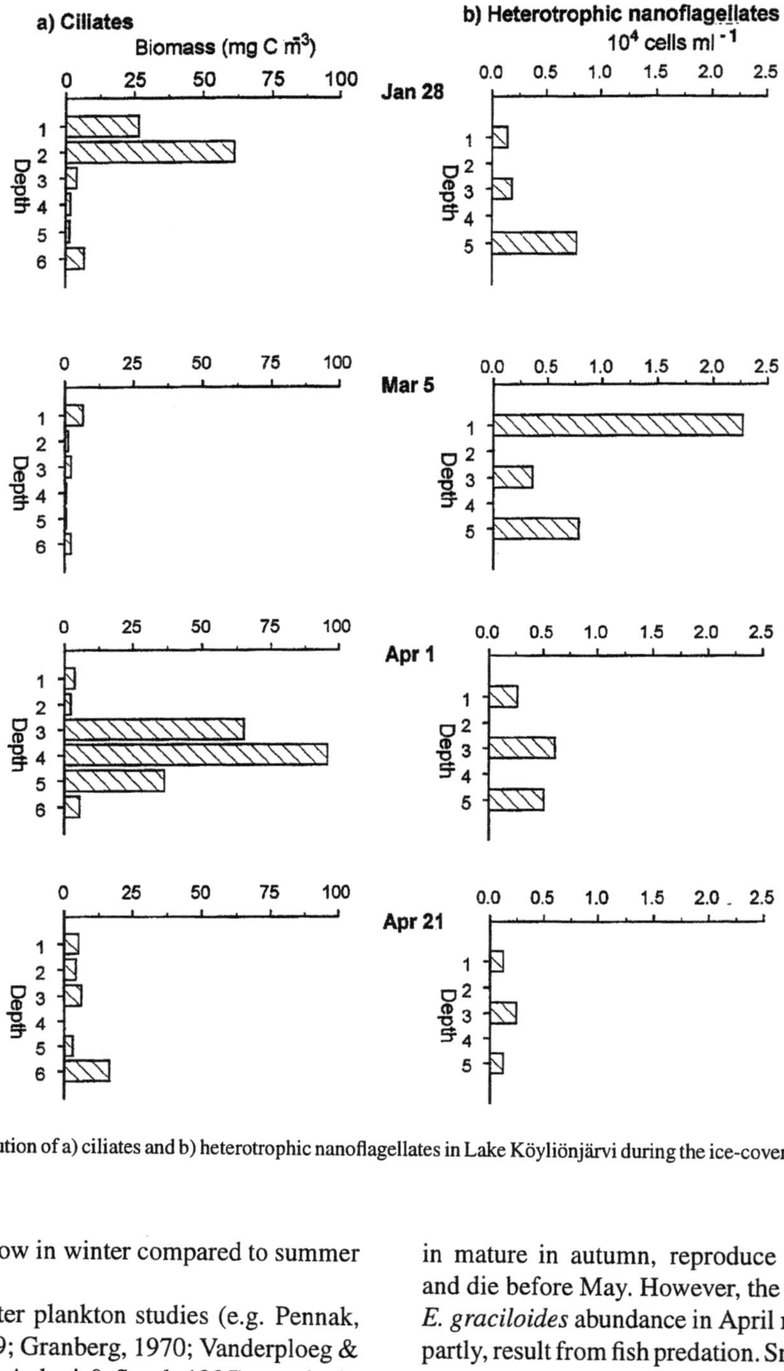

Figure 8. Vertical distribution of a) ciliates and b) heterotrophic nanoflagellates in Lake Köyliönjärvi during the ice-covered period, January-April 1996.

cell numbers were low in winter compared to summer levels.

In previous winter plankton studies (e.g. Pennak, 1968; Hakkari, 1969; Granberg, 1970; Vanderploeg & Bolsenga, 1992 and Agbeti & Smol, 1995) zooplankton was characterized by a low biomass of copepods, their juveniles, and rotifers. Although these observations were from oligotrophic lakes, the observed situation in the eutrophic Lake Köyliönjärvi seemed to be quite similar. However, *Eudiaptomus graciloides* was also relatively abundant during the ice cover. According to Papinska (1981), the adults of *E. graciloides* in mature in autumn, reproduce in February-March and die before May. However, the observed decline in *E. graciloides* abundance in April may also be, at least partly, result from fish predation. Smelt (*Osmerus eperlanus* L.), which is the second most abundant fish in Lake Köyliönjärvi, is known to feed actively on *Eudiaptomus* in late winter below the ice in Lake Syamozera in Russian Karelia (Sterligova 1979).

Our observation of a significant diatom bloom before the ice break is not a unique phenomenon. It is known from earlier reports that spring phytoplankton blooms may develop even under the ice-cover

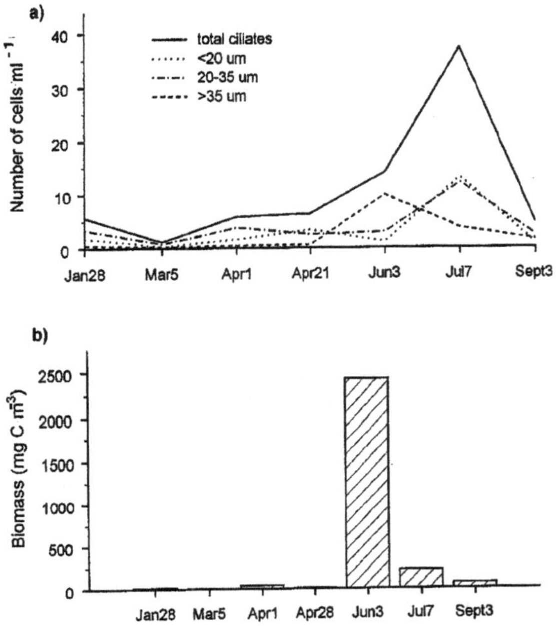

Figure 9. Seasonal variation in the (a) number and (b) biomass of ciliates. The ice-covered period lasted from January 28 to April 21. The winter time values are mean values including the samples taken in 1 m intervals from the depths 1–6 m from one sampling site. The summer time values are pooled samples including three sampling sites.

(Maede & Ichimura, 1973; Sommaruga & Psenner, 1995). We did not measure the amount of PAR at Lake Köyliönjärvi, but the snow cover had almost completely disappeared at the time of the bloom increasing the amount of available radiation. According to Cleve-Euler (1951) and Willén (1962), *Aulacoseira islandica* is a typical species in under-ice diatom blooms.

The ciliate numbers remained low during the ice cover but increased in spring. This kind of seasonal pattern has been observed also in more oligotrophic lakes (Laybourn-Parry et al., 1990; James et al., 1995; Sommaruga & Psenner, 1995) and in brackish water (Kivi, 1986). One of the possible reasons for the observed vertical patterns might be the predation by *E. graciloides*. The impact of calanoid copepods on ciliates is well documentated (Brett et al., 1994; Wiackowski et al., 1994; Burns & Schallenberg, 1996) and calanoids are also known to feed efficiently under the ice (Vanderploeg et al., 1992). The HNFs were more abundant in winter than in summer, which is well in accordance with previous studies (Wright, 1964; Arvola & Kankaala, 1989).

In their study of five Finnish lakes Arvola & Kankaala (1989) observed the peak in bacterial abundance during the phytoplankton spring maximum and also found a positive correlation between the densities of bacteria and algae. In Köyliönjärvi there was no similar correlation and rather the situation seemed to be opposite, as the number of bacteria decreased during the diatom bloom. High numbers of picoalgae could not be detected in Lake Köyliönjärvi, although Pennak (1968) found high numbers of small picoalgae in the eutrophic Tea Lake. The bacteria and picoalgae may be controlled by HNF, as flagellates can graze efficiently on bacteria (Sommaruga & Psenner, 1995) and autotrophic picoalgae (Christofferssen, 1994).

In this study we did not consider the horizontal or between-year variation in plankton community structure. Cloern et al. (1992) state that there may be significant horizontal variation in the winter phytoplankton population. Our own winter samples from Lake Köyliönjärvi in previous years also suggest that there may be substantial between-year differences in both phyto- and zooplankton communities (Vuorio & Saarikari, unpublished data), which is partly attributable to varying oxygen and weather conditions; the biomanipulation is also continuously adding to the between-year variation in plankton community structure.

Acknowledgements

We acknowledge the Laboratory of the Southern Finland Regional Environment Centre for making the chemical analyses, providing the field equipment and giving practical hints for winter sampling. We thank Jouko Sarvala for giving valuable comments throughout this work. We are also grateful to Jarkko Leka and Mika Ventelä, who helped with the hard winter-time sampling, to Asko Sydänoja and Ilkka Blomqvist who took the summer samples, to Krzyzstof Wiackowski who guided with ciliates and commented on the manuscript, to Marianne Moilanen who counted summertime picoalgae, to Martti Mäkinen and Dave Currie who helped with the English, and to Vesa Koivula, Erkki Korpimäki, Sami Kurki, Otso Suominen, Elina Vainio and Jaana Hietala who gave valuable suggestions to the manuscript and to Harri Kuosa and an anonymous reviewer who evaluated the first version of the manuscript. The study was supported by the Graduate School 'Biodiversity and ecological interactions' in University of Turku and the Academy of Finland.

References

Agbeti, M. D. & J. P. Smol, 1995. Winter limnology: a comparison of physical, chemical and biological characteristics in two temperate lakes during ice cover. Hydrobiologia 304: 221–234.

Arvola, L. & P. Kankaala, 1989. Winter and spring variability in phytoplankton and bacterioplankton in lakes with different water colour. Aqua Fenn. 19: 29–39.

Bolsenga, S. J. & H. A. Vanderploeg, 1992. Estimating photosynthetically available radiation into open and ice-covered freshwater lakes from surface characteristics; a high transmittance case study. Hydrobiologia 243/244: 95–104.

Brett, M. T., K. Wiackowski, F. S. Lubnow, A. Muellersolger, J. J. Elser & C. R. Goldman, 1994. Species-dependent effects of zooplankton on planktonic ecosystem processes in Castle Lake, California. Ecology 75: 2243–2254.

Burns, C. W. & M. Schallenberg, 1996. Relative impacts of copepods, cladocerans and nutrients on the microbial food web of a mesotrophic lake. J. Plankton. Res. 18: 683–714.

Christoffersen, K. 1994. Variations of feeding activities of heterotrophic nanoflagellates on picoplankton. Marine Microbial Food Webs 8: 111–123.

Cleve-Euler, A., 1951. Die Diatomeen von Schweden und Finnland. Kungliga Vetenskaps-akademiens Handlingarna 2/1. Bibliotheca Phycologica 5: 1–162.

Cloern, J. E., A. E. Alpine, B. E. Cole & T. Heller, 1992. Seasonal changes in the spatial distribution of phytoplankton in small, temperate-zone lakes. J. Plankton Res. 14: 1017–1024.

Granberg, L., 1970. Seasonal fluctuations in numbers and biomass of the plankton of Lake Pääjärvi, southern Finland. Ann. Zool. Fenn. 7: 1–24.

Hakkari, L., 1969. Zooplankton studies in the Lake Längelmävesi, south Finland. Ann. Zool. Fenn. 6: 313–326.

Hirvonen & Salonen, 1995. The first stage in restoring Lake Köyliönjärvi by fish removal. Vesitalous 36: 11–14.

Hirvonen, A., H. Helminen, V. Saarikari, S. Salonen, K. Vuorio & J. Sarvala, 1993. Effects of cyprinid reduction on water quality in Lake Köyliönjärvi, SW Finland. 5th International Conference on the Conservation and Management of Lakes 'Strategies for Lake Ecosystems beyond 2000', Proceedings 17–21 May, 1993 Stresa, Italy: 113–116.

James, M. R., C. W. Burns & D. J. Forsyth, 1995. Pelagic ciliated protozoa in two monomictic, southern temperate lakes of contrasting trophic state: seasonal distribution and abundance. J. Plankton Res. 17: 1479–1500.

Kivi, K., 1986. Annual succession of pelagic protozoans and rotifers in the Tvärminne Storfjärden, SW coast of Finland. Ophelia 4: 101–110.

Latja, R. & K. Salonen, 1978. Carbon analysis for the determination of individual biomasses of planktonic animals. Verh. Internat. Verein. Limnol. 20: 2556–2560.

Laybourn-Parry, J., J. Olver & S. Rees, 1990. The hypolimnetic protozoan plankton of a eutrophic lake. Hydrobiologia 203: 111–119.

MacIsaac, E. A. & J. G. Stockner, 1993. Enumeration of Phototrophic Picoplankton by Autofluorescense Microscopy. In P. F. Kemp, B. F. Sherr, E. B. Sherr & J. J. Cole (eds), Handbook of Methods in Aquatic Microbial Ecology. Lewis Publisher, Boca Raton.

Maeda, O. & S. Ichimura, 1973. On the high density of a phytoplankton population found in a lake under ice. Int. Rev. Ges. Hydrobiol. 58: 473–485.

Müller, H. & W. G. Geller, 1993. Maximum growth rates of aquatic ciliated protozoa: the dependence on body size and temperature reconsidered. Arch. Hydrobiol. 126: 126–327.

Papinska, K., 1981. Occurrence of filtering Crustacea in the near-bottom and pelagic waters of the Mikolajskie Lake. Hydrobiologia 83: 411–418.

Pennak, R. W., 1968. Field and experimental winter limnology of three Colorado mountain lakes. Ecology 49: 505–520.

Porter, K. G. & Y. S. Feig, 1980. The use of DAPI for identifying and counting aquatic microflora. Limnol. Oceanogr. 25: 943–948.

Putt, M. & D. K. Stoecker, 1989. An experimentally determined carbon-volume ratio for marine oligotrichous ciliates from estuarine and coastal waters. Limnol. Oceanogr. 34: 1097.

Salonen, K., 1981. Determination of carbon – an alternative method for the estimation of biomass of zooplankton. Lammi Notes 5: 7–11.

Sarvala, J., H. Helminen, V. Saarikari, S. Salonen & K. Vuorio, 1998. Relations between planktivorous fish abundance, zooplankton and phytoplankton in three lakes of differing productivity, Hydrobiologia 363: 81–95.

Sherr, E. B. & B. F. Sherr, 1993. Preservation and storage of samples for enumeration of heterotrophic protists. In P. F. Kemp, B. F. Sherr, E. B. Sherr & J. J. Cole (eds), Handbook of Methods in Aquatic Microbial Ecology. Lewis Publishers, Boca Raton: 207–213

Sommaruga, R. & R. Psenner, 1995. Trophic interactions within the microbial food web in Piburger See (Austria). Arch. Hydrobiol. 132: 257–278.

Sommer, U., Z. M. Gliwicz, W. Lampert & A. Duncan, 1986. The PEG-model of seasonal succession of planktonnic events in freshwaters. Arch. Hydrobiol. 106: 433–471.

Sterligova, O. P., 1979. Koryushka Osmerus eperlanus (L.) i ee rol v ikhtiofaune Syamozera. Voprosy Ikhtiologii 19: 793–800 (in Russian).

Tulonen, T., 1993. Bacterial production in a mesohumic lake estimated from (^{14}C) leucine incorporation rate. Microbiol. Ecol. 26: 201–217.

Vanderploeg, H. A., S. J. Bolsenga, G. L. Fahnenstiel, J. R. Liebig & W. S. Gardner, 1992. Plankton ecology in an ice-covered bay of Lake Michigan: utilization of a winter phytoplankton bloom by reproducing copepods. Hydrobiologia 243/244: 175–183.

Wiackowski, K., M. T. Brett & C. R. Goldman, 1994. Differential Effects of Zooplankton Species on Ciliate Community Structure. Limnol. Oceanogr. 39: 486–492.

Willén, T., 1962. Studies on the phytoplankton of some lakes connected with or recently isolated from the Baltic. Oikos 13: 169–199. Wright, R. T., 1964. Dynamics of a phytoplankton community in an ice-covered lake. Limnol. Oceanogr. IX: 163–178.

Hydrobiologia **363:** 241–251, 1998.
T. Tamminen & H. Kuosa (eds), Eutrophication in Planktonic Ecosystems: Food Web Dynamics and Elemental Cycling.

Biochemical composition of particulate organic matter and bacterial dynamics at the sediment–water interface in a Mediterranean seagrass system

Roberto Danovaro[1], Norberto Della Croce[2] & Mauro Fabiano[2]
[1] *Facoltà di Scienze, Università di Ancona, Via Brecce Bianche, 60131, Ancona, Italy*
[2] *Istituto Scienze Ambientali Marine, Università di Genova, Corso Rainusso 14, CP 79, 16038, Genova, Italy*
Address for correspondence: Dr Roberto Danovaro, Facoltà di Scienze, Università di Ancona, Via Brecce Bianche, 60131, Ancona, Italy; e-mail: danovaro@anvax1.unian.it

Key words: particulate organic matter composition, bacteria, sediment-water interface, *Posidonia oceanica*

Abstract

The seagrass *Posidonia oceanica* is the most productive system of the Mediterranean Sea. In order to gather information on the temporal and spatial variability of the suspended particulate matter in relation to bacterial dynamics, water samples were collected at 10 cm above the sediments over a period of 13 months in the Prelo Bay (Ligurian Sea, NW Mediterranean). Measurements of seston concentration, as well as the elemental (POC and PON) and biochemical composition (lipids, proteins, carbohydrates and nucleic acids) of particulate matter were carried out to assess the origin, composition and bacterial contribution to the food potentially available in the seagrass system to consumers. Lipids and proteins were the main biochemical classes of organic compounds, followed by carbohydrates. Despite the highly refractory composition of the seagrass leaves, particulate organic matter was mostly composed of labile compounds (69.9% of POC). POM temporal patterns were controlled by current speed at the sediment–water interface that resuspended only small particles largely colonised by bacteria after an intensive process of fractionation and aging. In the seagrass system, the POM appears to be dominated by bacteria (density ranging from 0.7 to 2.5 × 10^9 cells l^{-1}, representing more than 48.3% of POC and 68.7% of the biopolymeric carbon, as the sum of lipid, protein and carbohydrate carbon). This feature was characteristic of the seagrass system since much lower bacterial densities were found outside the *Posidonia* meadow. Bacteria were negatively correlated with the concentration of nitrite and nitrate suggesting a selective utilisation of inorganic nutrients to support their growth.

Introduction

Quantitative information on particulate organic matter and associated bacteria is central to carbon flow in marine ecosystems. Bacterial activity, indeed, is dependent, to a large extent, upon the availability of labile organic matter (Hermin, 1989). Organic matter availability is, in turn, the result of the interactions between physical and biological processes. Temporal variability and spatial scale play a key role in determining the extension of these interactions and the structure of the microbial food web. The sediment-water interface represents a frontier between two phases (bottom waters and surface sediments). This interface forms a complex system characterised by a living community dominated by a mineral phase and microorganisms influencing nutrient exchanges and detrital organic matter composition (Daumas, 1990). The importance of suspended matter in the planktonic and benthic food webs is well recognised, but only few studies have attempted to characterise the quantity and quality of the suspended particles at the water-sediment interface (Fichez, 1991). This is surprising since the most evident changes in the composition and characteristics of the particles occur in this transition layer. In coastal areas, the deposition of the suspended particulate matter to the sediment results in a rapid change in its biochemical composition. In fact, the biopolymeric fraction (*sensu* Fichez, 1990) of the organic carbon, dominant in the suspended POM (representing up to 90% of the organic carbon pool, Danovaro & Fabiano, 1997), becomes a minor fraction once has reached the sediments (usually ranging from 5 to 15%), so that the carbon pool becomes dominated by a geopolymeric fraction (Fabiano et al., 1995). The extent of such interactions depends on the amount and origin of the primary organic matter produced, the significance of the allochthonous inputs, the sinking rates and the resuspension effects.

In this study we focused our attention on the sediment-water interface of a seagrass bed (*Posidonia oceanica* Delile) of the Western Mediterranean. Despite the generally low inorganic nutrient concentrations, the seagrass meadows represent the most productive system of the Mediterranean with a production of 1000–3000 g DW m^{-2} y^{-1} (Ott, 1980). In autumn, large amounts of seagrass leaves and their epiphytes are deposited onto the sediments. Only a small fraction of this plant material (which is of highly refractory composition, Danovaro 1996) is directly consumed by benthic organisms and most of it must be fractionated before entering the food chain (Velimirov, 1986). Detrital material derived from the seagrass and 'exported' to the adjacent sediments serve as major source of organic matter supporting heterotrophic processes. Therefore the sediment–water interface of the seagrass bed represents a peculiar and highly productive system dominated by bacteria (Danovaro et al., 1994), that may provide useful information for comparison and for the understanding of the functioning of other highly productive systems characterised by much higher nutrient concentrations.

In the present study, the bacterial dynamics are investigated in relation to particulate organic matter composition and nutrient concentrations in order to point out: (1) origin, composition and dynamics of the particulate organic matter at the sediment-water interface; (2) the factors related to the dynamics of the bacterial community at the sediment-water interface; and (3) the role of bacteria in the composition of particulate organic matter; (4) the production cycled through bacterioplankton and the role of changes in particle availability on the size and activity of the microbial assemblages.

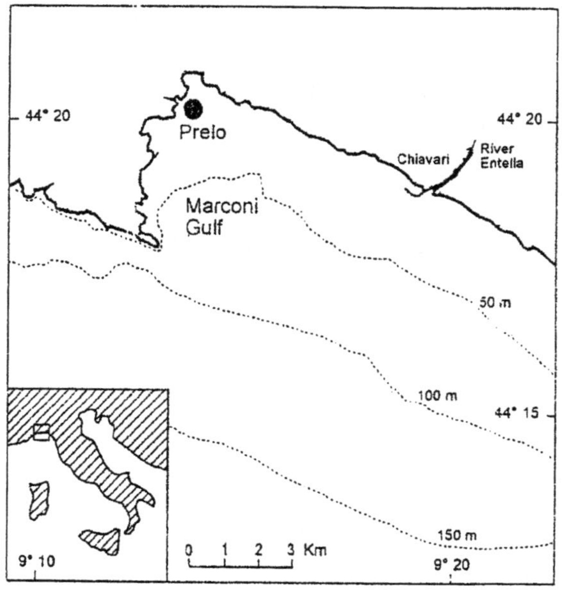

Figure 1. The sampling station in the Golfo Marconi, Ligurian Sea (NW Mediterranean).

Materials and methods

Study site and sampling

Samples were collected from January 1991 to January 1992: every 2 weeks in April and May and on monthly basis during the rest of the year. Sampling was carried out by SCUBA divers at Prelo cove (4 m depth) in the Gulf of Marconi, Ligurian Sea (NW Mediterranean, Figure 1). The study area is sheltered and characterised by a coarse-detritic bottom. The depth of the redox potential discontinuity layer ranged from 4 to 11 cm and the porosity in the top 2 cm of sediment varied between 22.1 and 55.7% (Danovaro, 1996). The Prelo cove is characterised by the presence of a wide, dense (375 leaves per m^2) and highly productive *Posidonia oceanica* meadow from a depth of about 0.4 to 9 m that reduces the current speed at the sediment-water interface (0.4–2.9 cm s^{-1}, Danovaro, 1996).

Water samples were collected using 5-litre PVC bottles previously washed with a 0.1 N HCl solution and placed horizontally 10 cm above the sediments. In order to avoid contamination with resuspended sediments, the bottles were kept at 1 m from the diver and closed manually. Water samples were sieved through a 200 μm mesh net and brought to the laboratory for subsequent filtration. For bacterial analyses water samples (10 ml) were collected on three replicates in sterile test tubes and fixed with a 2% buffered formalin solution (prefiltered with a 0.2 μm Puradisk, Nuclepore). For the analysis of total suspended matter, particulate organic carbon and nitrogen, particulate lipids, proteins and carbohydrates and particulate nucleic acids (DNA and RNA) 500–1000 ml of the water sample were filtered on precombusted (450 °C, 2 h) Whatman GF/C glass fiber filters. For phytopigment analysis, 1000 ml were filtered onto Millipore filters (0.45 μm pore size) and stored in the dark at –20 °C until analysis.

Total suspended matter

After filtration, total suspended matter retained on Whatman GF/C glass-fibre filters was rinsed with 20 ml of isotonic solution and weighed after desiccation (24 h, 60 °C) using a balance Mettler M3 (accuracy ± 1 μg). Particle concentration was reported to the volume filtered. Blank filters were subjected to the same treatment. Analyses were carried out on duplicate per sample.

Pigment content, elemental and biochemical composition of the suspended particles

The analyses of chlorophyll-*a* and phaeopigments were carried out according to Lorenzen & Jeffrey (1980) on a single replicate. Pigments were extracted with 90% acetone. After centrifugation, the supernatant was used to determine the functional chlorophyll-*a* and acidified with 0.1 N HCl in order to estimate the amount of phaeopigments (Plante-Cuny, 1974).

Particulate organic carbon (POC) and nitrogen (PON) were measured, after treatment on HCl vapour for 5 h, using a Carlo Erba CHN Analyzer (mod. EA 1108). Analyses were carried out on a single replicate. Cyclohexanone-2,4-dinitrophenilhydrazone was used as standard (Hedges & Stern, 1984). Particulate nucleic acids (DNA and RNA) were determined using the method of Lukavsky et al. (1973) slightly modified by Zachleder (1984). This method is based on the light-activated reaction of the extract in 0.5 N perchloric acid and the dyphenilamine reactive. DNA (thymus) solutions were used as standard. RNA concentrations were determined by difference. Particulate lipids were extracted by direct elution with chloroform and methanol. Analyses were carried out using the Bligh & Dyer (1959) and the Marsh & Weinstein (1966) methods. Tripalmitine solutions were used as standard. Particulate protein analyses were carried out according to Hartree (1972). Albumine solutions were used as standard. Particulate carbohydrates were analysed according to Dubois et al. (1956). D(+) glucose solutions were used as standard. Lipid, protein, carbohydrate and nucleic acid analyses were carried out on duplicate per sample.

Dissolved nutrients

Water samples were filtered through precombusted (450 °C, 2 h) Whatman glass fibre filters (GF/C) (pressure of filtration: –0.2/–0.4 bar). The filtrate was then placed into an acid-washed polyethylene bottle and stored frozen for the analysis of nitrogenous compounds: nitrate (NO_3^-), nitrite (NO_2^-) and ammonium (NH_4^+) and phosphate (PO_4^{3-}) by the standard automated nutrient analytical procedures (Strickland & Parsons, 1972) using a Technicon AutoAnalyzer mod. II.

Bacterial parameters

Each water replicate ($n=3$) was stained for 3 min. with Acridine Orange and filtered on black Nuclepore 0.2 μm filters and counted under epifluorescence microscopy (Zeiss Universal Microscope). At least 10 microscopic fields or > 400 cells were counted (Hobbie et al., 1977). The number of dividing bacteria (defined as cells with a clearly visible invagination) was determined. The contribution by different size classes of bacteria to the total biomass was followed by dividing bacteria into different size classes according to Palumbo et al. (1984). Bacterial biovolume was converted to carbon content assuming 310 fg C per μm^3 (Fry, 1990). The hourly growth rate was calculated as: $\mu = e^{(0.299FDC-4.916)}$ and the doubling time (D) was calculated from the daily μ as: $D = 0.693/\mu$ (Turley & Lochte, 1985). Bacterial DNA was estimated assuming the conversion factor 5 fgDNA $cell^{-1}$ (Simon & Azam, 1989). Bacterial contribution to the particulate organic carbon was estimated assuming that only bacteria attached to detrital particles or large free-living bacteria (> 1 μm) were retained by the GF/C filters.

Results

Total suspended matter (TSM)

Seasonal changes in TSM concentrations are reported in Figure 2. TSM concentrations varied significantly during the year. Highest concentrations were report-

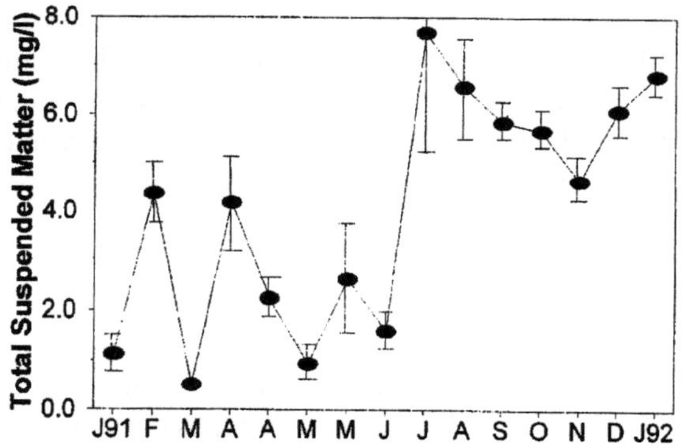

Figure 2. Seasonal changes in total suspended matter concentrations in the Golfo Marconi. Standard deviations are shown.

ed between July 1991 and January 1992 (range 4.63–7.68 mg l^{-1}) and lowest values were recored in March 0.51 mg l^{-1}.

Elemental and biochemical composition of the suspended particles

Changes in POC and PON concentration at the sediment-water interface are illustrated in Figure 3a and b while the C:N ratios are shown in Figure 3c. POC concentrations ranged from 82.1 to 490.6 μgC l^{-1} (in December and early April respectively) while PON concentrations varied from 10.30 to 40.09 μg l^{-1}. For both parameters the highest values were observed in late winter–early spring. The C:N ratio increased from February to April (up to 15.3). From June to the end of the year the values of the C:N ratio remained below 10.

Changes in chlorophyll-*a* and phaeopigment concentrations are shown in Figure 4. Chlorophyll-*a* showed two main peaks in April (3.74 μg l^{-1}) and September (5.95 μg l^{-1}) while phaeopigment content was always very low with a main peak in February (2.20 μg l^{-1}) and a smaller peak in November (0.23 μg l^{-1}). Particulate DNA and RNA concentrations are illustrated in Figure 5a and b. Highest DNA content in April (17.5 μg l^{-1}) corresponded to a chlorophyll-*a* peak. RNA concentrations at the sediment–water interface appeared more variable showing the highest value in July (54.5 μg l^{-1}) and lowest values in August 8.2 μg l^{-1}. As result RNA:DNA ratio showed high values in July and from September to November (range 2.4–4.1, Table 2). Particulate protein, carbohydrate and lipid concentrations are reported in Figure 6a, b and c respectively. The annual average concentration of

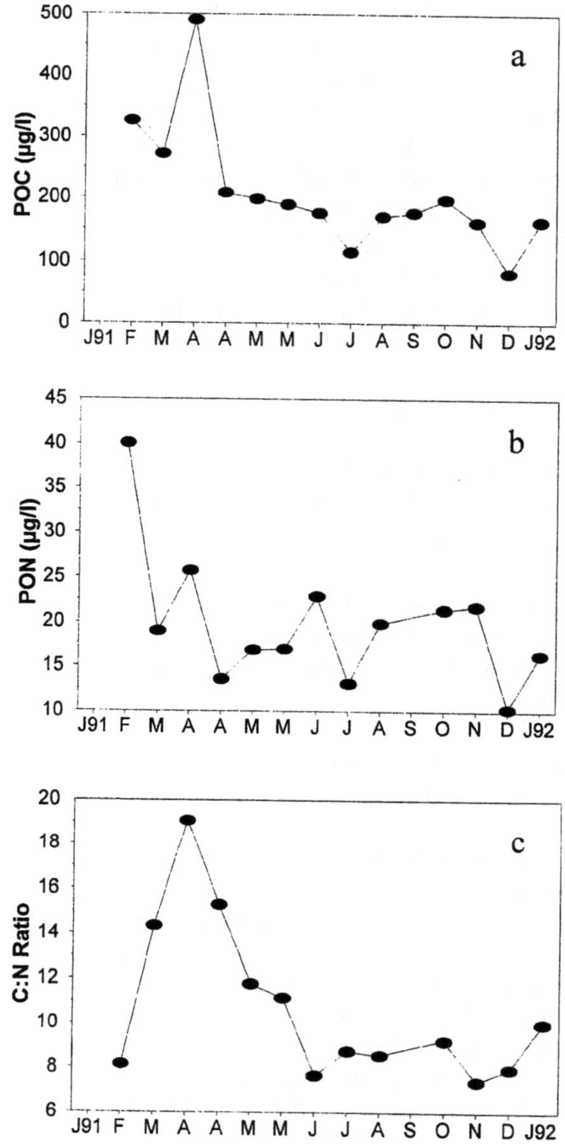

Figure 3. Seasonal variations of: (a) particulate organic carbon (POC); (b) particulate organic nitrogen (PON) concentrations and (c) C:N ratios.

the three biochemical components was quite similar (82.5, 72.2 and 89.0 μg l^{-1} for proteins, carbohydrates and lipids respectively). No significant seasonal trends were observed except a main peak of particulate lipids in April (179.1 μg l^{-1}). The biopolymeric fraction of the organic carbon (C-BPF), as the sum of lipid, protein and carbohydrate carbon (Table 1), accounted on annual average for 69.9% of POC concentrations, whilst protein nitrogen accounted for 67.4% of the PON.

Table 1. Concentrations of biopolymeric carbon (C-BPF) and chlorophyll-*a* carbon (C-CHL, obtained after conversion using the factor of De Jonge, 1980) and their contribution to POC (C-BPF/POC and C-CHL/POC). Contribution of chlorophyll-carbon to C-BPF (C-CHL/C-BPF). Contribution of the bacterial biomass retained by GF/C filters (BBMr) to POC and C-BPF

Sampling date	C-BPF μgC/l	C-Chl-*a* μgC/l	BBMr μgC/l	C-BPF/POC (%)	C-CHL/POC (%)	BBMr/POC (%)	BBMr/C-BPF (%)
17.01.91	174.57	56.07	50.30	nd	nd	nd	28.82
20.02.91	170.38	42.05	62.00	52.16	12.88	18.98	36.39
21.03.91	124.07	56.07	117.45	45.53	20.57	43.10	94.66
08.04.91	213.93	112.14	136.74	43.61	22.86	27.87	63.92
19.04.91	106.45	56.07	90.53	51.26	27.00	43.59	85.05
08.05.91	152.44	56.07	150.81	76.93	28.30	76.10	98.93
29.05.91	121.91	24.03	113.55	64.43	12.70	60.01	93.14
28.06.91	150.84	22.43	112.95	85.95	12.78	64.36	74.88
18.07.91	111.19	44.86	96.16	97.02	39.14	83.91	86.48
05.08.91	97.66	154.75	96.85	57.30	90.80	56.83	99.18
09.09.91	120.51	178.58	73.28	68.30	101.21	41.53	60.81
02.11.91	157.00	56.07	80.06	79.53	28.40	40.56	51.00
02.12.91	97.99	42.05	89.76	60.68	26.04	55.58	91.60
23.12.91	79.47	24.03	44.14	96.79	29.27	53.76	55.54
07.01.92	161.92	28.04	16.04	99.28	17.19	9.83	9.91

Table 2. Protein nitrogen concentrations (PRT-N, as protein content / 6.25) and protein contribution to total particulate organic nitrogen (PRT-N/ON). bacterial-DNA concentrations (B-DNA) and their potential contribution to the total DNA content of particulate matter and growth rates of bacterioplankton. RNA:DNA ratios, Growth rates (GR, d^{-1}) and Doubling time (d) and Bacterial production (PB as $\mu gC\ l^{-1}\ d^{-1}$)

Sampling date	PRT-N μgN/l	PRT-N/ON (%)	RNA:DNA	B-DNA μg/l	B-DNA/DNA (%)	GR (μ) d-1	D d	BP μC/l/d
17.01.91	nd	nd	2.03	3.54	22.41	0.419	1.65	39.05
20.02.91	14.24	35.53	1.67	5.33	47.43	0.308	2.25	37.09
21.03.91	17.42	91.81	1.05	6.69	50.36	0.595	1.17	123.96
08.04.91	13.17	51.19	1.71	7.15	43.01	0.301	2.31	71.80
19.04.91	12.27	90.42	0.88	4.90	28.04	0.534	1.30	84.99
08.05.91	15.02	89.26	2.38	6.69	47.36	0.206	3.37	52.78
29.05.91	12.99	76.67	1.46	7.07	54.14	0.289	2.39	59.33
28.06.91	15.20	66.38	1.13	4.93	34.92	0.368	1.88	70.58
18.07.91	10.59	80.87	4.41	7.40	59.81	0.277	2.50	50.31
05.08.91	10.56	53.07	0.64	6.06	47.64	0.252	2.74	44.19
09.09.91	12.33	nd	2.39	5.80	41.19	0.277	2.50	38.62
02.11.91	11.91	55.66	3.64	3.81	36.33	0.398	1.74	54.85
02.12.91	11.73	53.81	4.08	5.98	52.15	0.320	2.17	52.51
23.12.91	3.71	36.04	1.82	4.70	31.03	0.618	1.12	56.23
07.01.92	15.56	95.47	2.36	4.98	31.77	0.988	0.70	51.52

Dissolved nutrients

Dissolved nutrient concentrations are reported in Figure 7a, b, c and d respectively for ammonium, nitrite, nitrate and phosphate. Ammonium concentrations showed three main peaks in March, May and September whilst nitrite and partially nitrate concentrations showed a clear depletion in late spring and summer with highest values in autumn–winter. Highest values of nitrite, nitrate and phosphate were recorded in January, October and September (0.52, 2.32, 2.57 μm, respectively).

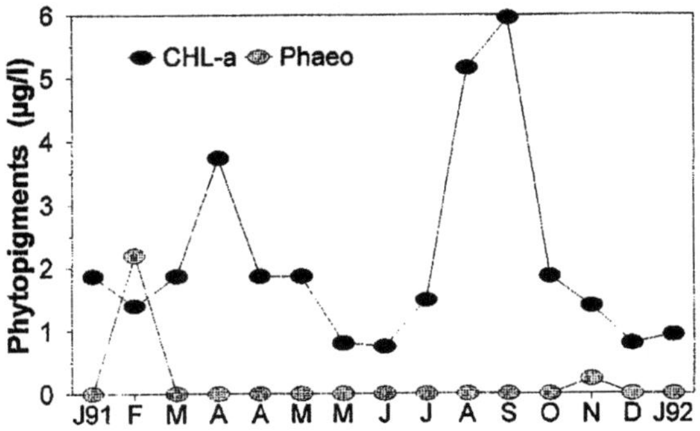

Figure 4. Seasonal variations of chlorophyll a (filled circles) and phaeopigment concentrations (empty circles) in the Golfo Marconi. Analyses on a single replicate.

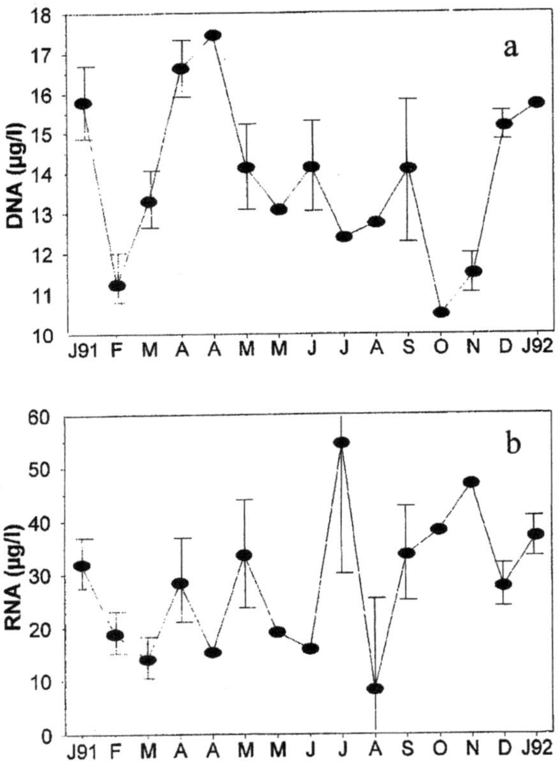

Figure 5. Seasonal variations of particulate nucleic acids: (a) DNA concentrations; (b) RNA concentrations. Standard deviations are reported.

Bacterial Parameters

Seasonal changes in bacterial abundance, biomass and productivity (as FDC) are reported in Figure 8a, b and c respectively. Bacterial parameters were characterised by clear seasonal changes with highest values in spring-summer and low values in winter. Highest bacterial

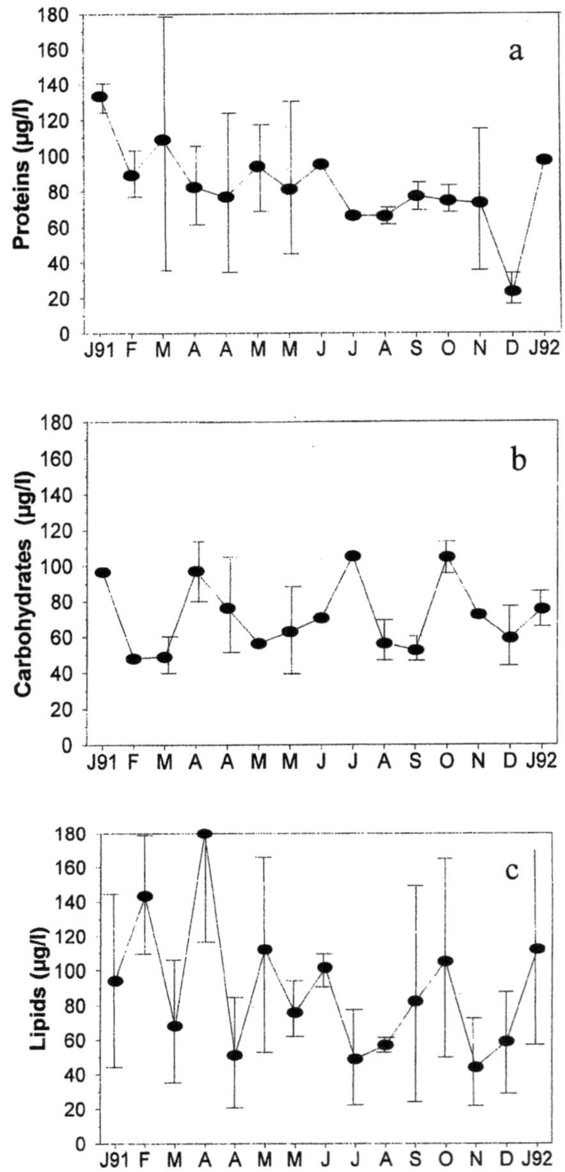

Figure 6. Seasonal variations of the biochemical composition of POM: (a) particulate proteins, (b) particulate carbohydrates, (c) particulate lipids. Standard deviations are reported.

density, reported in May (2.84×10^9 cells l^{-1}), was equivalent to a biomass of 177.8 μgC l^{-1}; a minor peak was reported in November (2.84×10^9 cells l^{-1}). FDC values ranged from 0.52 to 5.77% (May and January 1992 respectively). The fraction of bacterial biomass retained by the GF/C pore filters accounted on average for 77.1% of the total bacterial biomass (range 44.4–85.0 in January 1992 and June respectively). Average

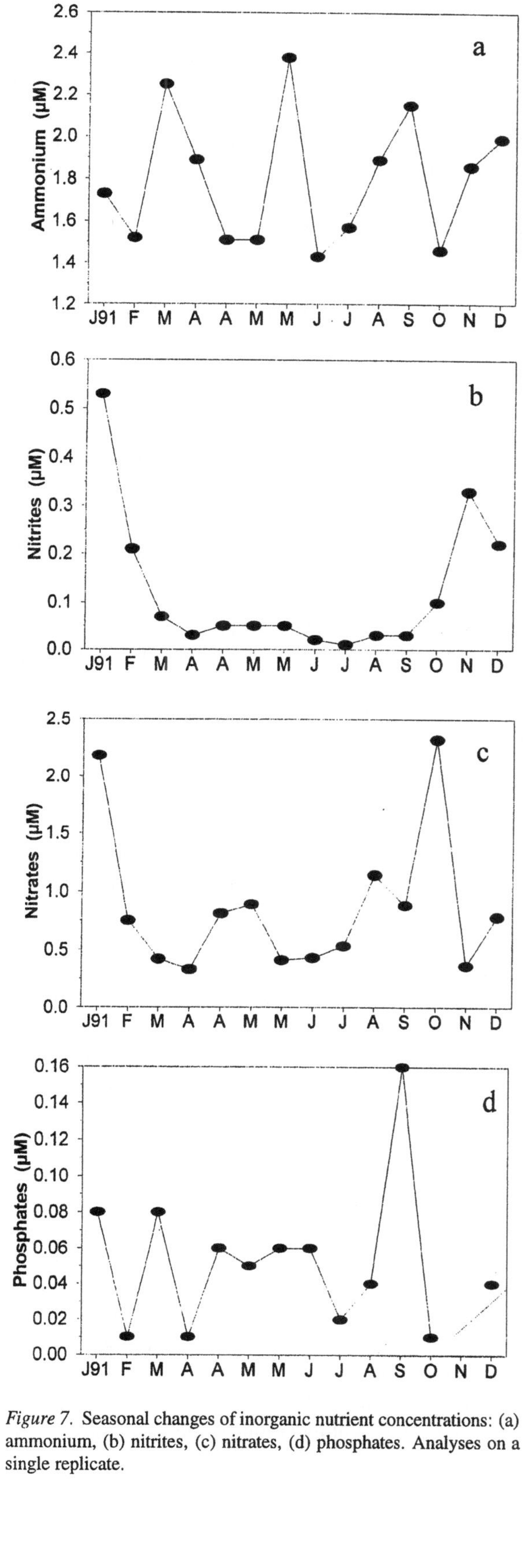

Figure 7. Seasonal changes of inorganic nutrient concentrations: (a) ammonium, (b) nitrites, (c) nitrates, (d) phosphates. Analyses on a single replicate.

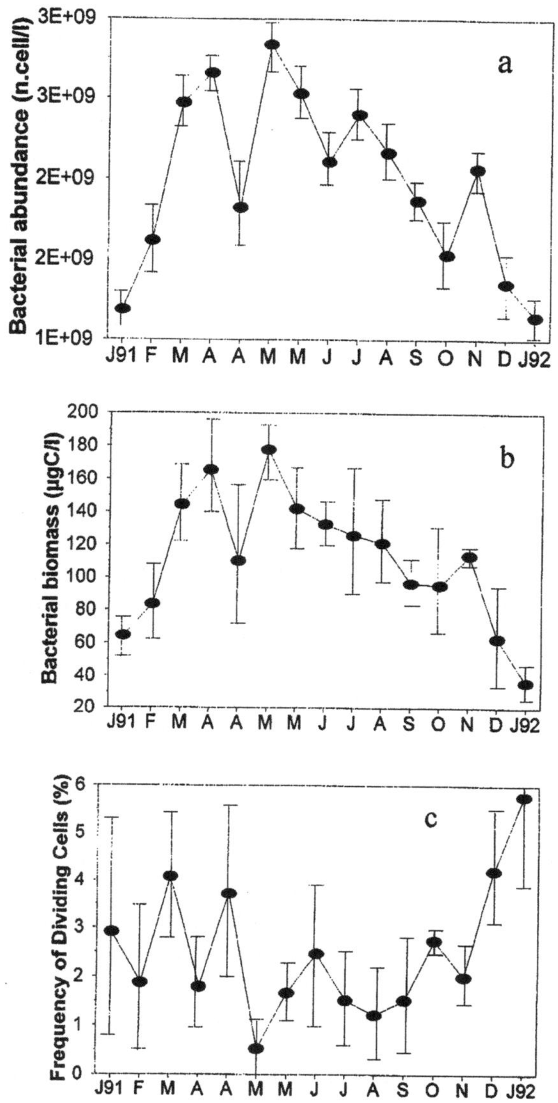

Figure 8. Seasonal changes of bacterial parameters: (a) bacterial abundance, (b) bacterial biomass, (c) frequency of dividing cells (FDC). Standard deviations are reported.

cell biomass ranged from 3.15 to 6.30 × 10^{-11} mgC $cell^{-1}$. High values of cell biomass were reported in spring (April–May: 6.04–6.24 × 10^{-11}mgC $cell^{-1}$) according to the phytoplankton peak. Table 1 shows the bacterial contribution to the particulate organic matter composition. Bacterial biomass accounted on average for 48.3% of POC concentrations and for 68.7% of the C-BPF content. Bacterial DNA accounted on annual average for 41.8% of the particulate DNA pool. Bacterial abundance was negatively correlated with nitrite and nitrate concentrations ($n=15$, $r=-0.694$

and $r = -0.664$, respectively). Data on bacterial growth rates (GR), doubling time (D) and bacterial production (BP) are reported in Table 2. Growth rates ranged from 0.206 to 0.988 d^{-1} (in early May and January 1992, respectively). Bacterial production from D values ranged from 37.09 to 123.96 $\mu gC\ l^{-1}\ d^{-1}$ showing a well defined seasonal pattern characterised by higher values in spring and summer lower values during the rest of the year.

Discussion

Quality and composition of the particles

The sediment–water interface of the seagrass meadow appears to be characterised by relatively high concentrations of particulate organic matter of high nutritional quality. Previous studies have shown that primary production cycles of *Posidonia oceanica* and its epiphytes were responsible for the large amounts of sedimentary organic carbon pools (Danovaro 1996), and that the inputs of plant material produced by the seagrass were characterised by a highly refractory composition (Velimirov, 1987; Lawrence et al., 1989). This is in contrast with the dominance of labile organic compounds found in the suspended particles, as, on average, about 70% of the organic carbon pool was composed by the biopolymeric fraction. Another interesting feature is that despite carbohydrates were dominant in this system, both in sedimentary organic matter and in the seagrass leaves (Lawrence et al., 1989; Danovaro, 1996), suspended particles displayed high lipid and protein concentrations. While the dominance of lipids appears to be a characteristic feature of the sediment-water interface (Fichez, 1990; Danovaro & Fabiano, 1997), the high protein content seems to be the effect of a protein enrichment due to attached bacteria, which for their low C:N ratio (from 3.5 to 5) and their high biomass may have contributed significantly to enhance the protein content of the suspended particles.

The large contribution of labile compounds (C-BPF) to POC (Table 1) provides further evidence of differences between the biochemical composition of the suspended and sedimentary organic matter, as in the sediments only about 15% of the organic pool was composed by biopolymeric compounds (Danovaro, 1996). However, the large dominance of labile compounds on POC was surprising also when compared to values reported for the same year in another subtidal station outside the Prelo meadow, where the biopolymeric fraction accounted for less than 60% of the POC concentrations and the concentrations of lipids, proteins and carbohydrates were 20–30% lower (Danovaro & Fabiano, 1997). Again the high bacterial density and biomass of the seagrass system, about one order of magnitude higher than outside the meadow, (Danovaro & Fabiano, 1997) are probably responsible for the large biopolymeric fraction and the high nutritional quality of the seagrass POM.

Source and dynamics of the suspended particles

The comparison between the suspended and sedimentary organic matter (data from Danovaro, 1996) in terms of composition and seasonality, provides some interesting indications about the nature, origin and the factors regulating particle dynamics at the sediment–water interface. Velimirov (1986a, 1987), studying the daily and seasonal variations of DOC and POC concentrations in seagrass beds, hypothesised that changes in DOC concentrations were mainly due to a shift from the dissolved to the particulate phase. The results of the present study indicate that other hypotheses, possibly complementary to one described, may be proposed. In fact we report that changes in POC concentrations at the sediment–water interface were coupled with those in the sediments. However, the amounts of total suspended matter were low and indicated, as expected from the coarse grain size of the sediments and the generally low current velocities, that sediment resuspension was limited or negligible. However, as shown in Figure 9, we observed that the significance of the biopolymeric fraction on POC at the sediment–water interface (Figure 9a) was inversely related to the percentage of the biopolymeric fraction in the sediments ($p < 0.05$; Figure 9b). This effect was more evident when higher current velocities at the sediment–water interface were recorded (such as in June, October and December, Figure 9c). These data lead us to hypothesise that the quantity, composition and origin of the suspended particles are regulated by a drift mechanism and it seems that this mechanism influenced primarily their composition. As suggested by the dominance of proteins and lipids in POM, this mechanism seems to induce the resuspension of small particles largely colonised by bacteria after an intensive process of fractionation and aging but did not, for instance, resuspend the fresh plant debris that, due to their high carbohydrate content, would have determined a different biochemical composition of the suspended parti-

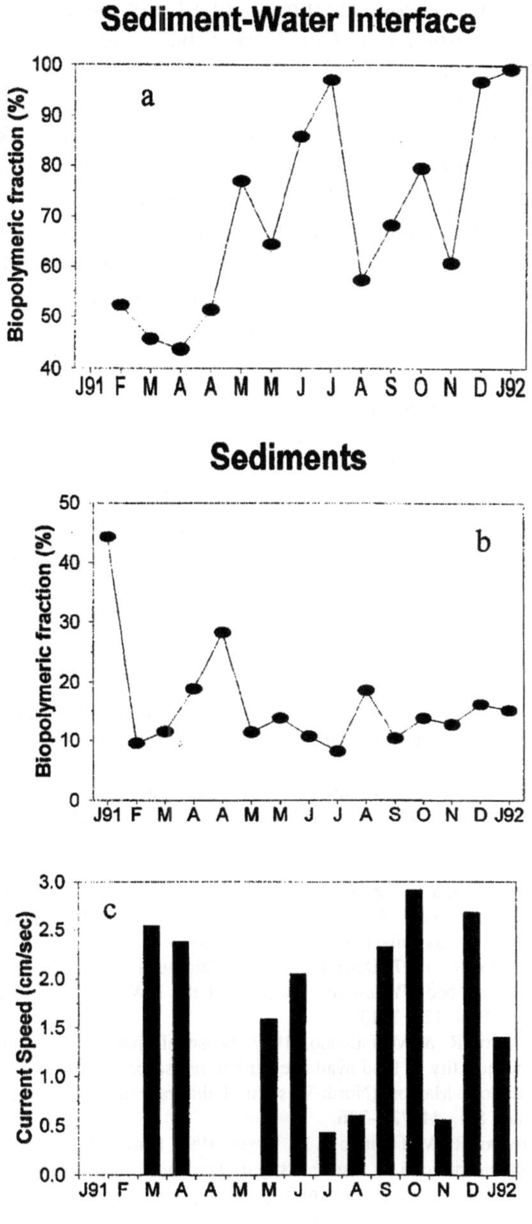

Figure 9. Resuspension of biopolymeric fraction induced by current speed at the sediment–water interface. (a) Reported is the increase of the percentage of the biopolymeric fraction on POC, (b) changes in the significance of the BPF in the sediments; (c) seasonal variation of the current speed at the sediment–water interface at the sampling dates (data from Danovaro, 1996).

cles. Such mechanism basically driven by the hydrodynamism would control the dynamics and composition of POM at the sediment–water interface. The hypothesis of a close coupling between suspended POM and the biopolymeric fraction of the organic matter in the sediments is also supported by Velimirov (1986) who provided indication that high DOC concentrations of the seagrass meadows result from DOC pulses from the sediments.

The observed patterns provide also additional information to the problem of the quality of the 'export' from the seagrass system. The resuspension of labile organic particles suggests that a large fraction of the material exported, calculated to reach about 1600–1700 g DW m^{-2} (Velimirov, 1987) doesn't consist of refractory plant debris but of POM of high nutritional quality able to support heterotrophic processes in areas around the seagrass meadows.

Bacterial and nutrient dynamics at the sediment-water interface

Velimirov (1986a, b) reported low numbers of free living and attached bacteria (below $1.2 \times 10^8\ l^{-1}$) typical of the seagrass system despite high DOC concentrations. Higher densities ($3.6–19.4 \times 10^8\ l^{-1}$) were reported by Velimirov & Walenta-Simon (1992, 1993) with a bacterial biomass ranging from 6.81 to 81.0 μgC l^{-1}. In the seagrass of the Prelo cove, bacterial biomass dominated the system reaching densities (up to $28.4 \times 10^8\ l^{-1}$) and a biomass on average 2–5 times higher than those reported by Velimirov & Walenta-Simon (1992, 1993). The significance of the bacterial biomass in the seagrass system, already reported for the benthic environment (Danovaro et al., 1994), is even more evident in the overlying waters. Phytoplankton accounted for a minor fraction of the biopolymeric carbon (about 30% of C-BPF, Table 1). The most striking result of the present study is that bacteria contributed for a significant fraction of the particulate organic carbon (representing more than 48% of POC) and dominated the biopolymeric fraction of the carbon pool (accounting for about 68% of C-BPF). These data are similar to those reported for highly oligotrophic areas (Furhman et al., 1989) but are unexpected for an highly productive system and are in contrast with findings reported by Velimirov (1986) in a seagrass system of the Gulf of Naples where bacteria accounted for a negligible fraction of POM (bacterial biomass accounting for 1–2% of POC and 1–6% of PON). However, it should be taken into account that those calculations were based on POM estimated from ignition loss and that POM estimates in the Prelo meadow showed values 3–4 times lower, on average 2.5 mg l^{-1} (Danovaro, 1993), than those reported by Velimirov (1986).

The significance of bacteria in the seagrass system is confirmed by the large contribution of bacterial DNA to the total particulate DNA pool (on average more than 40%). Since the amounts of particulate RNA appear to be dependent upon bacterial synthesis rates, the possibility of utilising the RNA:DNA ratio as indicator of bacterial production in 'bacterial-dominated systems' should be further investigated.

Bacterial dynamics at the sediment–water interface of the studied seagrass meadow did not appear to be controlled by temperature or by the amounts of labile organic compounds and were not related to the phytoplankton dynamics or to seagrass growth cycle. By contrast it appears that bacterial dynamics are related to inorganic nutrient availability (particularly nitrite and nitrate). The presence of a significant coupling between benthic bacterial density and inorganic nutrient concentrations has been previously reported in the same environment by Danovaro et al. (1994). To this regard, a few studies have demonstrated that heterotrophic bacteria uptake inorganic nutrients in order to complement internal N and P (Hörrigan et al., 1988; Tupas et al., 1994 and literature therein). The strong negative relationships between bacterial density and nitrite-nitrate concentrations in the seagrass of Prelo could be the result of the nutrient uptake by bacteria. This hypothesis is supported by the relatively high C:N (w:w) values of the seagrass POM (C:N comprised between 11 and 15) during the exponential bacterial growth (end of April-May) that indicated the low availability of detrital POM to bacteria so that they, probably, needed to complement the organic sources with inorganic N. Additional indirect evidence to support the hypothesis of the inorganic nutrient uptake by bacteria is provided by the phytoplankton and seagrass seasonal dynamics. Peaks of chlorophyll-*a* at the sediment–water interface, early April and September, and highest growth rates of the *Posidonia oceanica*, in winter, may not be responsible for the nutrient depletion in summer; but were related to the bacterial growth rates in late spring–summer. Our estimates of bacterial growth rates and doubling time well agree with those reported by Velimirov & Walenta-Simon (1992), although these authors found that growth rates, with highest values in summer, were dependent upon temperature and chlorophyll-*a* concentrations.

Bacterial secondary production (on annual average 59.19 μgC l^{-1} d^{-1}) and *Posidonia oceanica* primary production (311 gC m^{-2} y^{-1}, Boyer, 1991) allow to estimate the fraction of seagrass production potentially utilised by bacteria. Assuming a carbon conversion efficiency of 50%, bacterial production in 1 m^3 of the sediment–water interface would require 43.2 gC m^{-2} y^{-1} which is equivalent to about 14% of the seagrass primary production. Considering that more than 50% of the seagrass primary production is generally exported from the system, this value, even thought does not take into account the contribution given by phytoplankton and microphytobenthos to the total primary production (very low indeed when compared to primary production values due to the seagrasses and their epiphytes), suggests the relevance of bacterioplankton in the carbon cycling of the sediment–water interface of the Mediterranean seagrass systems.

Acknowledgements

Roberto Danovaro was supported by a grant of the *Ministero Università e Ricerca Scientifica e Tecnologica.* The authors gratefully acknowledge Dr Paolo Povero (University of Genoa) for help in POC and PON analyses.

References

Bligh, E. G. & W. Dyer, 1959. A rapid method for total lipid extraction and purification. Can. J. Biochem. Physiol. 37: 911–917.

Boyer, M., 1991. Variazione annuale di produzione primaria, biomassa e produzione batterica in una prateria di *Posidonia oceanica* del Golfo del Tigullio (Mar Ligure). Ph.D. Thesis, University of Genova, 160 pp.

Danovaro, R., 1993. Analisi della dinamica e della struttura trofica di comunità meiobentoniche in relazione al contenuto ed alla composizione della sostanza organica particellata (Mar Ligure). Ph.D. Thesis University of Pisa, 246 pp.

Danovaro, R., 1997. Detritus-Bacteria-Meiofauna interactions in a seagrass bed (*Posidonia oceanica*) of the NW Mediterranean. Mar. Biol. 127: 1–13.

Danovaro, R. & M. Fabiano, 1997. Seasonal changes in quality and quantity of food available for benthic suspension feeders in the Golfo Marconi (North-Western Mediterranean). Estuar. coast. Shelf Sci., 44: 723–736.

Danovaro, R., M. Fabiano & M. Boyer, 1994. Seasonal changes of benthic bacteria in a seagrass bed (*Posidonia oceanica*) of the Ligurian Sea in relation to origin composition and fate of the sediment organic matter. Mar. Biol. 119: 489–500.

Daumas, R., 1990. Contribution of the water-sediment interface to the transformation of biogenic substances: application to nitrogen compounds. Hydrobiologia 207: 15–29.

De Jonge, V. E., 1980. Fluctuations in the organic carbon to chlorophyll aa ratios for estuarine benthic diatom populations. Mar. Ecol. Prog. Ser. 2: 345–353.

Dubois, M., K. A. Gilles, J. K. Hamilton, P. A. Rebers & F. Smith, 1956. Colorimetric method for determination of sugars and related substances. Analyt. Chem. 28: 350–356.

Fabiano, M., R. Danovaro & S. Fraschetti, 1995. Temporal trend analysis of the elemental and biochemical composition of the sediment organic matter in subtidale sandy sediments of the Ligurian Sea (NW Mediterranean): a three year study. Cont. Shelf Res. 15: 1453–1469.

Fichez, R., 1990. Decrease in allochthnonous organic inputs in dark submarine caves, connections with lowering in benthic community richness. Hydrobiologia 207: 61–69.

Fichez, R., 1991. Composition and fate of organic matter in submarine cave sediments; implications for the biogeochemical cycle of organic carbon. Oceanol. Acta 14: 369–377.

Fry, J. C., 1990. Direct methods and biomass estimation. In Methods in Microbiology, Vol 22 Academic Press: 41–85.

Furhman, J. A., T. D. Sleeter, C. Carlson & L. M. Proctor, 1989. Dominance of bacterial biomass in the Sargasso sea and its ecological implications. Mar. Ecol. Prog. Ser. 57: 207–217.

Hartree, E. F., 1972. Determination of proteins: a modification of the Lowry method that give a linear photometric response. Analyt. Biochem. 48: 422–427.

Hedges, J. I. & J. H. Stern, 1984. Carbon and nitrogen determination of carbonate-containing solids. Limnol. Oceanogr. 29: 657–668.

Hermin, M.-N., 1989. Dégradation microbienne de la matière organique à l'interface eau sediment en milieu marin. Thèse Doct. Univ. Aix-Marseille II, 202 pp.

Hobbie, J. E., R. J. Daley & S. Jasper, 1977. Use of Nuclepore filters for counting bacteria by fluorescence microscopy. Appl. envir. Microbiol. 33: 1225–1228.

Horrigan, S. G., A. Hagström, I. Koike & F. Azam, 1988. Inorganic nitrogen utilization by assemblages of marine bacteria in sea water culture. Mar. Ecol. Prog. Ser. 50: 147–150.

Lawrence, J. M., Ch.-F. Boudouresque & F. Maggiore, 1989. Proximate costituents, biomass and energy in *Posidonia oceanica* (Potamogetonacea). PSZNI: Mar. Ecol. 10: 263–270.

Lorenzen, C. & J. Jeffrey, 1980. Determination of chlorophyll in sea water. Unesco Technical Papers in Marine Science 35: 1–20.

Lukavsky', J., K. Tetik' & J. Vandlova, 1973. Extraction of nucleic acid from the alga *Scenedesmus quadricauda*. Arch. Hydrobiol. Suppl. 9: 416–426.

Marsh, J. B. & W. J. Weinstein, 1966. A simple charring method for determination of lipids. J. Lip. Res. 7: 574–576.

Ott, J., 1980. Growth and production in *Posidonia oceanica* (L) Delile. PSZNI Mar. Ecol. 1: 47–64.

Palumbo R., J. E. Ferguson & P. A. Rublee 1984. Size of suspended bacterial cells and association of heterotrophic activity with size fractions of particles in eastuarine and coastal waters. Appl. envir. Microbiol. 48: 157–164.

Plante-Cuny, M. R., 1974. Evaluation par spectrophotometrie des teneurs en chlorophyll-*a* fonctionelle et en phaeopigments des substrats meubles marins. O.R.S.T.O.M. Nosy-Bé, Documn Techqs, 45 pp.

Simon, M. & F. Azam, 1989. Protein content and protein synthesis rates of planktonic marine bacteria. Mar. Ecol. Prog. Ser. 51: 201–213.

Strickland, J. D. H. & T. R. Parsons, 1972. A practical handbook of sea water analysis. Fish. Res. Bd Can. Bull. 167: 310 pp.

Tupas, L. M., I. Koike, D. M. Karl & O. Holm-Hansen, 1994. Nitrogen metabolism by heterotrophic bacterial assemblages in Antarctic coastal waters. Polar Biol. 14: 195–204.

Turley, C. M. & K. Lochte, 1985. Direct measurement of bacterial productivity in stratified waters close to a front in the Irish Sea. Mar. Ecol. Prog. Ser. 23: 209–219.

Velimirov, B., 1986. DOC dynamics in a Mediterranean seagrass system. Mar. Ecol. Prog. Ser. 28: 21–41.

Velimirov, B., 1987. Organic matter derived from a seagrass meadow: Origin, properties and quality of particles. PSZNI: Mar. Ecol. 8: 143–173.

Velimirov, B. & M. Walenta-Simon, 1992. Seasonal changes in specific growth rates, production and biomass of a bacterial community in the water column above a Mediterranean seagrass system. Mar. Ecol. Prog. Ser. 80: 237–248.

Velimirov, B. & M. Walenta-Simon, 1993. Bacterial growth rates and productivity within a seagrass system: seasonal variations in a *Posidonia oceanica* bed. Mar. Ecol. Prog. Ser. 96: 101–107.

Zachleder, V., 1984. Optimization of nucleic acids assay in green and blue-green algae: extraction procedures and the light-activated diphenylamine reaction for DNA. Arch. Hydrobiol. Suppl. 67: 313–328.

Hydrobiologia **363**: 253–260, 1998.
T. Tamminen & H. Kuosa (eds), Eutrophication in Planktonic Ecosystems: Food Web Dynamics and Elemental Cycling.

Development of microbial community during *Skeletonema costatum* detritus degradation

Kai Künnis
Tallinn Water Treatment Plant, Järvevana tee 3, Tallinn EE0001, Estonia

Key words: Skeletonema costatum, detritus, degradation, bacteria, flagellates, ciliates

Abstract

Microbial degradation of algal detritus was studied experimentally using the diatom *Skeletonema costatum* prekilled culture as a substrate for the marine microbial community. The qualitative and quantitative changes in the microbial community and the algal detritus structure were followed during 11 days of incubation at the water temperature of 20 °C.

Most of the bacterial parameters (epifluorescence microscopy counts of free-living and attached cells, mean cell volume, biomass and productivity) were the highest after 24 h from the algal detritus addition to the microbial community. The bacterial peak (3.6×10^6 cells ml^{-1}, biomass 3.1 $\mu gC\ ml^{-1}$, net production 66×4 $ngC\ ml^{-1}h^{-1}$) was followed by a precipitous increase of homogenous nanoflagellate population with a maximum number of 2.6×10^5 cells ml^{-1}, which in turn declined quickly after ciliates appeared in the community. The bacterial production, initiated by the supplement of algal detritus, was totally ingested by microzoans within 3 days. Changes of the structure of marine bacterial assemblage and relative increase of the amount of attached bacteria during the period of massive development of nanoflagellates emphasized the importance of small flagellates predation on the free-living average size (cell volume 0.1–0.2 μm^3) pelagic bacteria. After 11 days of incubation on algal detritus the initial bacterial assemblage was replaced by the mixed succession of bacteria, flagellates and ciliates at a proportion of 1000 : 1.5 : 0.2. The stabilization of microbial community and changes in algal detritus structure allow to expect that pelagic microbial utilization of *Skeletonema costatum* bloom (53×10^6 cells l^{-1}), could be finished within a period of 8–11 days at the summer water temperature.

Introduction

Planktonic bacteria play an important role in organic matter transformation and decomposition in aquatic environments. Great inputs of organic matter into marine ecosystems take place during the vernal and summer algal blooms. It have been demonstrated that approximately 80% of the primary production is decomposed (Lee & Cronin, 1982) or recycled (Eppley & Peterson, 1979) in the near surface water layer. Three main pathways of organic matter degradation exist in pelagic environments. These are based on predation, particle feeding, and dissolved organic matter uptake (Hoppe, 1988). Bacteria are involved in the latter two. The question is to which extent pelagic bacteria live on the organic matter exreted by living algae and to which extent on their dying and dead remnants (Jones, 1982).

The existing records on photosynthetically produced matter utilization by bacteria vary to a high extent. According to research results, bacteria can metabolise from 10 to 60% (Fuhrman & Azam, 1980; Lowell & Konopka, 1985), even to 98% (Cole et al., 1988) of primary produced organic carbon. Marine bacteria are not distributed evenly throughout the water mass, but have higher numbers around microsites of higher organic matter concentration, such as the phytoplankton cells and organic detrital aggregates (Fukami et al., 1981; Azam & Ammerman, 1984; Hoppe, 1988). The exoenzymatic activity of particle-associated bacteria, however, contributes to the pool of dissolved organic matter and benefits the production of free-living bacteria (Hoppe, 1988).

Modern studies of marine ecosystems have demonstrated that the production of bacterial carbon is equally important as the bacterial catabolic activity of organic matter mineralization. The production of pelagic bacteria, both free-living and particle-associated, is predominantly ingested by microzooplankters (Azam et al., 1983; Fenchel, 1980, 1984; Porter et al., 1985; Sherr et al., 1986, 1992; Sieburth, 1984). The aim of the present investigation was to estimate experimentally the effects which can bring about decaying diatom blooms by a marine microbial community. The post-bloom conditions were imitated by enriching natural seawater with prekilled culture of *Skeletonema costatum* (Grev.) Cleve. The reactions of the microbial community were detected by changes in the bacterial parameters (number of free-living and attached cells, mean cell volume, biomass, productivity) and in the composition of microbial community. The degree of algal detritus degradation was evaluated by microscope examinations.

Material and methods

The experiment of *Skeletonema costatum* decomposition by marine microbial community was carried out in the Helsingør Marine Biological Laboratory in April–May 1993.

Experiment design

Seawater for the microcosm experiment was taken from the open Kattegat below the euphotic water layer. The main parameters of the seawater were the following: salinity 30.1 ‰, temperature 12.6 °C and pH 7.97. The seawater (several cubic meters) was stored close to *in situ* temperature in the dark for 3 days before the start of the experiment. For experiment setting 3 l of seawater was passed through a 17-μm plankton net to remove natural detritus and larger organisms. The culture of *Skeletonema costatum* (Grev.) Cleve from the Helsingør Marine Biological Laboratory collection (batch culture, conc. $\sim 10^9$ cells l^{-1}) was diluted with particle-free seawater (1 : 1) and killed by the freezing-heating treatment (– 80/+ 50 °C). The cell suspension was resuspended in the seawater, after that the algal detritus concentration was approximately 53×10^6 cells per l. The seawater was thoroughly mixed (2 h) and aerated up to achieving oxygen saturation level (7.9 mgO_2 l^{-1}, at 20 °C). 100 ml plastic bottles, double for each sampling time, were filled with seawater-detritus mixture and incubated with control samples of seawater (without *Skeletonema* detritus addition) on a rotator-shaker at 20 °C in the dark. The follow-up examinations were performed after 2 h, 1 d, 2 d, 4 d, 6 d, 8 d and 11 d from the moment of *Skeletonema* detritus addition to the seawater. The parameters, which were determined in each microcosm both with and without algal detritus addition were following: the numbers of free-living bacteria, attached bacteria, flagellates and ciliates, bacterial biomass, bacterial productivity (by ^{3}H-thymidine incorporation), the number and size of algal detritus particles, and the density of bacterial attachment on the detritus particles.

Counting of bacteria, flagellates and ciliates

Bacteria, flagellates and ciliates were quantified using epifluorescence microscopy with acridine orange as a fluorochrome (Hobbie et al., 1977). Organisms were concentrated on 0.2 μm Nuclepore polycarbonate filters. Free-living and attached bacteria were counted separately. Bacteria were counted as 'attached' if they were seen in close contact with a particle, cells which were near, but not directly in contact with particles were counted as free-living bacteria. Duplicate preparations were observed per subsample and at least 30–50 optical fields were counted for bacteria and at least 50 fields for flagellates at 1250 × magnification. Ciliates were counted on an entire surface area of the filters at 500 × magnification.

Determination of mean cell volumes and bacterial biomass

Cell size measurements were done using a calibrated ocular graticule New Porton Grids (Graticules Ltd., England). Cell dimensions of at least 70 bacteria and 40 flagellates were measured in each subsample. The size parameters of ciliates were measured for all counted individuals. The bacterial biomass was calculated using the 0.35 pgC μm^3 conversion factor of carbon content(Bjørnsen, 1986).

Estimation of bacterial production

Bacterial production was measured according to the [^{3}H]thymidine incorporation method (Fuhrman & Azam, 1982). Triplicate subsamples of 10 ml and two formaldehyde killed controls were incubated with a final concentration of 20 nmol methyl-[^{3}H]thymidine (specific activity: 20 Ci mmol^{-1}) for 30 min at 20 °C in the dark. The incubation period was terminated by adding 25% formaldehyde. The subsamples were filtered onto 0.2 μm pore size polycarbonate filters (Nuclepore). After filtration the filters were washed 10 times with ice cold 5% trichloroacetic acid and radioassayed by liquid scintillation counting. Bacterial production was calculated using a conversion factor of 1.1×10^{18} cells produced per mole of thymidine incorporated (Riemann et al., 1987) and 0.35 pg C μm^3 cell carbon content (Bjørnsen, 1986).

Evaluation of bacterial attachment, algal detritus degradation and aggregation

The number of algal detritus particles, their size distribution and degree of being covered with bacteria (density of bacteria on a area unit of particle surface) were determined by epifluorescence microscopy on acridine orange stained filters. Bacteria were distinguished from particles by their morphology and intensity of staining. Changes in the detritus structure were evaluated visually.

The amount of bacterial microcolonies and the number of bacteria per colony was counted. Bacteria in microcolonies and bacteria in slime were included in the count of attached bacteria.

Results

The total number of bacteria ranged from 5.2×10^6 to 36.3×10^6 cells ml^{-1}. The bacterial number was the highest after one day of incubation from the moment of addition of algal detritus to the seawater (Figure 1). After 2 days of incubation the number of bacteria had decreased by six times (6.0×10^6 cells ml^{-1}), and stayed close to that level (5.2–9.5 $\times 10^6$ cells ml^{-1}) over the following incubation period. Decrease in the total bacteria number was accounted for mainly by the decline in the number of free-living bacteria. An intervening increase in the total bacteria number, observed at the forth day of incubation, coincided with decline in the number of nanoflagellates (Figures 1, 3).

The mean cell volume of bacteria ranged from 0.08 to 0.25 μm^3 (Figure 1). The mean cell size of bacteria, before algal detritus addition, was 0.23 μm^3 and did not change significantly after the seawater was supplied with algal detritus. The largest mean cell volume of bacteria was detected after one day of incubation. Decrease in the mean cell volume occured between the 1st and 4th day of incubation, parallel to a sharp increase in the abundance of nanoflagellates (Figure 3). After 4 days of incubation the mean cell volume had declined to 0.09 μm^3. The mean cell volume of bacteria in the control samples of seawater (without algal detritus addition) decreased during the incubation period from 0.23 to 0.08 μm^3. At the end of the 11 days incubation period the mean cell volume of the detritus-based microbial communities was rather similar to the mean cell volume of the detritus-free (control) samples of seawater.

The bacterial assemblage which was developed on the freshly killed algal material during the first hours of incubation, composed dominantly of free-living single cocci and coccobacilli (82% of the total bacterial assemblage). After one day of incubation the structure of bacterial assemblage began to change: the proportion of spherical cells declined and the percentage of rod-shape bacteria increased. During detritus degradation new bacterial forms developed and the bacterial assemblage became more diverse. When the incubation period had lasted 2 days a mixed community of cocci, rods, vibrions and spirilla with cell volumes reaching from 0.008 (cocci) to 1.70 μm^3 (spirilla) was associated with the algal detritus. The tendency of progressive differentiation of the bacterial community was observed during the incubation period: the proportion of large rods, filaments and spirilla increased, and at the same time a high number of very small cocci appeared in the community. After 4 days of incubation, 50% of the bacteria counted were cocci with a cell volume below 0.014 μm^3.

Bacterial biomass ranged from 0.17 to 3.11 μgC ml^{-1} (Figure 2), with the maximum value reached after one day and the minimum between the 6th and 8th day of incubation. Bacterioplankton increased in biomass until being followed by nanoflagellates (Figures 1, 3). After 11 days of incubation with algal detritus, the bacterial biomass had decreased to 0.25 μgC ml^{-1}, which was half of the initial bacterial biomass of natural seawater (0.48 μgC ml^{-1}), and approximately twice the mean bacterial biomass of the control without detritus incubated seawater samples (0.13 μgC ml^{-1}).

The attached bacteria formed 1–24% of the total amount of bacteria (Figure 1). Epifluorescence microscopy showed a dense bacterial attachment on *Skeletonema costatum* detritus already after 2 h from the moment of algal detritus contact with the microbial community. The highest number of attached bacteria (1.8×10^6 cells ml^{-1}) was detected after one day,

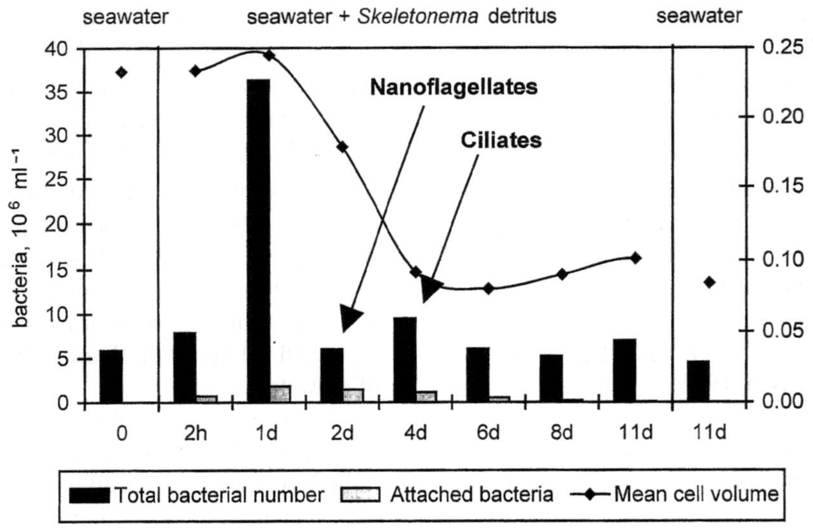

Figure 1. Changes in the total bacterial number, the number of attached bacteria and the mean cell volume of bacteria in the experimental microcosms of natural seawater and seawater with *Skeletonema costatum* detritus addition (average values of two replicate microcosms). The arrows indicate the appearance of nanoflagellates and ciliates in the microcosms.

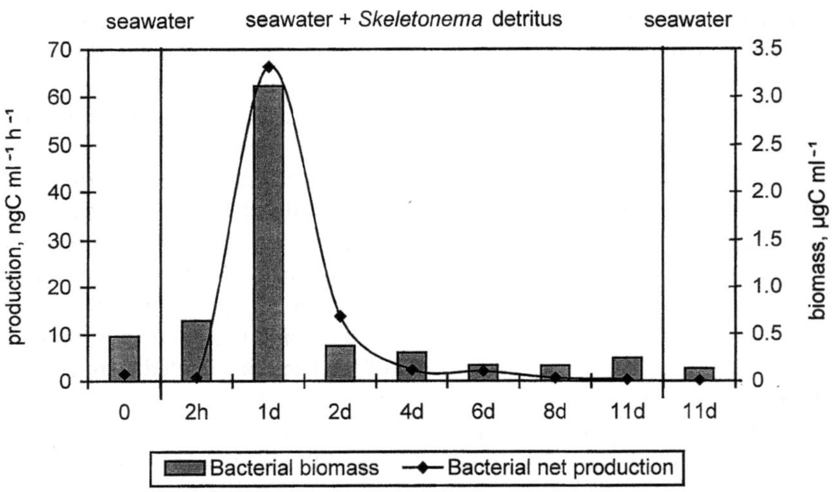

Figure 2. Changes in the bacterial biomass and the rate of bacterial production in the experimental microcosms of natural seawater and seawater with *Skeletonema costatum* detritus addition (average values of two replicate microcosms).

and their highest percentage (24%) appeared after two days of incubation. Later the proportion of attached cells began to decrease. After 11 days of incubation the percentage of attached bacteria in the total bacterial assemblage had dropped to 1%. The percentage of attached bacteria in the control samples of seawater (bacteria in microcolonies and bacteria associated with slime) did not exceed 0.5%.

Bacterial microcolonies were counted separately as a specific type of bacterial colonization. Two main forms of bacterial colonies appeared in the detritus-based marine microbial community. The first type of microcolonies (real colonies) formed during the first days of incubation from actively multiplied cells and looked like bacterial balls (average 13–30 cells per colony). That type of bacterial aggregates are typical to rapidly growing microbial populations (Andersen & Sørensen, 1986). The other type of bacterial formation, bacteria surrounded by a transparent mucus matrix, appeared later or after 6 days' incubation, and they

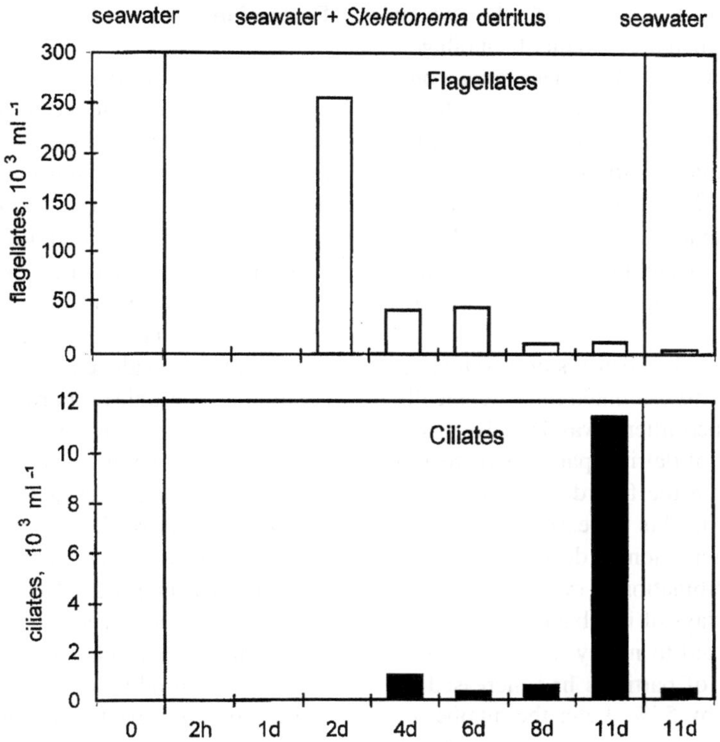

Figure 3. The number of nanoflagellates and ciliates in the experimental microcosms of natural seawater and seawater with *Skeletonema costatum* detritus addition (average values of two replicate microcosms).

were characteristic to the last stage of algal detritus destruction.

The rate of bacterial production in natural seawater, measured by [^{3}H]thymidine incorporation, was 18.6×10^3 cells ml^{-1}h^{-1} (1.52 ngC ml^{-1} h^{-1}). After the adding of algal detritus, the bacterial production first decreased to 10.2×10^3 cells ml^{-1} h^{-1} (0.83 ngC ml^{-1} h^{-1}) measured after 2 h of incubation, and then started to increase rapidly. After one day of incubation the rate of bacterial production had increased to a very high level 775×10^3 cells ml^{-1} h^{-1}(66.4 ngC ml^{-1}h^{-1}). One day later the bacterial production had dropped about 3.5 times (Figure 2). The bacterial production was obviously ingested by nanoflagellates, which appeared in large numbers in the community (Figure 3). From that moment bacterial production decreased gradually until the termination of the experiment. The average bacterial production of the control samples was 8.2×10^3 cells ml^{-1} h^{-1}(0.24 ngC ml^{-1} h^{-1}) after 11 days of incubation, being 2.3 times lower than the initial value.

The number of flagellates fluctuated to a large extent from 0 to 2.6×10^5cells ml^{-1} (Figure 3). At the beginning of the experiment the amount of flagellates was too low (< 500 cells l^{-1}) to be counted by the method used. A high number of nanoflagellates (2.6×10^5 cells ml^{-1}) appeared in the community after two days of incubation. The peak of homogenous nanoflagellate population lagged one day as compared to the peak of bacterial numbers. The flagellate population consisted of small 2.4–3.0 μm free-swimming naked cells with two flagella: 4 μm and 16 μm length. Over the period between the 2nd and 4th day of incubation the number of flagellates declined to a density of 41×10^3cells ml^{-1}, obviously due to grazing by the increasing population of ciliates (10^3 ind. ml^{-1}) (Figure 3). As a direct consequence of ciliate grazing on nono-flagellates, the pressure on bacteria decreased and this allowed a slight increase in the total bacterial number (Figure 1). The actively dividing healthy flagellate population had almost totally disappeared after 6 days from their maximum appearance.

The number of ciliates ranged from 0 to 11.4×10^3 ind. ml^{-1} (Figure 3). The amount of ciliates increased sharply after 4 days of incubation (Figure 3). The first peak of ciliates was detected 2 days and the second one 9 days later than the peak of nanoflagellates was observed. During the second peak the ciliates reached their maximum number of 11.4×10^3 ind. ml^{-1}, which was only slightly lower than the amount of nanoflagellates (12.4×10^3 cells ml^{-1}). The ciliate fauna composed of two main species: small ciliates similar to *Uronematidae* (diameter 5–8 μm, length 15–18 μm) and larger ciliates similar to *Strombidiidae* (diameter 20–22 μm, length 32–36 μm). The large ciliates appeared 4 days later than the small ones, and their proportion in the ciliate assemblage increased in time: after 8 days of incubation their percentage was 10% and after 11 days 65%.

The initial number of algal detritus units (algal filaments and their fragments), counted microscopically on acridine orange stained filters, was 18×10^3 particles ml^{-1}. The number of detritus particles increased to a certain extent within the first day of incubation (23×10^3particles ml^{-1}). This increase was obviously caused by the fragmentation of dead *Skeletonema* filaments due to the combination of mixing and bacterial activity. After two days of incubation the number of particles had decreased to nearly the initial level, whereas the mean size of particles had increased as a result of aggregation by 55%. Later the number of detritus particles continued to decrease until the end of the experiment. After 11 days of incubation the average number of detritus particles in experimental microcosms did not differ significantly from their number in the control samples of seawater (microcosms with *Skeletonema* detritus addition – 2.9×10^3 particles ml^{-1}; controls, without algae addition – 2.6×10^3 particles ml^{-1}). Two tendencies were observed in detritus transformation: degradation of the algal cells and progressive aggregation of more refractory algal material. After 6 days of incubation the measured average cross-section area of algal detritus conglomerates was 370 μm^2. During the following days intensive aggregation was observed, as a result of which the difference among the size of particles increased (from 70 to 20 000 μm^2).

Algal detritus was densely covered with bacteria already after 2 h of incubation with the marine microbial community. After one day of incubation the average measured level of bacterial coverage was 34 bacteria per 100 μm^2 of detritus surface area. A day later the attachment had decreased to 13 bacteria per 100 μm^2. After 11 days of incubation the detritus mostly included empty cells of *Skeletonema costatum* and large conglomerates of bacteria-free unstructured detritus material. In addition to those mixed aggregates of *Skeletonema* remains, dead or dying flagellates and bacteria (alive) were found in the microcosms.

Discussion

Skeletonema costatum was chosen for the particular source of organic matter because of its blooming nature and wide distribution in coastal and brackish waters. From late spring through early autumn, *Skeletonema costatum* is often one of the most common algae species in the above-mentioned ecosystems, frequently forming blooms with densities higher than 10^6 cells l^{-1} and accounted for 28 to 98% of the total phytoplankton cell numbers (Han et al., 1992). There is a good deal of evidence, however, that under natural conditions a variable proportion of bacteria in the community is metabolically inactive with the proportion of active cells ranging from 60% to as little as 6% depending, among other factors, on the organic matter available (Linley & Newell, 1984). The present experiment indicated that the bacterial number and productivity increased considerably after the substrate, algal detritus, appeared in seawater. The experiment demonstrated that a short adaptation period was needed for the bacterial productivity activation, which was not expressed in the bacterial numbers and cell volume measurements. The conducted laboratory experiment supported the assumption of field investigations that the larger cells within natural bacterial populations represent the rapidly growing segment of the population (Pritchard & Tempest, 1982), and that protozoan bacterivory control bacterial abundance by selective cropping of the actively growing and dividing cells (Sherr et al., 1992). It has been indicated that small ciliates (diameters < 20 μm) utilize the same food resources as heterotrophic microflagellates, i.e. bacteria and other picoplanktonic micro-organisms (Sherr et al., 1986, 1987). Fenchel (1980; 1984) has concluded from the theoretical considerations of ciliate filtering capabilities, as well as from the results of laboratory feeding experiments, that bacterivorous ciliates cannot maintain growth on bacterial suspensions if the bacterial concentration is lower than 5×10^6 – 5×10^8cells ml^{-1}. Sieburth (1984) has found that bacterivorous ciliates could exist even when the total bacteria count was low, if suspended bacteria were not randomly dispersed but aggregated in patches or around particles. These facts could explain why the zooplankters number achieved a countable level only after the bacterial number had increased from 5.9×10^6 to 36×10^6 cells ml^{-1}, which was sufficient concentration for intensive development of bacterivorous microzoans: first flagellates and then ciliates.

The detritus-induced bacterial production (bacterial peak) was ingested by nanoflagellates within 2–4 days of incubation. The peak of ciliates was detected 9 days later than the massive development of nanoflagellates had been observed. In natural marine conditions of similar water characteristics (in Limfjord, a shallow bay in Denmark) the flagellate peak appeared 3–8 days later and the ciliate peak 7–14 days later than the bacterial peak was detected (Andersen & Sørensen, 1986). The fact that the maximum percentage of attached bacteria appeared at the same time as the peak of nanoflagellates, allow to assume that small 2.4–3.0 μm cell size naked flagellates feed primarily on free-living bacteria. Free-living bacteria usually predominate in pelagic seawaters, while attached bacteria have been reported to be more active than free-living in most cases (Hodson et al., 1981; Kirchman & Michell, 1982; Iriberri et al., 1987).

Observations have shown that pelagic bacteria grow primarily on fresh, more labile algal material (Ducklow et al., 1985), and that the role of attached bacteria in particle decomposition is especially important during the phase of phytoplankton bloom decline, when attached bacteria may amount to 40% of the total bacteria number (Fukami et al., 1983), and as much as 50% of the total bacterial production (Simon, 1987). During the current experiment the percentages of attached bacteria remained on a moderate level and did not exceed 24%. It should be mentioned that as a result of the counting method used the amounts of attached bacteria could have been underestimated: firstly, because the bacteria on back sides of opaque particles were not possible to count, and secondly, because some amount of the loosely attached bacteria could have been detached during the filtration procedure. Progressive decrease of bacterial attachment after two days of incubation was obviously caused by protozoan grazing combined with bacterial production abatement.

The efficiency of free-living bacteria to utilize dissolved organic carbon has been measured to be 50–90% (Williams, 1981). Attached bacteria are probably less efficient of converting particulate organic carbon into biomass, because they have to first synthesize extracellular enzymes to degrade the more refractory substrates. Newell et al. (1981) found that bacteria growing on phytoplankton debris had conversion efficiencies averaging 10%. The need for bacterial exoenzymatic degradation of algal detritus is generally induced when the dissolved organic products are totally incorporated. As soon as the dissolved organic products are used, the efficiency of bacterial colonization and utilization of organic particles would depend on the existing bacterial population's capacity for growth (multiply) and spreading on the particle surfaces. Based on the mentioned considerations it is possible that the function of photosynthetically active algae exudates is to select and preactivate the local microbial community for the algal bloom utilization and thereby anable fast attack by bacteria when the bloom begins to decline. This could be one reason for the very quick and dense bacterial colonization of *Skeletonema costatum* detritus after it was added to seawater which had been taken for the experiment at time of diatoms (including *Skeletonema costatum*) blooming.

Conclusions

The conducted experiment indicated a close trophic coupling between bacteria, nanoflagellates and ciliates in the process of *Skeletonema costatum* detritus degradation. The microbial degradation of *Skeletonema costatum* detritus started very fast after the seawater had been supplied with prekilled algae. The algal detritus particles were totally covered by bacteria within first hours from the contact with natural marine bacterioplankton. Two tendencies were observed during the microbial degradation of algal detritus: breakdown and destruction (mineralization) of the algal material, and progressive formation of aggregates. High number of bacteria associated with the algal detrital material during the first few days of incubation demonstrated the importance of bacteria in algal detritus decay and in sedimentation through aggregate formation in pelagic marine ecosystems.

The main consumers of bacteria were small naked flagellates. These nanoflagellates probably not only control bacterial abundance but also affect the structure of bacterial assemblage via size-selective grazing mostly on free-living bacteria with cell sizes close to the average of pelagic bacterial population. The abundance of nanoflagellates was controlled by the ciliates, the peak of which was detected 9 days later than massive development of nanoflagellates had been observed.

The stabilization of the microbial community and changes in algal detritus structure allow to expect that 8–11 days period could be enough for pelagic utilization of a *Skeletonema* bloom (53×10^6cellc l^{-1}) at the the water temperature 20 °C.

Acknowledgements

This study was carried out at the Marine Biological Laboratory (University of Copenhagen) in Helsingør and supported by the grant of Nordic Academy for Advanced Study (NorFa). I thank Dr M. Olesen and Dr C. Lundsgaard for valuable practical and theoretical advice. I also thank Prof. T. Fenchel and all the friendly collective of the Helsingør Laboratory.

References

Andersen, P. & H. M. Sørensen, 1986. Population dynamics and trophic coupling of pelagic microorganisms in eutrophic coastal waters. mar. Ecol. Prog. Ser. 33: 99–109.

Azam, F., T. Fenchel, J. G. Field, J. S. Gray, L. A. Meyer-Reil & F. Thingstad, 1983. The role of water-column microbes in the sea. mar. Ecol. Prog. Ser. 10: 257–263.

Azam, F. & J. W. Ammerman, 1984. Cycling of organic matter by bacterioplankton in pelagic marine ecosystem: microenvironmental considerations. In M. J. R. Facham (ed.), Flows of Energy and Materials in Marine Ecosystems. Plenum Publishing Corp., New York: 345–360.

Bjørnsen, P. K., 1986. Automatic determination of bacterioplankton biomass by image analysis. Appl. envir. Microbiol. 51: 1199–1204.

Cole, J. J., S. Findlay & M. L. Pace, 1988. Bacterial production in fresh and saltwater ecosystems: a cross-system overview. mar. Ecol. Prog. Ser. 43: 1–10.

Ducklow, H. W., S. M. Hill & W. D. Gardner, 1985. Bacterial growth and the decomposition of particulate organic carbon collected in sediment traps. Cont. Shelf Res. 4: 445–464.

Ducklow, H. W., D. A. Purdie, P. J. LeB. Williams & J. M. Davies, 1986. Bacterioplankton: a sink for carbon in a coastal marine plankton community. Science 232: 865–867.

Eppley, R. W. & B. J. Peterson, 1979. Particulate organic matter flux and planktonic new production in the deep ocean. London. Nature 282: 677–680.

Fenchel, T., 1980. Relation between particle size selection and clearance in suspension – feeding ciliates. Limnol. Oceanogr. 25: 733–738.

Fenchel, T., 1984. Suspended marine bacteria as a food source. In M. J. Fasham (ed.), Flows of Energy and Materials in Marine Ecosystems. Plenum: 301–305.

Fuhrman, J. A. & M. Azam, 1980. Bacterioplankton secondary production estimates for coastal waters of British Columbia, Antarctica, and California. Appl. envir. Microbiol. 39: 1085–1095.

Fuhrman, J. A. & M. Azam, 1982. Thymidine incorporation as a measure of heterotrophic bacterioplankton production in marine surface waters: evaluation and field results. mar. Biol. 66: 109–120.

Fukami, K., U. Simidu & N. Taga, 1983. Change in a bacterial population during the process of degradation of a phytoplankton bloom in a brakish lake. mar. Biol. 76: 253–255.

Han, M.-S., K. Furuya & T. Nemoto, 1992. Species-specific productivity of *Sceletonema costatum (Bacillariophyceae)* in the inner part of Tokyo Bay. mar. Ecol. Prog. Ser. 79: 267–273.

Hobbie, J. E., R. J. Daley & S. Jasper, 1977. Use of Nuclepore filters for counting bacteria by fluorescence microscopy. Appl. envir. Microbiol. 33: 1225–1228.

Hodson, R. E., A. E. Maccubbin & L. R. Pomeroy, 1981. Dissolved adenosine triphosphate utilization by free-living and attached bacterioplankton. mar. Biol. 64: 43–51.

Hoppe, H. G., S. J. Kim & K. Gocke, 1988. Microbial decomposition in aquatic environments: combined process of extracellular enzyme activity and substrate uptake. Appl. envir. Microbiol. 54: 784–790.

Iriberri, J., M. Unanue, I. Barcina & L. Egea, 1987. Seasonal variation in population density and heterotrophic activity of attached and free-living bacteria in coastal waters. Appl. envir. Microbiol. 53: 2308–2314.

Jones, A. K., 1982. The interaction of algae and bacteria. In A. T. Bull & J. H. Slater (eds), Microbial Interactions and Communities, 1. Academic Press, London: 189–248.

Kirchman, D. L. & R. Michell, 1982. Contribution of particle-bound bacteria to total microhetero-trophic activity in five ponds and two marshes. Appl. envir. Microbiol. 43: 200–209.

Lee, C. & C. Cronin, 1982. The vertical flux of particulate organic nitrogen in the sea: decompo-sition of amino acids in the Peru upwelling area and the equatorial Atlantic. J. mar. Res. 40: 227–251.

Linley, E. A. S. & R. C. Newell, 1984. Estimates of bacterial growth yields based on plant detritus. Bull. mar. Sci. 35: 409–425.

Lovell, C. R. & A. Konopka, 1985. Primary and bacterial production in two dimictic Indiana lakes. Appl. envir. Microbiol. 49: 485–491.

Newell, R. C., M. I. Lucas & E. A. S. Linley, 1981. Rate of degradation and efficiency of conversion of phytoplankton debris by marine micro-organisms. mar. Ecol. Progress 6: 123–136.

Porter, K. G., E. B. Sherr, B. F. Sherr, M. Pace & R. W. Sanders, 1985. Protozoa in planktonic food webs. J. Protozool. 32: 409–415.

Pritchard, R. H. & D. W. Tempest, 1982. Growth: cells and populations. In J. Mandestam et al. (eds), Biochemistry of Bacterial Growth. Halsted Press, New York: 99–124.

Riemann, B., P. K. Bjørnsen, S. Newell & R. Fallon, 1987. Calculation of cell production of coastal marine bacteria based on measured incorporation of [^{3}H]thymidine. Limnol. Oceanogr. 32: 471–476.

Sherr, E. B., B. F. Sherr, R. D. Fallon & S.Y. Newell, 1986. Small, aloricate ciliates as a major component of the marine heterotrophic nanoplankton. Limnol. Oceanogr. 31: 177–183.

Sherr, B. F., E. B. Sherr & R. D. Fallon, 1987. Use of monodispersed, fluorescently-labeled bacteria to estimate *in situ* protozoan bacterivory. Appl. envir. Microbiol. 53: 958–965.

Sherr, B. F., E. B. Sherr & J. McDaniel, 1992. Effect of protistan grazing on the frequency of dividing cells in bacterioplankton assemblages. Appl. envir. Microbiol. 58: 2381–2385.

Sieburth, J. Mc. N., 1984. Protozoan bacterivory in pelagic marine waters. In J. E. Hobbie & P.J. Williams (eds), Heterotrophic Activity in the Sea. Plenum: 405–444.

Simon, M., 1987. Biomass and production of small and large free-living and attached bacteria in Lake Constance. Limnol. Oceanogr. 32: 591–607.

Williams, P. J. LeB., 1981. Incorporation of microheterotrophic processes into the classical paradigm of the planktonic food web. Kieler Meeresforsch 5: 1–28.

Hydrobiologia **363**: 261–270, 1998.
T. Tamminen & H. Kuosa (eds), *Eutrophication in Planktonic Ecosystems: Food Web Dynamics and Elemental Cycling.*

Nutrient composition, microbial biomass and activity at the air–water interface of small boreal forest lakes

U. Münster[1], E. Heikkinen[1] & J. Knulst[2]
[1] *Lammi Biological Station, University of Helsinki, FIN-16900 Lammi, Finland (Phone: +355-3-631-111; Fax: +358-3-631-1166; E-mail: Uwe.MunsterHelsinki.Fi)*
[2] *IVL, Swedish Environmental Research Institute, Aneboda Research Station, S-360030 Lammhult, Sweden*

Key words: boreal lakes, air–water interface, surface microlayer, nutrients, neuston, microbial biomass, activity

Abstract

Nutrient contents, microbial biomass and microbial activities were studied at the air–water interface in the surface microlayer (SM) and subsurface waters (10 cm depth) in small boreal forest lakes. Two different sampling techniques were used to evaluate differences in the nutrient composition in SM- and subsurface waters and to study their effects on microbial biomass and activities of neustonic and planktonic microbial communities. Inorganic nutrients were only slightly enriched in the SM compared to the subsurface water samples. P-PO^4 varied between 6–10 mg P-PO_4 m_{-3} in the SM and between 1–2 mg P–PO_4 m^{-3} in the subsurface waters. Dissolved inorganic nitrogen ($NO_2{}^-$ + $NO_3{}^-$ + NH_4^+) varied between 12–20 mg N m^{-3} in the SM and between 3–12 mg N m^{-3} in subsurface waters. Polymeric organic bound phosphorus and nitrogen were about 10 times enriched in the SM compared to the subsurface samples. However, microbial biomass like chlorophyll was by a factor of 8 to 280 times enriched in the SM of meso-polyhumic lakes and 2–25 time enriched in an acidified lake. Bacterioneuston biomass was by a factor of 1.5 to 2 times enriched in the SM compared to that of bacterioplankton. However, fungal biomass was 30 to 40 times higher in the SM than in the subsurface samples. Microbial activities was measured as [^{14}C]-UL-α-D-Glucose uptake and as microbial biopolymer processing measured with 4-methylumbelliferyl-α-D-Glucopyranoside as model substrate for tracing the enzymatic cleaving rate of α-glycosidic polymer bound glucose via microbial α-glucosidase (αGlucAse). [^{14}C]-UL-α-D-Glucose uptake was about 4–5 times higher in bacterioneuston compared to bacterioplankton and varied between 4–22 μm m^{-3} h^{-1} in bacterioneuston communities and between 1.5–2.5 μm m^{-3} h^{-1} in bacterioplankton. αGlucAse was about 1.5 to 8 times higher in SM samples and varied between 38–108 μm m^{-3} h^{-1} in the SM microbial communities compared to 12–35 μm m^{-3} h^{-1} in the subsurface water microbial communities. The ratios between αGlucAse activities and [^{14}C]-UL-α-D-Glucose uptake was about 3–5 times lower in the bacterioneuston than in the bacterioplankton communities which means a tighter metabolic coupling between biopolymer processing and substrate uptake in bacterioneuston than in bacterioplankton. Biofilms in surface microlayers at the air–water interface in small humic forest lakes can provide favourable microhabitats for the growth of neuston communities which may act as important sinks for allochthonous nutrient resources and may then generate new nutrient pools and prey for planktonic microbial food webs.

Introduction

It has been shown that phosphorus and nitrogen are the most limiting factors controlling the productivity and biomass developments of microplankton in the pelagic zone of boreal freshwaters (Jackson & Hecky, 1980; Jones, 1992; Schindler, 1977; Schindler & Bailey, 1993). However, inorganic nutrients in boreal freshwater are mostly low and the major P- and N- resources are combined to dissolved organic matter (DOM) which is largely composed of humic substances (Salonen et al., 1992a). The bioavailability of inorganic P- and N-nutrient sources in many of such waters can be influenced by humic substances

(Francko, 1986). Particularly in most Finnish forest lakes especially phosphorus is considered as the most limiting factor (Salonen et al., 1992b, 1994). Therefore osmotrophic organisms like algae and bacteria are forced to compete for limited P- and N-substrates (Francko, 1986, 1992; Jones, 1992). As bacteria are primarily sinks for P-resources (Jones, 1990; Salonen et al., 1994), micro- and macrozooplankton are considered as the most important phosphorus remineralizers in most humic lakes (Jones, 1992; Salonen et al., 1994).

In recent times studies on the chemistry and biology of biofilms are more under focus due to environmental and health aspects (Costerton, 1995). Biofilms at the air–water and on solid–water interfaces have been defined to provide favourable growth conditions for microbes, living in attached microcolonies and with cooperative metabolic activities (Costerton et al., 1987, 1994, 1995; Fletcher, 1986, 1991; Geesey et al., 1978; Maki, 1993; Marshall, 1984). However, in contrary to nutrient bioavailability patterns and food web interactions in pelagic water bodies, the prey as particulate polymeric P- and N-resources likely in biofirlms at the air–water interface are probable not so easily available for grazers like macro- and micro-zooplankton because of the high hydrophobicity of such biofilms. Also bacteria at the air–water interface may find not very favourable growth conditions because of higher environmental impacts (Schindler et al., 1996; Yan et al., 1996) and the strong hydrophobicity of these microenvironments. According to Lemke et al. (1995) hydrophobic conditions in microhabitats can effect the phosphorus utilization of bacteria.

Many surface waters have at the air–water interface a surface microlayer (SM) which is only 100 to 500 μm thick. This SM is acting as a biofilm and has exceptional environmental and ecological functions (Garrett, 1972; Hardy, 1982; Maki, 1993; Norkrans, 1980). The SM is considered as an important interface between the air and the subsurface water and it is an important regulator of heat and gases (e.g. CO_2, CH_4). It regulates also the nutrient, pollutant and corbon fluxes into deeper water layers (Garrett, 1972; Hardy, 1982; Maki, 1993). The SM is principally a microcosmos (Naumann, 1917) with primary producers (phytoneuston), decomposers (bacterioneuston) and micrograzers (e.g. flagellates, ciliates). Due to its chemical and biological composition (Liss, 1975; Williams et al., 1986) and its biofilm properties (Maki, 1993; Norkrans, 1980), it may favour nutrient enrichments and the growth of microorganisms on particles and in colloids in small microcolonies (Costerton et al., 1994; Marshall, 1984).

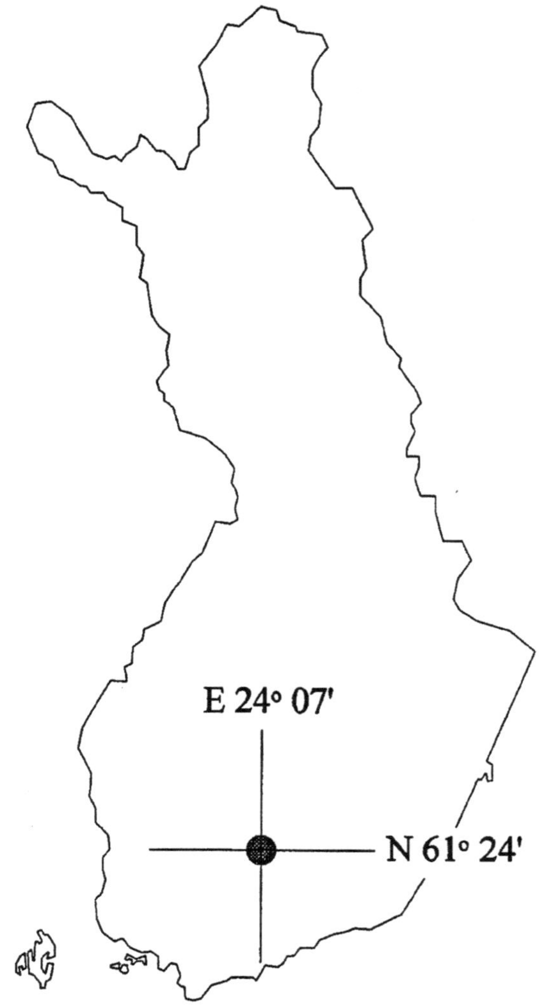

Figure 1. Location of sampled **Evo** lakes in southern Finland.

Due to such predicted scenarios of biofilm functions, we have started to study the composition of the SM and its subsurface waters in small humic forest lakes in the Evo research area in southern Finland (Figure 1). We were especially interested what is the most suitable sampling technique and whether we can measure significant differences in nutrient composition and microbial growth, biomass and activities in the SM and its subsurface waters. Due to the remarkable interactions of lakes and rivers in boreal environments with the littoral and its drainage areas (Kairesalo et al., 1992; Meili, 1992; Salonen et al., 1992a; Schindler et al., 1990, 1992, 1996) we hypothesized that especially small forest lakes may harbour significant amounts of nutrients and food resources located in the SM at the

Table 1. Name, location and general characterization of sampled lakes.

Lake name	Location and size	Catchment area (ha)	Max. depth (m)	Mean depth (m)	Humic contents	Comments
Mekkojärvi	Evo-Area; 0.35 ha	4	4	2.3	polyhumic	Small head-water lake with intensive studies on food web structures and nutrient cycling (Salonen et al., 1992a)
Valkea Kotinen	Evo-Area; 3.6 ha	30	6.5	3.0	mesohumic	Head water lake, subject of an International Integrated Monitoring Program (IMP) to study pollutants and catchment effects on ecosystem structure and function (Bergström et al., 1995)
Iso-Valkjärvi	Evo-Area; 3.8 ha	17	8	3.4	mesohumic	Acidified lake, subject of liming experiments on food web structure and fish populations (Rask & Järvinen, 1995)

air–water interface, which are not regulary measured, and not at all included in most nutrient budget studies and aquatic food web aspects. These nutrient pools and living biomasses in the SM may act as important primer in microbial food web processes at the air–water interface and may even provide important prey and nutrient resources for the food chain processes below and SM, likely for microbial communities in the subsurface waters (Södergren, 1993).

Material and methods

Three lakes with rather similar size and location but different chemical composition and environmental impact (Table 1) were selected for this SM-study: lake Iso-Valkjärvi, lake Mekkojärvi and lake Valkea Kotinen (Figure 1). The SM was sampled according to Knulst & Södergren (1994) with a surface slick sampler as already described by Södergren (1984, 1993). This is an especially constructed battery driven, remote radio-controlled operating catamaran (Figure 2a). The SM was collected by a rotating teflon coated drum (Figure 2a) in front of the surface slick sampler. The adsorbed SM is about 100–500 μm thick and is scraped off from the drum by a hydrophobic teflon razor (Figure 2b). Thereafter the slick was continously collected into pre-cleaned and sterilized glass bottles. For comparison studies an alternative SM-sampling technique was done with the glass-plate technique according to Harvey & Burzell (1972), which is in its principle shown in Figure 3. From both in parallel performed sampling techniques, water was kept in cooling boxes filled with crushed ice during sampling periods and transportation. Subsurface water samples were taken with an acid washed, pre-cleaned glass bottle, which was lowered exactly 10 cm below the water surface and kept in cooling boxes as described above. All samples were analyzed in the laboratory for basic physical-chemical parameters, nutrient concentrations, biomass determinations and microbial activities according to Koroleff (1979), Finnish standard methods (Mäkelä et al., 1992), Keskitalo & Salonen (1993), Hoppe (1993) and Wright & Hobbie (1966). The results from physical-chemical measurements are summarized in Table 2. Chlorophyll contents were measured with a spectrophotometer (Shimadzu model UV-2100) at 665 nm according to Keskitalo & Salonen (1993) after extraction with 94% ethanol. Bacterial numbers were counted as acriflavine direct counts (AFDC) with epi-fluorescence microscopy according to Bergström et al. (1985). Fungal biomass was measured as ergosterol after saponification with KOH/methanol and pentane extraction with a HPLC method according to Newell et al. (1986, 1988) and Newell (1993).

Microbial activities were measured as α-glucosidase activities and organic carbon assimilation as [^{14}C]-UL-α-D-Glucose uptake. We used 4-methylumbelliferyl-α-D-Glucopyranoside (MUFα-Gluc, from Sigma) as model substrate for tracing the enzymatic cleavage rate of α-glycosidic polymer bound glucose via microbial α-glucosidase (αGlucAse) activities according to the method recommendations described by Hoppe (1993) and modifications by Múnster et al. (1989). We prefiltered the lake water through 100 μm pore size plankton net to remove larger particles and zooplankton and incubated all MUFαGluc substrates with lake water at 15 °C in the laboratory in a constantly thermostated cuvette direct-

Table 2. Physical–chemical characterization of the studied lakes.

							ISVA					
	MKJ			VK			Basin-A			Basin-B		
Parameter	GP	RD	SSW	GP	RD	SSW	GP	RD	SSW	GP	RD	SSW
pH	6.75	6.6	6.2	5.45	5.33	5.36	nm	6.98	7.0	nm	5.91	5.96
Cond. (mS m^{-1})	52.9	50.2	49.9	31.9	33.7	30.9	nm	26.6	25.8	nm	13.2	13.0
Alkal (eq. m^{-3})	0.102	0.100	0.093	0.011	0.014	0.085	nm	0.147	0.128	nm	0.016	0.001
Colour (g Pt m^{-3})	252	262	237	162	165	143.5	nm	68	53	nm	43	37
Absorbance (254 nm)	0.72	0.77	0.70	0.477	0.497	0.440	nm	0.259	0.215	nm	0.156	0.135
P-PO_4 (mg P m^{-3})	2	2	2	6	10.8	1	nm	1	1	nm	2	1
N-NO_2^- +NO_3^- (mg N m^{-3})	7	7	7	3	3	2	nm	1	2	nm	2	1
N-NH_4^+ (mg N m^{-3})	13	6	5	7	4	3	nm	11	1	nm	10	3
P_{tot} (mg P m^{-3})	18.5	243.4	13	33	220	24	nm	200	10	nm	29	18
P_{org} (mg P m^{-3})	16.5	241.4	11	25	191.2	22	nm	188	8	nm	27	17
POP (mg P m^{-3})	13	68	3	26	102	22	nm	189	7	nm	26	7
DOP (mg P m^{-3})	0.5	173	8	1	8.2	1	nm	10	1	nm	3	10
N_{tot} (mg P m^{-3})	503	3315	518	666.8	2530	554	nm	3720	425	nm	445	480
DIN (mg P m^{-3})	20	13	12	10	7	5	nm	14	3	nm	12	4
PON (mg P m^{-3})	135	2452	24	260	1747	260	nm	2556	55	nm	411	61
N_{org} (mg P m^{-3})	483	3302	506	656.8	2523	549	nm	3706	422	nm	433	476
DON (mg P m^{-3})	348	850	482	396.8	776	289	nm	1150	367	nm	22	415

MKJ: Mekkojärvi. VK: Valkea-Kotinen. ISVA-A: Iso Valkjärvi, Basin-A-limed site. ISVA-B: Iso-Valkjärvi, Basin-B-acidified site. AFDC: Acriflavine Direct Counts. GP: Glass Plate Technique. RD: Rotating Drum Technique. SSW: Sub-Surface Water, nm. not measured.

P_{tot}: total phosphorus; P_{org} organic phosphorus; POP: particulate organic phosphorus; DOP: dissolved organic phosphorus. N_{tot}: total nitrogen; DIN: dissolved inorganic nitrogen; N_{org}: organic nitrogen; PON: particulate organic nitrogen; DON: dissolved organic nitrogen.

ly in the spectrofluorimeter during measurements. We used a direct *in situ* fluorescence reading method by adding 10–50 μM final MUFαGluc concentration to lake water in a 1.5 ml semi microcuvette and followed the increase of fluorescence with a spectrofluorimeter (Hitachi model F-4000) with 333 nm for excitation and 450 nm for emission wavelength settings. This MUFαGluc concentration was found to saturate in most cases the αGlucAse activities under *in situ* conditions. The relative fluorescence increase values were then used for calculating the α-glucosidase activities from corresponding 4-methylumbelliferyl (MUF) calibration curves made in each lake water samples from the same day as formerly described by Münster et al. (1989).

[^{14}C]-UL-α-D-Glucose uptake (302 mCi/mmol specific activity, Amersham, U.K.) was measured according to Hobbie & Rublee (1977) and Wright & Hobbie (1966). We added 50–100 nM [^{14}]-UL-α-D-Glucose final concentration to 30 ml of 100 μm prefiltered lake water samples in 50 ml glass bottles, which were acid-and Millipore-Q washed and preburned (450 °C for 4–5 h) before use. This concentration was found in pretests to saturate the uptake system. [^{14}C]-UL- α-D-Glucose uptake was measured at 15 °C under dark conditions in the laboratory for 1 to 2 h and uptake was terminated by adding 3 ml of prefiltered (Whatman GF/F) 35% formalin solution. Control samples were prepared in similar assays by adding 3 ml 35% formalin 10 min before [^{14}C]-UL-α-D-Glucose additions. Three parallel 10 ml formalin fixed water samples from each bottle were filtered through 0.2 μm cellulose nitrate membrane filters (25 mm ID, Sartorius, Germany). Filters were washed with 10 ml cold α-D-Glucose solution (500 μM, Merck, Germany), placed in 20 ml scintillation vials, amended with 250 μl ethylenglycolmonomethylether (Merck, Germany) and after 2 h 8 ml scintillation cocktail (OptiPhase High Safe 3, Wallac, Finland) was added. Vials were thoroughly mixed and after 10 h counted in a 1409 Liquid Scintillation Counter (Wallac, Finland). DPM values from control samples were substracted from sample values, parallel values were averaged to mean values and [^{14}C]-UL-α-D-Glucose uptake was calculated according to Wright & Hob-

Figure 2a. Surface Slick Sampler in Action at Lake mekkojärvi 1995 with Rotating Drum (**RD**) in Front.

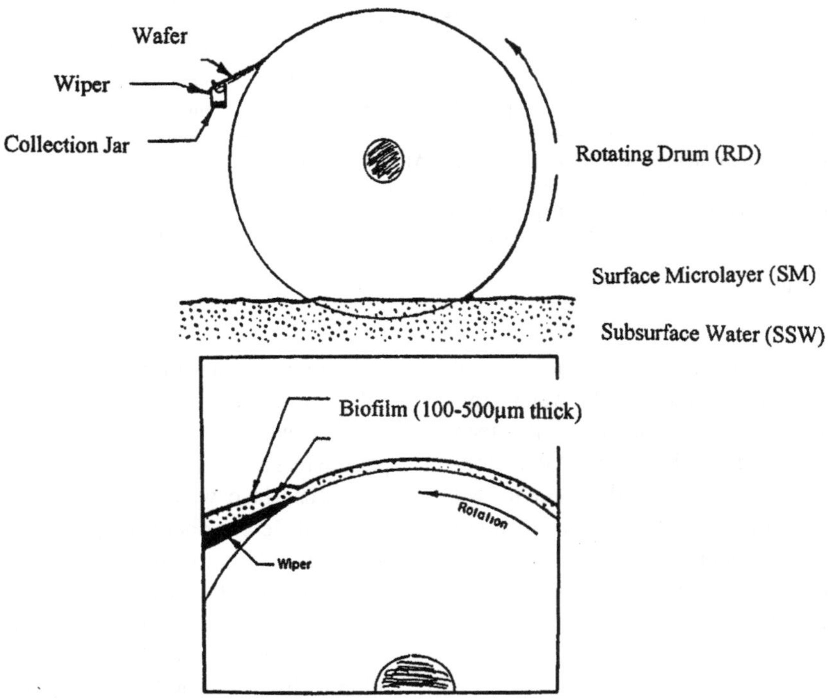

Figure 2b. Principles of Surface Microlayer (**SM**) Sampling with the Rotating Drum (**RD**) Tecnique (modified from Garrett, 1972).

bie (1966). Results from both microbial biomass and activity measurements are summarized in Table 3.

In addition to microbial activity measurements we have made growth experiments of bacteria and fungi on different agar plate growth media as summarized in Table 4. Lake water was prefiltered through a 100 μm size plankton net and diluted 1:10 with sterilized Millipore-Q water. Diluted lake water was

Table 3. Microbial biomass and activities in the studied lakes.

Parameter	MKJ			VK				ISVA Basin-A				ISVA Basin-B		
	GP	RD	SSW	GP	RD	SSW		GP	RD	SSW		GP	RD	SSW
Chlorophyll (mg m^{-3}	6	739.6	2.6	32.7	275.8	31.7		nm	117.2	4.6		nm	8.4	5
AFDC (cells 10^6 ml^{-1})	1.27	2.58	1.54	3.86	4.40	2.72		nm	9.50	1.99		nm	5.96	2.0
Ergosterol (μM)	23.7	33.7	0.84	11.34	33.7	0.72		nm	nm	nm		nm	nm	nm
[^{14}C]-α-D-Glucose uptake (μm m^{-3} h^{-1})	4.2	22.1	1.6	4.3	22.3	2.4	nm	nm	nm		nm	nm	nm	
α-Glucosidase Activity (μm m^{-3} h^{-1})	30.1	107.9	33.8	34.6	53.5	29.5		nm	92.5	12.4		nm	38.6	30.4
α-Glucosidase Activity/ [^{14}]-α-D-Glucose Uptake-Ratio	7.2	4.9	21.2	8.1	2.4	12.3		nm	nm	nm		nm	nm	nm

MKJ: Mekkojärvi. VK: Valkea-Kotinen. ISVA-A: Iso Valkjärvi Basin-A. ISVA-B: Iso-Valkjärvi, Basin-B. AFDC: Acriflavine Direct Counts. Gl Glass Plate Technique. RD: Rotating Drum Technique. SSW: Sub-Surface Water. nm: not measured.
* The data represent mean values from 6 sampling time measurements and with three parallels.

Table 4. Microbial growth as colony-forming units (CFU) on selective agarplate growth media.

Growth medium	MKJ				VK			ISVA Basin-A			ISVA Basin-B		
	GP	RD	SSW		GP	RD	SSW	GP	RD	SSW	GP	RD	SSW
CFU ml^{-1} on CPS	1680	9745	835		1625	4335	68	nm	5540	815	nm	6200	1505
CFU ml^{-1} on KAZ	1240	8660	1190		2650	4865	70	nm	6400	1034	nm	5500	786
CFU ml^{-1} on Malt	3475	7310	1640		1660	4640	140	nm	8620	133	nm	8935	340
CFU ml^{-1} on GP-agar plus	115	15		15	190	170	0	nm	245	0	nm	185	0

CPS: Casein, Peptone and Starch medium on agar plates, according to Moaledj (1975).
KAZ: Kaserer medium from A to Z on agar plates, according to Moaledje (1975)
Malt: Malt extract with peptone on agar plates, according to Deutsche Sammlung von Mikroorganismen und Zellkulturen GmbH, (DSM) Katalog 1989, Medium No 90 GP-Agar plus: Aureomycine/Rose Bengal: Aureomycine/Rose Bengal Glucose-Ppetone Agar, according to Maki et al. (1984) and Atlas & Parks (1993).
Other Abbreviations see Table 2–3!

inocculated on three parallel agar plates and control plates were prepared from autoclaved lake water samples according to standard methods under sterile clean bench conditions (Gerhardt et al., 1994). After inoculation the agar plates were incubated for one to two weeks at room temperature and the colony forming units (CFU) were counted manually. The CFU- counts were corrected for dilution with Millipore-Q water and the results are presented in Table 4.

Results and discussion

We have sampled the surface microlayer (SM) and the subsurface water (10 cm depths) from May until September in 1995 and have also compared during a one-week period the difference between the rotating drum (RD)- and the glass plate (GP)- sampling techniques. A summary of the physical and chemical parameters is given in Table 2. While there was no difference ($p > 0.05$) in the basic physical-chemical parameters and the inorganic nutrients between the SM and subsurface water values in all three lakes, and also no difference between the two sampling methods, we observed significant differences ($p < 0.05$) between total nutrients (P_{tot}, N_{tot}), and organic nutrients (P_{org}, N_{org}, POP, DOP, DON and PON) the SM and the subsurface waters in all lakes and between the two sampling techniques (Table 2). This observation was supported by the differences in biomass parameters like chlorophyll, bacteria number (AFDC) and fungal biomass (measured as ergosterol, Table 3). We could also recognize clear differences in between the lake samples with highest chlorophyll in Mekkojärvi (739.6 mg Chl m^{-3}), followed by Valkea Kotinen (275.8 mg Chl

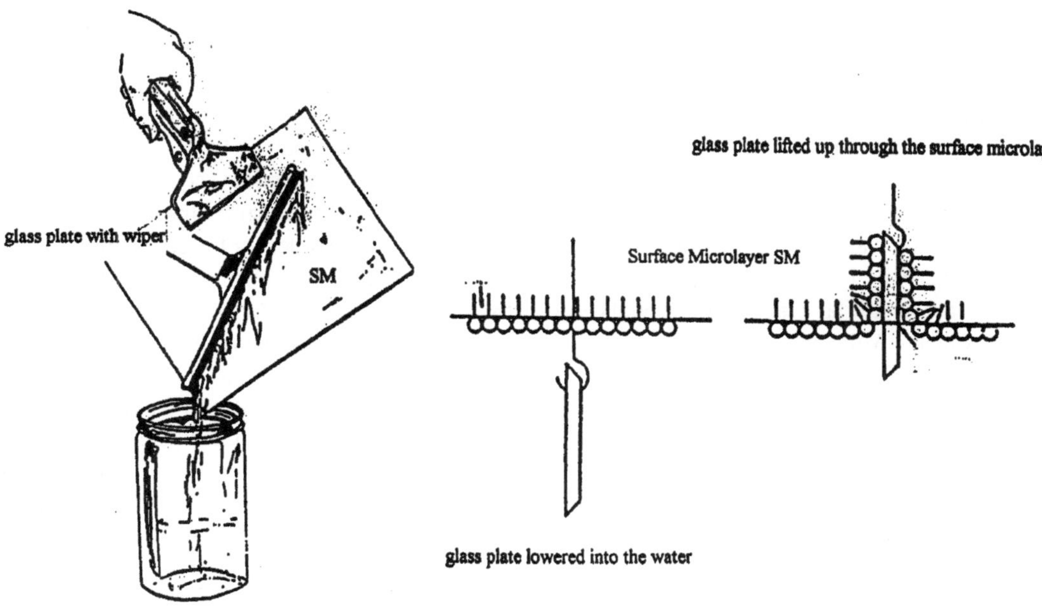

Figure 3. Scheme and principles of the Glass Plate (**GP**) sampling techniques (modified from Harvey & Burzell, 1972).

m^{-3}) and Iso Valkjärvi (117.2 mg Chl m^{-3} in Basin-A, and 8.4 mg Chl m^{-3} in Basin-B). Interestingly, Iso Valkjärvi showed much higher chlorophyll in the SM from the limed site (Basin-A) compared to the acidified, control site (Basin-B). This could be an indicator of acification or other stress effects on phytoneuston in the SM from the acidified site, because in the subsurface water there was little difference between both basins.

In contrary to the clear differences in phytoneuston/phytoplankton biomasses, we could observe only smaller differences in bacterioneuston and bacterioplankton biomasses based on numbers from acriflavin direct counts (AFDC). There were also little differences in bacterioplankton numbers between all lakes, but some differences in the AFDC counts in the SM and the subsurface waters from all lakes (Table 3). Amazingly we counted highest AFDC numbers in the SM from Iso Valkjärvi in the limed site (Basin-A), followed by the acidified Basin-B, Valkea Kotinen and the lowest numbers in the SM from Mekkojärvi. In general we found two to three times more bacterioneuston cells in the SM compared to bacterioplankton in the subsurface waters.

Microbial activity measurements provided however a more pronounced difference between the bacterioneuston and bacterioplankton community. αGlucAse showed in all sampled lakes clear and significant differences ($p < 0.05$) between the microbial community in the SM RD- samples compared to that in the subsurface water, but only small and insignificant differences ($p > 0.05$) between the GP-samples and the subsurface water. Highest αGlucAse of 108 μm m^{-3} h^{-1} was measured in the SM from Mekkojärvi followed with 92.5 μm m^{-3} h^{-1} from Basin-A in Iso Valkjärvi, and 53.5 μm m^{-3} h^{-1} in Valkea Kotinen and 38.6 μm m^{-3} h^{-1} in Basin-B in Iso Valkjärvi. As we have not differentiated between dissolved free αGlucAse and particle associated αGlucAse activity, and we have also not separated bacterioplankton and bacterioneuston from their predators like Protozoa, Rotatoria, Ciliata and Cladocera and from algae before the and αGlucAse activity measurements, we could not specify who were the main contributors to the αGlucAse activities. We have therefore preliminary defined the αGlucAse activity as microbial community activities, although we have used extremely short incubation times of 10 min, and assumed that bacteria were the main contributors.

This situation was however different and much clearer with the [^{14}]-UL-α-D-Glucose uptake measurements, which was mostly related to bacterial activities. Unfortunately we could measure [^{14}C]-UL-α-D-Glucose uptake only in two lakes due to technical reasons. Therefore we have no [^{14}C]-UL-α-D-Glucose uptake data for comparisons from Iso Valkjärvi. However, we clearly defined bacteria as the primary utilizer of glucose in our system. [^{14}C]-UL-α-D-Glucose uptake was significantly higher ($p < 0.05$) in bacterioneuston compared to bacterioplankton in Mekkojärvi and Valkea-Kotinen, however insignificant ($p > 0.05$) between the GP- and subsurface water samples and also not significantly different ($p > 0.05$) between both lakes. The rather similar values in bacterioneuston and bacterioplankton glucose uptake in both lakes could have been due to rather similar glucose pool sizes in the SM and subsurface waters from both lakes as has been reported formerly by Münster (1991, 1993). Due to the low and rather similar carbohydrate pools in the subsurface waters (Münster, 1991), we assumed that there should have been at least in an ideal case a close coupling between extracellular cleaving activities of biopolymers like polysacharides via e.g. αGlucAse and the released monomeric substrate uptake like α-D-Glucose. For this assumption we have calculated the ratios between αGlucAse activity and [^{14}C]-UL-α-D-Glucose uptake as shown in Table 3. From these data it was obvious that in both lakes the microbial community sampled by the RD-technique displayed the narrowest αGlucAse:[^{14}]-UL-α-D-Glucose-ratio (2.4–4.9) followed by the community in the GP-samples (7.2–8.1) and the widest ratio was observed in the planktonic community (12.3–21.2). We interpreted these results as an evidence that in the SM the microbial community was much more closely coupled in its catabolic/anabolic processes compared to that in the planktonic community, which probably lived under much less favourable growth conditions with much higher dilution and dissolution effects compared to the microbial community in biofilms in the SM at the air–water interface. Therefore, we assumed that due to the structure and function of microbial communities in biofilms, microbes formed and lived in microcolonies with cooperative metabolic network systems as has been recently described by Costerton et al. (1994, 1995; Fletcher, 1986, 1991; Marshall, 1984). Similar suggestions have been given in studies on Swedish lakes by Södergren (1979, 1984, 1993).

Under the assumption of better growth conditions due to more bioavailable organic nutrients and cooperative catabolic/anabolic network systems in biofilms of the SM at the air–water interface (Costerton et al., 1994, 1995; Marshall, 1984; Fletcher, 1991), we have made growth experiments and counted the number of colony forming units (CFU) on selective agar plate media as summarized in Table 4. In almost all experiments CFU values gave higher numbers in the SM samples with the RD- and the GP technique compared to the subsurface waters. From the four different applied growth media (conf. Table 4) we found for the nutrient rich casein-peptone-starch (CPS) medium in the RD-sample the highest CFU-numbers (9745 CFU ml^{-1}) in the polyhumic lake Mekkojärvi, followed by Iso Valkjärvi in Basin-B (6200 CFU ml^{-1}) and -A (5540 CFU ml^{-1}) and finally (4335 CFU ml^{-1}) in Valkea Kotinen, and with similar sequences but with 3 to 5 times lower CFU-numbers in the GP-samples. A rather similar sequence order for the RD-samples were found in the CFU-numbers on the pure mineral medium – Kaserer from A–Z (KAZ)- with highest values (8660 CFU ml^{-1}) in Mekkojärvi, followed by Iso Valkjärvi in Basin-A (6400 CFU ml^{-1}), Basin-B (5500 CFU ml^{-1}) and lowest numbers in Valkea Kotinen (4865 CFU ml^{-1}). Again the GP-values for KAZ were 2 to 7 times lower compared to the RD-samples. With Malt agar as growth medium we received a different sequence pattern in the RD-samples, now with highest numbers in Iso Valkjärvi in Basin-B (8935 CFU ml^{-1}), and Basin-A (8620 CFU ml^{-1}), followed by Mekkojärvi (7310 CFU ml^{-1}) and finally in Valkea Kotinen (4640 CFU ml^{-1}).

On our fourth growth medium (Agar-GP plus, see Table 4), which was especially prepared to grow only fungi (Maki et al., 1984), we counted highest values in RD-samples in Iso Valkjärvi in Basin-A (245 CFU ml^{-1}), then in Basin-B (185 CFU ml^{-1}), followed by Valkea Kotinen (170 CFU ml^{-1}) and finally in Mekkojärvi (150 CFU ml^{-1}). In this case there was only little difference between RD- and GP samples, and sometimes even higher CFU in the GP-samples (Table 4). We could not find any growth on our control plates although we have incubated them up to two to three weeks. Also the total counts after one and two weeks did not much differ, the second week counting provided only larger colonies, which cometimes overgrew each other. As we have a rather small data basis from these growth studies, we have to be careful with generalizations, and we have to approve and try to confirm these results in future studies.

In summary we found higher microbial biomass and activities in the SM with the RD-sampling technique and the surface slick sampler compared to the GP-sampling method. Probably with the GP-technique we have diluted the SM samples with water from the subsurface compared to those with the RD method. This would imply that the RD-technique obviously collected less with subsurface water contaminated biofilm material compared to the GP-technique. Unfortunately we have not tested all published SM-sampling techniques (Maki, 1993; Norkrans, 1980; Williams et al.,

1986) and our final conclusions regarding the best sampling techniques is probably to early.

Although the biofilm microbial communities in the SM at the air–water interface can be assumed to receive generally higher environmental impacts like UV-irradiance, heavy metals, pollutants and others (Frimmel, 1994; Garrett, 1972; Hardy, 1982; Herndl et al., 1993; Jokiel & York, 1984; Kirk, 1994; Maki, 1993; Salonen & Vähätalo, 1994; Schindler et al., 1996; Taalas et al., 1996; Yan et al., 1996), there were significantly higher microbial activities, bacterial and fungal growth potential in the SM samples compared to those in the subsurface waters. Probably the SM was indeed an exceptionally good microenvironment for growth and reproduction of aquatic microorganisms also in humic waters, though humic lake waters contain largely (90–99% of DOC) recalcitrant organic matter (Münster, 1991, 1993). Although quantitatively of minor relevance in most large lakes and oceans with larger mixing depths, in small forest lakes the SM may probably act as an important resource to trigger microbial growth due to limited nutrients. Such trigger effect can be important for phosphorus, nitrogen and labile organic carbon sources for subsurface water microorganisms especially in small boreal forest lakes as can typically be found in Finnish environments.

We concluded from our preliminary data that microbial communities in biofilms at the air–water interface can take profit from allochthonous nutrient and detritus inputs and can provide also to subsurface communities nutrients for osmotrophs like algae and bacterioplankton, and they can generate important prey for phagotrophs and macrograzers as has been also shown by Södergren (1993). However, more studies are needed to understand better the microbial processes within the SM and its ecophysiological role in humic and clear water ecosystem functioning in boreal freshwaters.

Acknowledgements

This project was funded during 1995 by the Maj and Tor Nessling Foundation, Helsinki, Finland. We acknowledge also the technical support from the IVL and the University of Lund, Department of Ecology for providing the surface slick sampler and Lammi Biological Station for technical support during sampling and transportations.

References

Atlas, R. M. & L. C. Parks, 1993. Handbook of Microbiological Media. CRC Press, Boca Raton, 91 pp.

Bergström, I., A. Heinänen & K. Salonen, 1986. Comparison of acridine orange, acriflavine and bisbenzimide stains for enumeration of bacteria in clear and humic waters. Appl. Envir. Microbiol. 51: 664–667.

Costerton, J. W., K.-J. Cheng, G. G. Geesey, T. I. Ladd, J. C. Nickel, M. Dasgupta & T. J. Marrie, 1987. Bacterial Biofilms in Nature and Disease. Ann. Rev. Microbial. 41: 435–464.

Costerton, J. W., Z. Lewandowski, D. DeBeer, D. E. Caldwell, D. R. Korber & G. James, 1994. Biofilms, the customized microniche. J. Bacteriol. 170: 2137–2142.

Costerton, J. W., Z. Lewandowski, D. E. Caldwell, D. R. Korber & H. M. Lappin-Scott, 1995. Microbial Biofilms. Ann. Rev. Microbiol. 49: 711–745.

Deutsche Sammlung von Microorganismen und Zellkulturen GmbH, (1989). Katalog, 286 pp.

Fletcher, M., 1986. Measurement of Glucose utilization by *Pseudomonas fluorescencs* that are free-living and that are attached to surfaces. Appl. Envir. Microbiol. 52: 672–676.

Fletcher, M., 1991. The physiological activity of bacteria attached to solid surfaces. Adv. Microb. Physiol. 32: 53–85.

Francko, D. A., 1986. Epilimnetic phosphorus cycling: influence from humic materials and iron coexisting major mechanisms. Can. J. Fish. aquat. Sci. 43: 302–310.

Frimmel, F., 1994. Photochemical aspects related to humic substances. Envir. Int. 20: 373–385.

Garrett, W. D., 1972. Impact of natural and man-made surface films on the properties of the air–water interface. In D. Dyrssen & D. Jagner, (eds), The Chemistry of the Ocean. Nobel Symposium 20: 75–91. Almquist & Wicksell, Stockholm.

Geesey, G. G., R. Mutch, J. W. Costerton & R. B. Green, 1978. Sessile bacteria: an important component of the microbial population in small mountain streams. Limnol. Oceanogr. 23: 1214–1223.

Gerhardt, P., R. G. E. Murray, W. A. Wood & N. R. Krieg, 1994. Methods for General and Molecular Bacteriology. American Society for Microbiology. Washington, D. C.: 135–292.

Hardy, J. T., 1982. The sea-surface film microlayer; biology, chemistry and anthropogenic enrichment. Prog. Oceanogr. 11: 307–328.

Harvey, G. W. & L. A. Burzell, 1972. A simple microlayer method for small samples. Limnol. Oceanogr. 17: 156–157.

Herndl, G. J., G. Müller-Niklas & J. Frick, 1993. Major role of ultraviolet-B in controlling bacterioplankton growth in the surface layer of the ocean. Nature 361: 717–719.

Hobbie, J. E. & P. Rublee, 1977. Radioisotope studies of heterotrophic bacteria in aquatic ecosystems. In J. Jr Cairns (ed.), Aquatic Microbial Communities. Garland Publ. Inc. New York: 441–476.

Hoppe, H.-G., 1993. Use of fluorogenic model substrates for extracellular enzyme activity (EEA) measurement of bacgteria. In P. Kemp, B. F. Sherr, E. B. Sherr & J. J. Cole (eds), Handbook of Methods in Aquatic Microbial Ecology. Lewis Publ.: 423–431.

Jackson, T. A. & R. E. Hecky, 1980. Depression of primary productivity by humic matter in lake and reservoir waters of the boreal forest zone. Can. J. Fish. aquat. Sci. 37: 2300–2317.

Jokiel, P. L. & R. H. York, 1984. Importance of ultraviolet radiation in photoinhibition of microbial growth. Limnol. & Oceanogr. 29: 192–199.

Jones, R. I., 1990. Phosphorus transformation in the epilimnion of humic lakes: biological uptake of phosphate. Freshwat. Biol. 23: 323–337.

Jones, R. I., 1992. The influence of humic substances on lacustrine planktonic food chains. Hydrobiologia 229: 73–91.

Kairesalo, T., A. Lehtovaara & P. Saukkonen, 1992. Littoral-pelagial interchange and the decomposition of dissolved organic matter in a polyhumic lake. Hydrobiologia 229: 199–224.

Keskitalo, J. & K. Salonen, 1993. Manual for the integrated monitoring subprogramme 'Hydrobiology of lakes'. Vesi- ja Ympäristöhallinnon Julkaisuja – Sarja B. Vol. 16: 40 pp.

Kirk, J. T. O., 1994. Optics of UV-radiation in natural waters. Arch. Hydrobiol. Ergebn. Limnol. 43: 1–16.

Knulst, J. & A. Södergren, 1994. Occurrence and toxicity of persistent pollutants in surface microlayers near an inceneration plant. Chemosphere 29: 1339–1347.

Koroleff, F., 1979. Meriveden yleisimmät kemialliset analyysimenetelmät. Meri 7, 60 pp.

Lemke, M. J., P. F. Churchill & R. G. Wetzel, 1995. Effect of substrate and cell surface hydrophobicity on phosphate utilization in bacteria. Appl. Envir. Microbiol. 61: 913–919.

Liss, P. S., 1975. Chemistry of the Sea Surface Microlayer. In J. P. Riley & G. Skirrow (eds), Chemical Oceanography, 2nd edn. 2: 193–243.

Mäkelä, A., S. Antikainen, I. Mäkinen, J. Kivinen & T. Leppänen, 1992. Vesi- ja Ympäristöhallinnon Julkaisuja- Sarja B. No. 10: Vesitutkimusten Näytteenottomenetelmät, 87 pp.

Maki, J. S., 1993. The air–water interface as an extreme environment. In T. E. Ford (ed.), Aquatic Microbiology; An Ecological Approach. Blackwell Sci. Publ. Oxford: 409–439.

Maki, J. S., S. C. Danos & C. C. Remsen, 1984. Quantitative changes in fungal colonyforming units in the surface microlayers of two freshwater ponds. Can. J. Microbiol. 30: 578–586.

Marshall, K. C., 1984. Microbial adhesion and aggregation. Dahlem Conference, Berlin, Springer Verlag.

Meili, M., 1992. Sources, concentrations and characteristics of organic matter in softwater lakes and streams of the Swedish forest region. Hydrobiologia 229: 23–41.

Moaledj, K., 1975. Qualitative analysis of an oligocarbophilic aquatic microflora in the Plußsee. Arch. Hydrobiol. 93: 287–302.

Münster, U., 1991. Enzyme activity in eutrophic and polyhumic lakes. In R. J. Chrost (ed.), Microbial Enzymes in Aquatic Environments. Brock/Springer Series, New York, 1991: 96–122.

Münster, U., 1993. Concentrations and fluxes of organic carbon substrates in the aquatic environment. Antonie van Leeuwenhoek 63: 243–274.

Münster, U., P. Einiö & J. Nurminen, 1989. Evoluation of the measurements of extracellular enzyme activities in a polyhumic lake by means of studies with methylumbelliferyl-substrates. Arch. Hydrobiol. 115: 321–337.

Naumann, E., 1917. Beiträge zur Kenntnis des Teichnannoplanktons. II. Über das Neuston des Süßwassers. Biol. Zentr. 37: 98–106.

Newell, S. Y., 1993. Membrane-containing fungal mass and fungal specific growth rate in natural samples. In P. F. Kemp, B. F. Sherr, E. B. Sherr & J. J. Cole (eds), Handbook of Methods in Aquatic Microbial Ecology. Lewis Publ.: 579–586.

Newell, S. Y., R. D. Fallon & J. D. Miller, 1986. Measuring fungal biomass dynamics in standing-dead leaves of a salt-marsh vascular plant. In S. T. Moss (ed.), The Biology of Marine Fungi. Cambridge University Press, New York: 19–25.

Newell, S. Y., T. L. Arsuffi & R. D. Fallon, 1988. Fundamental procedures for determining ergosterol content of decaying plant material by liquid chromatography. Appl. Envir. Microbiol. 54: 1876–1879.

Norkrans, B., 1980. Surface Microlayers in Aquatic Environments. Adv. Microb. Ecol. 4: 51–85.

Rask, M. & M. Järvinen, 1995. Neutraloinin vailkutukset happamoituneen metsäjärven ekosysteemiin – Iso Valkjärven kalkituskokeen tuloksia vuosilta 1990–1993. Kalatutkimuksia-Fiskundersökningar. No. 101: 1–90.

Salonen, K., T. Kairesalo & R. I. Jones, 1992a. Dissolved organic matter in lacustrine ecosystems: energy source and system regulator. Hydrobiologia 229, 291 pp.

Salonen, K., L. Arvola, T. Tulonen, T. Hammar, T.-R., Metaälä, P. Kankaala & U. Münster, 1992b. Planktonic food webs of a highly humic lake. I. A mesocosm experiment during spring primary production maximum. Hydrobiologia 229: 125–142.

Salonen, K. & A. Vähätalo, 1994. Photochemical mineralisation of dissolved organic matter in lake Skjervatjern. Envir. Int. 20: 307–312.

Salonen, K., R. I. Jones, H. De Haan & M. James, 1994. Radiotracer study of phosphorus uptake by plankton and redistribution in the water column of a small humic lake. Limnol. Oceanogr. 39: 69–83.

Schindler, D. W., 1977. Evolution of phosphorus limitation in lakes. Science 195: 260–262.

Schindler, D. W., K. G. Beatty, E. J. Fee, D. R. Cruishank, E. R., DeBruyn, D. L. Findlay, G. A. Linsey, J. A. J. Shearer, M. P. Stainton, 1990. Effects of climate warming on lakes of the central boreal forest. Science 250: 967–970.

Schindler, D. W., S. E. Bailey, P. J. Curtis, B. R. Parker, M. P. Stainton & C. A. Kelly, 1992. Natural and man-caused factors affecting the abundance and cycling of dissolved organic substances in precambrian shield lakes. Hydrobiologia 229: 1–21.

Schindler, D. W. & S. E. Bailey, 1993. The biosphere as an increasing sink for atmospheric carbon: estimates from increased nitrogen deposition. Global Biogeochem. Cycles 7: 717–733.

Schindler, D. W., P. J. Curtis, B. R. Parker & M. P. Stainton, 1996. Consequences of climate warming and lake acidification for UV-B penetration in North American boreal lakes. Nature 379: 705–708.

Södergren, A., 1979. Origin of ^{14}C and ^{32}P labelled lipids moving to and from freshwater surface microlayers. Oikos 33: 278–289.

Södergren, A., 1984. Small-scale temporal changes in the biology and chemical composition of surface microlayers in a eutrophic lake. Verh. int. Ver. Limnol. 22: 765–771.

Södergren, A., 1993. Role of aquatic surface microlayer in the dynamics of nutrients and organic compounds in lakes, with implications for their ecotones. Hydrobiologia 251: 217–225.

Taalas, P., T. Koskela, E. Kyrö, J. Damski & A. Supperi, 1996. Ultraviolet radiation in Finland. The Finnish Research Programme on Climate Change (SILMU), Final Report: 83–90.

Williams, P. M., A. F. Carlucci, S. M. Henrichs, E. S. Van Vleet, S. G. Horrigan, F. M. H. Reid & K. J. Robertson, 1986. Chemical and microbiological studies of sea-surface films in the southern gulf of California and of the west coast of Baja California. Mar. Chem. 19: 17–98.

Wright, R. T. & J. E. Hobbie, 1966. Use of glucose and acetate by bacteria and algae in aquatic ecosystems. Ecology 47: 447–464.

Yan, N. D., W. Keller, N. M. Scully, D. R. S. Lean & P. J. Dillon, 1996. Increased UV-B penetration in a lake owing to drought-induced acidification. Nature 381: 141–143.

Hydrobiologia **363**: 271–282, 1998.
T. Tamminen & H. Kuosa (eds), Eutrophication in Planktonic Ecosystems: Food Web Dynamics and Elemental Cycling.

Seasonal and spatial distribution of bacterial production and biomass along a salinity gradient (Northern Adriatic Sea)

A. Puddu[1], R. La Ferla[2], A. Allegra[2], C. Bacci[1], M. Lopez[1], F. Oliva[2] & C. Pierotti[3]
[1] *Istituto di Ricerca Sulle Acque – CNR, Via Reno 1, 00198 Roma, Italy*
[2] *Istituto Sperimentale Talassografico – CNR, Spianata S. Raineri, 98122 Messina, Italy*
[3] *Istituto Zooprofilattico dell'Umbria e delle Marche, V.le Adriatico 52, 61032 Fano, Italy*

Key words: bacterioplankton abundance, bacterial carbon production, microbial loop, carbon cycle, primary productivity coupling, Adriatic Sea

Abstract

The Adriatic Sea is a semi-enclosed ecosystem that receives in its shallow part, the northern basin, significant fresh-water inputs which markedly increase its productivity with respect to the oligotrophic features of the Mediterranean sea. In this area, especially on the western coast where river plumes diffuse, high physical (density) and chemical (nutrients) gradients occur on a small scale, both horizontal and vertical. Results of bacterial production as ^{3}H-thymidine incorporation, bacterial abundance as DAPI direct count, autotrophic biomass as chlorophyll *a* and total biomass as ATP from three areas in the Northern Adriatic Sea are reported. The three sites, differently influenced by the river water diffusion, were sampled seasonally over two days, every 24 h, in four surveys from April 1995 to January 1996. Bacterioplankton production, strongly correlated with primary production, was extremely high near the coast in low-salinity, high-nutrient waters, mostly as an indirect consequence of riverine inputs causing an increase in phytoplankton production stimulated by physically driven nutrient inputs. In the warm months bacterial activity was higher than in cold months. While bacteria abundance did not appear related to the salinity gradients, bacterial production (from 0.6 to 372 pM ^{3}H-thymidine h^{-1} incorporated, corresponding to 0.01–8.2 μ g C l^{-1} h^{-1}) and the relative generation times (from 0.2 to 35 days) showed a high range of values, representing a variety of situations, from estuaries to the ocean. The resulting role of the bacterial community in the carbon cycle is very consistent, processing amounts of carbon which have been estimated as high as the 80% and the 260% of those synthesized by autotrophs in summer and winter, respectively.

Introduction

The coastal belt of the Northern Adriatic, characterized in the western side by a strong gradient in salinity and nutrients, represents an area of enhanced biological activity. This is due principally to the river runoff that causes an increase in primary production driven by nutrient enrichment. Here, as in estuaries (Ducklow & Shiah, 1993), heterotrophic bacteria may increase in abundance and activity utilizing the large amount of dissolved organic matter deriving from autotrophs production or of allochtonous origin. Therefore the importance of the microbial loop could be enhanced in this situation, with important consequences for the total nutrient cycle.

Riverine inputs are principally due to the Po river that, with a mean flow of 1500 m^{3} s^{-1}, accounts for 51% of the total fresh-water flow into the Northern Adriatic (Degobbis & Gilmartin, 1990). The Adriatic circulation is driven primarily by strong thermohaline forces, caused by the fresh water input and by seasonal wind and heat forcing at the surface, producing a flow of less-dense surface waters and more-dense sub-surface waters south-easterly directed along the Italian coast (Artegiani et al., 1996). On the eastern side, an opposing remounting flow brings into the northern basin the oligotrophic waters entering the Adriatic from the Ionian Sea. This anticyclonic circulation pattern is active during most of the year except during the summer period, when a semi-closed circulation pattern and a strong vertical stratification allow the fresh-water to be advected eastward toward the Istrian peninsula (Revelante & Gilmartin, 1992). As a consequence, the dispersion of the fresh-water input principally controls the production, the stratification and the circulation of the entire basin.

Regarding the current knowledge of the biology of the Adriatic sea, as reported by Herndl et al. (1992), there is abundant literature describing species distributions and successional events which however do not contribute significantly to the functional understanding of the dynamics of the system. Primary production has been extensively studied (Franco, 1967; Kveder & Keckes, 1969; Kveder & Revelante, 1971; Revelante & Gilmartin, 1976; Gilmartin & Revelante, 1983; Faganeli et al., 1989; Zoppini et al., 1995) but information on bacterioplankton distribution and productivity in the water column is very scarce (Karner et al., 1992; Krstulovic, 1992; Krstulovic et al., 1995; Solic & Krstulovic, 1994). Moreover, among the reported papers very few considered the coastal ecosystem.

In the framework of a multidisciplinary national Project (PRISMA) investigating eutrophication phenomena in the Adriatic Sea, bacterial abundance and productivity in the Northern Adriatic were measured over a one year period, along with C, N and P, pico-, nano- and microplankton, mesozooplankton distribution, primary productivity and grazing. The objective of this paper is to describe the changes on bacterial production and biomass along a salinity gradient in different seasonal situations. From the obtained results, we will try to discuss from an ecological point of view the role of bacterial communities in the microbial loop and their implications in the carbon fluxes in this marine system.

Materials and methods

Location and sampling

The study was carried out on three sites (Figure 1): two in the coastal zone, one affected by the Po river outflow (S1), the other one by the Adige and Brenta rivers (S2), and the third one in the middle of the northern basin (S3), less influenced by riverine inputs. On these sites the main hydrological, chemical and biological parameters were determined on samples collected at various depths from the research vessel 'Lo Bianco' anchored in a fixed station. Four seasonal cruises, with subsequent sampling in the three sites, were performed on April 3–13, July 3–12, October 2–12, 1995 and January 10–19, 1996. Samples for bacterial production were collected twice in every station at 6 or 7 depths during the four surveys, at 10.00 am in the first and the second day. The same samples were used for primary productivity measurements and other correlated biological and chemical analysis carried out by other research groups. Samples for bacterial abundance, chlorophyll and ATP were collected every 4 h, but only data corresponding to bacterial production determinations will be presented. CTD measurements were performed with an Idronaut mod 401 probe. Reversing thermometers and bench salinometer Autosal Guildline were used to control CTD data.

Bacterioplankton production, abundance and specific growth rates

Bacterial carbon production (BCP) was measured with the ^{3}H-thymidine uptake method (Fuhrman & Azam, 1982). Samples (5 ml) and blanks in plastic disposable sterile vials were incubated *in situ* for 1h with 20nM of added labelled thymidine (80–90 Ci mmol^{-1}; NEN). This concentration gave maximum labelling in preliminary incorporation experiments with various concentrations of ^{3}H-thymidine added to samples collected in the study sites at various salinities. Two ice-cold trichloroacetic acid (TCA)- (5% final concentration) killed controls were run corresponding to every three live incubations of the same water samples. At the end of the incubation 5 ml of 10% TCA were added to extract the soluble pools from the cells. After 30 min of extraction at 0 °C the precipitate was collected by filtration through a 0.22 μm pore-size Millipore filter. The radioactivity was counted on a Packard Tri-Carb 4430 scintillation counter using Filter Count by Packard as a liquid scintillation cocktail. Blank values were subtracted from the corresponding sample. The conversion factor of 1.1×10^{18} cells produced *per* mole of thymidine incorporated into DNA was used as proposed by Riemann et al. (1987) for coastal marine bacteria, considering also the relative high concentration of ^{3}H-thymidine added in this study (Bell, 1990). Bacterial cell production was then converted in BCP assuming 2×10^{-14}g C cell^{-1} (Lee & Fuhrman, 1987).

Bacterial abundance (BAB) was measured on subsamples (100 ml) preserved at 4 °C in the dark, in polystyrene bottles containing prefiltered formalde-

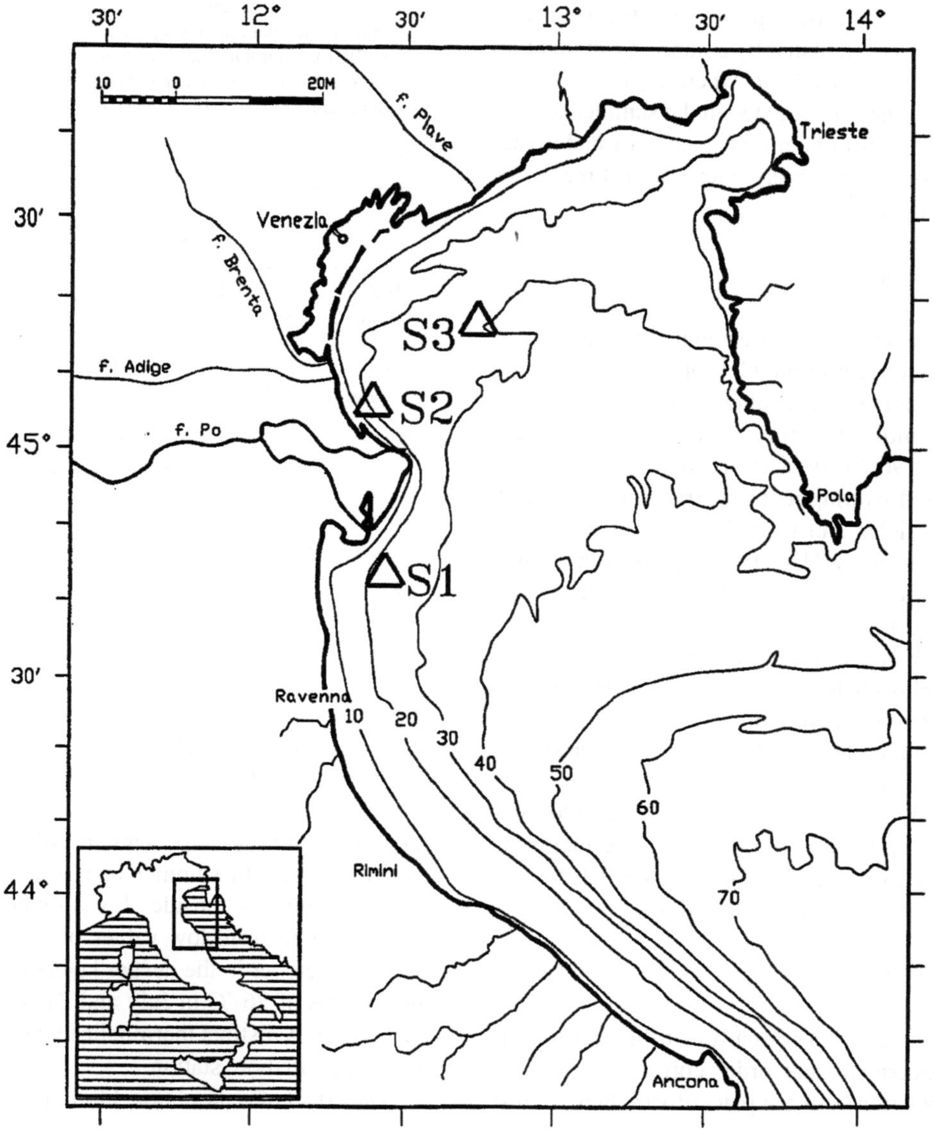

Figure 1. Location of sampling sites

hyde (final concentration, 2%) and stored a few weeks until lab treatment. Since the aim of the multidisciplinary research was to study the pico-, nano- and microplankton involved in the microbial network, samples from stations S2 and S3 were obtained from the picofraction, prefiltered through 2 μm pore-sized sterile membranes. Samples from S1, separately collected, were not prefiltered, so they refer to the total bacterial population. All samples were subjected to DAPI (4',6 diamidino-2 phenylindole) direct count method (Porter and Feig, 1980) and then filtered on to black Nuclepore polycarbonate membranes (0.2 μm pore size, 25 mm diameter). The wet filters were placed on glass slides and mounted in low-fluorescence immersion oil. The bacterial cells were counted using an Axioplan Zeiss microscope, equipped with an epifluorescent illumination system (50 W mercury lamp, G365 exciter filter, FT395 chromatic beam splitter, LP420 barrier filter). At least 20 randomly selected microscopic fields per filter were counted so that the standard deviation resulted less than 20% of the mean value. In this study we report only the abundances of heterotrophic bacteria since the phototrophic component was preliminarily subtracted.

For each observation the specific growth rate of bacterioplankton was calculated by assuming exponential growth as $\mu(d^{-1}) = \ln(Bt/B_0)/t$ where B_0 is the initial cell abundance, directly counted in the sample, and B_t is B_0 plus the number of cells produced in 1 h, derived from the ^{3}H-thymidine assay. Generation times (T_2) are equal to $\ln(2)\mu^{-1}$.

Biomass determination

The autotrophic biomass was estimated by measuring chlorophyll *a* concentration (Chl) on 0.5–1 l seawater filtered on Whatman GF/F filters. The samples were stored frozen until extraction, by grinding and extracting in 15 ml of 90% (v/v) acetone solution. After storing refrigerated overnight, the extracts were filtered to remove particles and their Chl content was measured according to APHA (1989) by using a Turner design fluorometer (model 10–005R) previously calibrated with a chlorophyll standard (Sigma).

The total biomass was estimated on the basis of adenosine triphosphate (ATP) cellular concentration on 0.2–0.3 l of sample filtered through 0.2 μm Nuclepore polycarbonate membranes. The filtered material was extracted at room temperature with dimethyl sulphoxide (Jakubczak & Leclerc, 1980; Zoppini, 1990) and the amount of ATP was measured by the luciferin-luciferase reaction with a luminometer LKB 1250 calibrated with a disodium ATP salt (Sigma).

Statistical analysis

Considering our data not normally distributed, we computed the Spearman rank order correlation to measure the association between pairs of variables. To test data means for differences, we computed the Mann-Whitney Rank Sum test, when comparing two groups, or the Kruskal-Wallis one way ANOVA on ranks, when comparing more than two groups.

Results

The physical characteristics of the water column were similar throughout the sampling period during every seasonal survey due to stable metereological conditions. CTD vertical profiles of salinity and temperature, *per* station and *per* survey, mediated over the two measurements made at 24 h interval, are shown in Figure 2.

Table 1. Mean values and coefficient of variation (cv%) of ^{3}H-thymidine incorporation rates (TdR), bacterial carbon production (BCP), chlorophyll *a* (Chl) and ATP concentration and bacterial abundance (BAB) grouped on the basis of the temperature and salinity of the sample (see text). LS = low salinity samples, IS = intermediate salinity samples, HS = high salinity samples.

		LS		IS		HS	
		mean	cv%	mean	cv%	mean	cv%
TdR	cold	27.5	69	6.6	84	2.4	74
(pmol l^{-1} h^{-1})	warm	126.7	76	15.9	57	9.5	28
BCP	cold	0.6	69	0.15	84	0.05	74
(μgC l^{-1} h^{-1})	warm	2.8	76	0.35	57	0.21	28
Chl	cold	2.45	105	1.28	83	0.67	44
(μg l^{-1})	warm	6.32	83	0.82	79	0.94	54
ATP	cold	0.34	125	0.18	70	0.09	52
(μg l^{-1})	warm	1.00	76	0.22	51	0.19	49
BAB	cold	5.6	44	6.3	130	4.4	48
(10^8 cells l^{-1})	warm	8.2	51	7.6	92	–	–

Considering all stations, the water temperature ranged from 12 °C (surface) to 10 °C (bottom) in April '95 and from 7 °C (surface) to 11.5 °C (bottom) in January '96 surveys, representative of early spring and winter seasons. In summer and autumn surveys the water column was considerably warmer, with temperatures ranging from low values of 12 °C, in July, or 15 °C in October in the deeper layers of S3 station, to high values of 26 °C in July and 21 °C in October at the surface. A distinct thermocline was evident only in the deeper offshore station.

The effect of river plume dilution was strongly evident at the upper layers of coastal stations, with the lowest salinities found in October at station S1. A mixed layer corresponding to salinities ranging from 34 to 36 psu was noticed in the coastal stations S1 and S2. This layer in some occasions reached deeper waters. Station S3 was only affected by a light dilution in the upper part of the water column. A totally homogeneous column was found in January at this station.

These observations allow two thermal periods and three water masses characterized by different salinity ranges to be identified. Samples collected in April and January are representative of the 'cold' period, while those of July and October belong to the 'warm' period of the year. It was possible to rank the 143 individual paired determinations of thymidine incorporation, bacterial abundance, chlorophyll and ATP concentrations

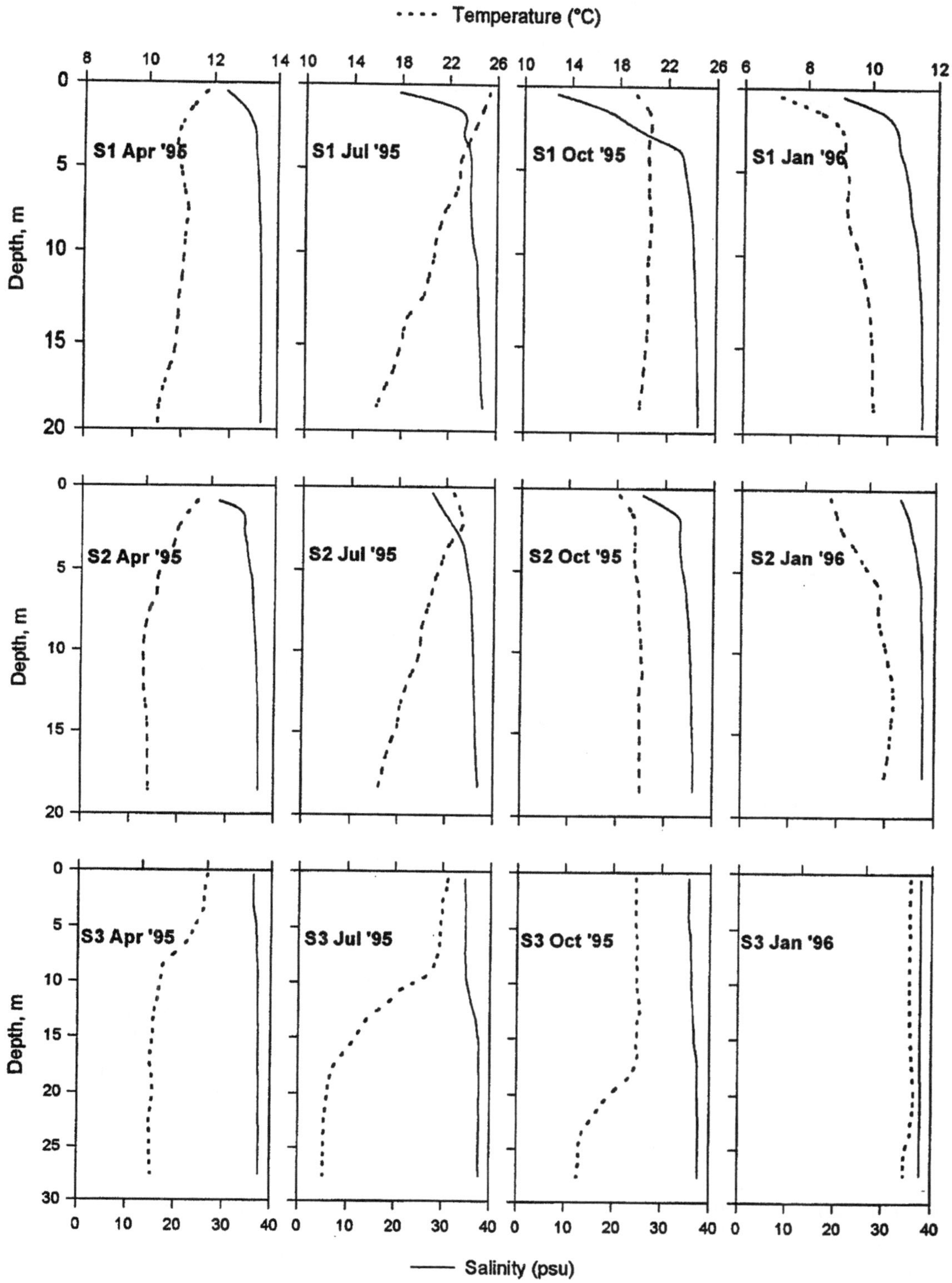

Figure 2. Vertical CTD profiles of salinity (solid line) and temperature (dashed line) mediated over two measurements.

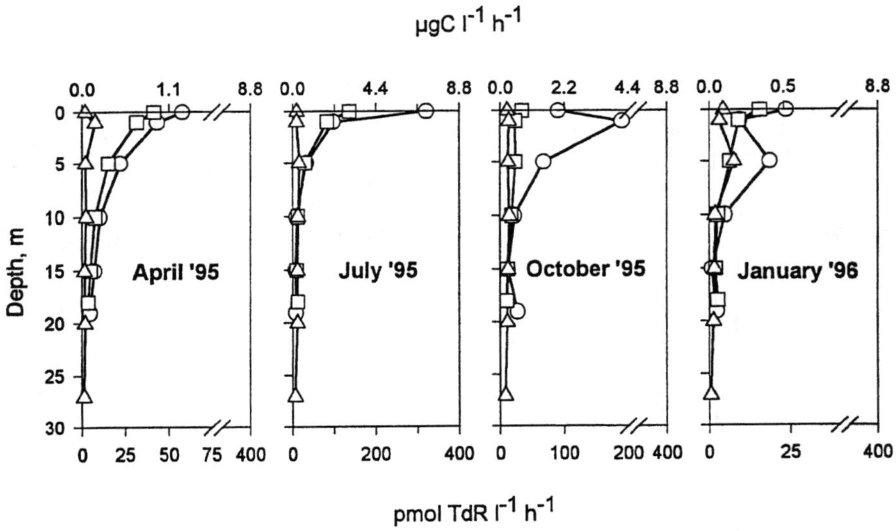

Figure 3. Vertical profiles of mean bacterial production as ^{3}H-thymidine incorporation rates and carbon production at the three sites: S1 (circles), S2 (squares), S3 (triangles).

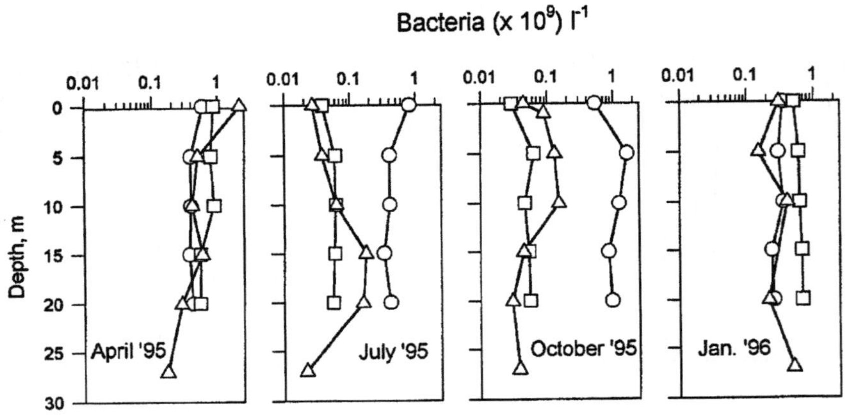

Figure 4. Vertical profiles of mean bacterial abundance at the three sites: S1 (circles), S2 (squares), S3 (triangles). Note that samples from S2 and S3 were prefiltered on 2 μm porosity filter.

into three groups with different temperatures and salinities. The low salinity group (< 35 psu in the 'cold' and < 33.4 psu in the 'warm' period) include 35 samples collected in the upper 5 meters of the water column in the two coastal stations. The high salinity group (> 37.0 psu) include 36 samples, most of them (26) collected from the offshore station at all depths and from the bottom layers at the coastal stations only in January. The intermediate salinity group include the remaining 72 samples equally distributed among all stations in April, July and October cruises.

Depth profiles of ^{3}H-thymidine (TdR) incorporation rate into the cold TCA are shown in Figure 3. Bacterial production varied in a wide range, from 0.6 to 372 pM TdR h^{-1}, corresponding to 0.01–8.2 μg C l^{-1} h^{-1}. Average values for the data organized as previously described are reported in Table 1.

The bacterial abundance profiles evaluated in the three stations during the four sampling surveys are shown in Figure 4. In station 1, the total bacterial counts appeared to be nearly homogeneous both in time (seasons) and space (throughout the water columns). In stations S2 and S3, during July and October, bacterial abundance resulted a factor of ten lower with respect to those obtained in April and January. Since prefiltration has probably removed a lot of bacteria not quantified, in the following discussion data from samples collected in July and October from S2 and S3 will not be

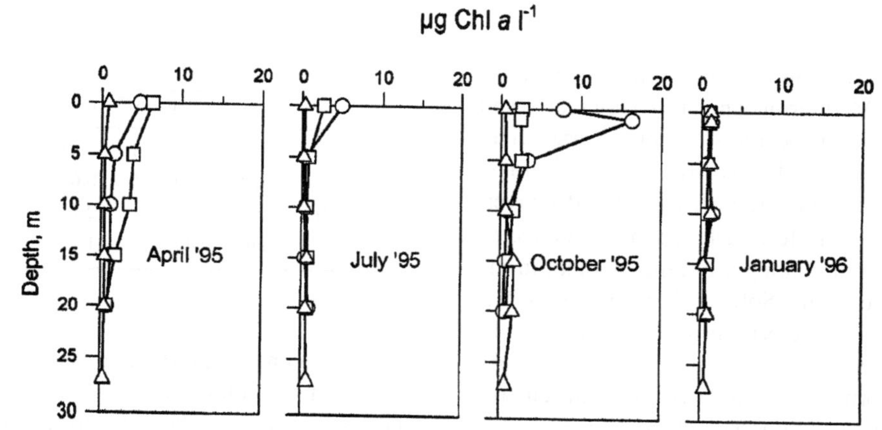

Figure 5. Vertical profiles of mean chlorophyll *a* concentration at the three sites: S1 (circles), S2 (squares), S3 (triangles).

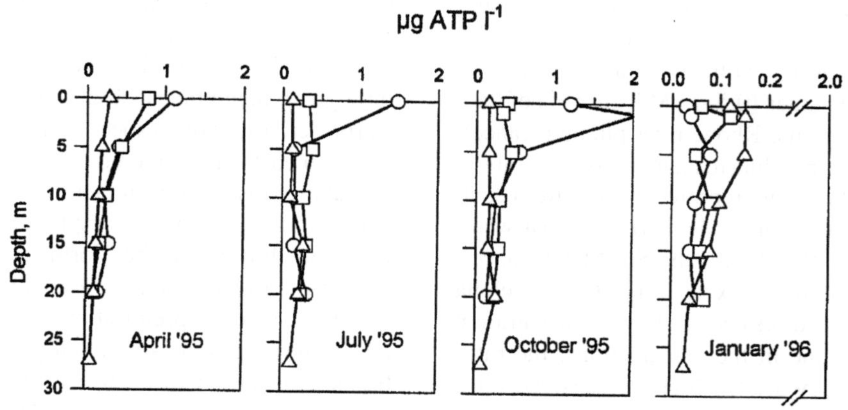

Figure 6. Vertical profiles of mean ATP concentration at the three sites: S1 (circles), S2 (squares), S3 (triangles).

considered. The whole quantitative trend of the bacterioplankton showed no high values, having an average of 6.0×10^8 cells l^{-1}. Bacterial abundance during the warm situation (only station S1) showed a mean value of 7.7×10^8 cells l^{-1}; during the cold period, the mean value of bacterial abundance was 5.4×10^8 cells l^{-1} considering all stations. We found inconsistent diel patterns of bacterial abundance during the four-month samplings.

Vertical profiles of chlorophyll *a* and ATP are shown in Figures 5 and 6, while average values are reported in Table 1. Chlorophyll *a* concentration varied from 0.2 to 19.8 $\mu g\, l^{-1}$ and ATP from 0.02 to 2.7 $\mu g\, l^{-1}$.

The analysis of variance (Kruskal-Wallis ANOVA) of bacterial production, chlorophyll and ATP showed that there is a significant difference ($P < 0.0001$) among the three salinity classes, while bacterial abundance shows on the contrary no significant differences among grouped data.

The comparison between the cold and the warm conditions (Mann-Whitney Rank Sum Test) showed that there is a significant difference ($P < 0.0001$) for the bacterial production and the chlorophyll and ATP mean concentrations in all the salinity classes, with the only exceptions of chlorophyll in the high salinity and ATP in the intermediate salinity group.

As for the ANOVA, the bacterial abundance is not significantly different between cold and warm periods. Bacterial production in low salinity water masses is ~12-fold higher than that in high salinity samples in all the cruise while average values in the warm period are ~4-fold higher than in the cold period. Biomass indicators show a similar behaviour between groups although differences are lower.

Discussion

Previous information on bacterial production in the Adriatic and the Mediterranean seas is very scarce and representative only of specific situations. Karner et al. (1992) reported a range of 10–100 pM TdR h^{-1} (transformed by us from their reported BCP and conversion coefficients) from few surface samples collected in spring and summer along a trophic gradient in the northern Adriatic basin. Solic & Krustolovic (1994) reported an average bacterial production of 26.2 pM TdR h^{-1}, with values ranging form 2.2 to 73 pM TdR h^{-1} (transformed from reported bacterial cell production and conversion coefficients as above) for the Kastela Bay, an enclosed shallow basin along the eastern coast in the middle Adriatic Sea. As regard to the Mediterranean Sea, Vives-Rego et al. (1988) measured bacterial production in front of Barcelona (Spain) as high as 126 pM TdR h^{-1} while Kirchman et al. (1989) reported very low values (0.5–7 pM TdR h^{-1}) of bacterial production for the Rhone river plume (France). Other values, corresponding to an incorporation rate of about 13 pM TdR h^{-1}, are reported by Hagström et al. (1988) for the oligotrophic Mediterranean sea water collected in front of Villefranche-Sur-Mer (France). Considering that, with the exception of the Barcelona study, all the cited authors added a lower concentration of labelled thymidine in their incubations, the rates reported in our study are the highest measured for a temperate coastal system. A similar very high range of thymidine incorporation rates was reported, to the best of our knowledge, only for the Chesapeake Bay (0.5–500 pM TdR h^{-1}, Ducklow & Shiah, 1993). Our average values, according to values reported by Ducklow & Carlson (1992) for the euphotic zones of major marine habitats, represents, in an area geographically reduced, a variety of situations from estuaries (1.8–3.3 μg C l^{-1} h^{-1}) to the ocean (0.1–0.2 μg C l^{-1} h^{-1}). Moreover, if we consider that the previous authors used in their review a standard thymidine conversion factor double than that we used, we may conclude that bacterial carbon production in coastal low saline waters of the Northern Adriatic is extremely high.

In this study case, the bacterial abundances were lower than those recorded in other estuarine areas and contrasted with the general understanding that the estuarine ecosystem, in consequence of the input of both allochtonous organic matter and autochthonous production, are characterized by high bacterial numbers (Heip et al., 1995). Moreover, bacterial densities determined at all three stations, did not reflect the different rates of secondary production occurring in relation to the various salinities. Perhaps the increase of bacterial density, due to the cell replication measured by the thymidine incorporation rate, was balanced by the bacterial loss (Gonzales et al., 1990, Shiah & Ducklow, 1995). We noted, instead, that the number of pico-sized bacteria was inversely related to temperature. The decrease in the density of bacteria passing through 2μ membranes, observed in S2 and S3 during warm periods, with respect to total bacterial density obtained from unfiltered samples at S1, suggests speculative discussion. In fact, from unpublished data concerning coastal water samples collected during the following July 96 in the Northern Adriatic sea, we observed a loss of about 47% of total bacterial population owing to the prefiltration on 2μm in warm period. Bacteria not only simply clogged on top of filters because the presence of detritus, but in fact were larger in size. Unfortunately, data from S1 with respect to S2 and S3 are not easily comparable but they consent a coherent hypothesis with previous published research. Since smaller size bacteria are generally characteristic of low nutrient waters and, according to Holligan et al. (1984) and Fuhrman et al. (1980), represent dormant or starved bacteria, while larger cells are typical of enriched environments, the occurrence of different trophic conditions may reasonably be associated not so much to the abundance but to the dimensional structure of the cells. The occurrence of diverse bacterial communities in the marine environment has been related to suspended particulate matter (Jacq & Prieur, 1986). Recently, the existence of two clearly distinct situations characterising the microbial community structure in warm and cold periods has been recognised (Unanue et al., 1992; Iriberri et al., 1993).

Table 2. Mean bacterioplankton specific growth rates (μ) and generation times (T_2)

		LS		IS		HS	
		mean	cv%	mean	cv%	mean	cv%
μ (d^{-1})	cold	0.89	41	0.33	71	0.17	107
	warm	1.70	46	0.61	81	0.40	24
T_2 (d)	cold	0.8		2.1		4.1	
	warm	0.4		1.1		1.7	

As a consequence of the different behaviour of bacterial activity and abundance, mean specific growth rates (μ) for the bacterial population (Table 2) show a wide range of values that correspond to mean generation times (T_2) from 0.4 days in low salinity July and October samples to more than 4 days in high salinity and 'cold' samples. The variability of generation times, considering all samples over the study period, is obviously wider, from 0.2 to 35 days. Slow growing populations ($T_2 > 20$ days) were collected in April and January prevalently in the offshore station (S3). Similar ranges of growth rates were reported in Kastela Bay (mid Adriatic), from 14 h to 7.8 days (Solic & Krstulovic, 1994) and at the mouth of the York river, in the Chesapeake Bay (Ducklow, 1982).

High values of biomass measured in low salinity water masses were due to phytoplankton blooms in the surface layers of coastal stations. Phytoplankton abundance, resulting from data of other authors collaborating to the same project, are consistent with our findings. Cell numbers reached values as high as 15 millions cells l^{-1} at station S1 in July (C. Totti, personal communication), with a dominance of small diatoms. Peak values of 19.8 μg Chl l^{-1} were measured in October '95 at the S1 station at - 1 m depth and correspond to the highest value of ATP (2.7 $\mu g\,l^{-1}$) and primary production (240.5 $\mu gC\,l^{-1}\,h^{-1}$) recorded over the study period (L. Alberighi, personal communication). Similar ranges of chlorophyll have been previously recorded in the most eutrophic coastal areas of the Northern Adriatic by many authors. On the contrary, data on ATP in this area are very scarce: only Krstulovic & Sobot (1982) reported values from 0.11 to 0.20 μg ATP l^{-1} for a coastal region of the eastern Central Adriatic (Split) and one high-sea station. Typical surface ATP concentrations range from > 0.5 μg ATP l^{-1} for eutrophic waters to 0.5–0.1 μg ATP l^{-1} for regions with moderate productivity to < 0.1 μg ATP l^{-1} for oligotrophic portions of the ocean (Karl, 1980). The ATP results of this study, characteristic of the different salinity groups, correspond, as they do for chlorophyll, to all the different regions previously described.

The correlations between bacterial carbon production and the other variables are presented in Table 3. Bacterial production appears strongly and inversely correlated with salinity and positively with chlorophyll *a* and ATP concentrations when data are considered all together or grouped with reference to the thermal periods, whereas the correlation degree is lower or nonexistent when different salinities are considered. No correlation exists at all between bacterial production and bacterial abundance. Bacterial production rates showed a very tight relationship with primary production rates measured by other authors. Having no information on the bioavailability of the allochtonous organic matter, we can just conclude from this observation that bacterioplankton activity in the study area shows an indirect link with the riverine inputs, modulated by the phytoplankton response. The latter is controlled by bottom-up forces reflecting the direct input of nutrients.

A tentative evaluation of the relative contributions of bacterial and phytoplankton biomass to the total planktonic biomass was carried out by transforming in carbon units bacterial density (20 fg C cell^{-1}), CHL ($\times$ 50) and ATP ($\times$ 250), respectively. Autotrophic carbon biomass so obtained resulted often higher than total biomass; this reflects the difficulty of applying a unique conversion factor to phytoplanktonic populations (Cho & Azam, 1990). Revelante & Gilmartin (1992, 1995) reported an extremely high variability of phytoplankton in size and physiological state and, consequently, in the chlorophyll content per cell and in the Chl *a*/carbon ratios in the Northern Adriatic. The percentage of bacterial biomass (Bact-C) to the total microbial biomass (ATP-C) is in the range 7–40%, with maxima in the high salinity waters. These results are consistent with previous findings of Krustolovic (1992) in the central Adriatic and with the reported dominance of bacterial biomass in the total microbial biomass in oligotrophic marine waters (Ducklow & Carlson, 1992).

In order to evaluate the magnitude of the flux of carbon from phytoplankton to bacteria, we estimated the relationship between bacterial and primary carbon production, integrated over the water column on a daily basis at the three studied sites (Table 4). Unpublished primary production data reported in the following discussion have been personally communicated by L. Alberighi, IBM-CNR Venice. Highest BCP/PP ratios, 66% and 36%, correspond to the lowest PP values, measured on January '96 at the coastal stations. After Ducklow & Carlson (1992) BCP values greater than 20% of the PP demand other DOM sources than phytoplankton exudation. Since the measured BCP includes only the quantity of organic carbon that remains as living C in the bacterial cells, the total bacterial carbon demand (BCD = BCP plus respiration) is consequently higher. The ratio BCP/BCD, which represents the bacterial growth efficiency, was, unfortunately, not measured in our study. Although literature values appear variable depending on community composition and environmental conditions (Jahnke & Craven, 1995), a realistic value in the open sea appear to be less than 20% (Ducklow & Carlson, 1992). To give a rough estimate of the BCD, we applied to our data the minimum and maximum growth efficiencies (26 and 55%), reported by Biddanda et al. (1994) for two stations in

Table 3. Probability levels of the correlation coefficient: bacterial carbon production (BCP) versus salinity (SAL), chlorophyll *a* (Chl) and ATP concentrations, primary production (PP) and bacterial abundance (BAB). Correlation was computed for all data grouped as in the text. Other symbols as in Table 1, n.s. = not significant ($P > 0.05$).

	All	Cold				Warm			
	data	all	LS	IS	HS	all	LS	IS	HS
BCP vs SAL	<0.001	<0.001	–	–	–	<0.001	–	–	–
BCP vs Chl	<0.001	<0.001	<0.1	0.001	<0.005	<0.001	n.s.	n.s.	n.s.
BCP vs ATP	<0.001	<0.005	n.s.	<0.005	n.s.	<0.001	<0.05	n.s.	n.s.
BCP vs BAB	<0.05	n.s.	n.s.	n.s.	n.s.	n.s.	n.s.	n.s.	–
BCP vs PP	<0.001	<0.001	n.s.	<0.005	n.s.	<0.001	<0.01	n.s.	<0.05

Table 4. Average values of bacterial carbon production (BCP) and primary production (PP) integrated on the water column, and their ratio expressed as percent BCP on PP.

Data	Station	BCP (mgC m^{-2} d^{-1})		PP (mgC m^{-2} d^{-1})*		BCP/PP
		mean	cv%	mean	cv%	(%)
April '95	S1	173	62	895	4	19
	S2	120	7	1510	27	8
	S3	31	41	500	9	6
July '95	S1	344	16	2237	60	15
	S2	276	43	1294	4	21
	S3	160	1	629	11	25
October '95	S1	553	15	3181	47	17
	S2	175	5	1241	37	14
	S3	169	8	853	21	20
January '96	S1	80	18	121	30	66
	S2	44	13	120	19	36
	S3	39	–	311	–	13

* L. Alberighi, personal communication.

the Gulf of Mexico, differently affected by the Mississippi plume. These stations appear to be similar for many physical, chemical and biological characteristics to our study area. Our BCD estimates would therefore correspond to 40–80% the amount of C produced by phytoplankton in periods of high and mean PP, April, July and October. On January, when PP reached its minimum levels, the amount of C needed to sustain bacterial production was, near the coast, 70 to 260% of that available from primary production. As reported by Chin-Leo & Benner (1992) for the Mississippi plume, we can suppose that in winter the BCD is partially supported by other substrates derived from riverine inputs. Our values for BCD/PP ratios should be considered as conservative estimates as they reflect the use of a thymidine conversion factor which was ~ 1.8 times lower that empirically determined by Biddanda *op. cit.* (1.89×10^{18} cells mol^{-1} thymidine) or that of 2×10^{18} cells mol^{-1} considered as a standard value (Ducklow & Shiah, 1993).

Conclusion

The reported results show that bacterioplankton production in the Northern Adriatic is extremely high near the coast in low-salinity, high-nutrient waters. This happens mostly as an indirect response to riverine inputs causing an increase in phytoplankton production fuelled by physically driven nutrients, but also as an effect of direct DOM discharges. The role of this community in the carbon cycle is very significant since it processes in productive periods a very high percentage (40–80%) of the carbon synthesized by autotrophs through a tight bacteria-phytoplankton coupling. Such a high bacterial carbon demand cannot be supported only by direct exudation of DOC by phytoplankton, but probably involves the availability of other DOM sources (Ducklow & Carlson, 1992). The need for external DOM sources is strengthened in winter when, near the coast, the bacterial carbon demand overcomes (70–260%) the relatively low amount of C produced by the phytoplankton. Therefore, we can conclude that the bacterial carbon demand in the Northern Adriatic area affected by riverine inputs is prevalently sustained by primary production in productive periods, while requires significant external carbon sources in less productive periods. Hence, after Azam et al. (1993), the term 'coupling' appears to be improper in this context due to the lack of knowledge on sources and sinks in the carbon transfer.

Acknowledgements

The authors wish to thank Prof. F. Azam for comments throughout the study and critical review of the manuscript. We are also grateful to C. Totti, IRPEM-CNR Ancona and to L. Alberighi and S. Rabitti, IBM-CNR Venice, for sharing data respectively on phytoplankton abundance, primary productivity and CTD, to A. Casiere, Ist. Zooprofilattico Fano, who collaborated on bacteria counting and to F. Bacciu, IRSA-CNR Rome, for assistance in sample collection and analysis. Thanks are also due to the crew of r/v 'Lo Bianco'. The manuscript was also improved by discussions with M. Pettine and A. Zoppini, IRSA-CNR Rome and by comments of an unknown rewier. This study is part of the PRISMA Project, supported by the Italian Ministry for University and Scientific and Technological Research.

References

APHA, AWWA, WPCF, 1989. Standard methods for the examination of water and wastewater. In L. J. Clesceri, A. E. Greenberg & R. R. Trussel (eds), American Public Health Association, Washington DC 17.

Artegiani, A., D. Bregant, E. Paschini, N. Pinardi, F. Raicich & A. Russo, 1996. The Adriatic Sea general circulation. Part II: baroclinic circulation structure. J. Phys. Oceanogr., in press.

Azam, F., D. C. Smith, G. F. Steward & A. Hagström, 1993. Bacteria-organic matter coupling and its significance for oceanic carbon cycling. Microb. Ecol. 28: 167–179.

Bell, R. T., 1990. An explanation for the variability in the conversion factor deriving bacterial cell production from incorporation of ^{3}H-thymidine. Limnol. Oceanogr. 35: 910–915.

Biddanda, B., S. Opsahl & R. Benner, 1994. Plankton respiration and carbon flux through bacterioplankton on the Luisiana shelf. Limnol. Oceanogr. 39: 1259–1275.

Chin-Leo, G. & R. Benner, 1992. Enhanced bacterioplankton production and respiration at intermediate salinities in the Mississippi River plume. Ecol. Prog. Ser. 87: 87–103.

Cho, B.C. & F. Azam, 1990. Biogeochemical significance of bacterial biomass in the ocean's euphotic zone. Mar. Ecol. Prog. Ser. 63: 253–259.

Degobbis, D. & M. Gilmartin, 1990. Nitrogen, phosphorus and biogenic silicon budgets for the northern Adriatic Sea, Oceanol. Acta 13: 31–45.

Ducklow, H. W., 1982. Chesapeake Bay nutrient and plankton dynamics. 1. Bacterial biomass and production during tidal destratification in the York River, Virginia, estuary. Limnol. Oceanogr. 27: 651–659.

Ducklow, H. W. & C. A. Carlson, 1992. Oceanic bacterial production. In K.C. Marshall (ed.), Advances in Microbial Ecology, Plenum Press, New York, 12: 113–181.

Ducklow, H. W. & F. K. Shiah, 1993. Bacterial production in estuaries. In T. E. Ford (ed.), Aquatic Microbiology: An Ecological Approach. Blackwell, Oxford 11: 261–287.

Faganeli, J., M. Gacic, A. Malej & N. Smodlaka, 1989. Pelagic organic matter in the Adriatic sea in relation to winter hydrographic conditions. J. Plankton Res. 11: 1129–1141.

Franco, P., 1967. Condizioni idrologiche e produttività primaria nel Golfo di Venezia. Arch. Oceanogr. Limnol. 15: 69–83.

Fuhrman, J. A., J. W. Ammerman & F. Azam, 1980. Bacterioplankton in the coastal euphotic zone: distribution, activity and possible relationships with phytoplankton. Mar. Biol. 60: 201–207.

Fuhrman, J. A. & F. Azam, 1982. Thymidine incorporation as a measure of heterotrophic bacterioplankton production in marine surface waters: evaluation and field results. Mar. Biol. 66: 109–120.

Gilmartin, M. & N. Revelante, 1983. The phytoplankton of the Adriatic sea: standing crop and primary production. Thalassia jugosl. 19: 173–188.

Gonzales, J. M., E. B. Sherr, B. F. Sherr, 1990. Size-selective grazing on bacteria by natural assemblages of estuarine flagellates and ciliates. Appl. envir. Microbiol. 56: 583–589.

Hagström A., F. Azam, A. Andersson, J. Wikner & F. Rassoulzadegan, 1988. Microbial loop in an oligotrophic pelagic marine ecosystem: possible roles of cyanobacteria and nanoflagellates in the organic fluxes. Mar. Ecol. Prog. Ser. 49: 171–178.

Heip, C. H. R., N. K. Goosen, P. M. J. Herman, J. Kromkamp, J. J. Middelburg & K. Soetaert, 1995. Production and consumption of biological particles in temperate tidal estuaries. Oceanogr. Mar. Biol. Ann. Rev. 33: 1–149.

Herndl, G. J., M. Karner & P. Peduzzi, 1992. Floating mucilage in the Northern Adriatic Sea: the potential of a microbial ecological approach to solve the mystery. Sci. Total Envir. Suppl.: 525–538.

Holligan, P. M., R. P. Harris, R. C. Newell, D. S. Harbour, R. N. Head, E. A. S. Linley, M. I. Lucas, P. R. G. Tranter & C. M. Weekley, 1984. Vertical distribution and partitioning of organic carbon in mixed, frontal and stratified waters of the English Channel. Mar. Ecol. Prog. Ser. 14: 111–127.

Iriberri, J., B. Ayo, M. Unanue, I. Barcina & L. Egea, 1993. Channeling of bacterioplankton production toward phagotrophic flagellates and ciliates under different seasonal conditions in a river. Microb. Ecol. 26: 111–124.

Jacq, E. & D. Prieur, 1986. Les associations bactéries-matière particulaire en milieu pélagique cotier: exemples de variations spatiales et temporelles. GERBAM II Colloque Int. de Bactériologie marine – CNRS, Brest 1–5 Octobre 1984 – IFREMER, Actes de Colloques 3: 229–236.

Jahnke, R. A. & D. B. Craven, 1995. Quantifying the role of heterotrophic bacteria in the carbon cycle: a need for respiration rate measurements. Limnol. Oceanogr. 40: 436–441.

Jakubczak, E. & H. Leclerc, 1980. Mesure de l'ATP bactérien par bioluminescence: étude critique des méthodes d'extraction. Ann. Biol. Clin. 38: 297–304.

Karl, D. M, 1980. Cellular Nucleotide Measurements and Applications in Microbial Ecology. Microbiol. Rev. 44: 739–796.

Karner, M., D. Fuks & G. J. Herndl, 1992. Bacterial activity along a trophic gradient. Microb. Ecol. 24: 243–257.

Kirchman, D., Y. Soto, F. Van Wambeck & M. Bianchi, 1989. Bacterial production in the Rhone River plume: effect of mixing on relationships among microbial assemblages. Mar. Ecol. Prog. Ser. 53: 267–275.

Krstulovic, N., 1992. Bacterial biomass and production rates in the central Adriatic. Acta Adriat. 33: 49–65.

Krstulovic, N., T. Pucher-Petkovic & M. Solic, 1995. The relation between bacterioplankton and phytoplankton production in the mid Adriatic Sea. Aquat. Microb. Ecol. 9: 41–45.

Krstulovic, N. & S. Sobot, 1982. Proportion of bacteria in total plankton of the central Adriatic. Acta Adriat. 23: 47–52.

Kveder, S. & S. Keckes, 1969. Hydrographic and biotical conditions in North Adriatic. V. Primary phytoplankton productivity. Thalassia jugosl. 5: 185–191.

Kveder, S., N. Revelante, N. Smodlaka & A. Skrivancic, 1971. Some characterstics of phytoplankton and phytoplankton productivity in the Northern Adriatic. Thalassia jugosl. 7: 151–158.

Lee, S. & J. A. Fuhrman, 1987. Relationships between biovolume and biomass of naturally derived marine bacterioplankton. Appl. envir. Microbiol. 53: 1298–1303.

Porter, K. G. & Y. S. Feig, 1980. The use of DAPI for identifying and counting aquatic microflora. Limnol. Oceanogr. 25: 943–948.

Revelante, N. & M. Gilmartin, 1976. The effect of Po river discharge on phytoplankton dynamics in the northern Adriatic sea. Mar. Biol. 34: 259–271.

Revelante, N. & M. Gilmartin, 1992. The lateral advection of particulate organic matter from the Po delta region during summer stratification, and its implications for the Northern Adriatic. Estuar. coast. Shelf Sci. 35: 191–212.

Revelante, N. & M. Gilmartin, 1995. The relative increase of larger phytoplankton in a subsurface chlorophyll maximum of the northern Adriatic Sea. J. Plankton Res. 17: 1535–1562.

Riemann, B., P. K. Bjornsen, S. Newell & R. Fallon, 1987. Calculation of cell production of coastal marine bacteria based on measured incorporation of ^{3}H-thymidine. Limnol. Oceanogr. 32: 471–476.

Shiah, F. K. & H. W. Ducklow, 1995. Regulation of bacterial abundance and production by substrate supply and bacterivory: a mesocosm study. Microb. Ecol. 30: 239–255.

Solic, M. & N. Krstulovic, 1994. Role of predation in controlling bacterial and heterotrophic nanoflagellate standing stocks in the coastal Adriatic Sea: seasonal patterns. Mar. Ecol. Prog. Ser. 114: 219–235.

Unanue, M., B. Ayo, I. Azùa, I. Barcina & J. Iriberri, 1992. Temporal variability of attached and free-living bacteria in coastal waters. Microb. Ecol. 23: 27–39.

Vives-Rego, J., J. Martinez & J. Garcia-Lara, 1988. Assessment of bacterial production and mortality in Mediterranean coastal water. Estuar. Coast. Shelf Sci. 26: 331–336.

Zoppini, A., 1990. Adenosintrifosfato cellulare (ATP). In S.I.B.M., Metodi nell'Ecologia del Plancton Marino. Nova Thalassia 11: 225–230.

Zoppini, A., M. Pettine, C. Totti, A. Puddu, A. Artegiani & R. Pagnotta, 1995. Nutrients, standing crop and primary production in western coastal waters of the Adriatic Sea. Estuar. Coast. Shelf Sci. 41: 493–513.

Hydrobiologia **363:** 283–288, 1998.
T. Tamminen & H. Kuosa (eds), Eutrophication in Planktonic Ecosystems: Food Web Dynamics and Elemental Cycling.

Incorporation of spatial structure into the analysis of relationships between environment and marine microbiological parameters

J. Truu[1,2], T. Nõges[3], K. Künnis[4] & M. Truu[1]
[1] *Environmental Protection Institute, Kreutzwaldi 5, Tartu, EE2400, Estonia*
[2] *Institute of Molecular and Cell Biology, Riia Street 23, Tartu, EE2400, Estonia*
[3] *Võrtsjärve Limnological Station, Institute of Zoology and Botany & Institute of Zoology and Hydrobiology of Tartu University, Rannu, Tartu County, EE2454, Estonia*
[4] *Water Protection Laboratory, Tallinn Technical University, Järvevana tee 5, Tallinn, EE0001, Estonia*

Key words: total bacterial number, plate count, bacterial production, partial multiple regression, variation partitioning

Abstract

During the RV *A. Veimer* cruise on 16 August 1990, the bacterial production, the bacterial total number and the plate count were estimated at nine sampling stations at four depths along a transect from the Estonian inshore area to the middle of the Gulf of Finland. To analyze the variability of microbiological parameters, the spatial structure was introduced into the statistical model and combined with the environmental variables. The total explained variation of the bacterial production was 74%, of which 62% was represented by the covariation between the environmental variables and the spatial structure. The variation in the bacterial production data that was explained by the environmental variables, independently of the spatial structure, was 9%, and variation due to spatial structure without environmental variables equaled 12%. The patterns of distribution of the bacterial total number and the plate count were related only to the spatial structure (explained variation 30% and 27%, respectively).

Introduction

Aquatic microbiological publications tend to employ increasingly more sophisticated methods of statistical analysis. Standard methods, like the computation of mean, spread, confidence limits, correlation and regression analysis have been supplemented by complex algorithms like ANOVA, factor analysis and multivariate statistics (Edgar & Laird, 1993; Holder-Franklin, 1992; van Tongeren, 1995).

Usually the data under study are related to a set of environmental data, for instance, to physico-chemical descriptors. While investigating aquatic ecosystems, large-scale and meso-scale horisontal gradients of surface concentrations in nutrients may be observed, as well as vertical (surface-to-bottom) gradients of temperature and dissolved oxygen, in small to medium scales (Karrasch et al., 1996; Roy et al., 1994). Such spatial heterogeneity may impair our ability to use standard statistical inference techniques, which were primarily designed for analysis of sets of observations drawn from independently and identically distributed random variables, and can affect the conclusions derived from statistical analyses performed which neglect such contributions (Dutilleul, 1993). It is likely that part of the biological variation is not explainable by any of the environmental variables considered, but rather by spatial variables alone. Legendre & Trousselier (1988) demonstrated in a study of planktonic heterotrophic and marine bacteria, that space may often act as a hidden variable and cause the appearance of spurious correlations.

Least-squares linear regression and multiple regression are among the most commonly used analytical tools of marine ecologists (White et al., 1991). Borcard et al. (1992) proposed a method to partition the variation of the target variable into independent components based on partial multiple regression. The aim of this method is to see if the environmental control model still holds if the spatial component is partialled out of the relationship between the target variable and the environment.

The purpose of this paper is to show that some assumed relationships between the marine microbiological parameters and the environmental variables can be spurious, implying a common spatial gradient, while others are real.

Material and methods

Study location and sampling

Material was collected during the cruise 37 of R.V. *Arnold Veimer* on 16.08.1990, at nine sampling stations and four depths along a transect from the Estonian coastal area to the middle of the Finnish Gulf (59° 40.1′N – 59° 59.1′ N, 25° 54.5′ E, stations 337–345). Sampling depths were 0.5 m, chlorophyll *a* maximum layer, above thermocline layer and 30 m, except for station 337 were water column was mixed and the second and third sampling depth were 13 m and 16 m, respectively. The collection of water samples started at 20.00 in the nearshore region and was finished at 24.00 in the offshore region. At the stations, the bacterial production derived from ^{3}H-thymidine incorporation, the total bacterial number by direct brightfield microscopy and the plate count were estimated. Bacterial production was measured only at 7 stations (337–343). The total count of bacteria was measured on erythrozine stained membrane filters (pore size 0.23 μm, Synpore) under the light microscope MBI-6 (magnification 1060) with oil immersion (Kuznetsov & Dubinina, 1989). Cell volumes were calculated using measured linear dimensions. The spread-plate technique with the peptone-yeast extract ZoBell medium 2216 E was used for bacterial plate counts (Oppenheimer & ZoBell, 1952). Bacterial production was measured by the tritiated thymidine incorporation according to Fuhrman & Azam (1982). The recommended scheme for Baltic Sea was followed (Guidelines…, 1988). The conversion factor 1.1×10^{18} cells mol^{-1} thymidine incorporated was applied (Riemann et al., 1987). The additional parameters included into the analysis were temperature, salinity, total nitrogen, total phosphorus, phosphate phosphorus, chlorophyll *a* and phytoplankton primary production.

Data analysis

For data analysis, the approach of partial multiple regression was used. This approach consists of expressing the spatial structure of the variables of interest as a linear combination of x- (distance from the coast) and y- (the sampling depth) coordinates. The spatial structure was included in the statistical model on an equal footing with the set of environmental variables. We used the third-order polynomial function of the spatial coordinates of the sampling locations and the stepwise variable selection. This approach allows to identify four fractions of variation:

(1) The nonspatial environmental variation,
(2) The spatially structured environmental variation,
(3) The spatial variation of the target variable that is not shared by the environmental variables,
(4) The unexplained, nonspatial variation.

For detailed explanation of the method, see Borcard et al. (1992). Environmental variables were analyzed using the Principal Component Analysis (PCA) after centring and standardization of the variables. The interpolation method used for producing graphs is the inverse squared distance weighting. The non-spatially structured environmental variation is mapped using the three-dimensional pin chart as suggested by Borcard & Legendre (1994). The multiple and partial linear regressions as well as the PCA was performed with the CANOCO program, release 3.12 (ter Braak, 1990).

Results

Table 1 reports the basic statistics for the physical, chemical, and microbiological variables measured along the studied transect. The total bacterial number and the plate count ranged from 0.24 to 4.76 million cells ml^{-1} and from 850 to 5400 CFU ml^{-1}, respectively. The highest values of both parameters were recorded at the inshore stations in the upper water layers. The bacterial production varied from 0.228 to 8.426 μgC l^{-1} h^{-1} with the highest values at the offshore stations. The interset correlation of the microbiological parameters showed positive relationship between the total bacterial number and the plate count ($r = 0.50$, $p < 0.01$, $n = 36$).

The Principal Component Analysis (PCA) of the environmental variables describes 87% of the variance in the data in first two axes. The passive analysis of the relationship with distance from the coast and the sampling depth indicates that the depth is correlated posi-

Table 1. Basic statistics for the variables under study ($n = 36$, in case of bacterial production $n = 28$)

Variable	Mean	Min	Max	CV[a] (%)
Temperature (°C)	13.9	2.7	20.3	48
Salinity (‰)	4.7	3.5	6.3	17
Total nitrogen (mg l^{-1})	25.7	20.2	35.0	15
Total phosphorus (mg l^{-1})	0.72	0.44	1.2	26
Phosphate phosphorus (mg l^{-1})	0.17	0	0.96	140
Chlorophyll *a* (mg m^{-3})	1.92	0.38	5.68	65
Primary production (mgC m^{-3} h^{-1})	5.2	0.02	16.3	88
Total bacterial number (10^6 cells ml^{-1})	1.26	0.24	4.76	76
Plate count (CFU ml^{-1})	2450	850	5400	52
Bacterial production (μgC l^{-1} h^{-1})	2.44	0.23	8.43	92

[a] CV, coefficient of variation

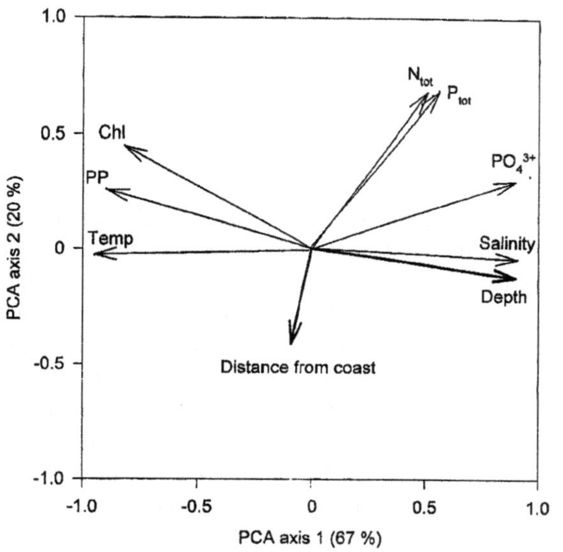

Figure 1. Principal Component Analysis of environmental variables with the depth and the distance from the coast as passive variables. Temp – temperature, PP – phytoplankton primary production, Chl – chlorophyll concentration, N_{tot} – total nitrogen, P_{tot} – total phosphorus, PO_4^{3+} – phosphate phosphorus.

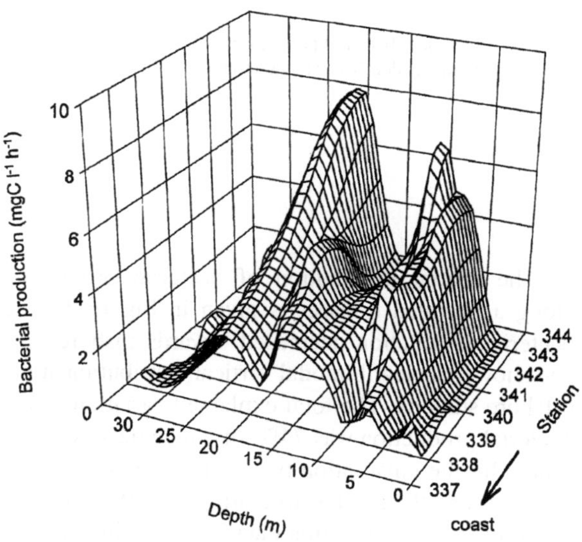

Figure 2. Spatial distribution of bacterial production along studied transect.

tively with the first PCA axis and the distance from the coast is related to the second PCA axis (See Figure 1). The depth was negatively correlated with such biological or related variables as temperature, phytoplankton primary production and chlorophyll concentration. The PCA shows that the inshore area is characterized by high total phosphorus and total nitrogen values.

The multiple regression analysis of microbiological parameters with environmental variables yielded significant results. After including the spatial structure into the model, only the relationships between the bacterial production and the environmental variables remained statistically significant ($p = 0.01$). All studied microbiological parameters were related to the spatial gradients, but with different patterns. The regression analysis of the bacterial total number constrained by 'space' matrix, independently of environmental variables, explained 30% of variability of this parameter. The analysis showed that the total bacterial number increased towards the inshore area and decreased with depth, but the depth effect was dominating. The spatial structure explained 27% of the variability of the plate count values, but in this case, the effects of depth and distance from the coast were equally important.

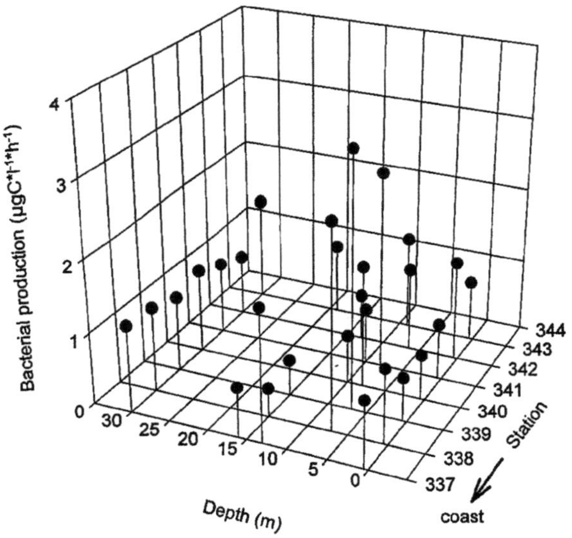

Figure 3. Fraction of bacterial production explained by environmental variables independently of spatial structure.

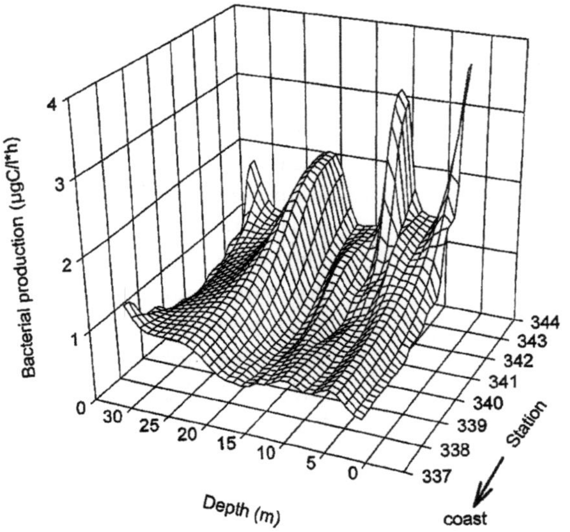

Figure 4. Map of bacterial production fraction explained by spatial structure independently of environmental variables.

The spatial distribution of bacterial production along a sampled transect is given in Figure 2. The bacterial production decreased towards inshore area and showed a clear bimodal vertical distribution at the offshore stations. The total explained variation of the bacterial production was 74%, of which 62% is represented by the covariation between the environmental variables and the spatial structure. 9% of the variation in the bacterial production data was explained by the environmental variables, independently of the spatial structure, while the variation due to the spatial structure without the environmental variables equaled 12%. The variables that were related to the variability of the bacterial production are salinity, chlorophyll concentration, total phosphorus and primary production.

Variation partitioning of the target variable allows us to visualize different fractions of variation. The graphs offer us an additional facility for interpreting the analysis results. Figure 3 shows that the values of high bacterial production, which were not related to the spatially structured environmental variability, were concentrated at the near coast stations in the horizon of chlorophyll maximum. The map of spatial fraction of bacterial production (See Figure 4) displays the increase in bacterial production starting below 15 m depth.

Discussion

Our results indicate that if the spatial structure is not partialled out, then the causality could falsely be attributed to the environmental variables, when in fact the correlation results from a common spatial structure present in both the dependent and the independent data sets. In our study this statement holds true in the case of the total bacterial number and the plate count. These two microbiological parameters were not truly related to the environmental variables which were included in our data set; although the usual regression analysis with environmental variables gave significant results.

When interpreting analysis results for the total bacterial number and the plate count, it should be taken in account that methods for determining these variables are not selective. The total bacterial number is derived from unspecific erythrozine staining technique, in which case only a minor and variable fraction of the total counts can be scored as living bacteria (Lovejoy et al., 1996; Zweil & Hagström, 1995). The obvious increasing trend in the plate count values towards the shore may reflect only the greater proportion of the aerobic heterotrophic bacteria in bacterioplankton, capable to grow on nutrient rich medium, in the nearshore stations. In coastal areas substrate sources for bacteria may come from allochthonous and autochthonous inputs, which means that not only substrate concentration, but also quality may change along the coast-offshore transect.

The incorporation of spatial structure into modeling of bacterial production shows that this process is, in great part, determined by the dynamics of spatially structured environmental variables. At the same time, the relationship between the bacterial production and the environmental variables could be the result of a strong dependence of the variability of the primary production on the spatially structured environmental variables (the spatially structured environmental variation equaled 79% in case of the primary production). It was revealed by the PCA analysis that the highest variation in the environmental data was related to depth.

Thus, the main constraint on bacterial production appeared to be the phytoplankton primary production. This result is supported by the findings of Baines & Pace (1991), who found that in coastal marine ecosystems the extracellular release of dissolved organic matter, as a good substrate for bacterioplankton growth, increases linearly with primary productivity, but is not related to the phytoplankton biomass. The increase in bacterial production with distance from the shore coincided with the decrease in the bacterial total counts. This result points to the difficulty in demonstrating positive correlation between the whole-community bacterioplankton production and abundance in the aquatic ecosystems. According to Letarte & Pinel-Alloul (1991), despite a relatively high abundance of small bacterioplankton cells, a few large bacteria, better adapted to nutrients that are relatively abundant at a time, could dominate production by reproducing at a higher rate and by producing much more biomass per cell.

As stated by Borcard & Legendre (1993), it is unclear how much of the spatial fraction is caused by the population or community dynamics, and how much is caused by the environmental variables; the biotic or abiotic, not measured in the study. In our case, the spatial fraction can be the reflection of the influence of the benthic processes, which were not captured by the environmental variables in present analysis, on pelagic microheterotrophic activity. The substrates for the bacterial growth change towards the deeper horizons. The predominantly living phytoplankton and their released products are replaced by sedimenting, partially decomposed organic matter. There is also a possibility that the spatial fraction reflects the effect of such biotic factor as predation.

Conclusions

We have shown that a strong relationship exists between the bacterial production and the spatially structured environmental variables, first of all, the phytoplankton primary production in the coastal marine ecosystem. The application of the variation partition technique was useful for revealing the variability patterns of the microbiological parameters, which otherwise could have remained hidden.

However, this raw data approach allows only to model trends. Therefore, a careful examination of these results could lay some groundwork for further analysis, perhaps using other statistical methods, in examining the ecosystem structure and the interrelations of processes.

Acknowledgements

This study was supported in part by grant from Norwegian Research Council. We thank U. Raudkivi for correcting English and two anonymous reviewers for helpful comments made on the manuscript.

References

Baines, S. B. & M. L. Pace, 1991. The production of dissolved organic matter by phytoplankton and its importance to bacteria: Patterns across marine and freshwater systems. Limnol. Oceanogr. 36: 1078–1090.

Borcard, D. & P. Legendre, 1993. Environmental control and spatial structure in ecological communities, with an example on Oribatid mites (Acari, Oribatei). J. envir. Stat. 1: 55–76.

Borcard, D. & P. Legendre, 1994. Environmental control and spatial structure in ecological communities: an example using Oribatid mites (*Acari, Oribatei*). Envir. ecol. Stat. 1: 31–61.

Borcard, D., P. Legendre & P. Drapeau, 1992. Partialling out the spatial component of ecological variation. Ecology 73: 1045–1055.

Dutilleul, P., 1993. Spatial heterogeneity and the design of ecological field experiments. Ecology 74: 1646–1658.

Edgar, R. K. & K. Laird, 1993. Computer simulation of error rates of Poisson-based interval estimates of plankton abundance. Hydrobiologia 264: 65–77.

Fuhrman, J. A. & F. Azam, 1982. Thymidine incorporation as a measure of heterotrophic bacterioplankton production in marine surface waters. Mar. Biol. 66: 109–120.

Guidelines for the Baltic monitoring programme for the third stage. 1988, Baltic Marine Environment Protection Commission. Helsinki Commission, 115 pp.

Holder-Franklin, M., 1992. Aquatic microorganisms: processes, populations, and molecular solutions to environmental problems. J. aquat Ecos. Health 1: 253–262.

Karrach, B., H.-G. Hoppe, S. Ullrich & S. Podewski, 1996. The role of mesoscale hydrography on microbial dynamics in the northeast Atlantic: Results of spring bloom experiment. J. mar. Res. 54: 99–122.

Kuznetsov, S. I. & G. A. Dubinina, 1989. Metody izuchenija vodnyh mikroorganizmov. Moskva, Nauka, 327 pp. [Methods for investigation of aquatic micro-organisms. In Russian.]

Legendre, P. & M. Troussellier, 1988. Aquatic heterotrophic bacteria: modeling in the presence of spatial autocorrelation. Limnol. Oceanogr. 33: 1055–1067.

Letarte, Y. & B. Pinel-Alloul, 1991. Relationships between bacterioplankton production and limnological variables: Necessity of bacterial size considerations. Limnol. Oceanogr. 36: 1208–1216.

Lovejoy, C., L. Legendre, B. Klein, J.-E. Tremblay, R. G. Ingram & J.-C. Therriault, 1996. Bacterial activity during early winter mixing (Gulf of St. Lawrence, Canada). Aquat. microb. Ecol. 10: 1–13.

Oppenheimer, C. H. & C. E. ZoBell, 1952. The growth and viability of sixty-three species of marine bacteria as influenced by hydrostatic pressure. J. mar. Res. 11: 10–18.

Riemann, B., P. K. Bjornsen, S. Newell & R. Fallon, 1987. Calculation of cell production of coastal bacteria based on measured incorporation of H-thymidine. Limnol. Oceanogr. 32: 471–476.

Roy, R., P. Legendre, R. Knowles & M. N. Charlton, 1994. Denitrification and methane production in sediment of Hamilton Harbour (Canada). Microb. Ecol. 27: 123–141.

Ter Braak, C. J. F., 1990. CANOCO- a FORTRAN program for canonical community ordination by [partial] [detrended] [canonical] correspondence analysis, principal components analysis and redundancy analysis. Scientia Publishing, 152 pp.

van Tongeren, O. F. R., 1995. Data analysis or simulation model: a critical evaluation of some methods. Ecol. Model. 78: 51–60.

White, P. A., J. Kalff, J. B. Rasmussen & J. M. Gasol, 1991. The effect of temperature and algal biomass on bacterial production and specific growth rate in freshwater and marine habitats. Microb. Ecol. 21: 99–118.

Zweil, U. L. & A. Hagström, 1995. Total counts of marine bacteria include a large fraction of non-nucleoid-containing bacteria (ghosts). Appl. envir. Microb. 61: 2180–2185.

Hydrobiologia **363:** 289–301, 1998.
T. Tamminen & H. Kuosa (eds), Eutrophication in Planktonic Ecosystems: Food Web Dynamics and Elemental Cycling.

Biomass and feeding activity of phagotrophic mixotrophs in the northwestern Black Sea during the summer 1995

T. Bouvier[1,2*], S. Becquevort[2] & C. Lancelot[2]
[1] *Laboratoire d'Hydrobiologie Marine et Continentale, UMR CNRS 5556, Université de Montpellier II, F-34095 Montpellier cedex 05, France*
[2] *Groupe de Microbiologie des Milieux Aquatiques, Université Libre de Bruxelles, Blvd du Triomphe, CP 221, 1050 Brussels, Belgium. E-mail: tbouvier@ulb.ac.be*
* *Address for correspondence*

Key words: Northwestern Black Sea, eutrophication, mixotrophy, ciliate, food web

Abstract

Biomass and activities of planktonic microorganisms (bacteria, nanoplankton and microplankton) were measured in the northwestern Black Sea during summer 1995. The method based on the uptake of fluorescently labeled prey was chosen to determine the ingestion rate of bacteria and nanoplankton by phagotrophic microorganisms. This method revealed the presence of mixotrophic organisms such as 'plastid-retaining ciliates' in the whole coastal area. Mixotrophic ciliates were dominated by micro-sized forms and maximum biomasses were recorded in the water masses characterised by low nutrient concentrations but high food particle concentrations. Mixotrophic nanoflagellates were absent and mixotrophic dinoflagellates were observed at one station only. Mixotrophic ciliates were shown to ingest preferably bacteria while mixotrophic dinoflagellates were grazing almost exclusively on nanoflagellates. Although the biomass of mixotrophic organisms were significantly lower than those of aplastidic protozoa, their feeding activity contributed to 14 and 24% of the ingestion of bacteria and nanoplankton, respectively. This is due to the high specific ingestion rate of mixotrophic micro-sized ciliates and dinoflagellates, which were two and three times higher, respectively, than the specific ingestion rate of bacteria and nanoplankton by aplastidic protozoa. This suggests a significant contribution of phagotrophic mixotrophs to the microbial network of the northwestern Black Sea.

Introduction

Mixotrophs are organisms that combine autotrophic and heterotrophic modes of nutrition (Gaines & Elbrächter, 1987; Sanders, 1991; Jones, 1994). Heterotrophic organisms may achieve photosynthesis through the sequestration of chloroplasts from ingested prey (Stoecker & Michaels, 1991; Laval-Peuto et al., 1986). On the other hand, autotrophic organisms like flagellates and dinoflagellates (Sander & Porter, 1988; Nygaard & Tobiesen, 1993; Bockstahler & Coats, 1993; Li et al., 1996) can sustain a phagotrophic (ingestion of entire prey) or an osmotrophic activity (ingestion of dissolved organic carbon). The competitive advantage of these organisms over strict autotrophs or heterotrophs rests in their ability to modify their nutritional mode in response to rapid changes in the environmental conditions such as light availability (Caron et al., 1993), inorganic and organic nutrient concentrations (Jones et al., 1994; Porter 1988; Sander et al., 1990), and food particle abundance (Sander et al., 1990; Caron et al., 1990).

Rapid changes in light and nutrient conditions are typical of the northwestern Black Sea submitted to the influence of the Danube river. On the long-term scale as well, the northwestern Black Sea has faced large ecological changes since 1960 as a consequence of modifications of environmental conditions driven by human activities (Mee, 1992; Gomoiu, 1992). The coastal environment has been enriched by terrestrial sources of inorganic nutrients and of dissolved and particulate organic matter delivered by the Danube river (Sapozhnicov, 1992), modifying the nutrient environment of ambient phytoplankton in both quantity and quality. Inorganic nutrients showed an excess of phosphorus over nitrogen with respect to phytoplankton requirements and over silicon with respect to diatom needs. Ratios of oxidised to reduced nitrogen and inorganic to organic nutrients were also modified with an increased presence of reduced and organic forms (Popa et al., 1985; Bodeanu, 1992). These changes in the coastal environment were stimulating, after 1970, the development of numerous phytoplankton blooms during summer, and generally the phytoplankton community was dominated by non-siliceous phytoplankton (Bodeanu, 1992). Among the bloom-forming species, the most frequently reported bloom-forming organism is the dinoflagellate *Prorocentrum* sp. causing blooms every summer since 1973. This primarily autotrophic dinoflagellate has been indeed documented as sustaining significant phagotrophy (Li et al., 1996). Besides this dinoflagellate, 5 to 30% (cell number) of the ciliate community, the dominant heterotrophic protozoa of the Black Sea, were reported as phagotrophic mixotrophs (plastid retaining ciliates) due to the presence of symbiotic algae in their body (Tumantseva, 1987). Among these, *Strombidium* (Blackbourn et al., 1973; Stoecker 1991), a member of the Strombidiidae family (Oligotrichida), has frequently been observed (Zaika & Averina, 1968). However, little is known of the abundance and biomass of mixotrophic ciliates in the Black Sea.

To which extent the recent occurrence of mixotrophs in the Black Sea is a direct consequence of eutrophication is not known yet. Their role in the ecological functioning of the Black Sea ecosystem has not been assessed due to the lack of quantitative information on their biomass and activity.

As a first step in this direction, the significance of phagotrophic mixotrophs in the microbial food web of the northwestern Black Sea was investigated during the EROS-2000 cruise of July 1995. Biomass and phagotrophic activity of mixotrophic organisms were determined over gradients of eutrophy-oligotrophy gradients crossing contrasting conditions of ambient light, nutrient concentration and prey abundance. Results are presented and discussed with respect to the biomass and phagotrophic activity of strictly heterotrophic protozoa (aplastidic organisms).

Material and methods

Study area, sampling and physico-chemical measurements

Field work was carried out during Leg 1 of the EROS2000-Black Sea expedition aboard RV 'Professor Vodyanitsky' from 17th July to 1st August 1995. The sampling grid (18 stations) crossed eutrophic-oligotrophic gradients in the coastal area influenced by the Danube river (Figure 1). At each station, seawater was collected at 6 different depths in the mixed layer, using 10 litres Go-Flo bottles mounted on a CTD (Mark III) rosette sampler. Incident PAR (Photosynthetically Active Radiation) was continuously recorded with a cosine Li-COR sensor set up on the upper deck of the ship. Vertical light attenuation and natural fluorescence were measured using the PNF300 underwater sensor (Biospherical instruments). Average ambient light in the upper layer was calculated from recorded incident PAR and vertical light attenuation using the equation of Riley (1957). Nitrate and nitrite concentrations were determined spectrophotometrically according to Grasshoff (1983).

Biomass of microorganisms

Biomass of microorganisms were determined by epifluorescence microscopy (Leitz, Laborlux D) after 4,6-diamidino-2-phenylindol (DAPI) staining. The experimental procedure was specific for each group of microorganisms studied:

Bacteria

After collection, seawater (5 ml) was fixed with 35% formalin (2%, final concentration) and stained with DAPI (0.01 μg cm^{-3}, final concentration) for 15 min, using the procedure of Porter & Feig (1980). Stained bacteria were collected by filtration onto 0.2 μm black polycarbonate filters (Nuclepore). Filters were mounted on slides and stored at +4 °C until examination. Bacteria were enumerated on a minimum of 20 fields at $\times$1250 magnification. Cell volume was calculated from the measurement of the shortest and longest axe, considering rods and cocci, respectively, as cylinders and spheres. Between 300–600 bacteria were measured for each sample. Biovolumes were converted to cell carbon, using the biovolume-dependent conversion factor established by Simon & Azam (1989) expressed as: $B = N\ V\ (130\ 10^{-15} + 350\ 10^{-15}\ e^{-V/0.0095})$ in which B is bacterial biomass; N is the bacterial number and V the bacterial volume.

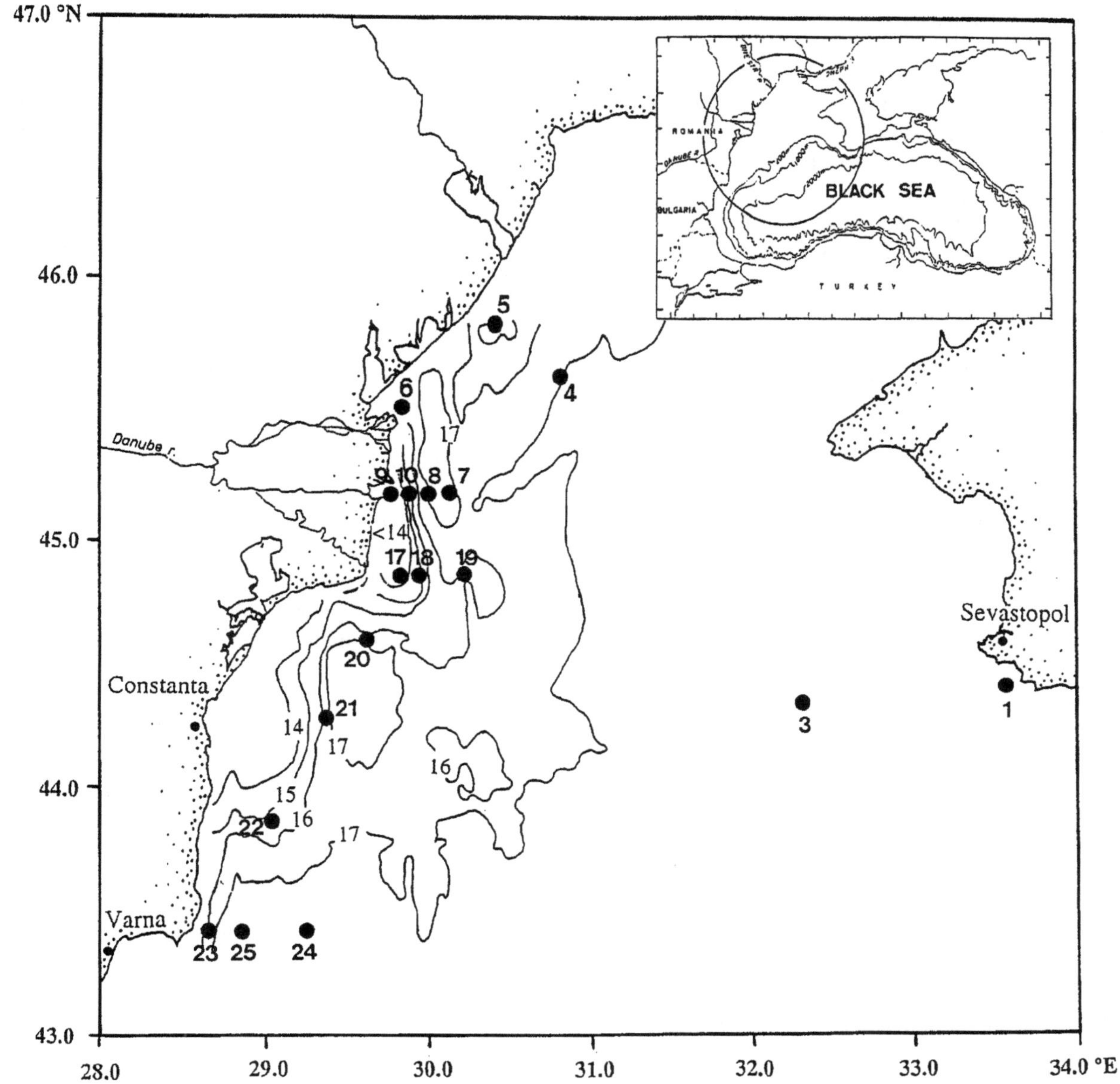

Figure 1. Map of the Black Sea indicating the salinity distribution (psu) in the surface mixed water layer (V. Egorov, pers. com., March 1996) and the geographical location of stations occupied during July, 1995 cruise.

Aplastidic protozoa

After collection, seawater (20 ml) was fixed with 25% glutaraldehyde (0.5%, final concentration) and stained for 15 minutes with DAPI (1 μg cm^{-3}, final concentration). Stained protozoa (ciliates, dinoflagellates and flagellates) were immediately collected by filtration on a 0.8 μm black polycarbonate filters (Nuclepore). Filters were mounted on slides and stored frozen until observation which took place within 2 days. Nano-sized (2–20 μm in diameter) protozoa were identified, counted and measured at a $\times$1250 and $\times$625 magnification while micro-sized (20–200μm in diameter) protozoa were observed at a $\times$125 magnification. For each size class, a minimum of 100 organisms per filter was counted. Plastidic (autotrophs) were distinguished from aplastidic (strict heterotrophs) organisms by the red autofluorescence of chlorophyll *a* observed under blue light excitation. Cell dimensions were estimated with an ocular micrometer. Cell biovolumes were calculated from measurement of dimensions and shapes of the cell. Recommended ratios to convert biovolume to biomass range between 0.08 and 0.22 pgC μm^{-3}

(Edler, 1979; Parsons et al., 1984; Børsheim & Bratbak, 1987; Choi & Stoecker, 1989) according to the taxon and the preservative and staining procedure. In this study, the conversion factor of 0.19 pgC μm^{-3} suggested by Putt & Stoecker (1989) for estimating the carbon biomass of ciliates fixed with Lugol's iodine was chosen, despite the fact that our samples were preserved with glutaraldehyde. A difference in cell shrinkage between fixation with Lugol's solution and glutaraldehyde (Choi & Stoecker, 1989; Wiackowski, 1994) has been observed. However, this difference becomes insignificant at the used concentration in glutaraldehyde (0.5%, f.c.) and the short fixation time (15 min.) (Choi & Stoecker, 1989; Børsheim & Bratbak, 1987). Biovolumes of flagellates and dinoflagellates were converted to organic carbon using the equation established by Smayda (1978), $\text{Log } B = 0.94 (\text{Log } V) - 0.6$, in which B is flagellate and dinoflagellate biomass and V flagellate and dinoflagellate volume.

Feeding activity

Ingestion rates of bacteria and nanoplankton by aplastidic protozoa were measured at 13 stations using a technique based on the uptake of monodispersed fluorescently labeled prey (Sherr et al., 1987a; Rublee & Gallegos, 1989). Fluorescently labeled prey were prepared prior to the cruise from microbial communities isolated from the coastal North Sea. Fluorescently Labeled Bacteria (FLB; mean volume of 0.08 μm^3, Equivalent Spherical Diameter, ESD: 0.55 ± 0.18 μm, $n = 100$) were prepared as described by Sherr et al. (1989). Fluorescently Labeled Algae (FLA) were obtained by the staining of cultivated *Phaeocystis* free-living cells (mean volume of 61.36 μm^3; ESD: 4.92 ± 0.41 μm, $n = 100$) using the procedure of Rublee & Gallegos (1989).

After collection, seawater was immediately prefiltered through a 200 μm mesh net to remove large zooplankton. Feeding experiments were performed in polycarbonate (Nalgene) bottles (500 ml) and run in the dark at *in situ* temperature under gentle shaking. Natural assemblages of protozoa were inoculated with tracer concentrations of FLA and FLB (10% of natural concentration of bacteria and nanoplankton) after a 30 minutes adaptation period to experimental conditions. Subsamples (6) were removed every 5 to 10 min during the one-hour incubation period. Feeding activity was stopped by the sequential addition of the three following preservatives: alkaline lugol (0.5%, final concentration), borate-buffered formalin (3%, final concentration) and a drop of sodium thiosulfate (Sherr et al., 1989). Fixed samples (20 ml) were stained with DAPI (1 μg cm^{-3}), and filtered through Nuclepore black polycarbonate membranes of 0.8 μm porosity. The filters were mounted on slides and stored at +4 °C until analysis.

The number of FLA and FLB ingested by ciliates, dinoflagellates and flagellates was determined by epifluorescence microscopy, using alternatively UV light and blue light (488 nm) excitation to identify protozoa and to determine the number of ingested FLA or FLB. Ingestion rates of bacteria and/or nanoplankton per organism were calculated from the linear slope of the time dependent curve of average FLB and FLA number ingested per organism and from the ambient concentration of bacteria and nanoplankton.

Biomass and activity of phagotrophic mixotrophs

The above method used for measuring feeding rate of aplastidic protozoa on bacteria and nanoplankton allowed to identify phagotrophic mixotrophic organisms and estimate their ingestion rate. Phagotrophic mixotrophs were thus defined as those microorganisms equipped with chloroplasts and having ingested FLB-bacteria and/or FLA-nanoplankton. Their abundance and biomass were determined from alkaline lugol-borate buffered formalin fixed samples under epifluorescence observations. Carbon biomass was calculated using the conversion factor chosen for calculating the biomass of aplastidic protozoa. Phagotrophic activity was estimated on the basis of the number of FLA and/or FLB ingested by mixotrophs during the incubation experiments described above.

Results

Environmental conditions

Experiments were conducted in the northwestern Black Sea, strongly influenced by the Danube river. The mixing of the Danube river with the marine waters (salinity: 19 psu) chiefly occurs at the river mouth (outer estuary). Accordingly, the investigated area crossed areas with salinity ranging between 9.37 and 19.3 psu (Table 1). Nitrate plus nitrite concentrations were the highest close to the river mouth (41.05 μm; Table 1), and decreased to phytoplankton growth-limiting values at salinities higher than 14 psu (L. Popa, A. Krastev, O. Ragueneau, J. Vervlimmeren, pers. com.). Ambi-

Table 1. Average depth-integrated salinity, photosynthetic active radiation (PAR) and nitrate plus nitrite concentrations in the upper mixed layer. N.D.: Not Determined

Stations	Salinity (psu)	PAR (μmol m^{-2} s^{-1})	NO_3+NO_2 (μmol l^{-1})
1	19.30	116	0.00
3	18.06	261	0.26
4	17.11	176	0.14
5	14.98	N.D.	0.25
6	15.34	114	0.20
7	17.50	N.D.	0.17
8	15.09	70	0.96
9	9.37	50	41.05
10	12.77	N.D.	15.32
17	12.55	N.D.	2.85
18	14.58	344	2.95
19	13.90	186	0.04
20	17.18	100	0.18
21	17.09	N.D.	0.78
22	16.59	120	0.00
23	15.05	N.D.	0.29
24	18.17	244	0.01
25	18.07	180	0.31

ent light in the surface layer varied between 50 and 344 μmol m^{-2} s^{-1} (Table 1) and was only exceptionally limiting photosynthetic activity (C. Lancelot & D. Van Eeckhout, unpublished data).

Biomass distribution of mixotrophic and aplastidic protozoa in relation with their potential prey.

Biomass of phagotrophic protozoa (mixotrophic and aplastidic protozoa) varied between 9.15 and 224.14 mg C m^{-3} in the investigated area of the northwestern Black Sea during the summer 1995 (Table 2). Phagotrophic mixotrophs were recorded in all investigated samples (Table 2). Their biomass ranged between 0.36 and 11.82 mg C m^{-3} and accounted for 1 to 15% of the total phagotrophic protozoan biomass. On an average, however, mixotrophic biomass was not very significant (4% of the total carbon-biomass of the phagotrophic organisms) except at two stations (stations 8 and 25) where the percentage of mixotrophic biomass reached 13 and 15%, respectively. Geographically with respect to the Danube river influence, biomass of aplastidic protozoa peaked around 13 psu salinity while the maximum carbon-biomass of phagotrophic mixotrophs was recorded in waters less influenced by the Danube river (15 psu) (Figure 2a–b).

The phagotrophic mixotroph community was composed of 'plastid retaining ciliates' and dinoflagellates (Table 3). No phagotrophic mixotrophy was evidenced among the flagellates. Phagotrophic mixotrophic dinoflagellates were, on the other hand, observed at one station only (station 25), south of the Danube river influence. At this station mixotrophic dinoflagellates were dominating the phagotrophic mixotrophic community. Their carbon biomass accounted for 82% of the total phagotrophic mixotrophic biomass. 'Plastid retaining ciliates' were observed at all stations. Their biomass varied between 0.36 and 11.82 mg C m^{-3} (Table 3). This group was dominating the mixotrophic community and was composed mainly of micro-sized cells. The biomass of nano-sized mixotrophs was low at all stations, varying between 0.04 and 0.27 mg C m^{-3} (Table 3). Nano- and micro-sized mixotrophic ciliates were not distributed similarly along a salinity gradient (Figure 2b). Maximum carbon biomass of micro-sized mixotrophic ciliates was observed at 15 psu salinity while the highest biomass of nano-sized mixotrophic ciliates was recorded in more marine waters (16–17 psu).

Nanoplankton (plastidic and aplastidic nano-cells) and bacteria are potential prey for mixotrophic and aplastidic protozoa. Relatively small cells with an average ESD of 5.77 ± 0.23 μm (Figure 3b) dominated the nanoplankton community. Their biomass ranged between 9.12 and 267.67 mg C m^{-3} (Table 2). Between 10 to 12 psu, about 85% of nanoplankton disappeared (Figure 2c) and was maintained at a low biomass of *ca.* 40 mg C m^{-3} at salinity above 12 psu. At a salinity of 15 psu (Station 5), however, a significant nanoplankton biomass was recorded. This maximum was mainly due to the development of *Prorocentrum* sp. (G. Ruta, pers. com.). Although this nano-dinoflagellate is currently reported as a mixotrophic organism in the literature (Li et al., 1996), our own feeding experiments did not detect any particle ingestion by *Prorocentrum* sp., suggesting that conditions for developing *Prorocentrum* phagotrophy were not fulfilled.

The average size of bacteria was significantly lower than 1 μm (0.60 ± 0.01 μm; Figure 3a). Bacterial biomass ranged from 14.20 to 53.94 mg C m^{-3} (Table 2). As a general trend biomass decreased along the salinity gradient (Figure 2d). However, significant accumulations of bacterial biomass were recorded between 14 and 15 psu (up to 53.94 mg C m^{-3}) and between 15 and 17 psu of salinity (up to 48.03 mg C m^{-3}).

Table 2. Average depth-integrated biomass of aplastidic (ciliates, dinoflagellates and flagellates) and mixotrophic (ciliates and dinoflagellates) protozoa, their total, and their potential prey (bacteria and nanoplankton). N.D.: Not Determined

Stations	Aplastidic sp. (mg C m^{-3})	Mixotrophs (mg C m^{-3})	Total Protozoa (mg C m^{-3})	Bacteria (mg C m^{-3})	Nanoplankton (mg C m^{-3})
1	24.10	0.79	24.89	19.66	33.10
3	66.79	N.D.	N.D.	19.20	30.79
4	8.12	1.03	9.15	18.66	9.12
5	37.20	N.D.	N.D.	53.94	135.20
6	111.94	N.D.	N.D.	22.00	24.94
7	97.78	1.19	98.97	23.70	55.78
8	82.35	11.82	94.17	20.13	18.35
9	134.67	5.04	139.71	39.83	267.67
10	223.78	0.36	224.14	27.36	58.78
17	104.02	2.20	106.22	32.76	64.02
18	124.29	2.10	126.39	39.66	50.29
19	33.00	N.D.	N.D.	29.06	61.00
20	11.11	0.76	11.87	20.34	20.11
21	12.41	N.D.	N.D.	26.67	49.41
22	128.23	1.48	129.71	39.15	57.23
23	80.65	5.32	85.97	48.03	49.65
24	46.00	0.54	46.54	14.18	37.00
25	66.23	11.77	78.00	18.30	52.23

Table 3. Biomass of nano- and micro-sized of mixotrophic ciliates and mixotrophic dinoflagellates. N.D.: Not Determined; –: Undetectable

	Mixotrophic ciliates			Mixotrophic dinoflagellates		
Stations	Nano-sized (mg C m^{-3})	Micro-sized (mg C m^{-3})	Total (mg C m^{-3})	Nano-sized (mg C m^{-3})	Micro-sized (mg C m^{-3})	Total (mg C m^{-3})
1	0.17	0.62	0.79	–	–	–
3	N.D.	N.D.	N.D.	N.D.	N.D.	N.D.
4	0.27	0.76	1.03	–	–	–
5	N.D.	N.D.	N.D.	N.D.	N.D.	N.D.
6	N.D.	N.D.	N.D.	N.D.	N.D.	N.D.
7	0.05	1.14	1.19	–	–	–
8	0.05	11.77	11.82	–	–	–
9	0.04	5.00	5.04	–	–	–
10	0.09	0.26	0.36	–	–	–
17	0.09	2.11	2.20	–	–	–
18	0.19	1.91	2.10	–	–	–
19	N.D.	N.D.	N.D.	N.D.	N.D.	N.D.
20	0.22	0.54	0.76	–	–	–
21	N.D.	N.D.	N.D.	N.D.	N.D.	N.D.
22	0.26	1.21	1.48	–	–	–
23	0.04	5.27	5.32	–	–	–
24	0.06	0.48	0.54	–	–	–
25	0.06	2.07	2.13	0.51	9.13	9.64

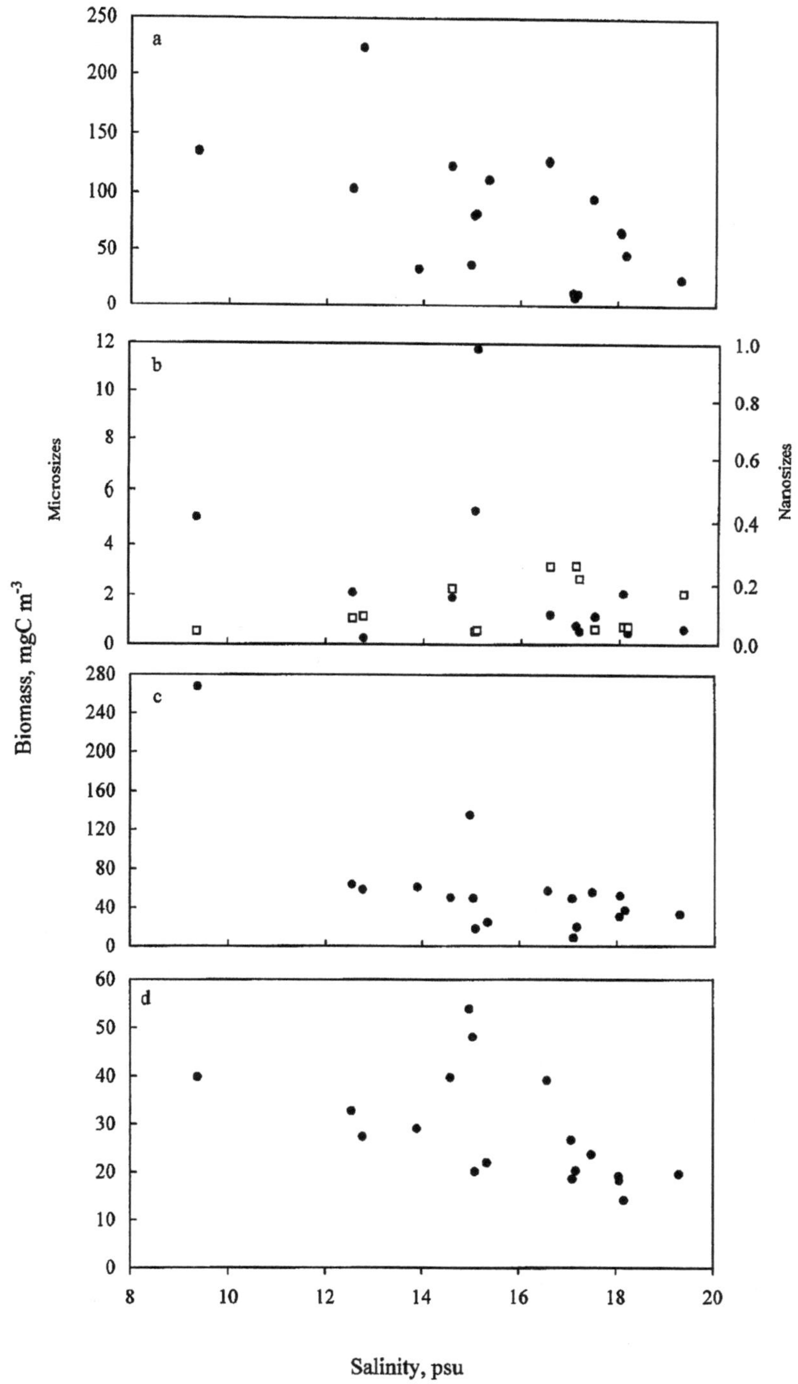

Figure 2. Microbial community biomass distribution along the salinity horizontal profile: (a) aplastidic protozoa (Ciliate, Dinoflagellate and flagellate); (b) mixotrophs, • micro-size and □ nano-sized; (c) nanoplankton and (d) bacteria.

Phagotrophic activity of mixotrophic organisms

Some authors (Gonzalez et al., 1990) reported that the method based on the addition of fluorescently labelled prey could induce under- or over-estimate of the ingestion rates by size selectivity owing to a mismatch between the size of natural and artificial prey. The high similarity observed between the size of natural

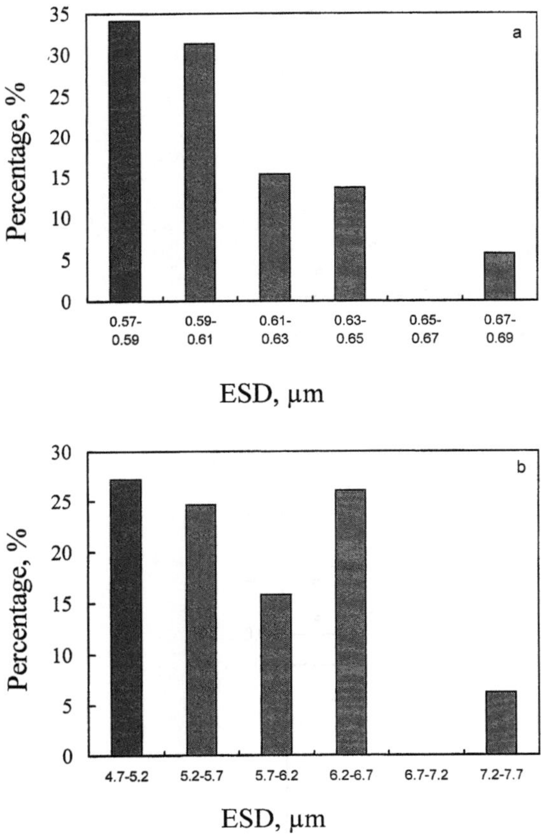

Figure 3. Size frequency distributions for bacterioplankton (a) and nanoplankton (b) from northwestern Black Sea (July 1995).

(bacteria and nanoplankton) and artificial (FLB and FLA) prey allows us to exclude size selective (positive or negative) ingestion induced by the addition of fluorescently labelled prey.

Specific ingestion rates of bacteria and nanoplankton by aplastidic and mixotrophic protozoa are summarised in Table 4. Diet greatly differed between ciliates and dinoflagellates. The latter were feeding almost exclusively on nanoplankton while ciliates were ingesting both bacteria and nanoplankton (Table 4). Interestingly enough, specific ingestion rates of nanoplankton by aplastidic ciliates and by mixotrophic nano- and micro-sized ciliates were almost identical, whereas specific ingestion of bacteria by mixotrophic micro-sized ciliates was more than two times higher than that of aplastidic ciliates.

In the nano-sized class, aplastidic ciliates were ingesting at higher rate than mixotrophs. The specific ingestion rate of nanoplankton by mixotrophic dinoflagellates was about three times higher than that by aplastidic dinoflagellates. Specific ingestion rates of nanoplankton by mixotrophic ciliates and dinoflagellates were identical, whereas a factor 3 was measured between those of aplastidic ciliates and aplastidic dinoflagellates (Table 4).

Mixotrophic daily ingestion rates of bacteria and nanoplankton ranged between 0.1–3.74 and 0.03–2.11 $mg\,C\,m^{-3}\,d^{-1}$, respectively. These were generally significantly lower than those estimated for aplastidic protozoa (Table 5a–b). Higher mixotrophic activity with respect to that of aplastidic protozoa was recorded in areas far of Danube river influence (stations 1, 18, 20, 24; salinity 19.3, 14.6, 17.2, 17.2 psu). On an average, summer daily mixotrophs activity contributed to 14 and 24% of total ingestion of bacteria and nanoplankton, respectively.

Discussion

This study reports the first measurements of the phagotrophic activity of mixotrophic microorganisms in the northwestern Black Sea during the summer period. The summer presence of mixotrophs in this coastal system is a well-known phenomenon since 1970 (Bodeanu, 1992; Tumantseva, 1987; Zaika & Averina, 1968). However, their feeding activity has not been measured yet. In this paper, the phagotrophic activity of mixotrophic organisms was measured using the method based on the use of fluorescently labeled prey (bacteria and nanoplankton) as first developed by Sherr et al. (1987) and Rublee & Gallegos (1989). This elegant method allows at the same time the recognition of phagotrophic mixotrophs and the measurement of their specific ingestion rate. Basically the method requires the use of fluorescently labeled, heat-killed bacteria and nanoplankton that mimic the size of ambient prey. In our experiments, the average size (ESD) of labeled bacteria and nanoplankton was 0.55 ± 0.18 and 4.92 ± 0.41 μm, respectively, i.e. quite comparable to that of ambient prey: 0.60 ± 0.01 μm and 5.77 ± 0.23 μm for bacterioplankton and nanoplankton, respectively.

These experiments revealed the presence of phagotrophic ciliates in the whole investigated area whereas mixotrophic nanoflagellates were absent and mixotrophic dinoflagellates were recorded at one station only. To which extent this absence or rarity could be the result of experimental biases is not known. It has been recently argued (Riemann, 1995) that the resolution of epifluorescence microscopy is currently

Table 4. Specific ingestion rate of bacteria and nanoplankton by aplastidic and mixotrophic ciliates and dinoflagellates

	Trophic prey	Specific ingestion rate in % of cell carbon per day (d^{-1})			
		Aplastidic		Mixotrophs	
		bacteria mean (S.D.[1])	Nanoplankton mean (S.D.[1])	bacteria mean (S.D.[1])	Nanoplankton mean (S.D.[1])
Ciliates	Nano-sized	2.37 (2.22)	0.71 (0.72)	0.90 (0.75)	0.72 (0.97)
	Micro-sized	0.15 (0.10)	0.23 (0.15)	0.36 (0.47)	0.22 (0.22)
		n : 13	n : 13	n : 13	n : 13
Dinoflagellates		2.10^{-4} (10^{-5})	0.08 (0.01)	0.001	0.21
		n : 2	n : 2	n : 1	n : 1

([1]) S.D. = Standard Deviation.
n : Experience number.

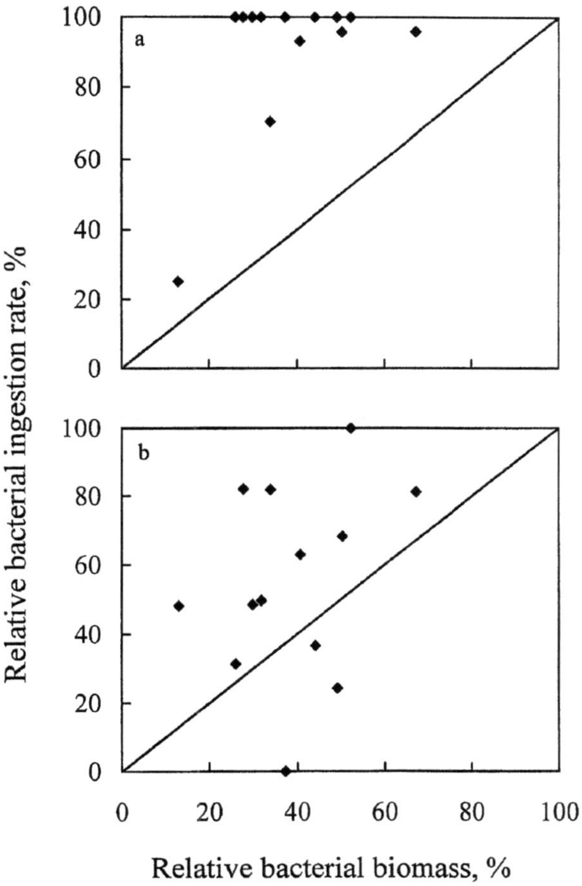

Figure 4. Relative ingestion rate of bacteria (the ratio of bacterial ingestion rate to the ingestion rate of bacteria and nanoplankton) by nano- (a) and micro-sized (b) mixotrophic ciliates in function of the relative bacterial biomass (the ratio of bacterial biomass to the biomass of bacteria and nanoplankton). The linear plot 1:1 corresponds to an absence of food preference between the prey.

too low to properly detect ingested fluorescent prey within pigmented nanoflagellate cells. Moreover, Sieracki et al. (1987) gave evidence that preservation of samples with formalin could induce some egestion of ingested particles by flagellates. Furthermore, experimental evidence has recently suggested that heat-killed prey are not ingested as readily as live prey by some dinoflagellates (*Ceratium* sp., *Prorocentrum* sp.; Li et al., 1996), suggesting that the importance of mixotrophic dinoflagellates could have been underestimated in the studied area.

Summarising, the chosen experimental procedure could have underestimated both the biomass and ingestion rate of mixotrophic dinoflagellates and nanoflagellates in the studied area, although the methods should be quite accurate to measure the biomass and activity of mixotrophic ciliates. Characteristics of ciliate mixotrophy are thus discussed in more detail.

Mixotrophic ciliates were recorded in the whole investigated area of the northwestern Black Sea and appeared to be a common component of the summer ciliate population. Accordingly, mixotrophy by ciliates has been reported in the literature for various marine systems including the Pacific Ocean (Blackbourn et al., 1973; Stoecker et al., 1996), Atlantic Ocean (Jonsson, 1987) and Mediterranean Sea (Rassoulzadegan, 1977; Laval-Peuto & Rassoulzadegan, 1988), where they represent a significant part of the ciliate population (Laval-Peuto & Rassoulzadegan, 1988).

In the northwestern Black Sea, however, the biomass of mixotrophs was not very significant, representing an average of 4% of the biomass of the whole mixotrophic community (95% carbon-biomass of the mixotrophic community was ciliates, Table 3). On

Table 5. Daily ingestion rates in mixed water layer for aplastidic and mixotrophic (a) bacterivorous and (b) nanoplanktovorous. N.D.: Not Determined; –: Undetectable

A

Stations	Ciliate aplastidic ($mg\ C\ m^{-3}\ d^{-1}$)	Mixotrophs ($mg\ C\ m^{-3}\ d^{-1}$)	Flagellate aplastidic ($mg\ C\ m^{-3}\ d^{-1}$)	Dinoflagellate aplastidic ($mg\ C\ m^{-3}\ d^{-1}$)	Mixotrophs ($mg\ C\ m^{-3}\ d^{-1}$)
1	0.07	0.12	–	–	–
3	N.D.	N.D.	N.D.	N.D.	N.D.
4	0.80	0.86	–	–	–
5	N.D.	N.D.	N.D.	N.D.	N.D.
6	N.D.	N.D.	N.D.	N.D.	N.D.
7	0.37	0.15	–	–	–
8	4.76	0.11	2.85	0.00	–
9	3.53	0.90	–	–	–
10	2.62	0.10	0.79	–	–
17	2.29	1.22	12.10	–	–
18	1.07	0.69	4.09	–	–
19	N.D.	N.D.	N.D.	N.D.	N.D.
20	0.16	3.74	0.00	–	–
21	N.D.	N.D.	N.D.	N.D.	N.D.
22	1.45	1.16	0.50	–	–
23	1.52	0.71	0.11	–	–
24	0.28	0.28	–	–	–
25	0.17	0.12	–	0.00	0.01

B

Stations	Ciliate aplastidic ($mg\ C\ m^{-3}\ d^{-1}$)	Mixotrophs ($mg\ C\ m^{-3}\ d^{-1}$)	Dinoflagellate aplastidic ($mg\ C\ m^{-3}\ d^{-1}$)	Mixotrophs ($mg\ C\ m^{-3}\ d^{-1}$)
1	0.17	0.03	–	–
3	N.D.	N.D.	N.D.	N.D.
4	0.67	0.11	–	–
5	N.D.	N.D.	N.D.	N.D.
6	N.D.	N.D.	N.D.	N.D.
7	0.27	0.14	–	–
8	4.27	0.00	1.76	–
9	3.60	1.03	–	–
10	4.69	0.06	–	–
17	1.52	0.28	–	–
18	0.24	0.76	–	–
19	N.D.	N.D.	N.D.	N.D.
20	0.09	0.57	–	–
21	N.D.	N.D.	N.D.	N.D.
22	0.43	0.30	–	–
23	1.47	2.11	–	–
24	0.12	0.06	–	–
25	0.95	0.22	1.40	2.04

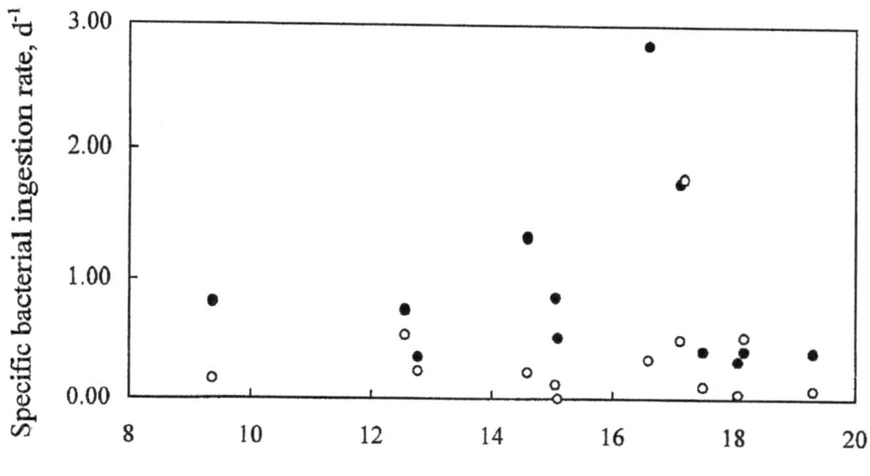

Figure 5. Specific ingestion rate of bacteria by nano- mixotrophic ciliates (•) and micro- mixotrophic ciliates (○) along the salinity horizontal profile.

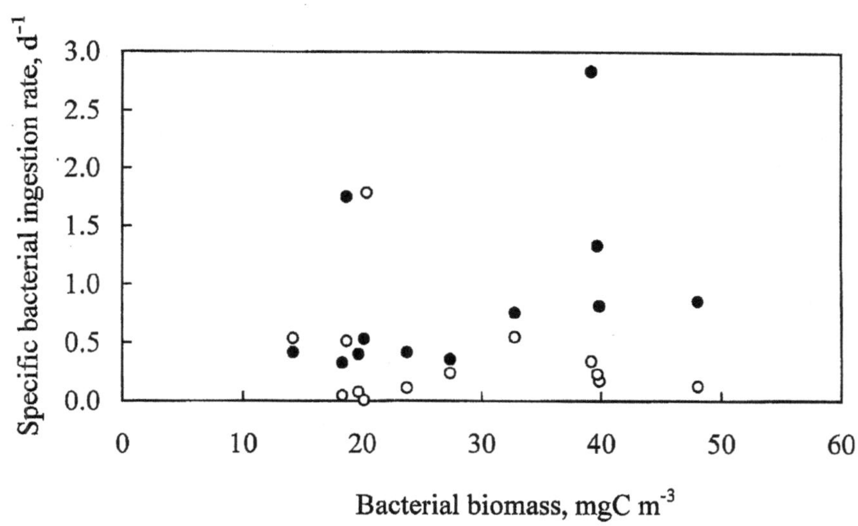

Figure 6. Specific ingestion rate of nano- (•) and micro-sized (○) mixotrophic ciliates in response of bacterial biomass.

the contrary, the feeding activity of mixotrophs was much more important, contributing to 14 and 24% of the mean ingestion rate of bacteria and nanoplankton, respectively, by the whole phagotrophic community. The mixotrophic ciliate population was mainly composed of micro-sized ciliates (71–99% carbon-biomass; Table 3). Individuals of both nano-sized and micro-sized mixotrophic ciliates were observed to ingest both bacteria and nanoplankton. However, as expected (Sherr et al., 1987), nano-sized mixotrophic ciliates exhibited a marked food preference for bacteria (Figure 4a). Food preference was estimated by relating the relative ingestion rate of bacteria to the relative prey biomass. In addition more than 60% of the nano-sized mixotrophic ciliates were feeding on the only bacteria (Figure 4a). On the other hand, food preference of the micro-sized mixotrophic ciliates was less distinct. Although the majority of the investigated populations exhibited a marked preference for bacteria, the populations were also feeding on nanoplankton (Figure 4b). Among them one was grazing exclusively on nanoplankton. This food preference of micro-sized ciliates for bacteria is surprising considering the small size of ambient bacteria (ESD less than 1 μm, Figure 3b) and other studies reporting a food preference for prey between 2 and 8 μm for micro-sized ciliates

(Jonsson, 1986; Rassoulzadegan et al., 1988; Bernard & Rassoulzadegan, 1990).

Protozoan ingestion rate has been shown to depend on the prey abundance following a hyperbolic function (Holling type II model, Fenchel, 1982, 1987). Although no specific food-enrichment experiments were conducted during cruise, the comparison between the distribution along salinity gradient of the specific bacterial ingestion rate of nano- and micro-sized mixotrophic ciliates with the biomass distribution of their favourite prey (Figure 5, Figure 2d) gave some indication that the feeding activity of the nano- and micro-sized mixotrophic ciliates would be positively correlated with bacterial biomass. However, no hyperbolic kinetics were found between the feeding activity of mixotrophic ciliates -either nano or micro-sized- and the bacterial biomass (Figure 6). This lack of correlation suggests the influence of other controlling factors. Light or the availability of inorganic and organic nutrients have been reported to control the balance between autotrophic-heterotrophic nutrition modes of pelagic mixotrophic organisms (Riemann, 1995). However the sampled stations – excepted in the close vicinity to the Danube mouth – were all characterised by a depletion of nitrate and the absence of physiological limitation by light. Other possible explanations are the variability of prey sizes distributions between different sampling locations, or the control of protozoan populations by metazooplankton grazing. The latter could regulate biomass and species composition and explain the variability observed in the specific ingestion rate in response to food biomass.

Conclusion

Despite their low biomass (4% of the total carbon-biomass of the phagotrophic organisms), phagotrophic mixotrophic organisms have been shown to contribute to an average of 14% and 24% of the feeding activity of phagotrophic protozoa due to their generally higher specific ingestion rate compared to that of strictly heterotrophic micro-grazers. The mixotrophic community was dominated by micro-sized ciliates; their specific ingestion rate on bacteria was two times higher than that of micro-sized aplastidic ciliates. Mixotrophic dinoflagellates grazing almost exclusively on nanoplankton appeared to be an important mixotrophic compartment as well, being characterised by a specific ingestion rate three times higher than that of aplastidic dinoflagellates. Considering the methodological limitations related to the use of fluorescently labeled prey for determining the phagotrophic activity of mixotrophs, it seems likely that the role of phagotrophic mixotrophs in the northwestern Black Sea during the summer period has been underestimated. Further work will seek to improve the resolution of current FLA/FLB methods to properly determine the phagotrophic activity of mixotrophs. When combined with simultaneous measurements of mixotrophs osmotrophic and photosynthetic activities, these measurements will allow to assess the importance of mixotrophs in the microbial food web as well as their controlling factors in the northwestern Black Sea.

Acknowledgements

We are grateful to two anonymous referees for their constructive remarks. We thank the crew of RV 'Dr Vodyanitsky' for their co-operation. We thank Pierre Duponcheel for providing nitrite and nitrate data. Special thanks are due to Denis Van Eeckhout for his comments on this manuscrit. This work is contribution N° 60 to the EC ELOISE Programme in the framework of EROS 2000/EROS 21 – Black sea projects funded by the Environment & Climate Programme of the European Commission under contract N° EV5V-CT940501 and ENV4-CT96-0219. T. Bouvier has been funded by a sectorial EC grant (contract N° EV5V–CT 94–5236).

References

Bernard, C. & F. Rassoulzadegan, 1990. Bacteria or microflagellates as a major food source for marine ciliates: possible implications for the microzooplankton. Mar. Ecol. Prog. Ser. 64: 147–155.

Blackbourn, D. J., F. J. K. Taylor & J. Blackbourn, 1973. Foreign organelle retention by ciliates. J. Protozool. 20: 286.

Bockstahler, K. R. & D. W. Coats, 1993. Spatial and temporal aspects of mixotrophy in Chesapeake Bay dinoflagellates. J. Euk. Microbiol. 40: 49–60.

Bodeanu, N., 1992. Algal blooms and development of the main phytoplanktonic species at the Romania Blak Sea littoral in conditions of intensification of the eutrophication process. In R. A. Vollenweider, R. Marchetti & R. Viviani (eds), Marin Coastal Eutrophication. Elsevier: 981–906.

Børsheim, K. Y. & G. Bratbak, 1987. Cell volume to cell carbon conversion factors for a bacterivorous *Monas* sp. enriched from seawater. Mar. Ecol. Prog. Ser. 36: 171–175.

Caron, A. C., K. G. Porter & R. W. Sanders, 1990. Carbon, nitrogen, and phosphorus budgets for the mixotrophic phytoflagellate *Poterioochromonas malhamensis* (Chrysophyceae) during bacterial ingestion. Limnol. Oceanogr. 35: 433–443.

Caron, D. A., R. W. Sander, E. L. Lim, C. Marrasé, L. A. Amaral, S. Whitney, R. B. Aoki & K. G. Porter, 1993. Light-dependent phagotrophy in the freshwater mixotropic chrysophyte Dinobryon cylindricum. Microb. Ecol. 25: 93–111.

Choi, J. W. & D. K. Stoecker, 1989. Effects of fixation on cell volume of marine planktonic Protozoa. Appl. Envir. Microbiol. 55: 1761–1765.

Edler, L., 1979. Recommendations for marine biological studies in the Baltic Sea. Baltic Mar. Biol. 4: 1–38.

Fenchel, T., 1982. Ecology of heterotrophic microflagellates II. Bioenergetics and growth. Mar. Ecol. Prog. Ser. 8: 225–231.

Fenchel, T., 1987. Flows of energy and materials. In O. Kinne (ed.), Ecology-Potentials and Limitations, 54 pp.

Gaines, G. & M. Elbrächter, 1987. Heterotrophic nutrition. In F. J. R. Taylor (ed.), The Biology of the Dinoflagellates. Oxford, Blackwell Scientific Publications: 224–268.

Gomoiu, M. T., 1992. Marine eutrophication syndrome in the northwestern part of the Black Sea. In R. A. Vollenweider, R. Marchetti & R. Viviani (eds), Marin Coastal Eutrophication, Elsevier: 683–692.

Gonzalez, J. M., E. B. Sherr & B. F. Sherr, 1990. Size-selective grazing by natural assemblages of estuarine flagellates and ciliates. Appl. Envir. Microbiol. 56: 583–589.

Grasshoff, K., 1983. Determination of nitrate. In K. Grasshoff, M. Ehrhardt & K. Kremling (eds), Methods of Seawater Analysis. Verlag Chemie. Basel: 143–150.

Jones, H. L. J., B. S. C. Leadbeater & J. C. Green, 1994. Mixotrophy in haptophytes. In J. C. Green & B. S. C. Leadbeater (eds). The Haptophyte Algae. Clarendon Press, Oxford 51: 247–263.

Jonsson, P. R., 1986. Particle size selection, feeding rates and grotwth dynamics of marine planktonic oligotrichous ciliates (Ciliophora: Oligotrichina). Mar. Ecol. Prog. Ser. 33: 265–277.

Jonsson, P. R., 1987. Photosynthetic assimilation of inorganic carbon in marine oligotrich ciliates (Ciliophora, Oligotrichina). Mar. Microb. Food Webs 2: 55–68.

Laval-Peuto, M. & F. Rassoulzadegan, 1988. Autofluorescence of marine planktonic Oligotrichina and other ciliates. Hydrobiologia 159: 99–110.

Laval-Peuto, M., P. Salvano, P. Gayol & C. Greuet, 1986. Mixotrophy in marine planktonic ciliates: ultrastructural study of *Tontonia appendiculariformis* (Ciliophora Oligotrichina. Mar. Microb. Food Webs 1: 81–104.

Li, A., D. K. Stoecker, D. W. Coats & E. J. Adam, 1996. Ingestion of fluorescently labeled and phycoerythrin-containing prey by mixotrophic dinoflagellates. Aquat. Microb. Ecol. 10: 139–147.

Mee, L. D., 1992. The Black Sea in crisis: a need for concerted international action, AMBIO 21(4): 278–286.

Nyggard, K. & A. Tobiesen, 1993. Bacterivory in algae: A survival strategy during nutrient limitation. Limnol. Oceanogr. 38: 273–279.

Parsons, T. R., Y. Maita & C. M. Lalli, 1984. A Manual of Chemical and Biological Mathods for Seawater Analysis. Pergamon Press, Oxford: 105–109.

Popa, A., A. Cociasu, L. Popa, I. Voinescu & L. Dorogan, 1985. Long-term statistical characteristics of several physico-chemical parameters of the near shore waters in Constantza zone. Cercetari mar. 18: 7–51.

Porter, K. G., 1988. Phagotrophic phytoflagellates in microbial food webs. Hydrobiologia 159: 89–97.

Porter, K. G. & Y. S. Feig, 1980. The use of DAPI for identifying and counting aquatic microflora. Limnol. Oceanogr. 25: 943–948.

Putt, M. & D. K. Stoecker, 1989. An experimentally determined carbon: volume ratio for marine 'oligotrichous' ciliates from estuarine and coastal waters. Limnol. Oceanogr. 34: 1097–1103.

Rassoulzadegan, F., 1977. Evolution annuelle des ciliés pélagiques en Méditerranee nordoccidentale. I. Cilies oligotriches (non-tintinnides) (Oligotrichina). Ann. Inst. Oceanogr. Paris 53: 125–134.

Rassoulzadegan, F., M. Laval-Peuto & R. W. Sheldon, 1988. Partitioning of the food ration of marine ciliates between pico- and nanoplankton. Hydrobiologia 159: 75–88.

Riemann, B., H. Havskum, F. Thingstad & C. Bernard, 1995. The role of mixotrophy in pelagic environments. In I. Joint (ed.). Molecular Ecology of Aquatic Microbes. NATO ASI Series g 38: 87–114.

Riley, A., 1957. Phytoplankton of the north central Sargasso Sea, 1950–1952. Limnol. Oceanogr. 2: 252–270.

Rublee, P. A. & C. L. Gallegos, 1989. Use of fluorescently labelled algae (FLA) to estimate microzooplancton grazing. Mar. Ecol. Prog. Ser. 51: 221–227.

Sanders, R. W., 1991. Mixotrophic protists in marine and freswater ecosystems. J. Protozoology 38: 76–81.

Sanders, R. W., K. G. Porter & D. A. Caron, 1990. Relationship between phototrophy and phagotrophy in the mixotrophic chrysophyte Poterioochromonas malhamensis. Microb. Ecol. 19: 97–109.

Sanders, R. W. & K. G. Porter, 1988. Phagotrophic phytoflagelattes. Adv. Microb. Ecol. 10: 167–192.

Sapozhnicov, V. V., 1992. Biohydrochemical causes of the changes of Black Sea ecosystem and its present condition. GeoJ. 27(2): 149–157.

Sherr, B. F., E. B. Sherr & R. D. Fallon, 1987a. Use of monodispersed fluorescently labeled bacteria to estimate in situ protozoan bacterivory. Appl. Envir. Microbiol. 53: 958–965.

Sherr, F. B., E. B. Sherr & C. Pedros-Alio, 1989. Silmutaneous measurement of bacterioplankton production and protozoan bacterivory in estuarine water. Mar. Ecol. Prog. Ser. 54: 209–219.

Sieracki, M. E., L. W. Haas, D. A. Caron & E. J. Lessard, 1987. Effect of fixation on particle retention by microflagellates: underestimation of grazing rates. Mar. Ecol. Prog. Ser. 38: 251–258.

Simon, M. & F. Azam, 1989. Protein content and protein synthesis rates of planktonic marine bacteria. Mar. Ecol. Prog. Ser. 51: 201–213.

Smayda, 1978. From phytoplankton to biomass. Monographs on oceanographic methodology, UNESCO 6: 273–279.

Stoecker, D. K., D. E. Gustafson & P. G. Verity, 1996. Micro- and mesoprotozooplankton at 140°W in the equatorial Pacific: heterotrophs and mixotrophs. Aquat. Microb. Ecol. 10: 273–282.

Stoecker, D. K., 1991. Mixotrophy in marine planktonic ciliates: physiological and ecological aspects of plastid retention by oligotrichs. In P. C. Reid, C. M. Turley & H. Burkill (eds). Protozoa and their Role in Marine Processes. Berlin, Heidelberg, NATO ASI series, Springer-Verlag 25: 161–179.

Tumantseva, N. I., 1987. Quantitative characteristics of protozinal plankton in the Black Sea in the spring of 1984. Modern condition of the Black Sea ecosystem. Nauka: 133–138.

Wiackowski, K., A. Doniec & J. Fyda, 1994. An empirical study of the effect of fixation on ciliate cell volume. Mar. Microb. Food Webs 8: 59–69.

Zaika, V. Ye. & T. Yu. Averina, 1968. Proportions of infusoria in the plakton of Stevastopol Bay, Black Sea. Oceanoolgiya 8: 843–.

Hydrobiologia **363:** 303–307, 1998.
T. Tamminen & H. Kuosa (eds), Eutrophication in Planktonic Ecosystems: Food Web Dynamics and Elemental Cycling.

Spatial structure and ecological efficiency in the summer zooplankton of a glacial lake

Galina A. Galkovskaya & Vadim V. Arapov
Institute of Zoology, Academy of Sciences of Belarus, Scorina St. 27, Minsk 220072, Belarus

Key words: zooplankton, spatial structure, ecological efficiency, weather conditions

Abstract

In June–July of 1993 and 1995, pelagic zooplankton dynamics of a glacial lake was investigated. In those years weather conditions difffered substantially. This resulted not only in a great difference in the water surface temperature (in 1993, 17.8 ± 0.47 °C; in 1995, 20.4 ± 0.33 °C), but also in an increase in the depth of metalimnion zone in 1995 in comparison with 1993. The ecological efficiency (K_e), the ratio of secondary to primary consumer productions, was 0.28 ± 0.031 in 1993 and 0.21 ± 0.034 in 1995. The higher K_e was accompanied by lowering of spatial structurization, primary and secondary consumer production.

Introduction

The trophic structure of glacial lakes is rather complicated and species diversity is much higher in comparison with shallow lakes (Yodris, 1981; Sprules & Bowerman, 1988). Since glacial lakes are evolutionally old, great species diversity of pelagic plankton and conservation of the relict complex may well be a consequence of a well developed internal stabilization system of the community.

The goal of the work was to compare pelagic zooplankton of a stratified glacial lake over the years that substantially differ in weather condition by studying vertical distribution of the species and by determining the ecological efficiency as the ratio of secondary to primary consumer production.

Materials and methods

South Voloso lake (55° 37′ N, 27° 08′ E, area 1.2 km^2, max. depth 40.4 m) of glacial origin is a mesotrophic lake with some oligotrophic traits (Yakushko, 1988).

Zooplankton samples were collected both with a plankton sampler (mesh size 100 μm, volume 10 litre) and with a Ruttner sampler (volume 1 litre) at the deepest station of the lake. The samples were taken in June–July, 1993 and 1995, for a month, every fourth day at noon. The samples were taken in the surface layer, the transparency depth (the visibility depth of Secchi disk), in the upper and lower boundary of metalimnion and in the hypolimnion (1 m above the bottom). At each sampling, the water temperature was measured with an electric thermometer from the surface to the bottom with a step of 1 m. The upper metalimnion boundary was defined as the depth at which the temperature was more than 1 °C lower as compared with the stratum located 1 m higher. The lower boundary of metalimnion was the depth at which the temperature was less than 1 °C lower as compared with the stratum located 1 m higher.

Niche overlaps were measured in terms of Pianka's index (Pianka, 1974):

$$\mathrm{ON}_{\mathrm{jk,kj}} = \frac{\Sigma p_{\mathrm{ij}} p_{\mathrm{ik}}}{\sqrt{\Sigma p_{\mathrm{ij}}^2 \Sigma p_{\mathrm{ik}}^2}},$$

where ON is the overlap index of j and k species; p_{ji} is the ratio of the species j density in sample i to the sum of this species density in all samples in the vertical water column, p_{ik} is the same ratio for species k. The values of ON vary between 0 and 1.

It was assumed that ON $\leq$ 0.1 gave evidence of niche nonoverlap. With this assumption we used the ratio of the number of $Q_{\mathrm{ON}\leq 0.1}$ to the total number of niche overlap versions $Q_{\text{ON}0-1}$ in the vertical water column for comparison of the zooplankton structurization degree.

The individual volume and dry weight were calculated for each species according to body length measurements and L–W equations of Bottrell et al. (1976).

The rotifer production was calculated from the equation:

$$P = N \cdot w \cdot (e^b - 1),$$

where P is the production, μg l^{-1} day^{-1}; b is the instantaneous birth rate, day^{-1}; N is the density, ind l^{-1}; w is the mean individual mass, μg. The birth rate was determined from the equation suggested by Paloheimo (1974):

$$b = 1/D_e \cdot \ln(1 + E/N),$$

where D_e is the duration of embryogenesis, days; E is the number of eggs in the population, eggs l^{-1}; N is the population density, ind l^{-1}.

The duration of embryogenesis was determined following Edmondson (1960). The specific production was calculated as the ratio of the daily production to the biomass. The production of *Limnocalanus macrurus* Sars was calculated with the data on absolute age gains (Vezhnovets, 1984).

Each day of sampling (t) the daily population production (P_t) was determined as the sum of production of all ages:

$$P_t = \Sigma P_i.$$

The production of cladocerans from *Daphnia* genus was calculated from the data on growth of individuals of different sizes as the sum of production of all size classes. The production of oviparous females was determined by:

$$P_{ov} = N \cdot N_{ov} \cdot w_{ov} \cdot 1/D_e,$$

where N is the number of oviparous females, ind l^{-1}; N_{ov} is the average number of eggs in brood chamber; w_{ov} is the egg mass, μg; D_e is the duration of embryogenesis, days.

The specific growth rate (c, day^{-1}) was defined as the ratio of the daily production to the biomass.

In calculation of production of the other crustacean species, we used mean daily specific growth rates (c) observed in similar lakes (Ivanova, 1985). The daily production (P) was determined as: where N is the population density, ind l^{-1}; w is the mean mass of an individual in the population, μg.

Zooplankton division into trophic levels (primary and secondary consumers) was performed according to literature. For example, copepods were divided into obligate and facultative predators following Monakov (1976). Five of the other species are known to be obligate predators (3 cladocerans *Leptodora kindtii* (Focke), *Polyphemus pediculus* (Linné) and *Bythotrephes longimanus* Leydig) and two rotifers: (*Asplanchna girodi* Guerne, and *Ploesoma truncatum* (Levander). One more rotifer species (*Asplanchna priodonta* Gosse) is a facultative predator. It is known that *Asplanchna sp. sp.* can be additional food for *L. kindtii*, but only in spring. In summer *L. kindtii* does not consume rotifers since there are many cladocerans which are the main food of *L. kindtii* (Karabin, 1974). The facultative predator production was divided in half between the primary and secondary consumers. We used the energy equivalent of zooplankton 20000 j g^{-1} dry mass (Peters Downing, 1984).

The ecological efficiency (K_e) was calculated as the ratio of the secondary to primary consumer production.

Differences in the values of parameters studied were checked with the t-test (Rokitsky, 1973).

Results and discussion

Weather conditions of the years differed much, which resulted in difference of dynamics of the surface water temperature and thickness of metalimnion (Figure 1). The average water surface temperature was 17.8 $\pm$ 0.47 °C in 1993 and 20.4 $\pm$ 0.33 °C in 1995.

The higher surface temperature in 1995 was accompanied by reduction of the epilimnion depth as the weather was sunny and windless. In 1995, the metalimnion was more than 2–3 times thicker than it was in 1993 (16 and 6 m, respectively).

The vertical zooplankton structurization was compared in the number of pairs of species with ON ≤ 0.1 and with ON $= 1$, in the total number of species pairs and in the ratio of these numbers. In 1995, $Q_{\text{ON}\leq 0.1}$ was larger than it was in 1993 with a significant difference ($P < 0.01$). The difference in the other parameters was insignificant (see Table 1).

In 1995 in the secondary consumer group, the average number of pairs of species with nonoverlapping niches ($Q_{\text{ON}\leq 0.1}$) was twice as large as that found in 1993. Similar results were obtained in comparison of the niche nonoverlap index of the secondary (Cs_2) and primary (Cs_1) consumers in the same parameter (Table 2).

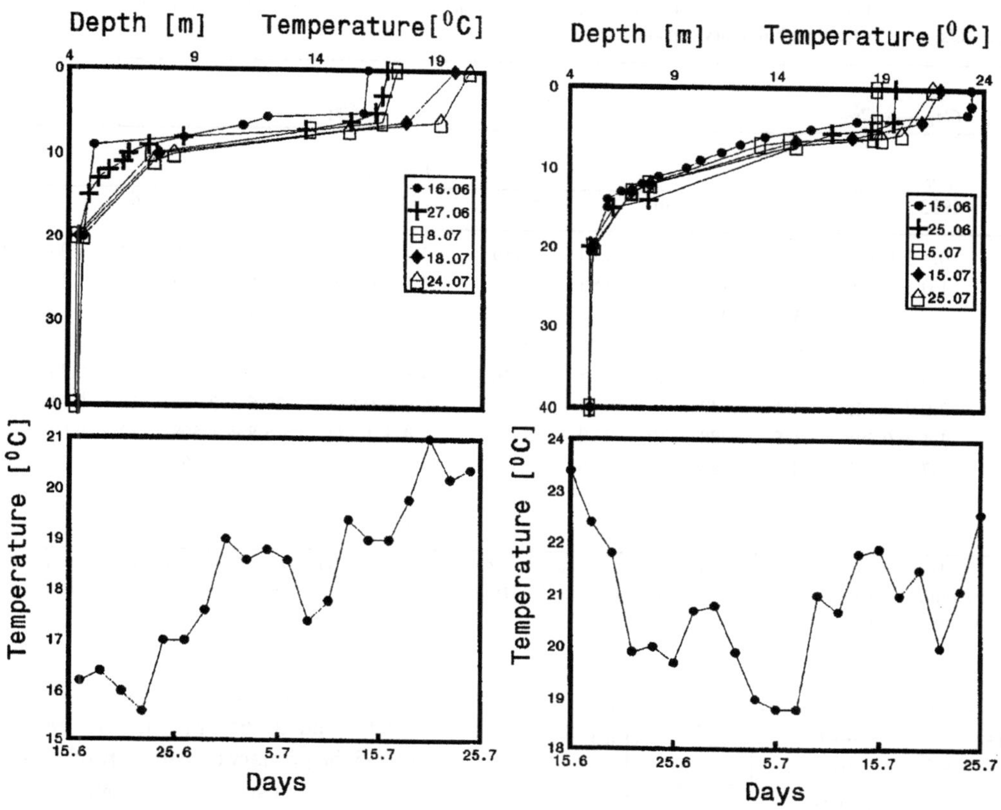

Figure 1. Temperature in vertical water column and average temperature in surface layer in 1993 (on the left) and 1995.

Table 1. Average niche overlap parameters in summer zooplankton of S. Voloso lake

Parameters	1993	1995	*t*	*P*
Numbers of species	38.5 ± 0.37	40.0 ± 0.76	1.770	<0.1
Q_{ON0-1}	722.7 ± 4.43	782.0 ± 29.90	1.786	<0.1
$Q_{ON\leq 0.1}$	162.0 ± 18.76	231.5 ± 16.50	2.780	<0.01
$Q_{ON\leq 0.1}/Q_{ON0-1}$, %	22.4 ± 2.59	29.2 ± 1.16	2.530	0.01
$Q_{ON=1}$	53.9 ± 6.28	69.1 ± 8.66	1.420	>0.1
$Q_{ON=1}/Q_{ON0-1}$, %	7.45 ± 0.85	8.8 ± 1.14	0.955	>0.1
$Q_{ON\leq 0.1}/Q_{ON=1}$	3.3 ± 0.53	3.86 ± 0.54	0.543	>0.1

Bearing in mind that increase in the niche nonoverlap index decreases the probability of interpopulation interactions between the primary and secondary consumers, it can be concluded that in 1995 this probability was half as high as that in 1993. Spatial niche nonoverlap is a real precondition for conservation of species populations since it assumes at least the absence of competitive displacement.

In the years compared, the species composition was almost the same, but different species were a basis of the production. In particular, in 1993 the primary consumer production was represented mainly by *Bosmina sp. sp.* and 'loricate' rotifers and in 1995, by *Daphnia cucullata* Sars and 'unloricate' rotifers. In 1993 *Heterocope appendiculata* Sars and *Mesocyclops leuckartii* (Claus) predominated in the production of the secondary consumers and *A. priodonta* and *L. macrurus* predominated in 1995.

As is evident from Figure 2, in 1995 the daily production of the primary consumers increased smoothly during the observation period from 150 j m^{-3} d^{-1} to 400 j m^{-3} d^{-1} and in 1993 this parameter varied in the

Table 2. Average number of pairs having nonoverlapping niches ($Q_{ON \leq 0.1}$) for every secondary consumer (Cs_2) with secondary consumers ($Cs_2 \rightarrow Cs_2$) and with primary consumers ($Cs_2 \rightarrow Cs_1$)

Secondary consumer, Cs_2^*	$Cs_2 \longrightarrow Cs_2$ $Q_{ON \leq 0.1}$ 1993	1995	df	t	P	$Cs_2 \longrightarrow Cs_1$ $Q_{ON \leq 0.1}$ 1993	1995	df	t	P
1	0.5 ± 0.179	4.0 ± 0.350	14	8.90	<0.001	4.1 ± 1.033	15.7 ± 0.755	14	9.06	<0.001
2	1.0 ± 0.479	1.6 ± 0.498	14	0.86	>0.1	3.1 ± 0.236	5.5 ± 1.736	14	1.37	>0.1
3	–	5.7 ± 1.006	–	–	–	–	22.7 ± 5.125	–	–	–
4	3.3 ± 1.020	4.4 ± 0.766	9	0.76	>0.1	17.4 ± 1.810	10.9 ± 1.673	9	2.64	<0.05
5	1.3 ± 0.243	3.0 ± 0.622	12	4.95	<0.001	4.8 ± 1.772	7.0 ± 1.836	12	0.86	>0.1
6	3.0 ± 0.626	3.4 ± 0.647	14	0.45	>0.1	20.0 ± 0.653	20.6 ± 2.415	14	0.24	>0.1
7	1.1 ± 0.295	1.5 ± 0.488	12	0.70	>0.1	4.4 ± 1.374	5.8 ± 1.508	12	0.68	>0.1
8	2.7 ± 0.250	2.9 ± 0.198	9	0.56	>0.1	14.3 ± 2.038	12.0 ± 2.562	9	0.7	>0.1
9	–	4.0 ± 0.000	–	–	–	–	16.0 ± 0.000	–	–	–
Σ	12.9	30.5				68.1	116.2			

* where 1 – *Asplanchna* sp., 2 – copepodids of *Cyclopoida*, 3 – *Cyclops lacustris* Sars, 4 – *Heterocope appendiculata* Sars, 5 – *Leptodora kindtii* (Focke), 6 – *Limnocalanus macrurus* Sars, 7 – *Mesocyclops leuckartii* (Claus), 8 – *Ploesoma truncatum* (Levander), 9 – *Polyphemus pediculus* (Linné).

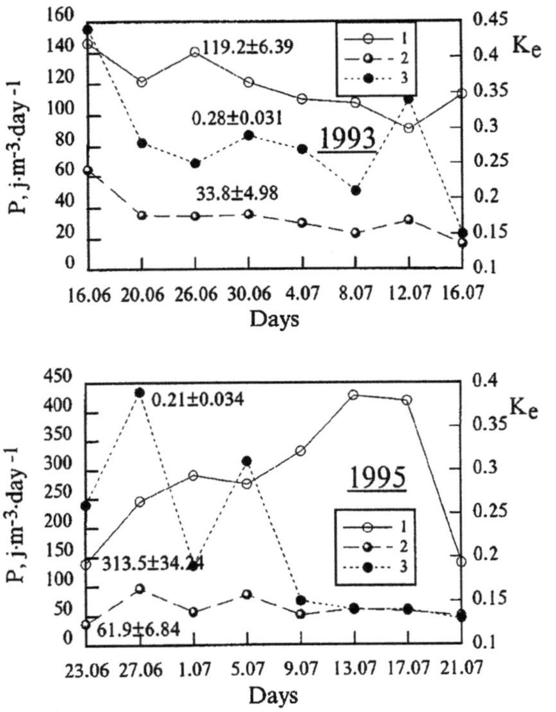

Figure 2. Production (P) of primary (1) and secondary (2) consumers and ecological efficiency (K_e) of secondary consumers (3) in 1993 and 1995 in S. Voloso lake.

range of 100–145 j m^{-3} d^{-1}. The range of the daily production of secondary consumers was 40–100 j m^{-3} d^{-1} in 1995 and 20–65 j m^{-3} d^{-1} in 1993.

In 1995 the primary consumer production increased substantially against the background of the number of species pairs with niche nonoverlap in the entire zooplankton (see Table 1) and in consideration of niche overlap between the primary and secondary consumers (see Table 2). In 1995 the number of pairs with niche nonoverlap among the secondary consumers was twice as large as that in 1993, just as the production of the secondary consumers.

Seasonal zooplankton succession is controlled by many factors, and temperature is as a considered factor that determines fluctuations of the species composition and zooplankton population density in the summer season (Sommer et al., 1986). However, the influence of temperature on spatial structure and production efficiency of summer zooplankton is practically unknown.

It is clear that the differences in temperature stratification of the water column in June–July 1993 and 1995 are related with weather conditions. The temperature stratification can restrict the habitat of some species and in this way affect the zooplankton structurality in the water column. Nevertheless, we can't define how the differences in the zooplankton structurality are caused by differences in temperature stratification. The hypothesis that temperature plays an important direct role in shaping zooplankton community is reasonable, but largely untested (Hu & Tessier, 1995).

The presented data do not allow us to state that there is a relation between the spatial structurality, the production efficiency in the trophic zooplankton system and changes in the weather conditions so additional studies are necessary.

One can only state with certainty that higher $Q_{\mathrm{ON} \leq 0.1}$ in the zooplankton decreases the probability of interpopulation interactions. On the one hand, it can result in weaker competition relations and can be a cause of changes in the population density and production, and on the other hand, a lower probability of interpopulation interactions can strongly affect the functioning of the ecosystem as a whole.

In the light of the data obtained it seems reasonable that the functioning of the ecosystem relates mainly with the number of species interactions, rather than with the number of species (King & Pimm, 1983; Lawton & Brawn, 1994).

Thus, the studies carried out in the years under different weather conditions in June–July have revealed substantial differences in the temperature conditions of a stratified lake, in the spatial zooplankton structure, in the production of primary and secondary consumers and in the K_e. These changes can be considered as an indication of internal mechanisms stabilizing the system, as they are not accompanied by changes in the species composition of zooplankton in which a relict crustacean complex is conserved.

Acknowledgements

We would like to thank the referees who gave us the opportunity to clarify our ideas. We thank Dr Vezhnovets for his help with sampling. This research was partially supported by National Science Foundation of Belarus, grants 13.065 and 001.003.

References

Bottrell, H. H, A. Duncan, Z. M. Gliwicz, E. Grygierek, A. Herzig, A. Hillbricht-Ilkowska, H. Kurasawa, P. Larsson & T. Weglenska, 1976. Review of some problems in zooplankton production studies. Norw. J. Zool. 24: 419–456.

Edmondson, W. T., 1960. Reproductive rates of rotifers in natural populations. Mém. Ist. ital. Idrobiol. 12: 21–77.

Hu, S. S. & A. J. Tessier, 1995. Seasonal succession and the strength of intra- and interspecific competition in a *Daphnia* assemblage. Ecology 76: 2278–2294.

Ivanova, M. B., 1985. Production of planktonic crustaceans in fresh water. USSR Academy of sciences, Zoological institute, 222 pp. (In Russian).

Karabin, A., 1974. Studies on the predatory rate of the cladoceran, Leptodora kindtii (Focke), in secondary production of two lakes with different trophy. Ecol. Pol. 22: 295–310.

King, A. W. & S. L. Pimm, 1983. Complexity, diversity, and stability: a reconcilliation of theoretical and empirical results. Am. Nat. 122: 229–239.

Lawton, J. H. & V. K. Brown, 1994. Redundancy in ecosystems. In E.-D. Schulze & H. A. Mooney (eds), Biodiversity and Ecosystem Function. Springer-Verlag, Berlin, Heidelberg: 265–268.

Monakov, A. V., 1976. Nutrition and food possibility of fresh-water copepods. Nauka, Leningrad, 170 pp. (In Russian).

Paloheimo, J. E., 1974. Calculation of instantaneous birth rate. Limnol. Oceanogr. 19: 692–694.

Peters, R. H. & J. A. Downing, 1984. Empirical analysis of zooplankton filtering and feeding rates. Limnol. Oceanogr. 29: 763–784.

Pianka, E. R., 1974. Niche overlap and diffuse competition. Proc. Natn. Acad. Sci. U.S.A. 71: 2141–2145.

Rokitsky, P. F., 1973. Biological Statistics. Minsk, University Press, 319 pp. (In Russian).

Sommer, U., Z. M. Gliwicz, W. Lampert & A. Duncan, 1986. The PEG-model of seasonal succession of planktonic events in fresh waters. Arch. Hydrobiol. 106: 433–471.

Sprules, W. G. & J. E. Bowerman, 1988. Omnivory and food chain length in zooplankton food webs. Ecology 69: 418–426.

Vezhnovets, V. V., 1984. Biology and production of *Limnocalanus grimaldii var. macrurus* (Sars). Abstracts of the candidate thesis. Minsk, 24 pp. (In Russian).

Yakushko, O. F. (ed.), 1988. Lakes of Belorussia, Minsk, Uradjai, 216 pp. (In Russian).

Yodris, P., 1981. The stability of real ecosystems. Nature 289: 674–676.

Hydrobiologia **363:** 309–321, 1998.
T. Tamminen & H. Kuosa (eds), *Eutrophication in Planktonic Ecosystems: Food Web Dynamics and Elemental Cycling.*

Nitrogen, phosphorus and *Daphnia* grazing in controlling phytoplankton biomass and composition – an experimental study

Heli Karjalainen[1]*, Satu Seppälä[1] & Mari Walls[2]
[1] *North Savo Regional Environment Centre, P.O. Box 1049, FIN-70101 Kuopio, Finland*
[2] *Department of Biology, University of Turku, FIN-20014, Turku, Finland* * *Present address: Department of Ecological and Environmental Sciences, University of Helsinki, Niemenkatu 73, FIN-15210 Lahti, Finland (E-mail: hekarjal@neopoli.helsinki.fi)*

Key words: nitrogen, phosphorus, phytoplankton biomass, *Daphnia* grazing

Abstract

The role of nitrogen as a factor controlling phytoplankton biomass was studied in nutrient enrichment incubations in the laboratory using water from pelagic region of two mesotrophic lakes in eastern Finland, Lake Kallavesi (in year 1994) and Lake Juurusvesi (in year 1995). We used different combinations of phosphorus and nitrogen additions in a total of eight experiments. Furthermore, we included *Daphnia* grazing treatment to the experimental design in Lake Juurusvesi experiments. The nitrogen treatments did not increase chlorophyll *a* concentration in any of the experiments compared with the controls. Chlorophyll *a* content was highest in those nutrient treatments where phosphorus was added with or without nitrogen. *Daphnia* grazing decreased chlorophyll *a* concentration compared with non-grazed treatments. In some cases grazing also caused higher ammonium concentrations. These experiments, as well as the nutrient ratio of the lake water used, suggest that phosphorus is likely to control the amount of phytoplankton biomass.

Introduction

The importance of phosphorus in the eutrophication process of originally oligotrophic freshwaters is well documented, whereas the role of nitrogen is more controversial. There is relatively little experimental work on the responses of freshwater phytoplankton communities to nitrogen additions. Studies performed in the Baltic Sea (i.e. Tamminen, 1990) reveal the key role of nitrogen in controlling primary productivity in brackish water ecosystems.

Effects of nutrients on the biomass of phytoplankton, commonly used to indicate eutrophication, are usually predicted on the basis of the absolute and relative amounts of nutrients in the watershed (Forsberg et al., 1978). However, these predictions do not take into account, for instance, the range of organisms and their biological demands for nutrients, or the ecological relationships present in the studied area. Furthermore, predictions on the effects of incoming nutrients focus on the responses of primary producers measured as changes in biomass, but not on the species composition of phytoplankton. Since the demands of inorganic nutrients vary greatly according to species and along changing environmental conditions, it is necessary to try to incorporate these latter factors in studies focusing on nutrient ratios.

Both abiotic and biotic environmental factors may influence the responses of a lake system to nutrient additions. Of the biotic factors, the impact of zooplankton grazing on phytoplankton may also influence nutrient conditions. Grazers may, for instance, produce small scale local patches of nutrients affecting the nutrient ratios and, consequently, the limiting nutrient if nutrients are excreted at a different ratio than they exist in the surrounding water (Elser & George, 1993) or in the material grazed (Sterner, 1990). It has been shown that some algae can directly utilize phosphorus excreted by grazers (Lehman & Scavia, 1982). Natural ecosystems are full of small scale patches. This

Table 1. Incubation experiments I–VIII were performed in years 1994 and 1995. Sampling site and time, and some basic chemical characteristics of the studied lake waters.

Year	Site	Experiment number	Date	chl-*a* $\mu g\ l^{-1}$	tot–N $\mu g\ l^{-1}$	NO_3–N $\mu g\ l^{-1}$	NH_4–N $\mu g\ l^{-1}$	PO_4–P $\mu g\ l^{-1}$	tot–P $\mu g\ l^{-1}$
1994	Lake	I	9 Jun	12	770	180	39	6	35
	Kallavesi	II	4 Jul	6	680	220	11	5	27
		III	1 Aug	9	680	140	37	3	23
		IV	29 Aug	8	770	170	108	<2	21
1995	Lake	V	6 Jun	8	690	150	22	<2	24
	Juurusvesi	VI	5 Jul	7	550	170	11	<2	19
		VII	2 Aug	9	520	79	15	<2	19
		VIII	30 Aug	6	610	140	14	<2	15

heterogeneity is, however, usually ignored in routine fieldwork.

Planktonic algae are capable of using both ammonium and nitrate as a nitrogen source. Ammonium is usually preferred because of its energetic advances (McCarthy et al., 1982; Wetzel, 1983; Miyazaki et al., 1989; Gu & Alexander, 1993), but the use of different nitrogen forms is dependent on their availability and abiotic conditions (Whalen & Alexander, 1986). We included the two nitrogen forms to investigate the possible differences in phytoplankton responses. We studied experimentally in the laboratory, (1) whether additions of inorganic nitrogen, ammonium and nitrate, have effects on phytoplankton biomass in waters from two mesotrophic lakes, and (2) whether the experimentally produced nitrogen enrichment causes changes in the phytoplankton composition.

We also took into account in our experiments the possible effects of zooplankton grazing and asked (3) whether the possibly strengthened recycling effects of *Daphnia* grazing change the responses of the phytoplankton in situation of relatively high nitrogen and low phosphorus content.

Materials and methods

Study sites

Lakes Kallavesi and Juurusvesi (Finland) are two different parts of the Vuoksi watersystem, which runs to the Gulf of Finland. Lake Kallavesi is 51 700 ha in area and has a maximum depth of 69 m. According to phytoplankton biomass, Lake Kallavesi is meso-eutrophic (measured as chlorophyll *a*) (Table 1). The sampling site was located in the pelagic region in the northern part of South-Kallavesi (area 31 500 ha) near a narrow strait, which separates the southern part of the lake from the northern part. Depth in the sampling site was 19 m. Lake Juurusvesi is mesotrophic (Table 1), it is 15 900 ha in area and the maximum depth is about 54 m. The study site was at Kuuslahti (maximum depth about 44 m, depth in the sampling site was 22 m), a narrow bay at the northern part of Lake Juurusvesi. Both Lake Kallavesi and Lake Juurusvesi are to some extent disturbed by industry.

Water samples for laboratory experiments were collected one day prior to the beginning of each incubation. Water was collected with a Limnos-sampler from 0–5 m so that the total water volume collected was 110–125 l in experiments I–IV and 250–287 l in experiments V-VIII. Water was sieved immediately through 150 μm filter (Elser & George, 1993) for the removal of the largest zooplankters (Suttle & Harrison, 1988; Dodds & Priscu, 1990), and stored in a dark and cool (+4 °C) place overnight.

Experimental design and treatments

In a series of experiments, the treatments used were nutrient additions and incubation time in experiments I–IV (year 1994, Lake Kallavesi), whereas in experiments V–VIII (year 1995, Lake Juurusvesi) we included an additional treatment, zooplankton grazing (with and without addition of *Daphnia*). The nutrient addition treatments consisted of six different levels: control, two different nitrogen additions (N), two (exp. V–VIII, year 1995) and three (exp. I–IV, year 1994) types of nitrogen + phosphorus additions (NP) (Table 2) and at Lake Juurusvesi (exp. V–VIII, year 1995) one manipulation where phosphorus was added alone (P) (Table 2). The levels of the additions, made once at the beginning

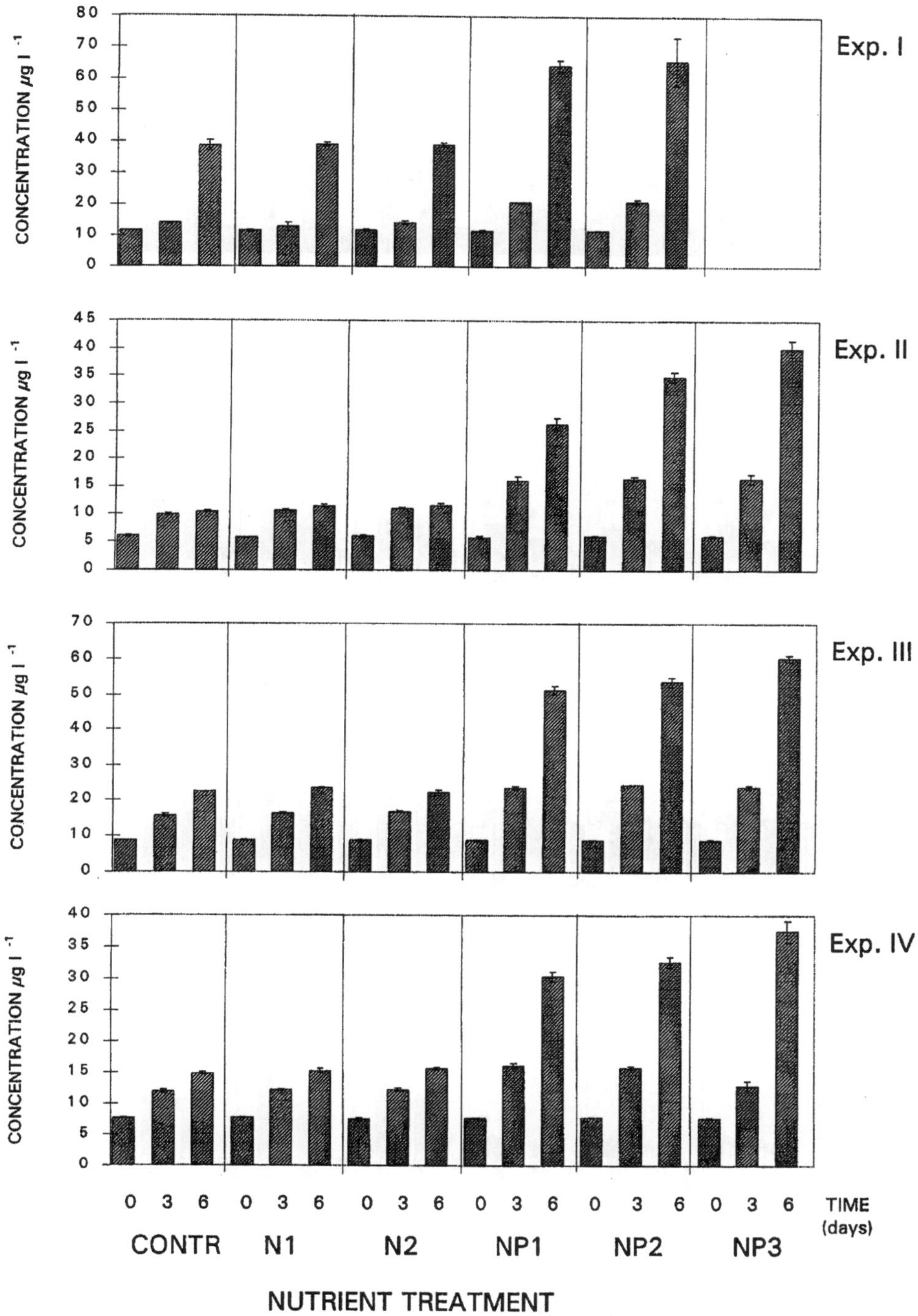

Figure 1. The mean chlorophyll *a* concentrations (± SD) in experiments I–IV in year 1994 in different treatments. Nutrient treatment abbreviations are explained in Table 2.

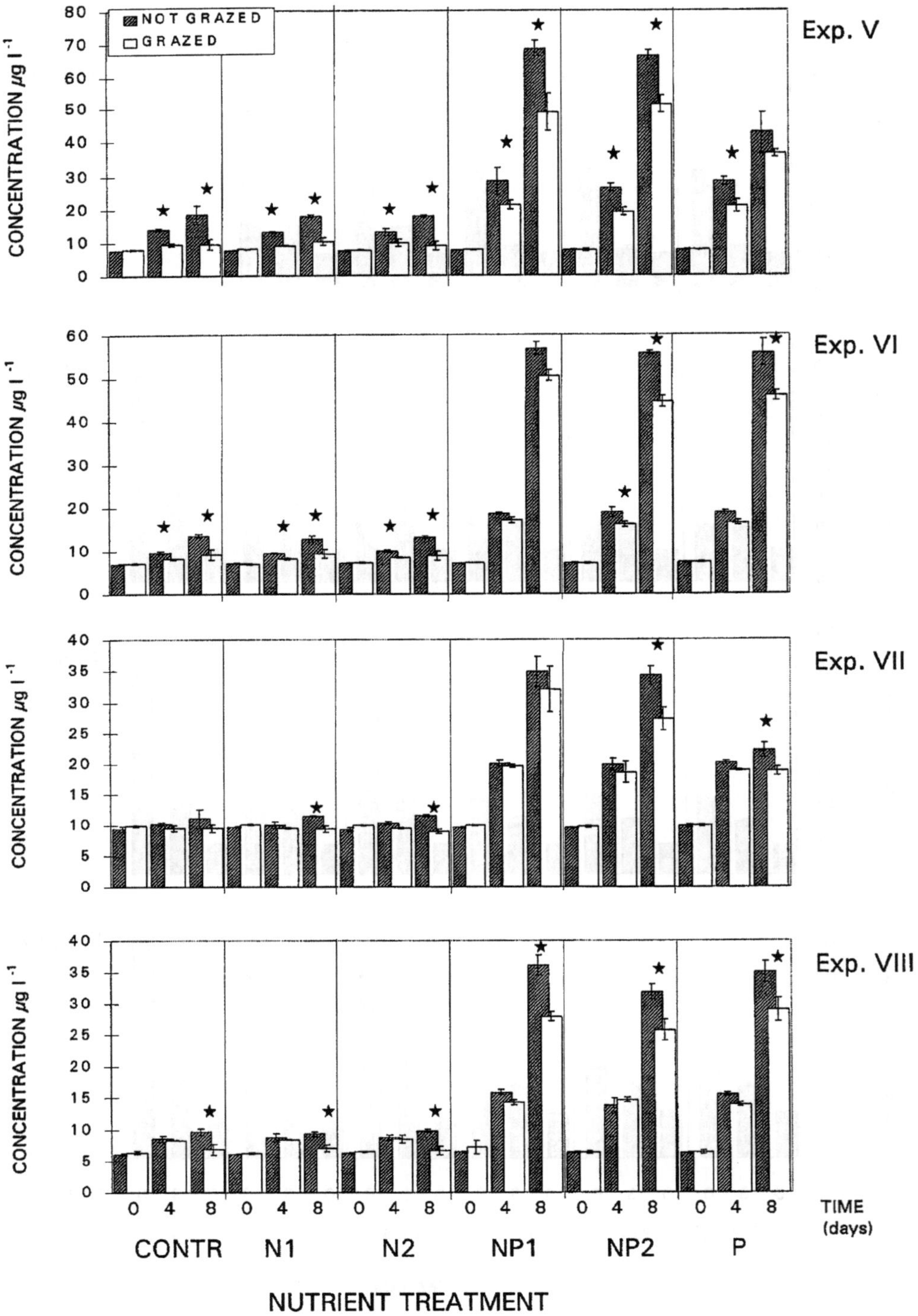

Figure 2. The mean chlorophyll *a* concentrations (± SD) in experiments V–VIII in year 1995 in different treatments. Nutrient treatment abbreviations are explained in Table 2. Statistically significant differences between *Daphnia* grazed and non-grazed treatments (Tukey, $p < 0,05$) are indicated with asterisks.

Table 2. Nutrient treatments of the experiments in years 1994 (Lake Kallavesi, experiments I–IV) and 1995 (Lake Juurusvesi, experiments V–VIII) and the abbreviations used in the figures and text.

Nutrient treatment		1994		1995	
		Nitrogen addition $\mu g\,l^{-1}$	Phosphorus addition $\mu g\,l^{-1}$	Nitrogen addition $\mu g\,l^{-1}$	Phosphorus addition $\mu g\,l^{-1}$
Control		–	–	–	–
Nitrogen (1)	N1	300	–	300	–
Nitrogen (2)	N2	1000	–	1000	–
Nitrogen + phosphorus (1)	NP1	300	10	300	15
Nitrogen + phosphorus (2)	NP2	1000	33	1000	15
Nitrogen + phosphorus (3)	NP3	300	125	(not performed in 1995)	
Phosphorus	P	(not performed in 1994)		–	15

of each experiment, were based on the existing nutrient loads to the areas studied. The incubation times were 0, 3 and 6 days in experiments I–IV and 0, 4 and 8 days in experiments V–VIII. In each treatment combination we had three replicate experimental units (incubation bottles). The total number of units was 45 in exp. I, 54 in exp. II–IV and 108 units in exp. V–VIII.

The experiments were conducted in the laboratory in 2 l Duran glass bottles under a light cycle of 16 h light/8 h dark ($8 \pm 0{,}8$ W m^{-2}, Osram daylight lamps 36 W/12,) and at constant temperature of 20 ± 1 °C. Four experiments were run during the summer of 1994 (experiments I–IV) with the water from Lake Kallavesi and in 1995 (experiments V–VIII) with the water from Lake Juurusvesi. Three replicate bottles for each treatment combination were filled with sieved lake water. For N and NP-treatments, the chemical used in experiments I–IV was NH_4Cl (Suttle & Harrison, 1988; Wehr, 1989) and in experiments V-VIII $NaNO_3$ (Mazumder & Lean, 1994). KH_2PO_4 (Suttle & Harrison, 1988; Aldridge et al., 1993) was used as a source of phosphorus in P and NP treatments. All chemicals were Merck's pro analyse quality. Incubation bottles were located randomly at the table and their places were altered daily to prevent any site effects. Water was mixed by stirring with a glass tube twice a day during the incubations.

Laboratory strain of *Daphnia pulex* De Geer, originating from small fishless pond Kupittaa from southern Finland was used as a grazer in the 1995 experiments. Group of *D. pulex* individuals from a laboratory strain were acclimated in sieved water from Lake Juurusvesi at the incubation temperature three weeks prior to the experiments. These individuals and their offspring were used as grazers in all experiments performed. Six to eight individuals, consisting of both juveniles and adults, were added to the respective units. With use of different age classes we wanted to insure that mortality in one age class would not diminish the grazing effect. At the end of each experiment, we collected the daphnids from the water with a 300 μm sieve for further analyses. During the incubations dead individuals were counted daily by visual observation. At the end of the incubations the numbers of egg-bearing adult *Daphnia* females were counted. In all experiments, the number of these females increased towards the end, indicating relatively good potential of population growth. The effect of sieving of *Daphnia*, done also for non-grazed units, on the phytoplankton content was at maximum 2.1% of the chlorophyll *a* concentration.

Phytoplankton and chemical analyses

All chemical analyses were performed at the laboratory of North Savo Regional Environment Centre according to Finnish Standard methods. NO_2–N + NO_3–N (further called as NO_3–N, reduction of NO_3–N to NO_2–N with Cu–Cd column, colorimetric determination of azo-colour, autoanalyzer), NO_2–N (only in 1994 experiments, colorimetric determination of azo-colour), NH_4–N (spectrophotometric determination with hypochlorite and phenol), total N (oxidation with $K_2S_2O_8$ to NO_3–N, measured as NO_2–N), PO_4–P (spectrophotometric determination as phosphomolybdate) and total P (preservated with H_2SO_4, oxidated with $K_2S_2O_8$, spectrophotometric determination as phosphomolybdate) were measured at the beginning, at the middle and at the end of the experiments. Chlorophyll *a* content, used as an indicator of phytoplankton biomass, was analysed spectrophotometrically using ethanol extraction method. Subsamples for identifying phytoplankton species composition were taken from each bottle prior to chemical analyses and were preserved with acetic Lugol solution. Phytoplankton $> 3\ \mu m$ in size was identified from day zero situation and controls, N2 and NP2 treatments from the last incubation day (day six) from two replicate experimental units in the year 1994 experiments, totalling 32 phytoplankton samples. In year 1995 experiments, just one phytoplankton sample from day zero and control, N2, NP2 and P from day eight with and without *Daphnia* grazing was analysed, totalling 36 phytoplankton samples. Utermöhl (1958) method was used for phytoplankton composition analyses. Phytoplankton cells were identified until 1500 units were counted. Cell volumes used in biomass estimations were either the measured volumes from the samples or the cell volumes for algae of corresponding size from Biological database (Finnish Environment Institute).

Statistical analyses

Multivariate analysis of variance was used in statistical testing of the results. In the experiments I–IV, all chemical parameters and chlorophyll *a* content, and in the experiments V–VIII, chlorophyll *a* and NH_4–N content were included in the model. Tukey's test was used as an *a posteriori* test. Logarithmic transformation was used if the data was not normally distributed or due to the heteroscedasticity of variances. The results of chemical parameters NO_3–N, PO_4–P, total N and total P, in experiments V–VIII were transformed to yield change from the mean at the beginning of the experiments. These transformations made the variables dependent and therefore they were tested with nonparametric methods (Kruskall-Wallis).

We present here the results of chlorophyll *a* data of the experiments performed in years 1994 and 1995, and the key data of the chemical analyses from the year 1995. Concerning the phytoplankton results, we present the data from one experiment in 1995 (exp. VII), which is a representative example of changes in phytoplankton composition. Other results from the experiments I–IV are presented and discussed in Karjalainen et al. (1996).

Table 3. Multivariate analysis of variance using factors nutrient treatment (NU) and incubation time (*T*) in the experiments I–VI in year 1994, showing only results of the variable chlorophyll *a*.

Exp.		Chlorophyll *a* df	*F*	*p*
I	NU	4	108.97	0.000
	T	2	3317.46	0.000
	$T \times$ NU	8	28.68	0.000
II	NU	5	834.33	0.000
	T	2	3855.65	0.000
	$T \times$ NU	10	466.37	0.000
III	NU	5	1294.55	0.000
	T	2	23293.26	0.000
	$T \times$ NU	10	468.04	0.000
IV	NU	5	881.40	0.000
	T	2	13094.65	0.000
	$T \times$ NU	10	377.92	0.000

Results

Chlorophyll a *data for years 1994 and 1995*

The single effects of incubation time, nutrient treatments (in years 1994 and 1995) and *Daphnia* grazing (in year 1995) on chlorophyll *a* concentration were statistically significant in every experiment I–VIII (Tables 3 and 4, Figures 1 and 2). Nutrient treatments and time in all experiments I–VIII and *Daphnia* grazing and time in experiments V–VIII had interactive effects on the chlorophyll *a*. Nutrient treatments and grazing had interactive effects in two (exp. V and VI) of the four experiments performed (Table 4 and Figure 2).

Phytoplankton biomass, measured as chlorophyll *a* content, increased during the incubation periods (Tables 3 and 4, Figures 1 and 2). At the end of the experiments the increase was the highest in those nutrient treatments where phosphorus was added with or without nitrogen. The phytoplankton biomass was higher in NP and P treatments than in the controls or N treatments at the end of the study in all experiments I–VIII (Tukey $p < 0{,}05$). Concerning phytoplankton biomass, N treatments did not differ from the controls in any of the experiments (Figures 1 and 2) and the effects of two N addition levels (N1 and N2) did not differ from each other. The NP treatments, NP1 and NP2, differed significantly (Tukey $p < 0{,}05$) just in experiments II, IV and VIII. NP3 treatment caused a slightly higher biomass than the other combined nutrient treatments in experiments II–IV (Figure 1). Phosphorus treatment

Table 4. Multivariate analysis of variance using factors nutrient treatment (NU), incubation time (*T*) and *Daphnia* grazing (*G*) in the experiments V–VIII in year 1995.

Exp.		Chlorophyll *a*			NH_4-N		
		df	*F*	*p*	df	*F*	*p*
V	NU	5;72	519.37	0.000	5;72	22.88	0.000
	T	2;72	2536.39	0.000	2;72	346.31	0.000
	G	1;72	329.21	0.000	1;72	7.47	0.008
	T × *G*	2;72	99.69	0.000	2;72	9.04	0.000
	T × NU	10;72	173.24	0.000	10;72	13.98	0.000
	NU × *G*	5;72	4.35	0.002	5;72	1.74	0.137
	T × NU × *G*	10;72	5.07	0.000	10;72	2.06	0.039
VI	NU	5;72	1729.77	0.000	5;72	38.27	0.000
	T	2;72	7354.45	0.000	2;72	451.73	0.000
	G	1;72	287.13	0.000	1;72	23.25	0.000
	T × *G*	2;72	100.15	0.000	2;72	10.87	0.000
	T × NU	10;72	629.61	0.000	10;72	15.22	0.000
	NU × *G*	5;72	5.00	0.001	5;72	4.99	0.001
	T × NU × *G*	10;72	2.69	0.007	10;72	4.69	0.000
VII	NU	5;71	756.07	0.000	5;71	117.11	0.000
	T	2;71	1169.14	0.000	2;71	338.80	0.000
	it G	1;71	52.70	0.000	1;71	64.67	0.000
	T × *G*	2;71	44.91	0.000	2;71	10.40	0.000
	T × NU	10;71	219.97	0.000	10;71	37.78	0.000
	NU × *G*	5;71	1.16	0.339	5;71	3.12	0.013
	T × NU × *G*	10;71	0.76	0.666	10;71	4.58	0.000
VIII	NU	5;72	773.05	0.000	5;72	32.33	0.000
	T	2;72	2585.01	0.000	2;727	66.08	0.000
	G	1;72	81.70	0.000	1;72	81.70	0.000
	T × *G*	2;72	86.12	0.000	2;72	10.78	0.000
	T × NU	10;72	279.00	0.000	10;72	17.50	0.000
	NU × *G*	5;72	0.85	0.519	5;72	5.12	0.000
	T × NU × *G*	10;72	2.18	0.029	10;72	5.26	0.000

(P, in year 1995) caused a lower chlorophyll *a* concentration compared with NP treatments in experiments V and VII (Tukey, $p < 0{,}05$) (Figure 2).

Chemical data for year 1995

The PO_4–P concentrations in the lakes studied, especially in Lake Juurusvesi, were low (Table 1). In the treatments where PO_4–P was added, amounts of it decreased significantly during the incubations (Table 5) and at the end of the experiments concentrations were $\leq 3\ \mu g\ l^{-1}$ in every treatment. Also the amount of inorganic nitrogen lowered towards the end (Table 4 and 5, Figures 3 and 4). The decrease was the highest in treatments where phosphorus was added with nitrogen. In non-grazed P treatments, both ammonium ($\leq 4\ \mu g\ l^{-1}$ NH_4–N) and nitrate ($\leq 10\ \mu g\ l^{-1}$ NO_3–N) concentrations were low at the end of the experiments. In experiments V and VII nitrate concentrations were relatively low ($< 60\ \mu g\ l^{-1} NO_3$–N) already at day four (Figure 3).

The effects of grazing

In experiments V–VIII, the chlorophyll *a* concentration was generally lower in the grazed treatments than in the treatments without added *Daphnia* (Table 4, Figure. 2). The difference in chlorophyll *a* was statistically significant (Tukey, $p < 0{,}05$) in almost all nutrient treatments at the end of each experiment (Figure 2). Grazing had also effects on the ammonium concentration of the water (Table 4, Figure 4). In some grazed controls and N treatments, ammonium concentrations were higher than in respective treatments without grazing (Tukey,

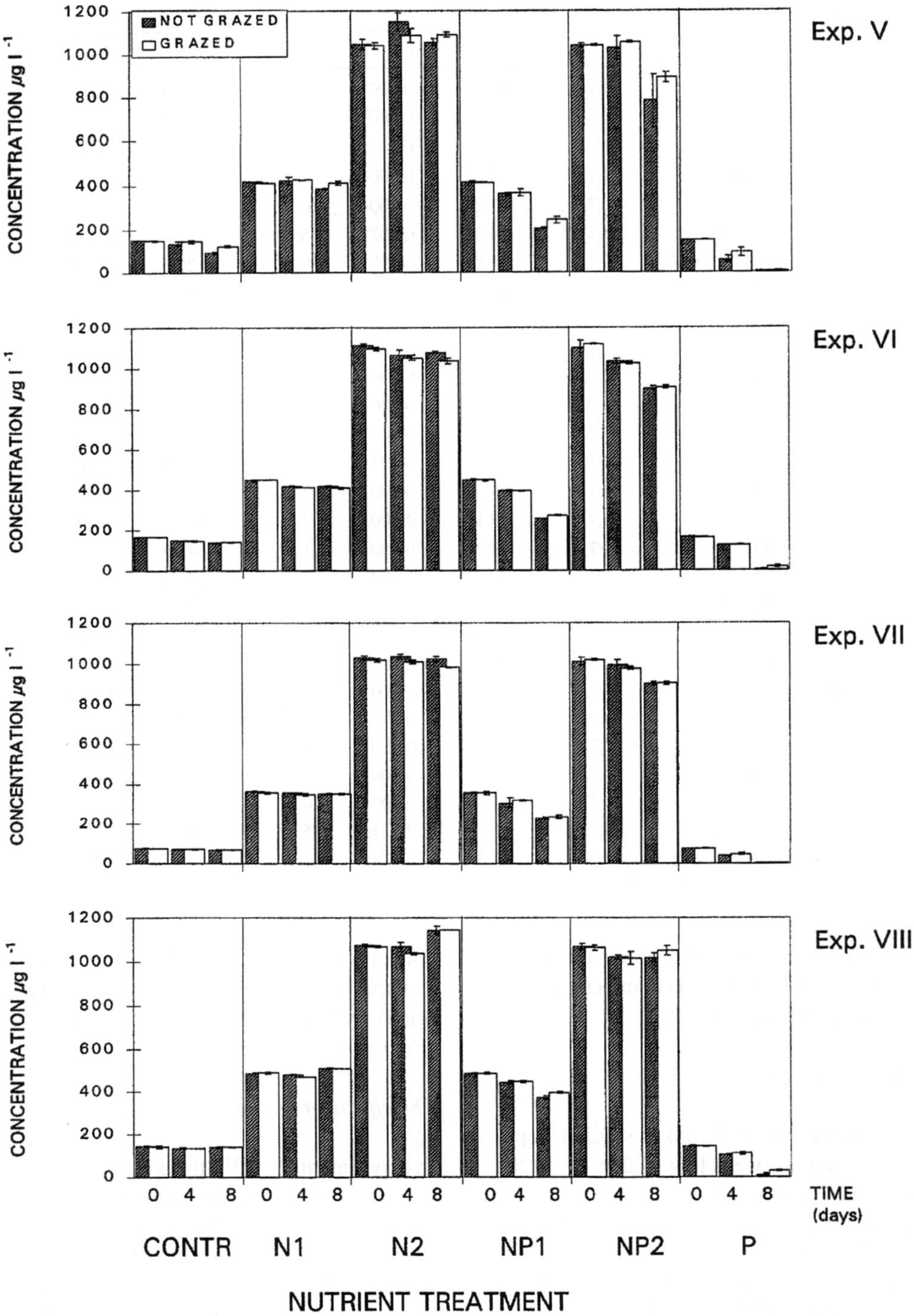

Figure 3. The mean NO_3–N concentrations (± SD) in experiments V–VIII in year 1995 in different treatments. Nutrient treatment abbreviations are explained in Table 2.

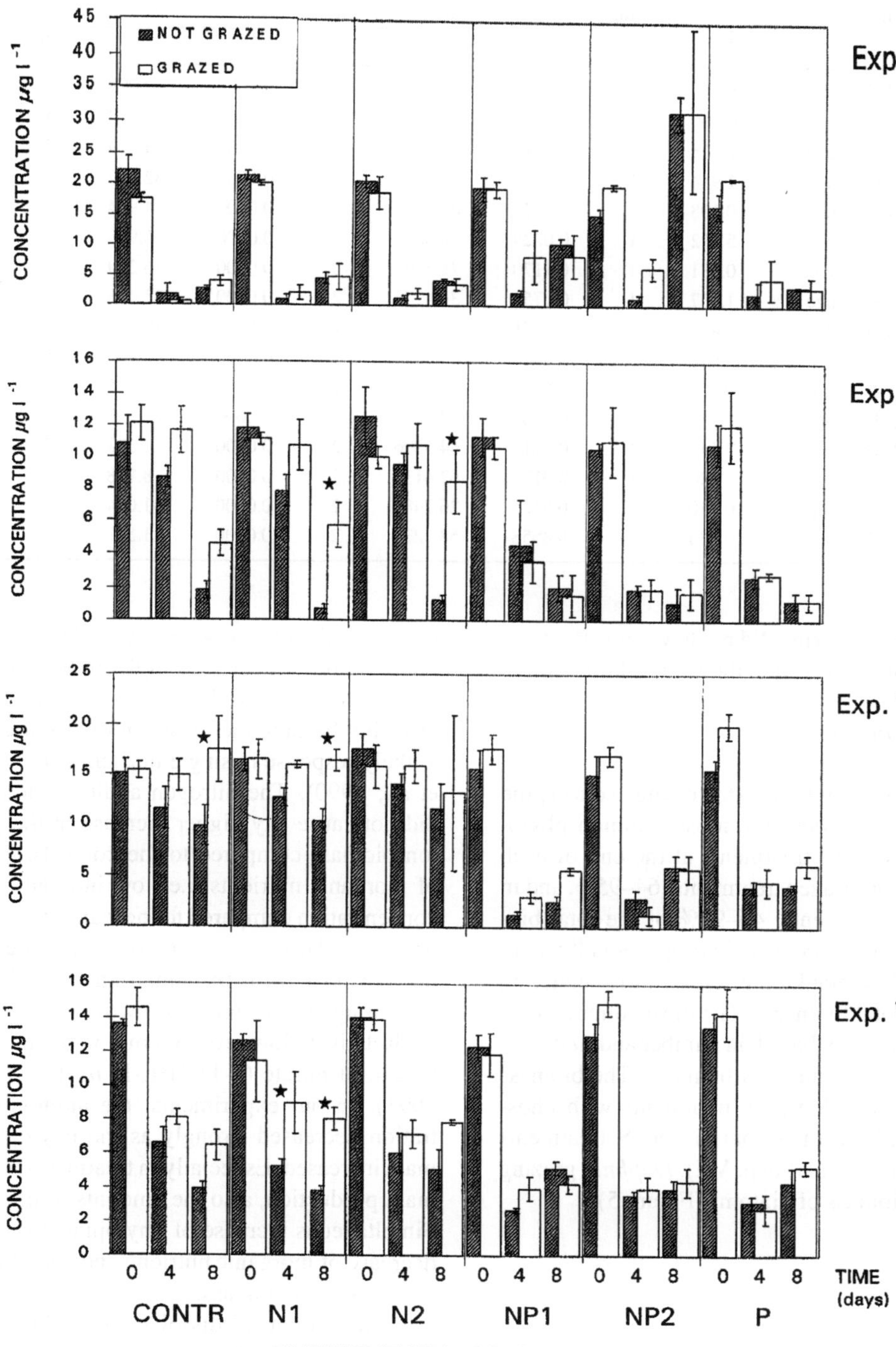

Figure 4. The mean NH_4–N concentrations (± SD) in experiments V–VIII in year 1995 in different treatments. Nutrient treatment abbreviations are explained in Table 2. Statistically significant differences between *Daphnia* grazed and non-grazed treatments (Tukey, $p < 0{,}05$) are indicated with asterisks.

Table 5. The effects of *Daphnia* grazing, incubation time and nutrient treatment on changes in total N (tot N), total P (tot P), NO_3–N ja PO_4–P in experiments V–VIII, (Kruskall-Wallis analysis of variance).

Exp	Parameter	*Daphnia* grazing			Incubation time			Nutrient treatment		
		χ^2	df	p	χ^2	df	p	χ^2	df	p
V	tot N change	9.326	1	0.0023	33.656	2	0.0000	9.009	5	0.1087
V	tot P change	10.566	1	0.0012	10.566	2	0.0000	5.159	5	0.3968
V	NO_3–N change	1.963	1	0.1612	24.842	2	0.0000	34.219	5	0.0000
V	PO_4–P change	1.914	1	0.1666	27.005	2	0.0000	32.037	5	0.0000
VI	tot N change	0.108	1	0.7423	65.271	2	0.0000	4.924	5	0.4253
VI	tot P change	5.212	1	0.0224	70.003	2	0.0000	6.943	5	0.2249
VI	NO_3–N change	0.131	1	0.7170	71.369	2	0.0000	16.219	5	0.0062
VI	PO_4–P change	1.837	1	0.1754	34.477	2	0.0000	33.498	5	0.0000
VII	tot N change	5.401	1	0.0201	31.553	2	0.0000	17.196	5	0.0041
VII	tot P change	8.948	1	0.0028	48.946	2	0.0000	6.718	5	0.2425
VII	NO_3–N change	0.221	1	0.6380	53.585	2	0.0000	17.825	5	0.0032
VII	PO_4–P change	0.569	1	0.4507	42.909	2	0.0000	26.539	5	0.0001
VIII	tot N change	2.533	1	0.1115	64.5396	2	0.0000	0.383	5	0.9958
VIII	tot P change	5.686	1	0.0171	62.4138	2	0.0000	8.788	5	0.1178
VIII	NO_3–N change	0.050	1	0.8225	26.7485	2	0.0000	31.094	5	0.0000
VIII	PO_4–P change	0.286	1	0.8658	58.1237	2	0.0000	21.523	5	0.0006

$p < 0{,}05$). *Daphnia* grazing did not have any effects on the concentration of PO_4–P in the water (Table 5).

Phytoplankton composition

The composition of phytoplankton changed during the incubation. Diatoms were the most common phytoplankton species in all treatments at the end of each experiment (in non-grazed treatments, 64–95%, and in grazed nutrient treatments, 45–96% of the total biomass). Diatoms increased usually proportionally more than the other phytoplankton groups (in the beginning 26–47% of the total biomass). The most common taxa were *Rhizosolenia* spp. based on number and *Acanthoceras zachariasii* measured as biomass. The biomass of chrysophytes was lower in treatments with phosphorus compared with the control and N treatments (see Figure 5 on results of exp. VII). *Daphnia* grazing decreased the biomass of diatoms (Figure 5).

Discussion

In our experiments, the increase in phytoplankton biomass (measured as chlorophyll *a* concentration) was to a large extent determined by the amount of inorganic phosphorus in the water. Inorganic nitrogen regulated biomass only in those cases where incubation had been going on already for a couple of days, and the phosphorus additions caused a high production of biomass. In cases,where phosphorus alone was added, nitrogen probably became the most limiting nutrient, which is a likely response in single nutrient enrichments (Elser et al., 1990). The nitrogen addition alone, however, did not cause any higher increase in the phytoplankton biomass compared to the controls. The amounts of inorganic nutrients, i.e. low inorganic phosphorus concentration compared to inorganic nitrogen concentration at the beginning of the experiments supports our conclusion on the importance of phosphorus in regulating phytoplankton growth.

Relatively low level of ammonium may inhibit the uptake of nitrate (7-11 NH_4-N $\mu g\ l^{-1}$, Dodds et al., 1991). In our experiments, the ammonium concentration decreased strongly as the phytoplankton biomass increased. Especially in treatments with high biomass production, also the amounts of nitrate lowered. Simultaneous increase of phytoplankton biomass and decrease of inorganic nutrients indicate their uptake in building algal biomass.

The amount of dissolved inorganic nutrients in waterbody is usually much smaller than the amount of nutrients bound in particles (Dodds, 1993). In short time scales, biological processes strongly determine the amounts of inorganic nutrients in the water (Dodds et al., 1991). Ammonium is the basic form of nitrogen

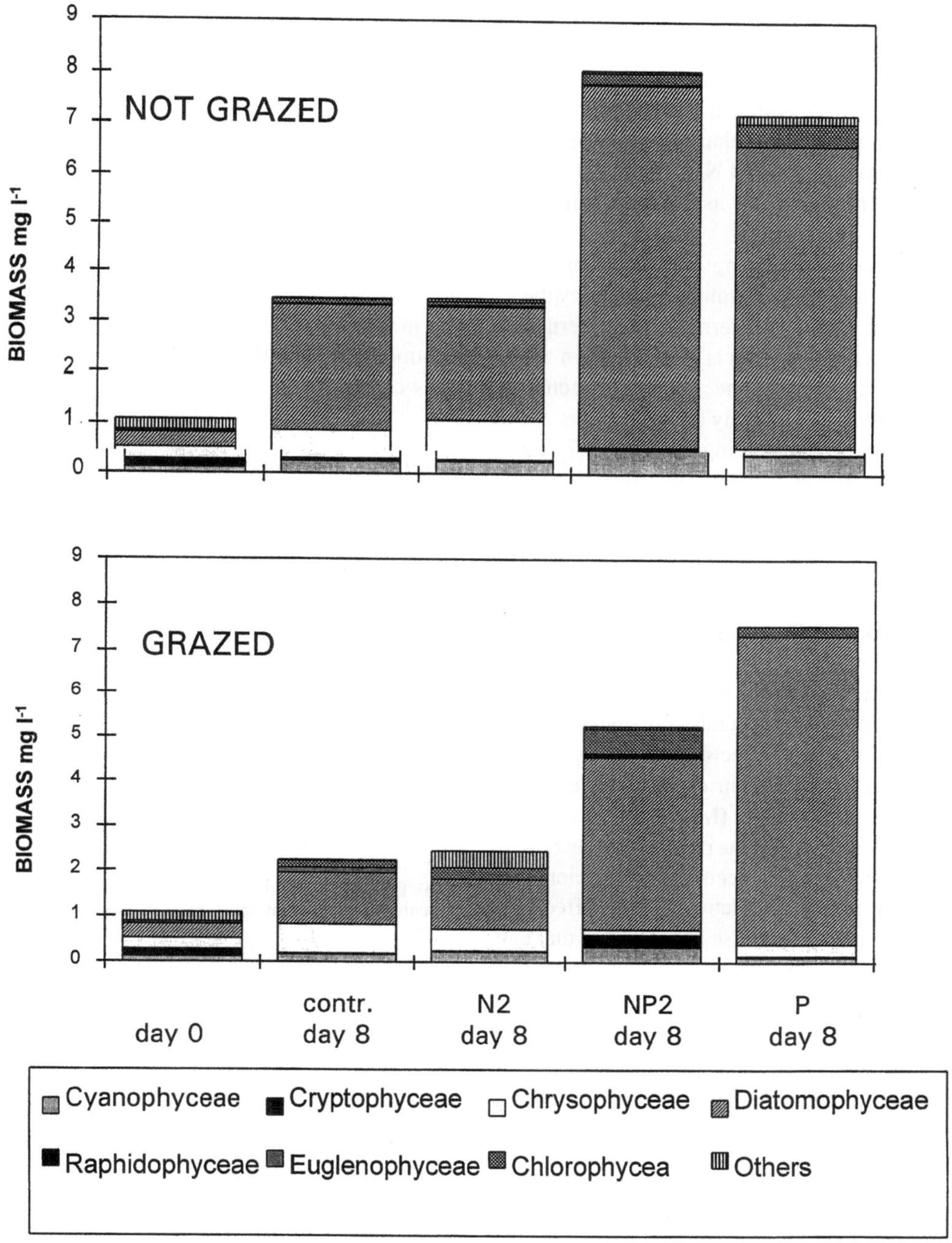

Figure 5. The group composition of phytoplankton in experiment VII in year 1995 in different treatments. Nutrient treatment abbreviations are explained in Table 2.

produced by excretion (Lehman, 1980). Therefore, in the case of low ammonium content and high nitrate content grazing may enhance significantly the availability of nitrogen for phytoplankters by excretion. In our study, the ammonium concentrations were higher in some of the grazed controls and nitrogen treatments when compared with the respective units without grazing (Figure 4). In treatments where phosphorus was added and the increase of phytoplankton biomass was high, no differences were found. It is probable that *Daphnia* grazing cycled nitrogen as ammonium and in the treatments where the production was relatively low, surplus ammonium stayed in the water column. It is possible that the effects of *Daphnia* grazing were indirect via changes in smaller size zooplankton (i.e. protozoans, ciliates), which were not affected by the presieving of the water and were obviously present in all experiments. This was not, however, examined here. When the increase of phytoplankton biomass was high and extra phosphorus (P and NP treatments) was added, the excreted ammonium was probably utilized by phytoplankton.

The negative effects of grazing on phytoplankton biomass can be, to some extent, compensated for by the better availability of nutrients (Sterner, 1986; Sterner, 1990). The excretion of nutrients is dependent on the nutrient demands of the grazer. The ratio of nitrogen to phosphorus of *Daphnia* is relatively low (Sterner et al., 1992), with *Daphnia magna* being around 12.7 (by atoms) according to Hessen (1990). Therefore, *Daphnia* have probably acted more like phosphorus traps rather than effectively recycling phosphorus in the systems studied.

In the studied systems, the biomass of phytoplankton was generally lower in grazer treatment combinations compared with non-grazer treatments. This indicates a major impact of zooplankton grazing in addition to nutrient control (especially phosphorus) of phytoplankton. The negative effects of large *Daphnia* on phytoplankton, in spite of a nutrient load, have also been found in *in situ* experiments (Mazumder & Lean, 1994). In addition to decreasing the phytoplankton biomass, grazing may also have effects on the efficiency of phytoplankton primary production. These effects, however, remain unclear, because we did not make any direct production measurements.

The sieving rate of *Daphnia pulex*, controlled for instance by temperature (Mourelatos & Lacroix, 1990), density of food particles (Wiedner & Vareschi, 1995) and age of the individual, has been measured to be about 2 ml ind.$^{-1}$ h^{-1} in natural cell densities (Wiedner & Vareschi, 1995). Based on these estimates, eight *D. pulex* individuals may sieve three litres in eight days, which exceeds the volume of one experimental unit used in our experiment.

Different phytoplankton species have different demands of inorganic nutrients. These demands and the capability to take up nutrients differ according to abiotic conditions (Rhee & Gotham, 1981), the physiological status of phytoplankton (Lean & Pick, 1981) and the ecological strategy concerning the nutrition of the algae (autotrophy-mixotrophy) (Bird & Kalff, 1987; Nygaard & Tobiensen, 1993). Some experiments have shown that both herbivory and changes in the nutrient loading can change diversity of algae dramatically (Proulx et al., 1996). In our study, the phytoplankton composition at the group level became more homogenous during the incubation. The relatively high amounts of diatoms in all the nutrient treatments was probably partly due to our incubation conditions. Diatoms are known to be good competitors at low light levels (Willen, 1991) and may, therefore, have gained advantage due to the relatively low light levels used. Some algae, including many diatoms, have also good capability of storing phosphorus, which explain the increasing algal biomass in control and nitrogen treatments, where the amounts of available phosphorus were low.

Acknowledgements

We will thank the staff of the North Savo Regional Environment Centre laboratory for analysing all nutrient samples (totally 720) and the field workers for a patient sampling of several hundred litres of lake water. T. Hammar, H. Tanskanen and J. Kangasjärvi kindly commented the manusript and made also valuable proposals during the studies. These studies were financially supported by Savon Sellu Oy and Kemira Chemicals Oy.

References

Aldridge, F. J., C. L. Shelske & H. J. Carrick, 1993. Nutrient limitation in a hypereutrophic Florida lake. Arch. Hydrobiol. 127: 21–37.

Bird, D. F. & J. Kalff, 1987. Algal phagotrophy: Regulating factors and importance relative to photosynthesis in *Dinobryon* (Chrysophyceae). Limnol. Oceanogr. 32: 277–284.

Dodds, W. K., 1993. What controls levels of dissolved phosphate and ammonium in surface waters? aquat. Sci. 55: 132–142.

Dodds, W. K. & J. C. Priscu, 1990. A comparison of methods for assessment of nutrient deficiency of phytoplankton in a large oligotrophic lake. Can. J. Fish. aquat. Sci. 47: 2328–2338.

Dodds, W. K., J. C. Priscu & B. K. Ellis, 1991. Seasonal uptake and regeneration of inorganic nitrogen and phosphorus in a large oligotrophic lake: size-fractionation and antibiotic treatment. J. Plankton Res. 13: 1339–1358.

Elser, J. J. & N. B. George, 1993. The stoichiometry of N and P in the pelagic zone of Castle Lake, California. J. Plankton Res. 15: 977–992.

Elser, J. J., E. R. Marzolf & C. R. Goldman, 1990. Phosphorus and nitrogen limitation of phytoplankton growth in the freshwaters of North America: A review and critique of experimental enrichments. Can. J. Fish. aquat. Sci. 47: 1468–1476.

Forsberg, C., S-O. Ryding, A. Claesson & A. Forsberg, 1978. Water chemical analyses and/or algal assay? – Sewage effluent and polluted lake water studies. Mitt. int. Ver. Limnol. 21: 352–363.

Gu, B. & V. Alexander, 1993. Seasonal variations in dissolved inorganic nitrogen utilization in a subarctic Alaskan lake. Arch. Hydrobiol. 126: 273–288.

Hessen, D. O., 1990. Carbon, nitrogen and phosphorus status in *Daphnia* at varying food conditions. J. Plankton Res. 12: 1239–1249.

Karjalainen, H., S. Seppälä & M. Walls, 1996. Ammoniumtypen merkitys kasviplanktontuotantoa säätelevänä tekijänä – esimerkkinä Kallavesi. Suomen ympäristö 19. In Finnish with English abstract.

Lean, D. R. S. & F. R. Pick, 1981: Photosynthetic response of lake plankton to nutrient enrichment: A test for nutrient limitation. Limnol. Oceanogr. 26: 1001–1019.

Lehman, J. T., 1980. Nutrient recycling as an interface between algae and grazers in freshwater communities. In W. C. Kerfoot (ed.), Evolution and Ecology of Zooplankton Communities. The University Press of New England, Hanover (N.H.): 251–263.

Lehman, J. T. & D. Scavia, 1982. Microscale patchiness of nutrients in plankton communities. Science 216: 729–730.

Mazumder, A. & D. R. S. Lean, 1994. Consumer-dependent responses of lake ecosystems to nutrient loading. J. Plankton Res. 16: 1567–1580.

McCarthy, J. J., D. Wynne & T. Berman, 1982. The uptake of dissolved nitrogenous nutrients by Lake Kinnert (Israel) microplankton. Limnol. Oceanogr. 27: 673–680.

Miyazaki, T., M. Watase & K. Miyake, 1989. Daily changes of uptake of inorganic carbon and nitrogen, and their relation to phytoplankton blooms in late spring-early summer in Lake Nakanuma, Japan. Hydrobiologia 185: 223–231.

Mourelatos, S. & G. Lacroix, 1990. In situ filtering rates of Cladocera: Effect of body length, temperature, and food concentration. Limnol. Oceanogr. 35: 1101–1111.

Nygaard, K. & A. Tobiensen, 1993. Bacterivory in algae: A survival strategy during nutrient limitation. Limnol. Oceanogr. 38: 273–279.

Proulx, M., F. R. Pick, A. Mazumder, P. B. Hamilton & D. R. S. Lean, 1996. Experimental evidence for interactive impacts of human activities on lake algal species richness. Oikos 76: 191–195.

Rhee, G-Y. & I. J. Gotham, 1981. The effect of environmental factors on phytoplankton growth: Temperature and the interactions of temperature with nutrient limitation. Limnol. Oceanogr. 26: 635–648.

Sterner, R. W., 1986. Herbivores' direct and indirect effects on algal populations. Science 231: 605–606.

Sterner, R. W., 1990. The ratio of nitrogen to phosphorus resupplied by herbivores: zooplankton and the algal competitive arena. Am. Nat. 136: 209–229.

Sterner, R. W., J. J. Elser & D. O. Hessen, 1992. Stoichiometric relationships among producers, consumers and nutrient cycling in pelagic ecosystems. Biogeochemistry 17: 49–67.

Suttle, C. A. & P. J. Harrison, 1988. Ammonium and phosphate uptake rates, N:P supply ratios, and evidence for N and P limitation in some oligotrophic lakes. Limnol. Oceanogr. 33: 186–202.

Tamminen, T. 1990: Eutrophication and the Baltic Sea: Studies on phytoplankton, bacterioplankton, and pelagic nutrient cycles. Academic dissertation. Helsinki, 22 pp.

Utermöhl, H., 1958. Zur Vervollkommnung der quantitativen Phytoplankton-Methodik. Mitt. int. Ver. Limnol. 9: 1–38.

Wehr, J. D., 1989. Experimental tests of nutrient limitation in freshwater picoplankton. Appl. envir. Microbiol.: 1605–1611.

Wetzel, R. G., 1983. Limnology. Saunders College Publishing, Philadelphia, 767 pp.

Whalen, S. C. & V. Alexander, 1986. Seasonal inorganic carbon and nitrogen transport by phytoplankton in an arctic lake. Can. J. Fish. aquat. Sci. 43: 1177–1186.

Wiedner, C. & E. Vareschi, 1995. Evaluation of fluorescent microparticle technique for measuring filtering rates of *Daphnia*. Hydrobiologia 302: 89–96.

Willen, E., 1991. Planktonic diatoms – an ecological review. Algolog. Stud. 62: 69–106.

Hydrobiologia **363:** 323–332, 1998.
T. Tamminen & H. Kuosa (eds), Eutrophication in Planktonic Ecosystems: Food Web Dynamics and Elemental Cycling.

In situ grazing pressure and diel vertical migration of female *Calanus euxinus* in the Black Sea

S. Besiktepe, A. E. Kideys & M. Unsal
Institute of Marine Sciences, Middle East Technical University, Erdemli-Icel, 33731 Turkey

Key words: grazing pressure, vertical migration, copepod, Calanus, Black Sea

Abstract

Gut pigment and abundance of the female *Calanus euxinus* (Hulsemann) were measured from several water layers (defined by density values), with 3–5 h intervals during 30 h and 21 h at a station in the southwestern Black Sea in April and in September 1995, respectively. The female *C. euxinus* was observed to begin migration to the upper phytoplankton-rich layer approximately 3 or 4 hours before the sunset. Only a fraction of the female *Calanus* population (0.2% in April and 3.6% in September) did not migrate but remained at the depth of the oxygen minimum zone during the nighttime. The migrating population was determined to have spent 7.5 h in the euphotic zone in April and 10.5 h in September. The grazing rate of female *Calanus euxinus* was measured from the gut content data collected from the layers which contain the euphotic zone. The percentage of primary production grazed by the female *C. euxinus* was calculated as 14.5% in April and 9.5% in September.

Introduction

There is an intriguing relationship between diurnal migration and feeding activity in zooplankton. There may be several reasons for the vertical migration of zooplankton, but in any case this would interact with grazing. The herbivorous zooplankton grazing impact represents an essential part of phytoplankton and zooplankton interactions in the ecosystems. The gut fluorescence method has been increasingly applied to determine *in situ* algal grazing rates of planktonic copepods in recent years (e.g. Boyd et al., 1980; Dagg & Grill, 1980; Kiorboe & Tiselius, 1987; Dagg et al., 1989; Morales et al., 1993; Tsuda & Sugisaki, 1994).

In this paper, the diel vertical migration of female *Calanus euxinus* was described by means of time series sampling in the Black Sea. The *in situ* grazing pressure of these copepods was also determined. Since our aim was specifically to determine the share of female *Calanus* grazing as a percentage of primary production, we analysed the gut pigment content only for the copepods collected from the euphotic zone.

Physical and biological properties of the Black Sea

The Black Sea, with an average depth of ~ 1240 m, contains the world's largest anoxic water volume (4.6×10^5 km^3, ~ 87% of the sea volume) below a thin layer (about 150 m) of oxygenated surface waters (Oguz et al., 1992). A pycnocline (or halocline) separates the oxic and anoxic waters and there is a well defined oxygen minimum zone (OMZ; with less than 10μM O_2) between these waters. The depth of the OMZ varies both seasonally and from one part of the Black Sea to another depending on the circulation and the intensity of eddies. However, recent investigations have shown that the OMZ (as well as other chemical and physical characteristics of the water column) could be explained better by water density rather than depth (Tugrul et al., 1992; Saydam et al., 1993; Murray et al., 1993). If the OMZ (and the other characteristics) is explained as a function of water density then spatial and temporal variations disappear. For example, the lower boundary of the OMZ changes from 120 m to 175 m at different locations during the April and September 1995 cruises in the Black Sea when it is expressed in terms of length units (e.g. meter), however it is always situated at the density of sigma-theta = 16.2 (Gokmen, 1996).

The distribution of the pelagic fauna is related to the boundaries of the OMZ (Vinogradov et al., 1992; Wishner et al., 1995). Vinogradov et al. (1992) has divided Black Sea pelagic ecosystem into two parts; the aerobiotic and the chemobiotic.

The aerobiotic waters of the Black Sea are biologically productive because of high run-off from rivers around the basin. The monthly primary production of water column for offshore areas from 1960 to 1991 was compiled by Vedernikov & Demidov (1993). In April the mean primary production was 520 mgC m^{-2} day^{-1} (range 50–990 mgC m^{-2} day^{-1}). In September the average value was 200 mgC m^{-2} day^{-1} (range 35–360 mgC m^{-2} day^{-1}). The euphotic zone depth (1% light depth of surface light) in the anticyclonic regions was around 35–40 m deep and in the cyclonic region around 30–35 m in May in the Black Sea (Vidal, 1995). However, the depth of the photosynthetic layer usually may extend down to 50–60 m with the optimum photosynthetic intensity being measured at a depth of 5–10 m (Bologa, 1985/1986). The seasonal primary production pattern in the Black Sea is bimodal. Diatoms predominantly bloom in spring, while *Emiliana huxleyi*, and to a lesser degree dinoflagellates, predominantly bloom in summer and fall (Hay & Honjo, 1989).

The chemobiotic environment includes the OMZ and the anaerobic layer. Only very few species can survive in the OMZ. Of these the principal one is *Calanus euxinus* (Vinogradov et al., 1992), which achieves the maximum biomass amongst Black Sea copepods (Ergun, 1994). The late copepodite stages and the adults of *C. euxinus* undertake diel vertical migrations and they spend the daytime in the OMZ. They decrease their oxygen consumption rate in the OMZ (Vinogradov et al., 1992).

Methods

Diel vertical migration

Samples were collected from a 1250 m deep station chosen in a dynamically non-active region (Lat: 41.54 Long: 29.50) during 26–28 April and 27–28 September 1995 (Figure 1). Vertical distribution of copepods was determined by vertical hauls using a Nansen Closing Net of 70 cm mouth opening and 112 m mesh size. The water column was sampled over five depth strata (in terms of density levels; see Introduction) according to major biogeochemical characteristics of the water column which may affect the distribution of mesozooplankton in the Black Sea (Figure 2). These depth strata are:

(1) from the depth of thermocline to the surface;
(2) from the depth of sigma-theta 14.6 to the thermocline; sigma-theta 14.6 roughly corresponds to lowest boundary of the euphotic zone (1% light level) (Oguz et al., 1996).
(3) the depth range between sigma-theta 15.4 and 14.6; at this stratum, the majority of nitrification and remineralization of organic matter take place (Lipp & Kempe, 1993).
(4) the depth range between sigma-theta 15.8 and 15.4; during the increase in the sinking of particulate organic matter, the upper boundary of the OMZ raises to the sigma-theta 15.4 (Basturk et al., 1994)
(5) and the depth range between sigma-theta 16.2 and 15.8; sigma-theta 16.2 corresponds to the bottom of the OMZ or the beginning of the anoxic water layer. According to Vinogradov et al. (1992), this is the daytime aggregation layer for late copepodite stages and adult of *Calanus*.

A series of vertical tows were taken in April and in September 1995 at 3–5 h intervals through a 30 h and 21 h cycle respectively. Samples were preserved in sodium-borate buffered 4% formalin-seawater solution, then subsampled with a Folsom splitter and identified by a stereomicroscope.

Determination of gut pigment content and grazing pressure

Following each vertical tow, female *Calanus euxinus* specimens were immediately separated on GF/F filters under the microscope for gut pigment content analysis, then frozen on dry-ice within 15 min after collection. Phytoplankton pigments in the guts of 10–15 freshly collected individuals of female *Calanus* (collected from the two uppermost layers in April and the uppermost layer in September encompassing the euphotic zone) were measured via whole animal fluorescence (Mackas & Bohrer, 1976; Boyd et al., 1980). For the background fluorescence of the female *Calanus*, a portion of the collected individuals were exposed to starvation in the GF/C filtered sea water during 24 h. Pigment analyses from each net tow were made in duplicate or triplicate depending on abundance of the organism.

For the gut evacuation rate experiment, copepods were collected at night from 50 m to the surface, where the maximum feeding is assumed to occur. After col-

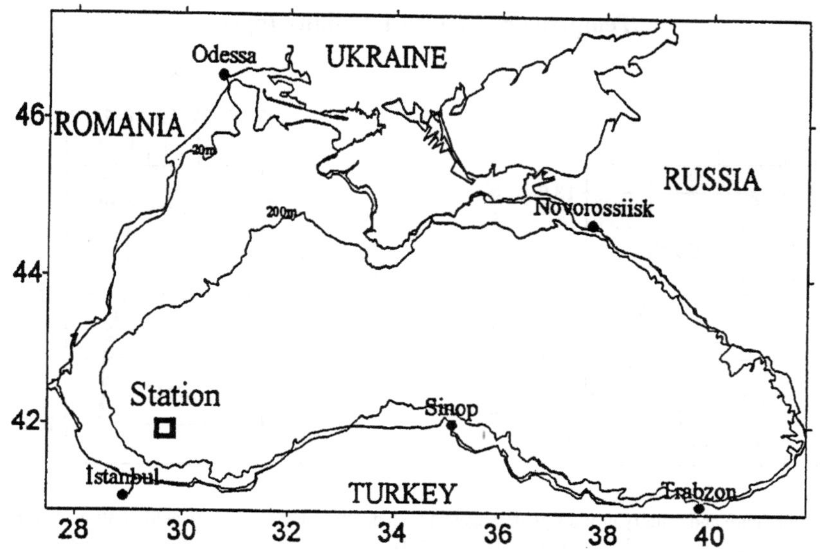

Figure 1. The sampling station in the Southwestern Black Sea.

lection, the cod end contents were rapidly retrieved and immediately sieved through a 2000 μm mesh to remove jelly organisms and sieved through 1000 μm mesh to remove smaller organisms. Some part of the filtrate was poured into soda/seawater solution (1:5, v/v) to anaesthetize the animals for initial gut fluorescence, whilst the rest was transferred into an aquarium filled with filtered seawater and kept in dark on deck. The decline rate of gut content was determined by periodically analyzing copepods at a time from aquarium over a 2 to 6 h period.

Gut fluorescence samples were homogenized in 10 ml of 90% aqueous acetone and the fluorescence of the filtrate was measured before and after acidification with 10% HCl using Hitachi model F-3000 Fluorescence Spectrophotometer. The chlorophyll and phaeopigment content of each copepod was calculated using the equations of Strickland & Parsons, (1972, cf. Boyd et al., 1980).

The studies of gastric evacuation in copepods indicate that the evacuation rate of food from the gut is exponential (Mackas & Bohrer, 1976; Dagg & Grill, 1980). However our gut evacuation rate experiments were unsuccessful because values of gut pigment analysis did not decrease regularly but fluctuated greatly with time. Therefore, we instead used the equation of Dam & Peterson (1988) for the estimation of the evacuation rates which is based on the relationship between gut clearance time and temperature for a wide variety of copepods:

$$R = 0.0117 + 0.001794T,$$

where R = Instantaneous evacuation rate; T = temperature.

Gut evacuation rate can be used to convert each measure of gut contents to an ingestion rate. Ingestion rate is calculated from;

$$I = SR$$

where S = level of gut contents (ng pigment copepod^{-1}); R = instantaneous evacuation rate (min^{-1}). By extrapolation over 60 min this equation can be used to estimate hourly ingestion rates.

Since recent studies have revealed that ingested chlorophyll-*a* is converted to phaeopigments during passage through the copepod gut (Head & Harris, 1992), the gut content pigment was expressed as chlorophyll-*a* + phaeopigments in a chlorophyll-*a* equivalent weight.

Grazing pressure on primary production

The average abundance of female *Calanus* for the whole water column was calculated by combining the data from all performed tows. These were carried out at different hours (8 times in April and 6 in September) for the 5 different layers. Then the daily grazing pressure of the female *C. euxinus* was estimated by taking into account its average abundance with ingestion rate. Grazing pressure, which was calculated as total pigment, was converted to carbon (C) by using (Phytoplankton Carbon) PC:Chl-*a* ratio. The PC values were estimated from the cell volume measurements of phytoplankton using carbon-volume relationship of

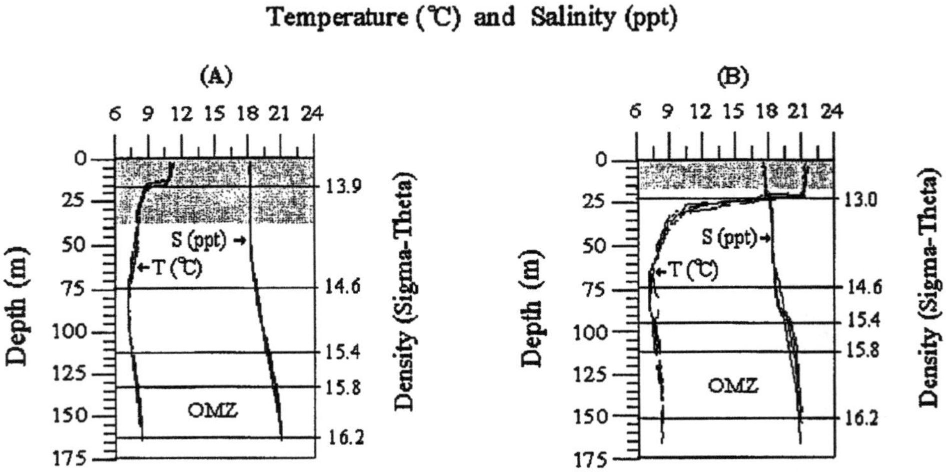

Figure 2. The potential temperature and salinity profiles plotted against both depth and water density at the station in April (A) and in September (B) 1995. Measurements were repeated several times during the stay at the station. The sampling layers are shown with horizontal solid lines corresponding to different sigma-theta values (see Methods for detailed explanation). The shaded area shows the euphotic zone. The water column between the surface and the depth of 15.8 density surface is oxic. The depth below 16.2 density surface is anoxic. OMZ = Oxygen Minimum Zone.

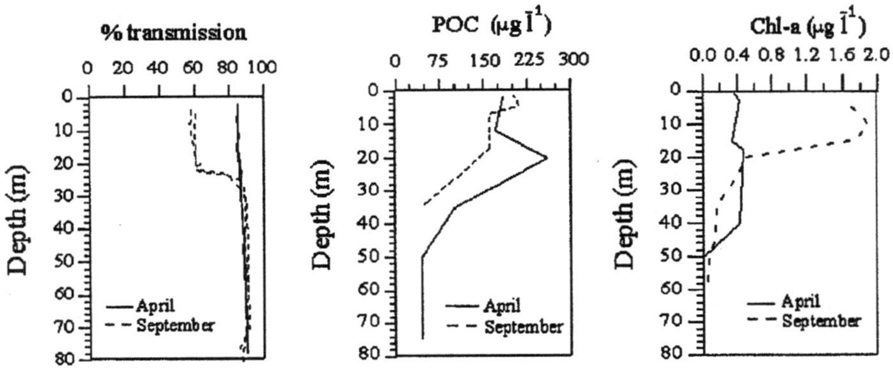

Figure 3. Vertical profiles of light tranmission, POC (Particulate Organic Carbon) and Chl-*a* (Chlorophyll-*a*) at the station in April and September 1995.

Strathmann (1967, Miss E. Eker unpublished data). The grazing pressure was obtained by dividing the calculated consumption of grazing rate by the integrated primary production of the water column (Yilmaz et al., 1996) .

Results

Hydrography and phytoplankton composition of the station

Salinity, potential temperature, Particulate Organic Carbon (POC), chlorophyll-*a*, and light transmission profiles obtained during the biological measurements are presented in Figures 2 and 3 to illustrate hydrochemical properties of the water column (down to beginning of the anoxic water) of the region. In April the surface mixed layer is a relict feature of the winter time convective mixing and due to the spring time solar heating a new layer was formed in the first 10–15 meters of the water column (Figure 2). The depth of chlorophyll-a maximum was between 18–40 m (Figure 3). In September there was a sharp stratification at around 25 m depth due to the temperature. The Chl-*a* profiles showed maxima at 12–13 meters. Both in April and in September the light transmission measurements displayed minimum values near the surface indicating the presence of higher amounts of suspended matter at this layer compared to lower depths, however the difference between the surface and deeper layers is much more pronounced in September. In September the transmission in the surface mixed layer was much lower than in April which was partly due to a higher abundance of both alive and broken gelatinous organisms observed visually in the former sampling period. In the top few meters, POC concentrations in September were also higher than that in April (Figure 3). The euphotic zone depth was 38m in April (corresponding to the two uppermost sampling layers) and 17 m in September (corresponding to the uppermost sampling layer).

In April there were 31 species of phytoplankton in the euphotic zone, and among these, the coccolithophorid *Emiliana huxleyi* was the most abundant species (Miss Elif Eker, unpublished data). *Heterocapsa triquetra* (synonym *Peridinium triquetrum*) and *Scrippsiella trochoidea* (synonym *P. trochoideum*) were the main species in Peridinea, and *Nitzschia delicatula* and *N. closterium* were the most dominant ones in Diatomea. Average phytoplankton concentration at the surface was 14×10^4 cells l^{-1}in April. This abundance was made up by 53% coccolithoporids, 19% dinoflagellates and 1.3% diatoms. In September both species number (56 species) and abundance (11×10^5 cells l^{-1}) were higher than those in April. Coccolithoporids represented almost entirely by *E. huxleyi* made up the bulk of phytoplankton abundance with 92%. The share of dinoflagellates and diatoms in phytoplankton was same with 2.8 and 2.9% respectively. *Exuviella cordata*, *E. compressa* and *Glenodinium paululum* among dinoflagellates and *Rhizosolenia calcar-avis* and *Nitzschia delicatula* among diatoms were the most abundant species (Miss Elif Eker, unpublished data).

During both sampling periods the oxygen minimum zone (OMZ) was observed between sigma-theta 15.8 and 16.2 (Figure 4).

Diel vertical migration

The vertical distributions of the female *Calanus euxinus* in April and in September are presented in Figures 5 and 6. The sun sets at 19:52h in April and at 17:43h in September. In both sampling periods the female *Calanus* began to migrate upward to the phytoplankton rich upper layers towards the evening. At night they concentrated in the first layer between the thermocline and the surface. After midnight the majority of copepods started to migrate downward. During

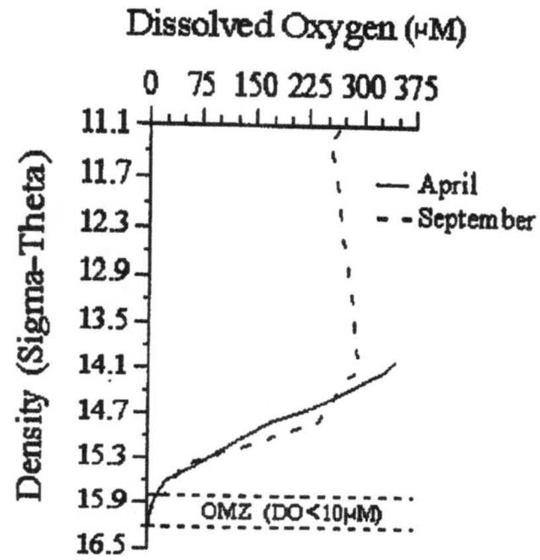

Figure 4. Vertical profiles of DO (Dissolved Oxygen) concentration at the station in April and in September 1995. OMZ = Oxygen Minimum Zone.

nighttime only a very small percentage of the female *Calanus* were present in the OMZ in both sampling periods: 3.6% in September and 0.2% in April. In the early morning (in April and in September the sun rises at 06:06h and 05:47h respectively) the bulk of the copepods was already in the OMZ. During daytime, female *C. euxinus* concentrated in the OMZ and this cycle was repeated throughout the sampling period. During the study, the abundance of the female *Calanus* in the water column showed some variations with different sampling time resulted most probably from physical dynamism as lateral intrusion of the water masses.

It was difficult to define accurately the extent of feeding period (in the sampling layers corresponding to the euphotic zone) due to the long (3–5 h) sampling interval; however, we can estimate roughly the duration that female *Calanus* spends at the euphotic zone from its vertical distribution figures (Figures 5 and 6). In April as some copepods are already in the upper layers at 17:00 and 18:00 hours, it could be assumed that they can begin to migrate around at 16:00h. The majority of the individuals must reach to the euphotic zone depth at around 21:00h. The lower boundary of the OMZ located at 162 m and thickness of the euphotic zone was 38 m so they should migrate at least 125 m to reach the euphotic zone. They spend 5 h to transit this 125 m resulting an upward speed of around 25 m h^{-1}. At 02:00h, just over the half of the copepods were still in the euphotic zone (Figure 5). According to the estimation of Hardy & Bainbridge (1954, cf. Marhall & Orr, 1972) the downward speed of *Calanus* is about 3 times higher than its upward speed over a long period experiment (i.e. 1 h). If this ratio is taken into account

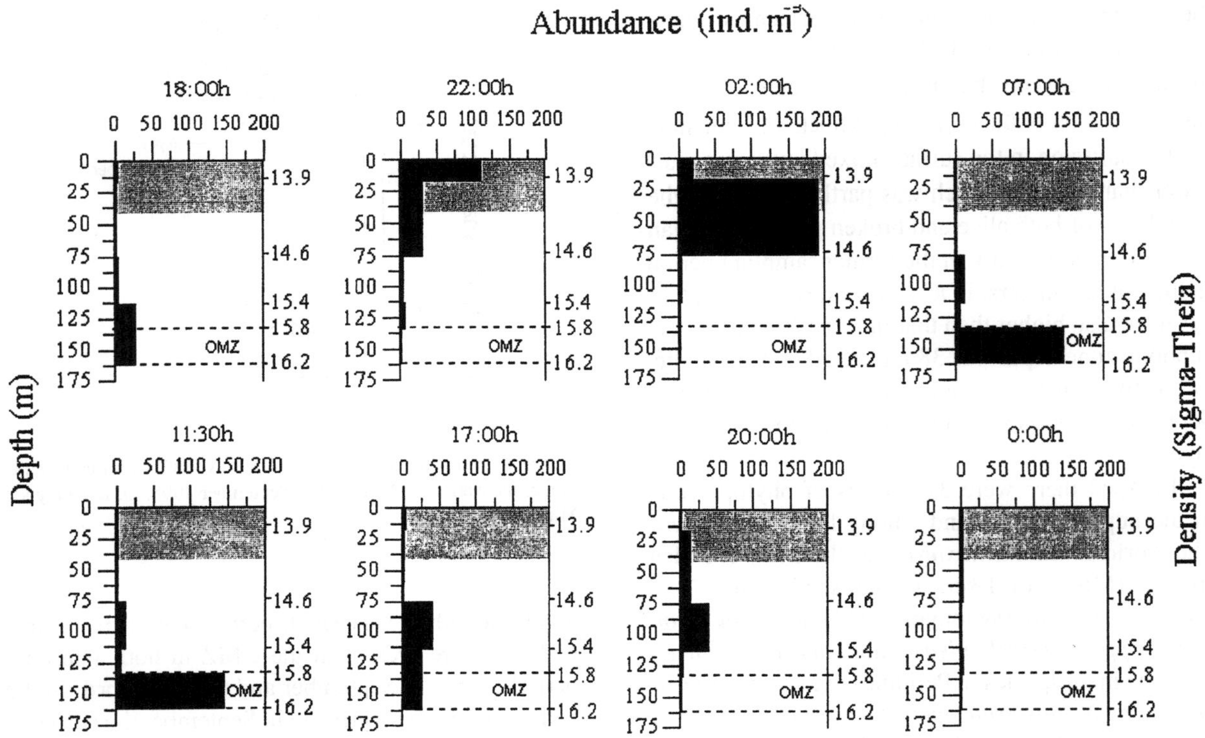

Figure 5. Diel changes in the vertical distribution of female *C. euxinus* abundance (dark area) in April 1995. The shaded area indicates the euphotic zone. OMZ = Oxygen Minimum Zone.

the downward speed of *Calanus* will be 75 m h^{-1}, then it must take approximately 2 h to reach the OMZ. Since they were already in the suboxic layer at 07:00h, they must begin downward migration sometime between 02:00 and 05:00h. Consequently the duration of stay of female *Calanus euxinus* in the euphotic zone was 7.5 h in April. With a similar calculation, the duration of stay in the euphotic zone in September would be around 10.5 h, with an upward speed of around 34 m h^{-1} and downward speed of 102 m h^{-1}.

Gut pigment content

The background fluorescence for the starved individuals was estimated as 0.68 ± 0.39 ng pigment copepod^{-1}. After this value was accounted for, the overall average gut pigment concentration of female *Calanus* was calculated to be 10.1 ng pigment copepod^{-1} in April and 14.0 ng pigment copepod^{-1} in September (Table 1). Although gut content values were similar, the overall average ingestion rate varied due to the differences in the gut evacuation rate constant between the sampling periods. Using the equation of Dam & Peterson (1988) the gut evacuation rate constant was 1.86 h^{-1} in April and 3.0 h^{-1} in September.

Grazing pressure on primary production

The ingestion rate of *Calanus* was calculated as 18.7 ng pigment copepod^{-1} h^{-1} in April and 42.0 ng pigment copepod^{-1} h^{-1} in September (Table 1). The average abundance of female *Calanus* in the whole water column was 3365 ind m^{-2} in April and 1343 ind m^{-2} in September (Table 2). These made up 11.4% of total copepod abundance (including copepodite stages) in April and 6.7% in September. Almost all female *Calanus* belonged to the migratory group. Only 3.6% in September and 0.2% in April were observed during the night time in the OMZ which could belong to a non-migrating population. By ommitting these non-migrating fractions, we assumed that each female *Calanus* in the water column migrate to the euphotic layer and spend 7.5 h day^{-1} in the euphotic zone in April and 10.5 h day^{-1} in September for feeding. Daily consumption by the female *Calanus* was estimated by taking into account the feeding duration, the number of individuals in the whole water column and

Table 1. Gut pigment content (S, ng pigment copepod^{-1}) and ingestion rate (ng pigment copepod^{-1} h^{-1}) of female *Calanus euxinus* in the layers encompassing the euphotic zone during April and September 1995. Gut evacuation rate constants (R) were 1.86 h^{-1} in April and 3 h^{-1} in September

Sampling hours	Gut Pigment Content in the Layers from the thermocline to the surface	Gut Pigment Content in the Layers from sigma-theta 14.6 to the thermocline	Ingestion Rate ($I = SR$)
	April		
18:00	–	-	–
22:00	4.77	12.63	16.18
02:00	6.76	6.86	12.67
07:00	–	–	–
11:30	–	–	–
17:00	–	–	–
20:00	21.28	3.42	22.97
24:00	8.46	16.2	22.93
Overall Average =		10.1 ± 5.8	18.7 ± 4.4
	September		
19:00	8.7		26.2
23:30	8.7		26.1
03:30	18.8		56.3
07:30	–		–
12:00	–		–
16:00	19.9		59.6
Overall Average =	14.0 ± 5.3		42.0 ± 15.9

– there was not sufficient number of female copepods for the analysis.

Table 2. Number of female *C. euxinus* in the oxic water column and grazing pressure on primary production in the euphotic zone during April and September 1995. Standard deviations are shown in paranthesis

	April	September
No. of Ind. (m^{-2})	3365 (3340)	1343 (853)
POC (μg l^{-1})	178.9 (55.8)	188.6 (23.4)
PC (μg l^{-1})	25.8 (18.3)	85.0 (50.0)
Chl-*a* (μg l^{-1})	0.34 (0.05)	1.31 (0.51)
PC:Chl-*a*	76	65
Average Phytoplankton Conc. (cell l^{-1})	14×10^4 (2×10^4)	11×10^5 (54×10^5)
Consumption (as μg pigment m^{-2} day^{-1})	471.6 (112.1)	592.8 (224.8)
Consumption (as mgC m^{-2} day^{-1})	35.9 (8.5)	38.5 (14.6)
Primary Prod. (mgC m^{-2} day^{-1})	247.0	405.4
Grazing Pressure (%)	14.5 (3.5)	9.5 (3.6)

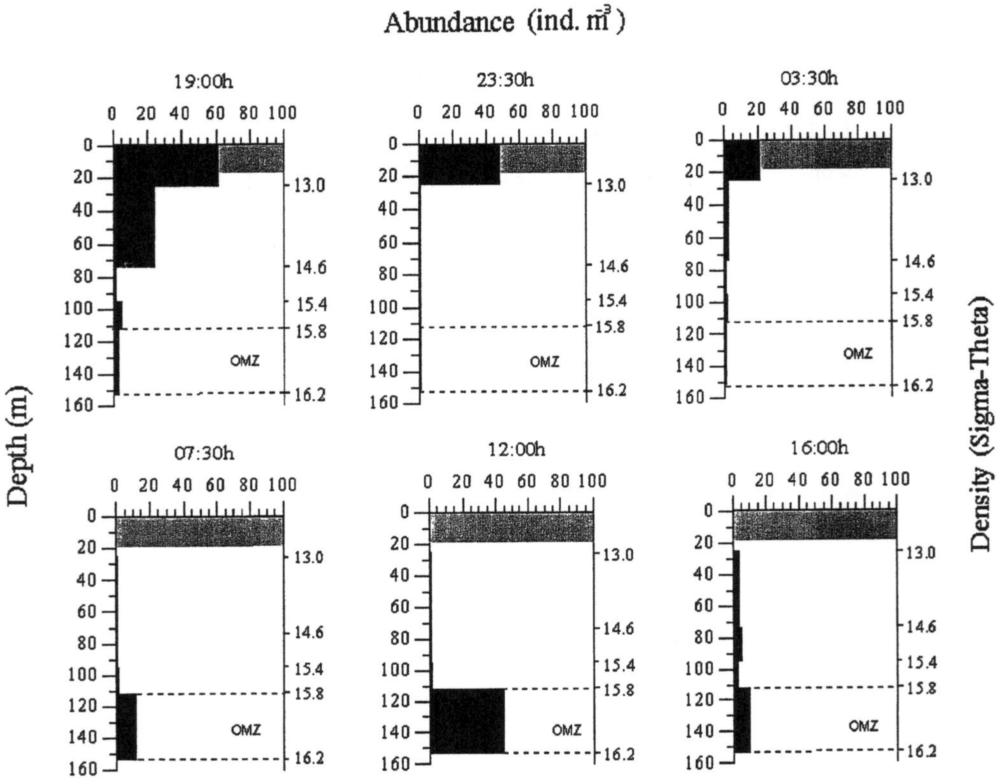

Figure 6. Diel changes in the vertical distribution of female *C. euxinus* abundance (dark area) in September 1995. The shaded area indicates the euphotic zone. OMZ = Oxygen Minimum Zone.

gut pigment concentrations. In April 472 μg pigment m^{-2} day^{-1} and in September 593 μg pigment m^{-2} day^{-1}was found to be consumed in the euphotic zone by the female *Calanus* (Table 2). The PC:Chl-*a* ratios are 76 in April and 65 in September. These PC:Chl-*a* ratios were used to convert the consumed gut pigment to carbon. The estimated primary production values at this station were 247 mgC m^{-2} day^{-1} in April and 405 mgC m^{-2} day^{-1} in September (Yilmaz et al., 1996). So in April the consumption rate of female *C. euxinus* was calculated as 35.9 mgC m^{-2} day^{-1} representing 14.5% of the primary production, and in September the consumption was 38.5 mgC m^{-2} day^{-1}, equal to 9.5% of the primary production.

Discussion

Besides the higher abundance of jelly substances, the high suspended matter (which is deduced from the light transmission profile in Figure 3) observed above thermocline (the mixed layer) in September is due to two other additional factors; the effect of Danube river and the phytoplankton. A higher input from Danube river to the sampling area is clear from the salinity profile (Figure 2) which is lower in this period than that in April (as was previously observed by Sur et al., 1994). Since rivers in general and Danube river in particular (Balkas et al., 1990) contribute a considerable amount of terrestrial material, a higher river input into the area would cause a decrease in light transmission observed in September. Perhaps as a result of this high river input, the phytoplankton abundance in this sampling period (11 × 10^{5} cells l^{-1}) was also much higher than that in April (14 × 10^{4} cells l^{-1}). As a result the high concentration of phytoplankton did not only contribute to a decrease in light transmission, but also to a higher content of chlorophyll-*a* in September compared to April (Figure 3).

In the Black Sea a great portion of copepods migrate daily from the OMZ containing little or no phytoplankton into surface waters where phytoplankton is abundant (Vinogradov et al., 1985 and 1992; Vinogradov & Nalbandov, 1990), and this was also observed for the female *C. euxinus* in this study. In both sampling periods, excepting a very small percentage, almost all female *C. euxinus* were observed to undertake diel migration. This result is in a good agreement with that found by Vinogradov & Nalbandov (1990) and Vinogradov et al. (1992). Therefore we can conclude that despite the existence of a non-migrating population of copepodite V of *Calanus*, females of this copepods are migratory. We estimated the feeding duration of female *Calanus* at the euphotic zone to be about 7.5 h in April and 10.5 h in September. Morales et al. (1993) found that feeding lasted at least 6 to 8 h for large fraction (1000–2000μm) of copepods. Our rough estimation of speed was 25 m h^{-1} upwards and 75 m h^{-1} downwards in April. In September, the speed values were 34 m h^{-1} upwards and 102 m h^{-1} downwards. These differences in speed between both sampling periods may come from several environmental conditions, most notably being the temperature. The values found here are in the range of experimental results of Hardy & Bainbridge (1954, cf. Marshall & Orr, 1972). They observed that *Calanus* migrated with a speed of 66 m h^{-1} upwards and 107 m h^{-1} downwards over a short experimental period (2 minutes) while the speeds were 15 m h^{-1} upwards and 47 m h^{-1} downwards over a longer experimental period (1 hour).

There may be several reasons for the higher grazing pressure value in April (14.5%) compared to that in September (9.5%). First of all, there is a higher abundance of copepods with lower primary productivity in April than that in September. Besides these, phytoplankton composition in April could suit better for *Calanus* feeding. Petipa (1964) observed that Peridinea was the dominant group and Diatomea was the second most abundant group in the gut of *Calanus*. Coccolithophorids were rarely observed as a food item. While in both seasons coccolithophorids are more important than the other groups of phytoplankton, in April the abundance percentage of Peridinea was 19% while that of Diatomea was 1.3% in phytoplankton. In September the percentages of these two groups of phytoplankton were equal (3%).

The values on grazing pressure of the present study are comparable with those in the literature. Morales et al. (1993) estimated copepod community grazing < 10% of the daily primary production in the northeast Atlantic. Tsuda & Sugisaki (1994) found that the grazing rate of the copepods was 1.4 to 2.0% of the measured primary production in the western subarctic North Pacific during spring. Their results indicate that the copepod community was unimportant as a primary consumer where nano and picoplankton had dominated over phytoplankton. For the grazing intensity, the size of phytoplankton is very important, nano and picophytoplankton being too small for grazing by copepods (Tsuda & Sugisaki, 1994). In contrast to many other regions, the contribution of picophytoplankton to the Black Sea phytoplankton is low (Stelmakh, 1988). Arinardi et al. (1990) estimated the grazing intensity of 27 species of female copepods to be between 5 and 26% of the primary production in the upwelling site in the Banda Sea, Indonesia.

According to Ergun (1994), in the Southern Black Sea the bulk of copepod abundance is made up of only five species; *Calanus euxinus, Acartia clausi, Pseudocalanus elongatus, Centropages kroyeri* and *Paracalanus parvus*. His results suggested that *C. euxinus* constituted the biggest proportion of the biomass in each sampling period; June 1991, January and July 1992. During these sampling periods the average percentage of *C. euxinus* was 85% as biomass and 22% as number among all 5 common copepods. Therefore the considerable grazing pressure by the female *C. euxinus* found in this study is not suprising. Finally it can be concluded that female *Calanus euxinus* has a major importance in the transfer of organic matter from primary producers to the higher taxa, including pelagic fish.

Acknowledgements

We thank to Dr D. Ediger and Mrs M. Sur for their help with spectrophotometer and fluorometer measurements, to Drs H. Ducklow and S. Tugrul for their critical reading of the manuscript, and to Miss E. Eker for allowing us to use her unpublished data on phytoplankton. We are grateful to the crew of the R/V 'Bilim' for their assistance at sea and to Alison M. Kideys for the correcting the English of the text. This study was supported by the Turkish Scientific and Technical Research Council (TUBITAK) and NATO TU-Black Sea Project.

References

Arinardi, O. H., M. A. Baars & S. S. Oosterhuis, 1990. Grazing in tropical copepods, measured by gut fluorescence, in relation to seasonal upwelling in the Banda Sea (Indonesia). Neth. J. Sea Res. 25: 545–560.

Balkas, T., G. Dechev, R. Mihnea, O. Serbanescu & U. Unluata, 1990. State of the marine environment in the Black Sea region. UNEP regional seas reports and studies. No. 124, 41 pp.

Basturk, O., C. Saydam, I. Salihoglu, L. V. Eremeeva, S. K. Konovalov, A. Stoyanov, A. Dimitrov, A. Cociasu, L. Dorogan & M. Altabet, 1994. Vertical variation in the principle chemical properties of the Black Sea in the Autumn of 1991. Mar. Chem. 45: 149–165.

Bologa, A. S., 1985/1986. Planktonic primary productivity of the Black Sea: A review. Thalassia Jugosl. 21/22: 1–22.

Boyd, C. M., S. L. Smith & T. J. Cowles, 1980. Grazing patterns of copepods in the upwelling system off Peru. Limnol. Oceanogr. 25: 583–596.

Dagg, M. J. & D. W. Grill, 1980. Natural feeding rates of *Centropages typicus* females in the New York Bight. Limnol. Oceanogr. 25: 597–609.

Dagg, M. J., B. W. Frost & W. E. Walser, 1989. Copepod diel migration, feeding and the vertical flux of phaepigments. Limnol. Oceanogr. 34: 1062–1071.

Dam, H. G. & W. T. Peterson, 1988. The effect of temperature on the gut clearance rate constant of planktonic copepods. J. exp. mar. Biol. Ecol. 123: 1–14.

Ergun, G., 1994. Distribution of five calanoid copepod species in the southern Black Sea. Ms. Thesis in Middle East Technical University, Turkey, 134 pp.

Gokmen, S., 1996. A comparative study for the determination of hydrogen sulfide in the suboxic zone of the Black Sea. Ms. Thesis in Middle East Technical University, Turkey, 156 pp.

Hay, B. & S. Honjo, 1989. Partical deposition in the present and holocene Black Sea. Oceanography 2: 26–31.

Head, E. J. H. & L. R. Harris, 1992. Chlorophyll and carotenoid transformation and destruction by *Calanus* spp. grazing on diatoms. Mar. Ecol. Prog. Ser. 86: 229–238.

Kiorboe T. & P. T. Tiselius, 1987. Gut clearance and pigment destruction in a herbivorous copepod, *Acartia tonsa*, and the determination of in situ grazing rates. J. Plankton Res. 9: 525–534.

Lipp, A. & S. Kempe, 1993. The Black Sea. A summary of new results. Presented to International Advanced Study Course on Biogeochemical processes, environment/development interactions and the future for the Mediterranean Basin. Nice, August 30–September 17.

Mackas, D. & R. Bohrer, 1976. Fluorescence analysis of zooplankton gut contents and an investigation of diel feeding patterns. J. exp. mar. Biol. Ecol. 25: 77–85.

Marshall, S. M. & A. P. Orr, 1972. The biology of a marine copepod. Springer-Verlag, Heidelberg/New York, 195 pp.

Morales, C. E., R. P. Harris, R. N. Head & P. R. G. Tranter, 1993. Copepod grazing in the oceanic northeast Atlantic during a 6 week drifting station: the contribution of size classes and vertical migrants. J. Plankton Res. 15: 185–211.

Murray, J. W., L. A. Codispoti & G. E. Friederich, 1995. Oxidation-Reduction Environments: The suboxic zone in the Black Sea. In C. P. Huang, C. R. O'Melia & J. J. Morgan (eds), Aquatic Chemistry. Am. Chem. Soc., Washington, DC., 157–176.

Oguz, T., P. E. La Violette & U. Unluata, 1992. The upper layer circulation of the Black Sea: Its variability as inferred from hydrographic and satellite observations. J. Geophys. Res. 97: 569–584.

Oguz, T., H. Ducklow, P. Malanotti-Rizzoli, S. Tugrul, N. P. Nezlin & U. Unluata, 1996. Simulation of annual plankton productivity cycle in the Black Sea by a one-dimensional physical-biological model. J. Geophys. Res. 101(C7): 61,585–16,599.

Petipa, T. S., 1964. Diurnal rhythm of food and diurnal diet of *Calanus helgolandicus* in the Black Sea. Works of Sevastopol Biological Station. Academy of Sciences, USSR, (In Russian).

Saydam, C., S. Tugrul, O. Basturk & T. Oguz, 1993. Identification of the oxic/anoxic interface by isopycnal surfaces in the Black Sea. Deep Sea Res. 40: 1405–1412.

Stelmakh, L. V., 1988. The contribution of picoplankton to primary production and the content of chlorophyll *a* in euphotic waters as exemplified by Sevastopol Bay. Oceanology 28: 95–99.

Strathmann, R. R., 1967. Estimating the organic carbon content of phytoplankton from cell volume or plasma volume. Limnol. Oceanogr. 12: 411–418.

Sur, H. I., E. Ozsoy & U. Unluata, 1994. Boundary current instabilities, upwelling, shelf mixing and eutrophication processes in the Black Sea. Prog. Oceanogr. 33: 249–302.

Tugrul, S., O. Basturk, C. Saydam & A. Yilmaz., 1992. Changes in the hydrochemistry of the Black Sea inferred from density profiles. Nature 359: 137–139.

Tsuda, A. & H. Sugisaki, 1994. In situ grazing rate of the copepod population in the western subarctic North Pacific during spring. Mar. Biol. 120: 203–210.

Vedernikov, V. I. & A. B. Demirov, 1993. Primary production and chlorophyll in the deep regions of the Black Sea. Oceanology 33: 193–199.

Vidal, C. V., 1995. Bio-optical characteristics of the Mediterranean and the Black Sea. Ms. Thesis in Middle East Technical University, Turkey, 134 pp.

Vinogradov, M. E., M. V. Flint & E. A. Shushkina, 1985. Vertical distribution of mesoplankton in the open area of the Black Sea. Mar. Biol. 107: 89–95.

Vinogradov, M. E. & Y. R. Nalbandov, 1990. Effect of changes in water density on the profiles of physicochemical and biological characteristics in the pelagic ecosystem of the Black Sea. Oceanology 30: 567–573.

Vinogradov, M. E., E. G. Arashkevich & S. V. Ilchenko, 1992. The ecology of the *Calanus ponticus* population in the deeper layer of its concentration in the Black Sea. J. Plankton Res. 14: 447–458.

Wishner, K. F., C. J. Ashjian, C. Gelfman, M. M. Gowing, L. Kann, A. L. Levin, S. L. Mullineaux & J. Saltzman, 1995. Pelagic and benthic ecology of the lower interface of the Eastern Tropical Pacific oxygen minimum zone. Deep Sea Res. 42: 93–115.

Yilmaz, A., C. Polat, D. Ediger & S. Tugrul, 1996. On the production, elemental composition (C, N, P) and distribution of photosynthetic organic matter in the southern Black Sea. Presented to International PELAG symposium. August 26–30 1996, Helsinki, Finland.

Hydrobiologia **363:** 333–339, 1998.
T. Tamminen & H. Kuosa (eds), Eutrophication in Planktonic Ecosystems: Food Web Dynamics and Elemental Cycling.

Zooplankton-phytoplankton interactions: a possible explanation of the seasonal succession in the Kuršių Marios lagoon

Zita Rasuolė Gasiūnaitė[1] & Irina Olenina[2]
[1]*Centre for System Analysis, Klaipėda University, Manto 84, LT- 5808, Klaipėda, Lithuania*
[2]*Centre for Marine Research, Taikos 26, LT- 5802, Klaipėda, Lithuania*

Key words: zooplankton, phytoplankton, trophic interactions, seasonal succession, Cyanobacteria blooms

Abstract

The zooplankton-phytoplankton interactions in the Kuršių Marios lagoon (southeastern Baltic Sea) were investigated in 1995. The objective was to evaluate the role of herbivores (crustacean zooplankton) in the seasonal succession of phytoplankton, as well as the influence of food conditions on structure and dynamics of zooplankton community. Our results demonstrated that the crustacean grazing pressure may restrict the development of small Chlorophyta and Diatomophyceae and, in turn, favouring growth of Cyanobacteria. Blooms of filamentous Cyanobacteria possibly has an inhibitory effect for *Daphnia,* decreasing their biomass as well possibly explaining the shift of dominant zooplankton species. The influence of planktivory on seasonal plankton succession remains unclear because of lack of fish data.

Introduction

It is well known that the density, species composition and size distribution of phytoplankton are regulated not only by nutrient conditions, but also by zooplankton grazing (Sarnelle, 1993; Kivi et al., 1993). Grazers may affect phytoplankton communities through several mechanisms, including direct suppression of edible algae, enhancement of inedible algae and shifts in the outcome of competition caused by grazer effects on nutrient supply ratios and regeneration rate (Sterner, 1989; Kivi et al., 1993). Algae smaller than 35 μm are better competitors for light and nutrients than larger ones but they are often suppressed by grazing. Larger algae, between 35 and 250 μm, may be unaffected by herbivores (Riegman et al., 1993; Edgar & Green, 1994).

Algal biomass and primary production should be lower in *Daphnia*-dominated lakes than in lakes dominated by small zooplankton such as *Bosmina*, small calanoid copepods or rotifer and algal biomass and production should be inversely related to mean zooplankton size (Kitchell & Carpenter, 1996). Nevertheless, at times, increased body size or biomass of zooplankton and grazing pressure on small algae shifts the phytoplankton community to one dominated by larger, 'resistant' species, such as filamentous cyanobacteria (Porter, 1977; Sarnelle, 1993; Carpenter et al., 1996a).

Zooplankton depend on a complex of various physical, chemical and biotic environmental factors. Planktivory and food supply often are responsible for community structure as well as seasonal succession (DeMott, 1989; Gliwicz & Pijanowska, 1989). Small-bodied zooplankton will be more abundant in the presence of planktivorous fish, while invertebrate predation can limit the smallest size (Brooks & Dodson, 1965). In addition, predation in conjunction with effects of food limitation could also alter the results of competition. For instance, in the absence of fish, large-sized zooplankton species will monopolise resources as they are competitively superior (Hall et al., 1976). On the other hand, large daphnids are more affected by inhibitory particles (e.g. filamentous cyanobacteria) than small daphnids. This inhibitory effect can reverse the outcome of competition between species with different body sizes (Gliwicz & Lampert, 1993). Thus, the complexity of the mechanisms may explain the multifarious and individualistic outcomes of studies on zooplankton–phytoplankton interactions (Carpenter et al., 1996a).

Here we present the results of field studies carried out during 1995 in the Kuršių Marios lagoon. The objective was to evaluate the interdependency between seasonal succession of phytoplankton and herbivorous crustacean zooplankton community structure. We focused on the functional relations between zooplankton and phytoplankton communities as well as key groups of crustaceans and planktonic algae, examining shifts in dominant species, size structure and biomass dynamics of the plankton.

Study site

The Kuršių Marios lagoon is a shallow transitory freshwater basin, situated in the southeastern Baltic Sea and connected to the Sea through the narrow Klaipeda straight (Figure 1). The area of the lagoon is 1584 km^2, its mean depth is 3.8 m; maximum depth is 5.8 m in the southern region and 10–12 m in the artificially deepened Klaipeda Strait harbour area. The southern and central portions of the lagoon are freshwater due to discharge from the Nemunas River and other smaller rivers, while the northern area is slightly influenced by the Baltic. The predominant flow of water is from the south to the north discharging approximately 22 km^3 into the Baltic Sea and maintaining freshwater conditions in the northern part of the lagoon. Brackish water intrusions are most common during August to October when 70% ($\approx$5 km^3) of the total annual input occurs. (Pustelnikovas, 1994, Stankevičius, 1996).

In 1995, water temperature began to increase at the end of April (13°C), reaching 16 °C temperature in May and 16–22 °C in June–August. Water temperatures decreased to 4–6 °C at the beginning of November (Figure 2).

The salinity in the northern part varied from 0 to 7.9 psu, dependent on wind activity affecting brackish water inflow from the Baltic Sea.

Materials and methods

Phytoplankton sampling for estimation of biomass dynamics was conducted two times per month during the vegetative phytoplankton period (end of February–October) at one station in the northern part of Kuršių Marios lagoon (see Figure 1). Samples (200 ml) from surface layer (0.5–1 m depth) were preserved with Lugol's solution, made with glacial acetic acid. The settling chamber procedure (Utermöhl, 1958; Hällfors & Niemi, 1990), was used to identify and enumerate phytoplankton with a Leitz Fluovert microscope at a magnification of 600 $times$ and 300 ×. The biovolume of algae cells was calculated using taxon specific stereometrical formulas and converted to biomass assuming the specific gravity of phytoplankton to be 1.0 (HELCOM, 1988). 'Edible' phytoplankters were defined as having a cell size less than 35 μm (see Sterner, 1989; Edgar & Green, 1994).

Zooplankton was quantitatively sampled at the same site, approximately every three-four days from March to November 1995. Pooled duplicate vertical hauls were taken with a 150 μm mesh plankton net (13 cm diameter) from 4 m depth. Samples were fixed with 4% formalin. Individuals were counted and measured (total length excluding spines) in a Bogorov tray using a binocular scope. Lengths were converted to the wet weight using length-weight relationships from Salazkin et al. (1984) and Jorgensen et al. (1991). Average length of Crustacea was determined for each sample. Rotifers were not quantitatively evaluated.

Temperature and salinity were also measured. Samples, where salinity was over 1 psu were not considered for analysis since they contained mostly marine species, transported by passive inflow from the Sea. Therefore, data discussed here represents only freshwater zooplankton and phytoplankton communities.

Correlation and time series (cross-correlation) analysis were performed using STATISTICA 4.5 software.

Results

Phytoplankton

Diatomophyceae and Cyanobacteria domination in spring and summer-autumn periods respectively, and was characteristic for seasonal phytoplankton succession in the Kuršių Marios lagoon (Figure 3). The spring bloom was dominated by small diatoms, *Stephanodiscus* spp. (cell size 8 × 6 μm–14 × 12 μm), and cryptophytes (*Chroomonas acuta*). Biomass of the Diatomophyceae increased from 0.9 g m^{-3} at the end of February to 2.9 g m^{-3} in the April and was responsible for more than 90% of total phytoplankton

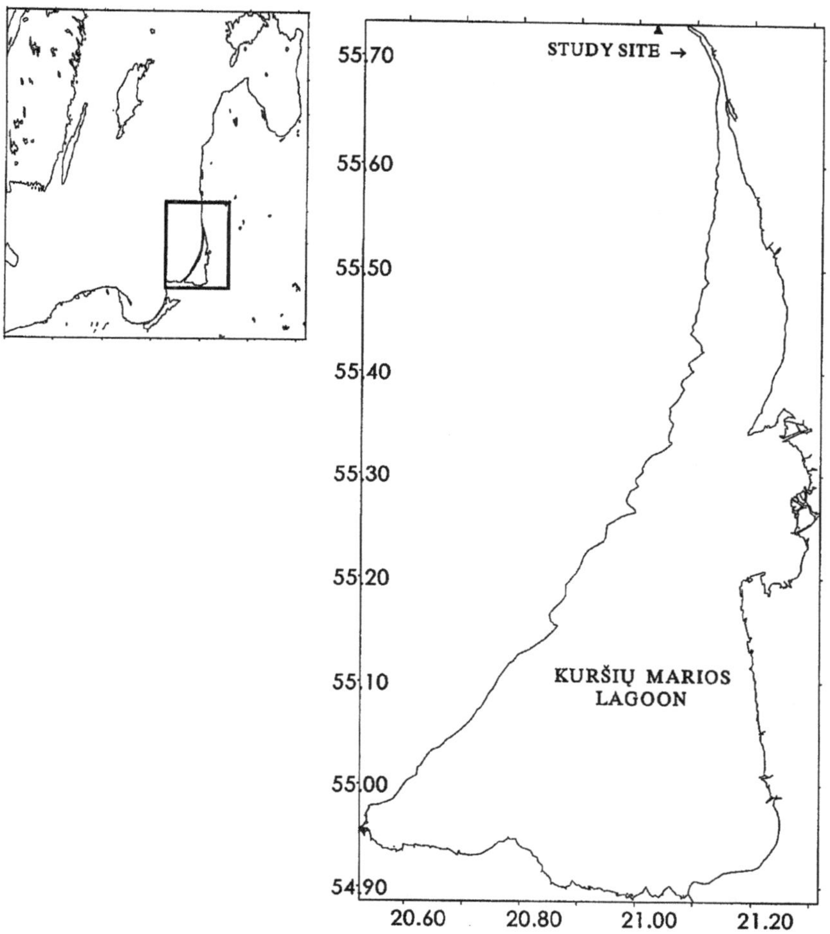

Figure 1. Study site.

biomass. At the end of May, diatoms reached 4.9 g m^{-3}(51% of total biomass).

During summer and autumn phytoplankton biomass increased and dominance shifted to large filamentous cyanobacteria *Aphanizomenon flos-aquae*. This species biomass reached 33.1 g m^{-3}, or 77% of the total in July. Relatively high biomass of *A. flos-aquae* was observed also in August (14.4 g m^{-3}, 56% of total biomass) and in the middle of September (24.4 g m^{-3}, or 83% of the total).

Chlorophyta (*Scenedesmus* spp., *Planktonema lauterbornii*, *Pediastrum boryanum*, *Pandorina morum*) were less important in the Kuršių Marios lagoon phytoplankton community. Only at the end of May they reached 38% of total phytoplankton biomass (3.7 g m^{-3}). Their biomass maximum was observed at the end of September (6.6 g m^{-3}, 25% of total biomass), when biomass and density of herbivorous zooplankton decreased (see Figure 4).

Zooplankton

Cladocera dominated zooplankton among the study period, with *Daphnia longispina*, *Chydorus sphaericus*, *Diaphanosoma brachyurum*, *Bosmina coregoni*; Copepoda (*Mesocyclops leuckarti*, *Cyclops strenuus*, *Eudiaptomus graciloides*) were important only in spring, forming up to 98% of total biomass (Figure 4). Rotatoria (densities and biomass) were not evaluated quantitatively in this study, but other studies indicated that their contribution is considerably lower than the crustaceans, only approximately 6% of overall zooplankton biomass (Naumenko, 1996). *Keratella* and *Brachionus* were usually the dominant genera.

The first maximum in zooplankton density and biomass occurred at the end of May when *D. longispina*

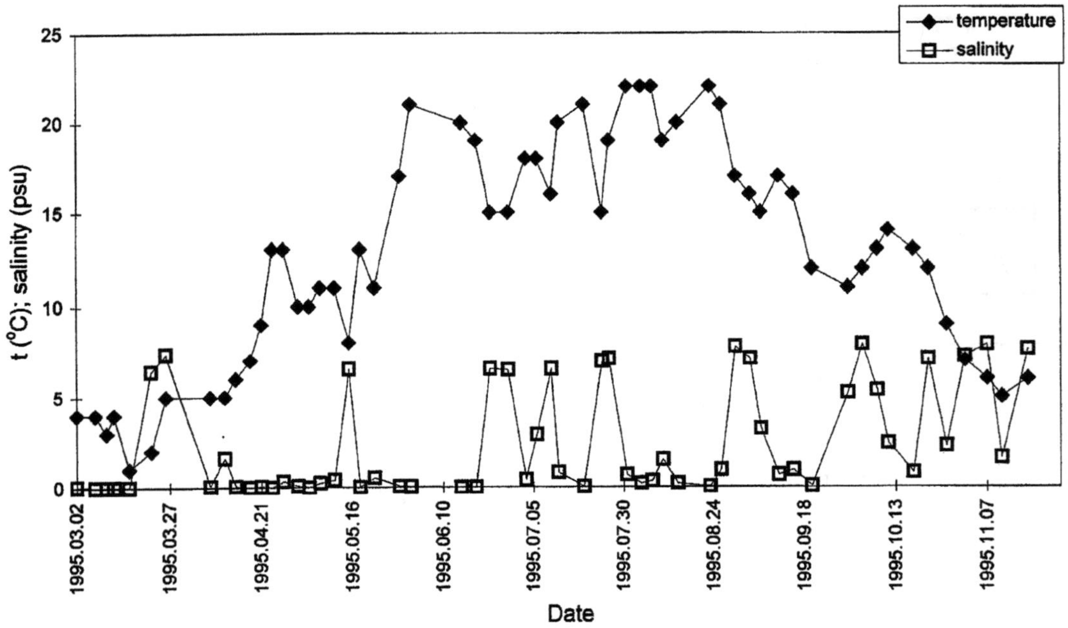

Figure 2. Changes in salinity and temperature in the northern portion of Kuršių Marios lagoon in 1995.

appeared and total biomass of planktonic crustaceans reached 4.6 g m^{-3}. Thereafter zooplankton biomass increased rapidly and reached a second maximum at the beginning of July when *D. longispina* reached approximately 10 g m^{-3} (92% of total zooplankton biomass). The mean length of the planktonic crustaceans also increased from 0.4 mm in the spring to 1 mm during the second peak. It is important to note that the dominance of large daphniids coincided with a bloom of filamentous cyanobacteria.

The biomass of *Chydorus sphaericus* increased at the end of summer and reached approximately 2.3 g m^{-3} and 50% of overall zooplankton biomass. The mean length of zooplankters at this time had declined to 0.3–0.4 mm.

Since some delay was anticipated on interresponse between different groups of zooplankton and phytoplankton, we used time series analysis (cross-correlation) for capturing such relationships. Some significant relationships between zooplankton biomass and biomass of different groups of phytoplankton were detected (Table 1). High positive correlations with time lag of 2 weeks was observed between overall zooplankton biomass and overall phytoplankton biomass (cross-correlation coefficient (CC) = 0.78, SE = 0.28), Cladoceran biomass and overall phytoplankton biomass (CC = 0.74, SE = 0.27), Cladoceran biomass and Cyanobacteria biomass (CC = 0.83, SE = 0.27,), and *D. longispina* and Cyanobacteria biomass (CC = 0.61, SE = 0.16). Zooplankton were also significantly correlated with biomass of small edible phytoplankton (CC = 0.46, SE = 0.16), but no significant correlation were founded between *D. longispina* and edible phytoplankton biomass.

Discussion

The results of the statistical analyses suggest that both phytoplankton and zooplankton community structure and dynamics in Kuršių Marios lagoon are strongly interdependent. Increases in zooplankton biomass and body size were followed by increases in phytoplankton biomass, particularly in biomass of filamentous and non-edible algae *Aphanizomenon flos-aquae*. According to Sommer (1987), the most probable explanation is that large cyanobacteria are weak competitors for nutrients and their bloom may, at least partly, be caused by grazing pressure on small algae, changing the outcome of competition in the phytoplankton community.

When the phytoplankton reached their summer biomass maximum, the zooplankton biomass had already declined. A shift of dominants in the zooplankton community followed. Possibly, filamentous cyanobacteria have an inhibitory influence on large

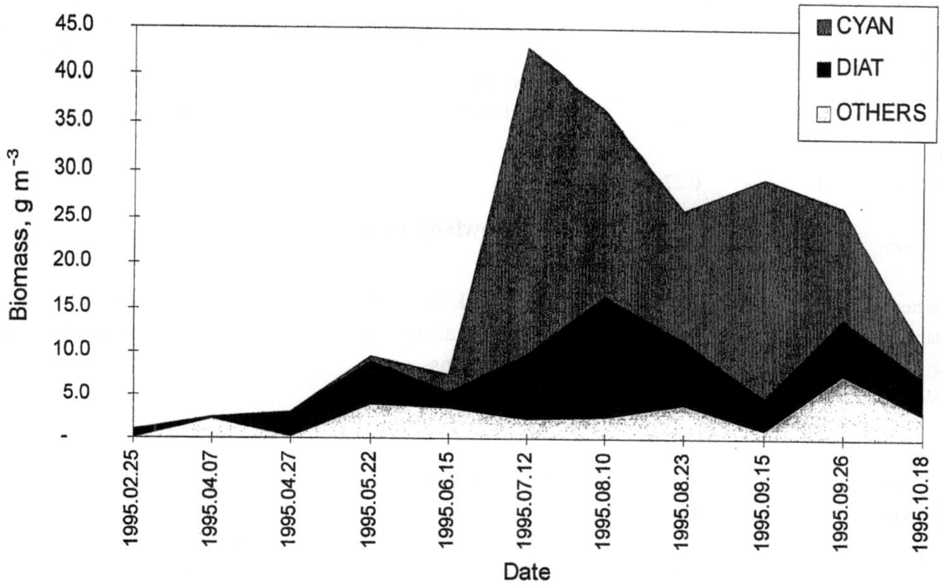

Figure 3. Phytoplankton community biomass dynamics. DIAT – Diatomophyceae, CYAN – Cyanophyta.

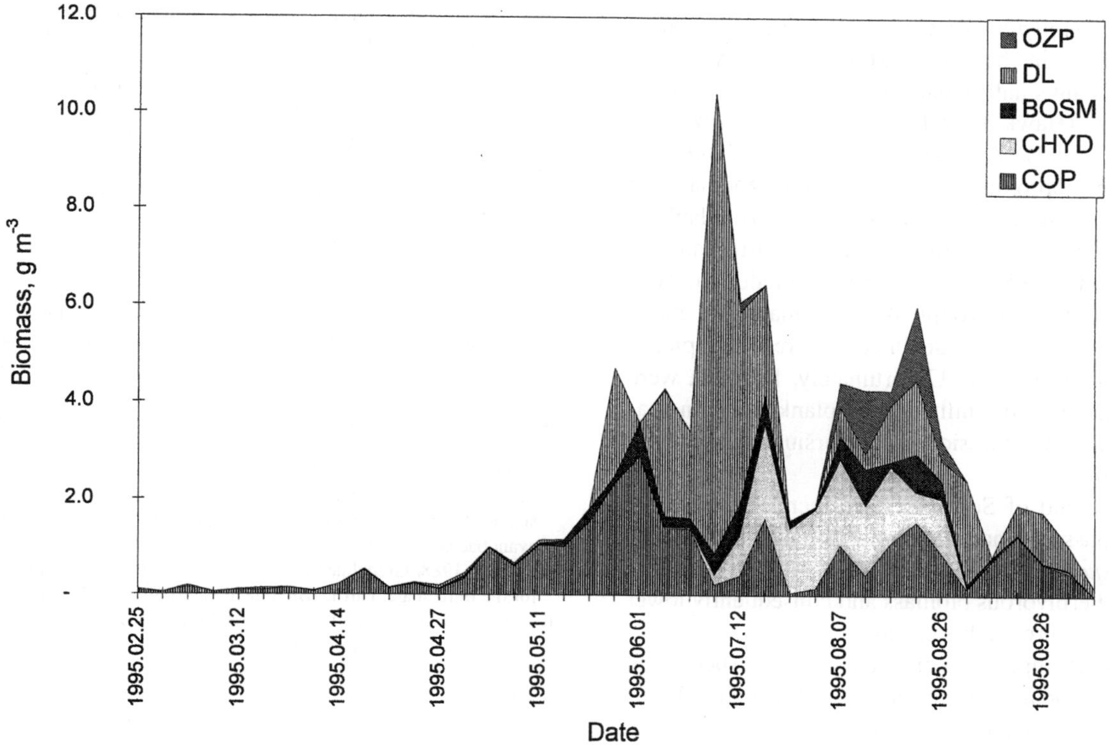

Figure 4. Zooplankton community biomass dynamics. DAPH – *Daphnia longispina*, BOSM – *Bosmina coregoni*, CHYD – *Chydorus sphaericus*, COP – Copepoda, OZP – other zooplankters.

Table 1. Time series analysis of key groups of crustaceans and planktonic algae. D – delay (weeks), CC – cross correlation coefficient, SE – standard error. Correlations, significant at $p<0.05$, are marked by $*$.

First	Lagged	D	CC	SE
Zooplankton biomass	Phytoplankton biomass	2	0.78*	0.28
Cladocera	Phytoplankton biomass	2	0.69*	0.27
Zooplankton biomass	Cyanobacteria	2	0.82*	0.26
Cladocera	Cyanobacteria	2	0.83*	0.27
Daphnia longispina	Phytoplankton biomass	2	0.57*	0.16
Daphnia longispina	Cyanobacteria	2	0.61*	0.16
Daphnia longispina	Chlorophyta+ Diatomophyceae	2	0.30	0.16
Zooplankton biomass	Chlorophyta+ Diatomophyceae	2	0.46*	0.3
Chydorus sphaericus	Chlorophyta+ Diatomophyceae	2	0.58*	0.16

Daphnia longispina species, explaining the change of dominants in the zooplankton community toward more resistant small-bodied species such as *Chydorus sphaericus* (Hawkins & Lampert, 1989; Gliwicz & Lampert, 1993). As mentioned above, *C. sphaericus* formed approximately 50% of overall zooplankton biomass during the late summer peak in zooplankton and mean body size of the planktonic crustaceans decreased to 0.3–0.5 mm. We cannot exclude, however, the possibility that zooplankton biomass and mean body length may have declined as a result of planktivorous fish pressure. Unfortunately, fish data were not available so the influence of planktivory on seasonal plankton succession in the Kuršių Marios lagoon remains unclear.

By the end of September, an increase in density and biomass of small phytoplankton and decrease of *A. flos-aquae* was observed. The increase may be due to lower herbivorous biomass and consequently lower grazing pressure on this group.

Summarising, we conclude that the succession of the phytoplankton community in Kuršių Marios lagoon depends on interactions with higher trophic levels, as well as the food conditions has an influence on structure and dynamics of zooplankton community. Grazing pressure of daphnids may restrict the development of small Chlorophyta and Diatomophyceae, giving opportunity for Cyanobacteria. In turn, the bloom of Cyanobacteria is unfavourable for large *Daphnia* and might explain the decrease in their biomass and shift of dominant zooplankton species to a chydorid-dominated community.

Acknowledgments

We thank A. Razinkovas and two anonymous reviewers for reading the manuscript, helpful comments and suggestions.

References

Brooks, J. L. & S. I. Dodson, 1965. Predation, body size and composition of plankton. Science 150: 28–35.

Carpenter, S. R., J. A. Morrice, J. J. Elser, A. S. Amand & N. A. MacKay, 1996a. Phytoplankton community dynamics. In S. R. Carpenter & J. F. Kitchell (eds), The Trophic Cascade in Lakes. Cambridge University Press: 189–209.

Carpenter, S. R., J. A. Morrice, P. A. Soranno, J. J. Elser, N. A. MacKay & A. S. Amand, 1996b. Primary production and its interactions with nutrients and light transmission. In S. R. Carpenter & J. F. Kitchell (eds), The Trophic Cascade in Lakes. Cambridge University Press: 189–209.

DeMott, W. R., 1989. The role of competition in zooplankton succession. In Plankton Ecology: Succession in Plankton Communities, U. Sommer (ed.), Berlin. Springer-Verlag: 195–252.

Edgar, N. B. & J. D. Green, 1994. Calanoid copepod grazing on phytoplankton; seasonal experiments on natural communities. Hydrobiologia 273: 147–161.

Gliwicz, Z. M. & W. Lampert, 1993. Body-size related survival of cladocerans in a trophic gradient: an enclosure study. Arch. Hydrobiol. 129: 1–23.

Gliwicz, Z. M. & J. Pijanowska, 1989. The role of predation in zooplankton succession. In Plankton Ecology: Succession in Plankton Communities, U. Sommer (ed.), Berlin. Springer-Verlag: 253–296.

Hall, D. J., S. T. Threlkeld, C. W. Burns & P. H. Crowley, 1976. The size-efficiency hypothesis and the size structure of zooplankton communities. Annu. Rev. Ecol. Syst. 7: 177–208.

Hawkins, P. & W. Lampert, 1989. The effect of *Daphnia* body size on filtering rate inhibition in the presence of a filamentous cyanobacterium. Limnol. Oceanogr. 34: 1084–1089.

HELCOM, 1988. Guidelines for the Baltic monitoring programme for the third stage. Part D. Biological determinants, 27D: 164.

Hällfors, G. & A. Niemi, 1990. Proposal for standartization of the way of presenting phytoplankton results. Fin. Mar. Res. 257: 29–36.

Jorgensen, S. E., S. N. Nielsen & L. A. Jorgensen (eds), 1991. Handbook of Ecological Parameters and Ecotoxicology. Elsevier Science Publishers B.V.: 1263.

Kitchell, J. F. & S. R. Carpenter, 1996. Cascading trophic interactions. In S. R. Carpenter & J. F. Kitchell (eds), The Trophic Cascade in Lakes. Cambridge University Press: 1–14.

Kivi, K., S. Kaitala, H. Kuosa, J. Kuparinen, E. Leskinen, R. Lignell, B. Marcussen & T. Tamminen, 1993. Nutrient limitation and grazing control of the Baltic plankton community during annual succession. Limnol. Oceanogr. 38: 893–905.

Naumenko, E., 1996. Species composition, seasonal and long-term dynamics of zooplankton abundance and biomass in the Currish lagoon of the Baltic Sea. ICES C.M. 1996/ L: 12 Ref. J.: 19.

Porter, K. G., 1977. The plant-animal interface in freshwater ecosystems. Am. Sci. 65: 159–170.

Pustelnikovas, O., 1994. Transport and accumulation of sediment and contaminants in the lagoon of Kuršių Marios (Lithuania) and Baltic Sea. Neth. J. Aquat. Ecol. 28: 405–411.

Riegman, R., B. R. Kuipers, A. A. M. Noordeloos & H. J. White, 1993. Size differential control of phytoplankton and the structure of plankton communities. Neth. J. Sea Res. 31: 255–265.

Salazkin, A. A., M. B. Ivanova & V. A. Ogorodnikova, 1984. Metodicheskie rekomendacyi po sboru i obrabotke materialov pri hydrobiologicheskikh issledovanijakc na presnovodnykh vodojemakh: zooplankton i ego produkcija. [Methodical recommendations for data collecting and analysis in hydrobiological studies of fresh waters: Productivity of zooplankton]. Leningrad: 33 (in Russian).

Sarnelle, O. , 1993. Herbivore effects on phytoplankton succession in a eutrophic lake. Ecol. Monographs 63: 129–149.

Sommer, U., 1987. Factors controlling the seasonal variation in phytoplankton species composition – a case study for a deep, nutrient rich lake. Prog. Phycol. Res. 5: 123–178.

Stankevičius, A. (ed.), 1996. Annual report of scientific work. Center of Marine Research, Klaipeda: 92 pp.

Sterner, R. W., 1989. The role of grazers in phytoplankton succession. In U. Sommer (ed.), Plankton Ecology: Succession in Plankton Communities, Springer-Verlag, Berlin: 70–107.

Utermöhl, H., 1958. Zur Vervollkommnung der quantitativen Phytoplankton- Methodic. Fur theoretishe und angewandte Limnologie 9: 1–39.

Hydrobiologia 363: 341–344, 1998.
T. Tamminen & H. Kuosa (eds), Eutrophication in Planktonic Ecosystems: Food Web Dynamics and Elemental Cycling.

Closing address: perspectives on basic and applied, freshwater and marine pelagic research

Jouko Sarvala
Department of Biology, University of Turku, FIN-20014 Turku, Finland (e-mail jouko.sarvala@utu.fi)

Key words: hydrodynamics, pelagic ecosystem, freshwater, marine, basic research, applied research

When asked to make some concluding comments about this symposium, I did not hesitate long, because I have had the privilege of following more or less closely the development of the various Pelag projects during the years. In fact, I have seen the Pelag in its germ line phase, when I was still working myself at Tvärminne Zoological Station in the early 1980s. We then constructed the first-ever carbon budget for the pelagic and benthic system off Tvärminne, which was submitted to the ICES ecosystem symposium in Kiel in 1982. The pelagic system structure of our carbon flow model was admittedly crude compared with the present knowledge, but, nevertheless, the paper was accepted as a lecture, and we got it published (Kuparinen et al., 1984). Many things have changed since those days, but I have been happy to notice that the good humour and untiring scientific curiosity of the group have persisted through the years.

The present symposium was organized by the Pelag group to mark the completion of their third project, and it is no wonder that the selection of symposium topics mirrors the interests of the organizers. As a consequence, during this week we became familiar with different kinds of material fluxes, often vertical (e.g. Wassmann, 1998), but what we especially experienced was an extensive horizontal flux of information, the source of which could partly be traced to long-distance transport, although part of the flux was clearly of local origin.

I think that the general structure of the symposium was successful. We have heard nine excellent summary papers setting out the state-of-the-art in several subtopics under the general title of the symposium. In addition, there were many high-quality contributed papers and posters with really impressive data. I am not going to review the whole symposium – to summarize almost fifty papers and a similar number of posters is simply impossible in a few lines. Instead, I shall pick up some thoughts that came to my mind while I was listening the presentations during the week.

The symposium had a wide geographical coverage from the Sargasso Sea in the west to the Black Sea in the east, and from the Norwegian fjords and the Baltic Sea in the north to the warm Mediterranean in the south. The salt content of study waters also varied considerably, from the very soft brown freshwaters of the Lammi area in southern Finland to the higher than ocean salinities of the Mediterranean Sea.

It was refreshing to see marine biologists and freshwater scientists in the same symposium. The presentations here amply confirmed that the planktonic systems have very similar structure and function throughout the whole salinity gradient. The variations that occur have little to do with salinity as such. Instead, the major differences between fully marine and lacustrine systems probably arise from the different amount and distribution of physical energy (Nixon, 1988b); in addition, the largest space and time scales are only present in the oceans. Yet in the past marine ecology and limnology have tended to be quite separate disciplines, marine scientists not reading freshwater papers, and vice versa. During the recent years, there have been a few conscientious attempts at a better unification, e.g. some symposia have been intentionally organized for both groups combined (Nixon, 1988a; Giller et al., 1994). Having myself done research both in the Baltic Sea and in lakes, I especially welcome this kind of joint meeting. From my own experience I know that switching between environments is scientifically rewarding, although it is also very demanding.

A deeper knowledge of the tight coupling between the physical and biological processes is necessary for a better understanding of aquatic life. The physical processes are especially important in the sea, although hydrophysics is equally important in large lakes. The integration of hydrophysics and biology has already long been in the research agenda (e.g. Archer, 1995; Mann & Lazier, 1996). One of the central ideas of the Pelag III project was the elucidation and combination of mesoscale physical patterns with microscale biological data. However, in spite of good intentions, little has been yet achieved in this respect. I think that one of the problems weakening the applicability of our basic science still lies here.

In this symposium, I was impressed at listening how the current and mixing patterns regulate phytoplankton productivity in the Black Sea (Yılmaz et al., 1998). This immediately reminded me about quite parallel phenomena in the lakes Tanganyika and Malawi in East Africa, where the wind-driven regular and episodic mixing pumps nutrients from the permanently anoxic hypolimnion to the euphotic zone (Hecky et al., 1991; Bootsma, 1993; Patterson & Kachinjika, 1993, 1995; Huttula, 1998; Plisnier et al., 1996). These lakes also provide an excellent example of the importance of the environmental physics over even larger scales. The El Niño Southern Oscillation (ENSO) events in the Pacific area affect tropical climate in the southwestern Indian Ocean (e.g. Charles et al., 1997), and influence the north-south location of the trade-wind belt in eastern Africa, where e.g. rainfall and maize yield in Zimbabwe have been shown to be predictable from ENSO events (Cane et al., 1994). The timing and extent of these seasonal winds in turn has dramatic consequences for the water mixing regimes in Malawi (Patterson & Kachinjika, 1995) and Tanganyika (Huttula et al., 1994; Savijärvi, 1995, 1997; Verburg et al., 1997; Huttula, 1998). Mixing regulates vertical nutrient transport, controlling primary production of phytoplankton (Patterson & Kachinjika, 1995) and probably ultimately affecting fish catches in the area (Plisnier, 1997, 1998). The nutrient upwellings and mixing events themselves are relatively small-scale phenomena both in space and time, as are even more the production processes, but they are controlled from distance by incidents at the largest global scale.

These examples highlight the importance of taking the hydrodynamic environment into account in trying to understand biological phenomena in whole lakes and sea areas. I fully endorse the statement heard in this symposium that improved time and space resolution does not need to mean lower resolution of trophic processes – in contrast, to understand correctly the trophic processes, it is necessary to improve on our description of the physical environment. This is particularly important when trying to apply our basic understanding of aquatic biology to water quality management.

The title of the symposium was 'Eutrophication in planktonic ecosystems: food web dynamics and elemental cycling'. Although not explicitly stated, the title implies that practical problems of water quality management come into the scene. There is an urgent need to improve the use of ecological knowledge in resource management (e.g. Underwood, 1995). The symposium was thought to help build a bridge between water quality managers and basic scientists. Yet most of the papers dealt with basic research, only few presentations directly tackling the water quality management issues.

One of the leading principles in the Pelag III project was that successful management in the long-term requires sound knowledge of the real mechanisms acting in nature (e.g. Tamminen, 1990). To find out these requires theoretical work, rigorous experimental testing, and modelling, in addition to and in combination with the more traditional observational field studies. These basic research aspects were all well represented in this symposium.

There was a wide consensus about the general structure of the planktonic ecosystem, especially about its biological components and the nature of network control (Thingstad, 1998). Concepts that were heretical some years ago seem now firmly established paradigms. I think that this is a good basis for future development of understanding. However, the symposium also clearly witnessed that the aquatic world is rather complicated, maybe, as Peter Kofoed Bjørnsen put it in his presentation, too complicated for the purposes of applied science.

The level of sophistication of most management models is as yet insufficient to yield realistic predictions. Future development is clearly needed in order to cope with the complexity of life in a satisfactorily simple way to produce solutions to applied problems. I am not sure whether it will ever be possible to achieve the required balance between complexity, realism, and practical applicability. It is, among others, necessary to realize that the actors in the ecological theatre are the outcomes of a long-term evolutionary history, and the evolution is not all behind, we are living with it today. What this means is that the characteristics and reactions of organisms are not fixed, but flexible and adaptable. This pertains as well to zooplankton, algae and fish, to protists and multicellular organisms alike. Quite similar considerations have become necessary in the terrestrial ecology when trying to manage pollution (e.g. Eeva et al., 1997).

Our best understanding is restricted to small-scale phenomena. Basic research proceeds in small steps, making very detailed measurements of invisibly small things, typically in small experimental chambers over short time scales. The problem here is how this detailed small-scale information, often working at the behavioural scale of individuals, can be scaled up to the larger-scale patterns that are of interest for the water manager, to the scale where our environmental problems occur. It is still very difficult to understand how it is possible that the organisms that reproduce, grow and die in a few hours or days and develop short-lived erratic population peaks, anyhow can persist in fairly similar numbers from year to year, or, how the very dynamic seasonal plankton successions translate into gradual long-term trends. Kiørboe (1998) noted this problem of scale in the case of copepod population regulation.

In order to understand aquatic life and especially to be able to manage it, we must perform studies at all scales, not only in the small bottles or mesocosms, but also at the scale of whole lakes, bays, seas and oceans, and not only over minutes, hours or days, but also over years, decades, and even longer time scales utilizing the methods of palaeoecology. And we must be able to link the phenomena at different scales. I think this is one of the great challenges ahead.

I think that this symposium was a useful step forward in organizing our existing knowledge and preparing ground for future progress. I congratulate the organizers for creating the framework for a successful symposium, and I extend my thanks to all participants, who finally filled the framework with science and life.

References

Archer, D., 1995. Upper ocean physics as relevant to ecosystem dynamics: a tutorial. Ecological Applications 5: 724–739.

Bootsma, H. A., 1993. Spatio-temporal variation of phytoplankton biomass in Lake Malawi, Central Africa. Verh. int. Ver. Limnol. 25: 882–886.

Cane, M. A., G. Eshel & R. W. Buckland, 1994. Forecasting Zimbabwean maize yield using eastern equatorial Pacific sea surface temperature. Nature 370: 204–205.

Charles, C. D., D. E. Hunter & R. G. Fairbanks, 1997. Interaction between the ENSO and the Asian monsoon in a coral record of tropical climate. Science 277: 925–928.

Eeva, T., E. Lehikoinen & T. Pohjalainen, 1997. Pollution-related variation in food supply and breeding success in two hole-nesting passerines. Ecology 78: 1120–1131.

Giller, P. S., A. G. Hildrew & D. G. Raffaelli, 1994. Aquatic Ecology. Scale, Pattern and Process. Blackwell Science, Oxford, 649 pp.

Hecky, R. E., R. H. Spigel & G. W. Coulter, 1991. The nutrient regime. In G. W. Coulter (ed.), Lake Tanganyika and its Life. British Museum (Natural History) & Oxford Univ. Press, London, Oxford & New York: 76–89.

Huttula, T. (ed.), 1998. Flow, thermal regime and sediment transport studies in Lake Tanganyika. Kuopio Univ. Publ. C. nat. envir. Sci., in press.

Huttula, T., V. Podsetchine, A. Peltonen, P. Kotilainen & H. Mölsä, 1994. Hydrology of Lake Tanganyika. In C. Kern-Hansen, D. Rosberg & R. Thomsen (eds), Proc. of Nordic Hydrological Conference, Faroe Islands 2–4 Aug. 1994. NHK-Report (København) 34: 34–52.

Kiørboe, T., 1998. Population regulation and role of mesozooplankton in shaping marine pelagic food webs. Hydrobiologia, 363: 13–27.

Kuparinen, J., J.-M. Leppänen, J. Sarvala, A. Sundberg & A. Virtanen, 1984. Production and utilization of organic matter in a Baltic ecosystem off Tvärminne, southwest coast of Finland. Rapp. P.-v. Réun. Cons. int. Explor. Mer 183: 180–192.

Mann, K. H. & J. R. N. Lazier, 1996. Dynamics of Marine Ecosystems: Biological and Physical Interactions in the Oceans. Blackwell Science, Oxford, 394 pp.

Nixon, S. W. (ed.), 1988a. Comparative ecology of freshwater and marine ecosystems. Limnol. Oceanogr. 33: 649–1025.

Nixon, S. W., 1988b. Physical energy inputs and the comparative ecology of lake and marine ecosystems. Limnol. Oceanogr. 33: 1005–1025.

Patterson, G. & O. Kachinjika, 1993. Effect of wind-induced mixing on the vertical distribution of nutrients and phytoplankton in Lake Malawi. Verh. int. Ver. Limnol. 25: 872–876.

Patterson, G. & O. Kachinjika, 1995. Limnology and phytoplankton ecology. In A. Menz (ed.), The Fishery Potential and Productivity of the Pelagic Zone of Lake Malawi/Niassa. Chatham, UK: Natural Resources Institute: 1–67.

Plisnier, P.-D., V. T. Langenberg, L. Mwape, D. Chitamwebwa, K. Tshibangu & E. Coenen, 1996. Limnological sampling during an annual cycle at three stations on Lake Tanganyika (1993–1994). FAO/FINNIDA Research for the Management of the Fisheries on Lake Tanganyika GCP/RAF/271/FIN–TD/46 (En), 136 pp.

Plisnier, P.-D., 1997. Climate, limnology and fisheries changes of Lake Tanganyika. FAO/FINNIDA Research for the Management of the Fisheries on Lake Tanganyika. GCP/RAF/271/FIN-TD/73 (En), 50 pp.

Plisnier, P.-D., 1998. Lake Tanganyika: recent climate changes and teleconnections with ENSO. Proceedings of the International Conference on Tropical Climatology, Meteorology and Hydrology, Royal Academy of Overseas Sciences and Royal Metereological Institute of Belgium, Brussels, May 1996 (in press).

Savijärvi, H., 1995. Sea-breeze effects on large-scale atmospheric flow. Contrib. atmos. Physics 68: 281–292.

Savijärvi, H., 1997. Diurnal winds around Lake Tanganyika. Q. J. r. meteorol. Soc. 123: 901–918.

Tamminen, T., 1990. Eutrophication and the Baltic Sea: studies on phytoplankton, bacterioplankton, and pelagic nutrient cycles. Ph.D. thesis, University of Helsinki.

Thingstad, T. F., 1998. A theoretical approach to structuring mechanisms in the pelagic food web. Hydrobiologia 363: 59–72.

Underwood, A. J., 1995. Ecological research (and research into) environmental management. Ecological Applications 5: 232–247.

Verburg, P., T. Huttula, B. Kakogozo, A. Kihakwi, P. Kotilainen, L. Makasa & A. Peltonen, 1997. Hydrodynamics of Lake Tanganyika and meteorological results. FAO/FINNIDA Research for the Management of the Fisheries on Lake Tanganyika, GCP/RAF/271/FIN-TD/59 (En). Bujumbura, Burundi.

Wassmann, P., 1998. Retention vs. export food chains: processes controlling sinking loss from marine pelagic systems. Hydrobiologia 363: 29–57.

Yılmaz, A., S. Tuğrul, Ç. Polat, D. Ediger, Y. Çoban & E. Morkoş, 1997. On the production, elemental composition (C, N, P) and distribution of photosynthetic organic matter in the southern Black Sea. Hydrobiologia, 363: 141–155.

FSC
www.fsc.org
MIX
Papier aus verantwortungsvollen Quellen
Paper from responsible sources
FSC® C105338

If you have any concerns about our products,
you can contact us on
ProductSafety@springernature.com

In case Publisher is established outside the EU,
the EU authorized representative is:
Springer Nature Customer Service Center GmbH
Europaplatz 3, 69115 Heidelberg, Germany

Printed by Libri Plureos GmbH
in Hamburg, Germany